WORK DESIGN: INDUSTRIAL ERGONOMICS

WORK DESIGN: INDUSTRIAL ERGONOMICS

SECOND EDITION

Stephan Konz
Kansas State University

Grid Publishing, Inc., Columbus, Ohio

Printed in the United States

1 2 3 4 5 6 ☒ 8 7 6 5 4 3

Library of Congress Cataloging in Publication Data

Konz, Stephan A.
  Work design.
  (The Grid series in industrial engineering)
  Includes index.
  I. Work design. I. Title. II. Series.
T60.8.K66 1983   658.5'4   82-3041
ISBN 0-88244-249-X   AACR2

CONTENTS

PART V RECOMMENDED WORK DESIGN PRINCIPLES

FOREWORD

Work Design is an introductory text for engineers and technicians who design and evaluate jobs and working environments in industry. It is, in my opinion, an excellent presentation which combines the "What to do," "How to do it," and "Why" features. To my knowledge, it is the first of its kind. Concepts and techniques from work-time measurement, human factors (ergonomics), and industrial safety and hygiene are merged to provide a comprehensive viewpoint. However, the required background need be no more than a basic course in statistics. Most of the specialized material on work physiology and industrial psychology relative to the design of equipment layouts, workplaces, handtools, and job training is included within the text. Principle techniques and variations for job design are covered in detail. These scientific principles and operational aids serve to augment common sense under the author's philosophy that "one ought to work smart rather than hard." This philosophical theme is distributed throughout the book and highlighted by numerous examples. I am sure that readers of the intended audience will find this book interesting and highly useful in preparing them for the technical jobs of making industry more human, safe, and productive.

The organization of this text is straightforward. A historical and sociological overview is presented initially to set the stage and to show how this text's material fits into this perspective. Also shown in this initial part is the "work smart not hard" principle. Next the text turns to some notions, ideas, and principles of engineering design which can be used along with the design arts. This background on design quickly fades into several specific technique classes which are used as aids in the design process. Techniques of recombining or eliminating job tasks and evaluating these tasks are shown along with the more traditional techniques of charting macro or micro operations, predetermined time studies, and work sampling. The next two sections of this book turn to recommended design principles; section IV concentrates on the scientific background of psychological, social, and physiological concepts applied to job design, whereas section V carries this background and these principles to specific areas of design and operations of work stations and handtools. In the next section there are principles and concepts for creating safe and productive working environments with regard to visual and auditory needs as well as atmospheric conditions of heat, humidity, and protection from toxic elements. These topics of job design are important features of human factors (ergonomics) and occupational safety and health. Section VII of this book turns to the traditional topic of work-time measurement, which plays an important part in industrial engineering and serves many needs in planning, designing, and controlling industrial operations. The final section of this book addresses the problem area of implementing the design in operation through job instructions and overcoming resistance to change. Many of the ideas shown in this section are particularly important to technical personnel who must sell their ideas to nontechnical management and operating personnel for a successful design. These eight sections span useful design concepts and arts of this field from inception to execution.

For many years in the past there have been books available which addressed either human factors (ergonomics), occupational safety and health, or work-time measurement, nearly excluding the other two related areas. This book is an exception and the beginning of a long-needed change. Professor Konz is to be congratulated for this undertaking. I am pleased to have the honor of playing a minor role on the sidelines of what I feel will become an important historical event in our profession.

James R. Buck
School of Industrial Engineering
State University of Iowa
Iowa City, Iowa

PART I
OVERVIEW AND
HISTORY

INDUSTRIAL SOCIETY

HISTORICAL BACKGROUND

Starting about 1750 in England, about 1800 in the United States, about 1825 in northwestern Europe, and later in the rest of the world, a profound change began—the Industrial Revolution. The key word is *change*. Prior to that time there had been good times and poor times, but the overall standard of living was no better in 1700 than it had been in 1000 or in Roman times. Roman houses in Britain had bathrooms; it was during the time of Queen Victoria before bathrooms again were built in private houses. The rising wind of change has continued until today it is a firestorm. Slow progress, although steady, is no longer acceptable. The standard of living improved about 2%/year in England between 1950 and 1970; this improvement was about half the improvement of England's neighbors and has been termed "the English disease."

KEY ASPECTS OF IMPROVED LIVING STANDARDS

The following aspects of our technological society do not have any particular rank order but are all interrelated.

Increased Knowledge

The nature of knowledge is that it is cumulative and that it diffuses. The human race has gradually increased its store of knowledge. Occasionally the reservoir has declined (400 to 1500 in Europe) or stagnated (Arabic world from 1250 to 1900; China from 1350 to 1900), but the general trend was upward. Many developments require developments in related fields. For example, the steam engine required development of a machine which could bore an accurate cylinder; both the steam engine and the boring machine required a coal technology for fuel and metal.

Diffusion of Knowledge

Literacy and education seem essential. One of our great inventions is the development of libraries spread throughout the country and the world rather than just a few isolated libraries such as the famous library of Alexandria. In addition, due to the invention of printing (China in 950, Europe in 1400), copies are widely available so that wars and revolutions no longer can destroy the only copy of a text. The system of professional journals ensures that specific knowledge in, say, industrial engineering, biochemistry, or electronics becomes available within a month of publication to specialists in that field in Johannesburg, Bombay, Eindhoven, and Manhattan, Kansas. The total impact of television on our society is just beginning to be felt. In the USA in 1970 the median schooling completed by the population over age 25 was 12.2 years; 30,000 Ph.D.'s were graduated each year.

Freedom from War

Throughout history man has torn down prosperity through war, revolution, and military expenses. It is not a coincidence that the Industrial Revolution began in the United Kingdom and the USA. There was political stability; neither country was invaded; wars, when occurring, were fought on foreign soil (with the exception of the US Civil War) and military spending was a low percent of gross national product. Promising civilizations in the Arab world and China were stopped by the Mongol invasions in the thirteenth century. During the period 1770 to 1945, Germany and France have "eaten each other up." The three countries increasing their standard of living most rapidly since 1950 (excluding the unusual situation of the Arab oil countries) have been Germany, Japan, and Brazil—all with low military expenditures.

In the past, prosperous but weak countries were soon overrun by warlike neighbors. What seems to be needed is sufficient military strength to deter attack (e.g., Sweden, Switzerland) or a "big brother" (USA acting as policeman for the world).

Capital

It takes money to make money. In England, the capital came out of the hides of the poor and the colonies; stories of the poverty of the working class at this time are well known (for example, Dickens or Marx). But this capital did permit "getting over the hump" so that future generations could have a better life. In the USA the country itself was virgin; the natural resources had not been picked over for tens of generations. (The number of Indians in 1770 has been estimated as less than 1,000,000 for the entire area now covered by the USA.) In addition, as a country welcoming emigrants, the USA did not have the expense of raising those workers to adulthood.

Capital (deferred consumption) is needed for education, machine tools, buildings, bridges, and research not only at the start of industrialization but also to improve living standards in advanced countries. Low capital availability equals low advances. Naturally the capital must be used on productive projects or it will not be beneficial.

A Development Orientation

A necessary condition of development is that the country must want to be developed. There must not be an entrenched dominant social class eager to maintain the status quo. Japan remained undeveloped by choice until the 1850s. The colonies of England encouraged the merchants to develop and market new products. It must be admitted that the spark of the Industrial Revolution may have begun in England from sheer good luck. For example, one British mine brought out coal, iron ore, *and* clay for firebrick from the *same* shaft. The USA especially had a minimum of restrictions of an entrenched group trying to avoid change.

There are many countries illustrating Churchill's saying "The inherent vice of capitalism is the unequal sharing of blessings while the inherent virtue of socialism is the equal sharing of miseries."

A Large Market

The profitability of many developments depends upon having a large number of customers. England had the British empire and her ships for transportation, while the USA had a reasonably large population and excellent transportation (first railroads and then highways). (There were 35,000 miles of railroad track in the USA in 1860.) Much of Europe was politically fragmented which restricted the market for, say, an Italian development. Land transportation was poor in Africa, Asia, and South America while sea transportation was dominated by the ships of Europe and America.

KEY CONCEPTS OF AN INDUSTRIAL SOCIETY

Separate from the larger social context, the industrial society has six key concepts:

1. Specialization of labor
2. Mechanical and electrical energy
3. Standardization and interchangeable parts
4. Precision from machines, not men
5. Mass production and mass demand
6. The assembly line

Specialization of Labor

Specialization is not new. In the *Livre de Metiers (Book of Trades)* issued in 1254 concerning the 101 Paris guilds, there were separate guilds in the leather industry for skinners, tanners, cobblers, harness makers, saddlers, and makers of fine leather goods; in carpentry there were guilds for chest makers, cabinetmakers, boatbuilders, wheelwrights, coopers, and twiners. What is new is the degree of specialization throughout society. When we think of specialization we think of an industrial employee whose sole function is to add three screws and three nuts to each assembly on the line. But other jobs also have become specialized. In the USA most farmers no longer raise their own food but specialize in one or two crops; in fact a rancher who raises beef probably buys the beef for his own table from the grocery store. Teachers formerly covered all grades and subjects in the "one room school"; now they specialize by grade and subject. Physicians specialize in radiology, obstetrics, dermatology, and brain surgery. Even cleaning of blackboard erasers has become specialized, as at Kansas State University where one person cleans all erasers on campus once/week.

Mechanical and Electrical Energy

Before the Industrial Revolution, power came from muscles (human or animal), wind, or falling water. A horse, for example, can carry 120 kg, pull a cart of 500 kg on a rough road and 900 kg on a paved road, and pull 40,000 kg in a canal barge. It was Watt's steam engine, Faraday and Henry's electric motor, and Otto's internal combustion engine which permitted the enormous magnification of physical effort. The computer probably is accomplishing the same magnification of mental effort. The telephone has magnified our ability to communicate. It has been well said that "steam has liberated more men than the Declaration of Independence has."

Standardization and Interchangeable Parts

With galleys "as alike to another as swallows' nests," the 3,000-man Venetian shipyard could ready 100 galleys for battle in 60 days in the sixteenth century. The concept did not catch on in manufacturing due to the lack of accurate machine tools at that period. Maudslay made some special machines to make standardized ship components in England around 1800 but the real development of interchangeable parts had to await accurate machine tools. It was put into practice in the USA during the 1820s to 1850s in a few industries (gun manufacture, textiles, clocks) and spread gradually. At one time it was feared that standardization would mean that there would be no choices—for example, Ford's Model T was available only in black. It hasn't worked out that way, since clever engineers have designed many options for the basic product line. In fact, people long for the "good old days" when there weren't so many choices to make.

Precision from Machines, Not Men

Craftsmen are expensive and their parts are not identical so parts are not interchangeable. By putting the precision in the machine, the parts can be interchangeable and mass production of standardized units is possible. In addition, machines can be given capabilities (power, strength, repeatability, etc.) which man with his muscles, eyes, and human variability cannot match.

Mass Production and Mass Consumption

These really are chicken and egg—you can't have one without the other. Without mass production the cost is too high, so you don't get mass consumption. For example, in the automotive industry, output of some accessories is 30,000/day from a single-plant complex. In other industries, such as soap, light bulbs, and food, the rate is even higher. A large number of poor consumers is not sufficient as the customer must be able to afford the purchase. Henry Ford realized that without purchasing power the masses cannot buy and thus the manufacturer cannot sell. That's why he doubled his employees' wages.

The Assembly Line

In construction you move the workers and the machines to the product; in manufacturing you move the product past stationary machines and workers. Although conveyors are commonly used to move the product, they are not a requirement since transportation can be by hand, chute, or vehicle. The assembly line is the culmination of the previous five concepts into a production system.

SUMMARY

The developments that make entire nations rich (rather than just a few individuals) are the pivotal developments of history. These developments have occurred both in the larger social structure and within the area of economic development (standard of living).

Within the area of economic development, the following six characteristics—specialization of labor, mechanical and electrical energy, standardization and interchangeable parts, precision from machines not men, mass production and mass consumption, and the assembly line—are so much a part of our culture that most people don't realize how recent these developments are.

SHORT-ANSWER REVIEW QUESTIONS

1.1   In 25-50 words for each concept, discuss the six key concepts of industrial society.

1.2   The industrial revolution seems to have five key factors: increased knowledge, freedom from war, capital, development orientation, and a large market. Discuss each in 10-25 words.

1.3   Is specialization of labor a development of the Industrial Revolution?

1.4   When did the Industrial Revolution begin?

1.5   How is the emigration of an engineer from India to the USA related to capital formation in the USA?

THOUGHT-DISCUSSION QUESTIONS

1. The steam engine required development of an accurate boring engine. Discuss how another specific development depended upon a development in another field.

2. Should immigration of skilled individuals be encouraged as a matter of national policy?

INDIVIDUAL CONTRIBUTORS

THE CONCEPT OF TWO CULTURES

C.P. Snow wrote of the "two cultures": one of scientists and engineers and the other of poets, artists, and writers. This chapter is written so that engineers not only can understand their culture but also appreciate the contribution their culture has made to society. As Galbraith commented: "Insist on the priority of dams, ditches, and fertilizer plants—it is these that feed poets."

Emerson said, "There is properly no history; only biography." However, the nontechnical historian ignores the people who have really changed our lives while paying extraordinary attention to the trivial behavior of politicians. For example Barbara Tuchman's acclaimed history, *The Proud Tower: A Portrait of the World 1890-1914*, covers Europe and the USA during this period without *once* mentioning that during this time the automobile was developed, the radio was invented, people flew, and the telephone and electric illumination became common. Ford, Marconi, the Wright brothers, Bell, and Edison are not mentioned once. Another example is the *Oxford History of the American People*, which uses 1,150 pages to cover America from prehistoric times to 1963. Whitney is covered in 26 words on the cotton gin and slavery; Colt is described in 10 words as the inventor of the "equalizer"; Bell is dismissed in 8 words as "Alexander Graham Bell, the inventor of the telephone"; and the sole mention of Edison is: "the American notion of a scientist still remained, as before the [Civil] War, a practical inventor such as Thomas Edison." Henry Ford and the auto industry are covered in 3 pages. There is no mention whatsoever of Taylor, Hollerith, Gilbreth, or de Forest. In contrast, the author Oliver Wendell Holmes is mentioned on 9 different pages.

Make your own list of contributors. You might consider the following in addition to these described in this chapter: Richard Arkwright, Alexander Graham Bell, John Bardeen, William Shockley, Walter Brattain, Henry Gantt, Gugliemo Marconi, Walter Shewhart, Louis Pasteur, Johan Gutenburg, Orville and Wilbur Wright, Edwin Land, Frank Whittle, Wilhelm Roentgen, Leo Baekeland, Elias

Howe, Christopher Sholes, Cyrus McCormick, Vladimir Zworykin, and Ronald Fisher. Who can deny the importance of (in a random order) the transistor, mechanical spinning, analysis of variance, the typewriter, "plastics," the reaper, the sewing machine, the airplane, instant photography, the jet engine, printing, the germ theory of disease, the telephone, statistical quality control, radio, TV, production scheduling, and X-rays? Asimov lists over 1,000 contributors (and some of the above names are not included), so there is plenty of room for discussion.[1]

The following vignettes give a brief view of some of the movers and shakers that made the world what it is.

> The reasonable man adapts to the world
> The unreasonable man tries to adapt
>   the world to himself
> Therefore all progress comes from
>   unreasonable men
>
> G. B. Shaw

VIGNETTES

James Watt: Key Concept—A New Prime Mover, the Steam Engine

Before Watt there were four basic sources of power—prime movers: human, animal, wind, and water. Watt's steam engine permitted the power source to be located anywhere and be of unlimited magnitude. It permitted implementation of the factory system (where machines are concentrated into one central location), cheap and fast transportation (steam-powered ships and trains), and urbanization as people forsook the rural idyll of the poet's imagination and the commuter's dream for a new life of working and living in the factory and city.

Watt was an instrument maker for the University of Glasgow and worked for several years on Newcomen steam engines. In 1764, he conceived the condenser. In the Newcomen engine the steam cylinder itself was cooled with water; then the metal of the cylinder had to be brought up to temperature again for the next stroke. In Watt's design, this inefficient cooling and heating was eliminated so fuel efficiency was doubled. The condenser was patented in 1769 but Watt ran out of money, and there was little progress until Mathew Boulton supported Watt in 1774. The first engine was installed in Wilkinson's foundry in 1776. Wilkinson's invention of an accurate boring mill, upon which Watt's cylinders were machined, had made possible Watt's engine. (Wilkinson's boring mill could make a cylinder that "erred from a true circle not more than the thickness of an old shilling." Since his mill could bore a 50 inch diameter and a worn shilling was .050 inches thick, this is a precision of .001 inch.)

Before Watt retired, rich and famous, in 1800 he made two additional improvements. First he made the engine double acting, which made rotary power practical as well as improving efficiency. Then he modified the ball governor used in flour milling by adding the critical concept of feedback so that the speed of the engine was self-regulating rather than just automatic. (Watt also invented steam heat when he heated his office in 1784.) Mining engineers in Cornwall, led by Trevithick, improved the energy efficiency of steam engines by a factor of six between 1810 and 1840; meanwhile, Carnot, in France, published his work on the ideal heat engine.

Using present day concepts, application was slow. The carding mill in the New Salem village of Abraham Lincoln (built in 1832) was powered by oxen; a census of steam engines in France in 1833 reported 947 in use; the Secretary of the Treasury reported to Congress in 1838 that there were 3,010 steam engines in the USA (800 on steamboats, 350 in locomotives, and 1,860 in factories and public works).

Henry Maudslay: Key Concept—Accurate Machine Tools

It has already been mentioned that Watt's steam engine could not be manufactured until Wilkinson had a boring mill that could produce reasonably concentric cylinders. In a similar manner the concept of interchangeable parts is based on the premise of each part being alike; in practice, each part will not be alike unless the machines that make the parts are accurate.

Maudslay's fame rests on two foundations: his improvement of the lathe (described by Farey in 1810 as "the most perfect of its kind") and his influence on the next generation of machine-tool builders. Clement, Roberts, Whitworth, and Nasmyth all worked at one time for Maudslay.

Clement improved lathe and planer design and worked on Babbage's mechanical computers. Roberts built a metal-planing machine in 1817, improved lathe gearing, and built many drilling machines. Whitworth introduced a standard screw thread and by manufacturing and selling standard gauges made his influence felt worldwide. Nasmyth invented a special purpose milling machine, the shaper, and the steam forge hammer.

In Nasmyth's words:

"Illustrating his often repeated maxim 'that there is a right way and a wrong way of doing everything,' Maudslay would take the shortest and most direct cuts to accomplish his objects. The grand result of thoughtful practice is what we call experience: it is the power or facility of seeing clearly, before you begin, what to avoid and what to select."

His "innate love of truth and accuracy" led him to develop a bench micrometer which he called "the Lord Chancellor" after the one from which there is no appeal.

Eli Whitney: Key Concept—Interchangeable Parts

Eli was born in 1765 and spent his youth on the family farm in Massachusetts. The Revolutionary War caused shortages and Eli manufactured nails at the age of 14. At 23 he entered Yale. Graduating at 27 and wanting to study law, he intended to take a job as a tutor in the South. He visited a southern plantation and within two weeks had invented the cotton gin. Unfortunately for Eli, his factory burned down just as he was going into production. The gin was simple in design so any mechanic could make one—and they all did. Eli never made much money from the cotton gin. However, the cotton gin made slavery profitable—an "environmental impact" never dreamt of by Eli.

He decided his primary problem had been financing so for his next project he turned to the federal government. In 1798 he signed a contract for 10,000 muskets for $13.40 apiece (normal price was $9.40) all to be delivered in 2 years. The key concept was that of interchangeable parts.

An inventive French mechanic, Honore Blanc, had developed a system of interchangeable parts for gun-making. (Moveable type probably is the first example of completely standardized and interchangeable parts.) Thomas Jefferson had visited his shop in the 1780s, seen parts "gaged and made by machinery" and had tried to get him to move to America. Jefferson was president from 1801 to 1809, which was fortunate for Eli as it took him 10 years to complete the 10,000 muskets. Although Whitney emphasized interchangeable parts and even staged a demonstration in 1801 of a musket on which he could fit any of several locks, all of the 10,000 muskets had identifying marks on each part, something which truly interchangeable parts do not need. The Whitney muskets in the Smithsonian Museum have parts that cannot be interchanged. Simeon North (pistols in 1799) and John Hall made truly interchangeable parts. Whitney, although never really achieving his objective, had good public relations and so receives the credit for the "American system of manufacture." Yet, his cotton gin probably caused the Civil War; that should be fame enough for anyone.

Michael Faraday and Joseph Henry: Key Concept—the Dynamo, the Source of Electricity; the Motor, a New Prime Mover

Michael Faraday, one of a blacksmith's 10 children, did not attend school. Fortunately he was apprenticed to a bookbinder where he not only bound books but looked inside. In 1812 he attended a public lecture by Davy. Michael took careful notes and made colored illustrations and sent them to Davy, asking to be his assistant. Davy hired him in 1813. By 1825, Faraday was director of the laboratory.

In 1823, innocent of mathematics but the "prince of experimenters," he liquified gases under pressure, in 1825 he discovered benzene, and in 1831 he had his greatest discovery, the dynamo and the electric motor. When Queen Victoria heard of the invention of the dynamo she asked "What good is it?" Faraday replied, "Madam, someday you will tax it!"

Joseph Henry came from a poor family. He left school at 10 and was apprenticed as a watchmaker. At the age of 16 he found a book, *Lectures on Experimental Philosophy,* in an abandoned church, read it, and became fired with the desire for education. He returned to school and by 1826 was teaching math and science.

In 1829, he made a vastly improved electromagnet which lifted 750 lbs; in 1835 he invented the electrical relay (in effect, the telegraph); in 1846 he became head of the Smithsonian Museum; and during the Civil War he founded the National Academy of Sciences. At the Smithsonian he began the policy of "publishing original research in a series of volumes and giving a copy to every major library on earth." Technology includes social concepts such as public libraries, technical societies, and technical journals as well as physical devices such as electric motors. If technology is an engine, then knowledge is its fuel; the fuel is becoming richer. Henry had a habit of not patenting his ideas as he thought that discoveries of science should be for the benefit of humanity. He discovered the principle of induction in 1831 but put his work aside at the end of August to be completed the following summer. Faraday published his results in November, 1831.

The electric motor has three major advantages over the steam engine: it can be made any size (especially smaller), it can be started and stopped quickly, and it can be powered at a distance by use of wires. Somewhat surprisingly, motors were not applied until about 1880 but then their growth was rapid.

Samuel Colt: Key Concept—Assembly Line

Sam designed a repeating pistol "the six shooter," "Colt's Patent Pacifier," "The Difference."

He opened a factory in Patterson, New Jersey in 1835. In Texas, where shooting was serious business, they liked it; the Texas Rangers ordered 100. The army thought it was "too complicated." His factory closed in 1842.

In 1847, the Mexican War began. General Zachary Taylor ordered 1,000 Colt 44s, and the weapon that tamed the Plains was back in production.

> The first workman would receive two or three . . . important parts and would affix them together and pass them on to the next who would add a part and pass the growing article to another who would do the same . . . until the complete arm is put together.

The National Bureau of Standards reports that the conveyor belt for assembly was not used until 1908.

Nikolaus Otto: Key Concept—the Internal Combustion Engine, a New Prime Mover

Reuleaux wrote in 1875 of the need for a small engine due to the high capital cost of steam engines: "How to make power independent of capital? . . . Engineers must provide small engines with low running costs . . . . These little engines are the true power units of the people."

In steam engines, combustion takes place outside the engine; in 1860 Jean Lenoir built the first engine where the combustion was internal. It was powered with illuminating gas and had a very poor efficiency, but it was a start.

Then in 1862, Nikolaus Otto (with his partner Eugen Langen in a relationship much like Watt and Boulton) began work on internal combustion engines. In 1876 (100 years after the steam engine) he produced the *silent Otto*, the first 4-cycle engine.

Improvements by fellow Germans Karl Benz, Gottlieb Daimler, William Mayback, and Rudolf Diesel came before 1900. The advantages over steam were low capital cost, quick starts, and high power/weight. The advantage over the electric motor was the elimination of the tether—the wire. The age of the automobile and flight began.

Thomas Edison: Key Concept—Electric Illumination

Edison represents the classic tale of the self-made man: the poor boy who, without schooling or influence, made his way to fame and fortune by intelligence and hard work. Enrolled at birth in the school of hard knocks, later with 1,093 patents to his name, he was the most productive inventor in the history of the United States—probably in the history of the human race. He also is very quotable:

> There is no substitute for hard work. Genius is 99% perspiration and 1% inspiration. [To a job applicant] Well, we don't pay anything and we work all the time. We will make electric light so cheap that only the rich will be able to burn candles.

Early inventions of an improved telegraph and the stock market ticker permitted Edison (age twenty-nine) to found Menlo Park (the first industrial research laboratory in the world—in itself one of Edison's many inventions) in 1876. He hoped to produce a new invention every 10 days! In fact, during one four-year stretch before he became involved in finance, he obtained 300 patents or one every five days.

In 1876 he improved the telephone and made it practical; in 1877 he invented the phonograph; in 1878 he announced he would tackle the problem of producing light by electricity.

Edison, who scorned theory, used the research method since then known as the *Edison method*—a patient trial of all possible alternatives. In 1879, he produced the first practical light bulb.

To apply the concept, he had to invent a host of auxiliary inventions—devices for sealing the bulbs, screw-in sockets, light switches, electric meters, safety devices—and then found the first electric utility, Consolidated Edison, which opened in 1882.

His one contribution to science, *the Edison effect,* which led to the vacuum tube and electronics, he ignored.

Herman Hollerith: Key Concept—Machine Computation

After receiving a degree in mining engineering in 1880, Herman worked on statistics for the census office. Marking tallies with dip-pen and ink just was too slow. At first he tried edge-marked cards. Then he tried holes in paper tape but was discouraged at the amount of tape that would have to be wound just to find the small amount of information that probably would be at the end of the roll. There is no evidence that he knew of the punched paper rolls used by French weavers since the mid-1700s (player pianos would use this concept in 1900) or weaver Jacquard's punched card (adapted from the roll) or even the punched card concept of Charles Babbage's *analytical engine*—a device that was never finished.

One day Herman took a trip to St. Louis. At this time train robbers were posing as passengers, and the government asked the railroads to keep track of everyone aboard. The conductors punched the ticket in specific places to indicate specific body characteristics—brown hair, blue eyes, medium weight—a *punch-photograph.*

Herman adopted the idea into the *Hollerith card,* the size of an 1890 dollar bill so it could fit existing file drawers. He developed a keyboard punching machine, rented his machines, and astounded the world with the speed of his *census machine.* IBM had been born.

Frederick Taylor: Key Concept—Scientific Study of Work

Frederick, the son of a wealthy Philadelphia family, attended Phillips Exeter prep school. Although he passed the Harvard entrance exams, his eyesight was impaired; so at 18 he became an apprentice machinist and patternmaker. Starting work at 22 during the recession of 1878 as an ordinary laborer at the Midvale Steel Co., he was successively time clerk, journeyman, lathe operator, gang boss, and foreman of the machine shop before being appointed chief engineer in 1887. He studied engineering at night and received a B.S. in mechanical engineering from Stevens Institute in 1883. The combination of practical experience and theory was to prove fruitful.

Taylor asked the question: "What is the best way to do this job?"

He did not accept opinions; he wanted facts—evidence. The questioning scientific approach (hypothesis, experiment, evaluation of data to prove or disprove the hypothesis) is well known today. It was known then and applied to chemistry and physics but not to the design of everyday jobs. This is Taylor's primary contribution—the application of the principles of science to improving jobs.

A famous example of this approach was his study of shoveling. In 1898, Taylor worked for Bethlehem Steel. In those days before mechanized material handling, when you wanted material moved you shoveled it by hand. In this one plant, 400 to 600 men spent most of their time shoveling. A variety of materials was shoveled each day by each man; each man furnished his own shovel: there was no training in shoveling techniques; pay was $1.15/day.

One thing Taylor noticed was that with a constant volume shovel the load was only 3.5 lb when shoveling rice coal but 38 lb when shoveling ore. The question was: "What was the best size shovel?" For the *experiment*, he had material shoveled with large shovels (i.e., heavy loads); then cut a little off the end of the shovel and had the same material shoveled; cut off a little more the next day, etc. *For subjects*, he used two good experienced shovelers (i.e., he replicated his data). Taylor used as his *criterion* the amount shoveled per day. The *results* indicated that maximum material was shoveled per day when the load on the shovel was 21.5 lb (for distances up to 4 ft and heights less than 5 ft). (For more on shoveling, see Muller and Karrasch, Wyndham et al., and Wyndham et al.[8, 13, 14])

To *apply* this knowledge Taylor had a toolroom established and special shovels purchased. The foreman was required to notify the toolroom of the work his gang would do that day. A large scoop was provided for shoveling ashes, a middle-size shovel for coal, and a small shovel for ore.

Taylor instituted another concept that was quite radical for that day; he did not believe that all the benefits from increased productivity should be retained by the organization; the worker also should benefit. Therefore he specified a standard tonnage to be shoveled for each type of material. The worker was trained in the proper work method, given the proper tools, and put to work. When he achieved standard he received a 60% bonus above the day wage rate. If he could not achieve standard even after training he was put on a different job. The concept of pay-by-results (incentive wages) and personnel selection (select the best people for each job) are commonplace today; they were radical innovations then.

After these methods were applied to the yard at Bethlehem, the same amount of work was done with 140 men; material handling cost for the company (including cost of the study, toolroom, and bonuses) was reduced from $.08/ton to $.04. Employee wages, of course, were 60% higher.

Taylor's *scientific management* is a good example that technology includes techniques as well as devices.

The concept that the job is to be designed by experts (engineers) and the worker's duty is just to follow instructions is called "Taylorism." It has resulted in dramatic improvements in productivity and the standard of living for many cultures for approximately 100 years. However, the educational level of the workforce has risen enough in some countries so that greater participation of the workers in job design has become feasible. See chapter 27 for some specific techniques.

Henry Ford: Concepts—High Pay for Workers, Low Cost Auto, Mass Consumption

Just as Whitney is incorrectly credited with interchangeable parts, Ford is incorrectly credited with the auto assembly line. Specialization of labor combined with use of conveyors had been used earlier in slaughterhouses (a "disassembly" line) but the auto was the product in the public eye (see Figure 2.1). Random Olds applied the assembly line concept to building Oldsmobiles in 1899, 10 years before Ford installed an assembly line. By 1904, Oldsmobile production had reached 5,000/yr. Ford, however, dominated auto production; in 1924 Model Ts accounted for one-half of the world's motor vehicles. Ford's assembly line cut throughput time/car from 13 hours to 1. Formerly one man took 20 minutes to assemble a magneto by himself; using a subassembly line with 29 operators, time was cut to 13 man-minutes/unit.

High pay for workers and a low cost auto were implemented with Ford's decision in 1914 to pay his workers $5 for an 8-hour day when they had been getting $2.50 for a 9-hour day. In addition he set up a $30,000,000 profit-sharing fund. How could he double wages while reducing costs? Through productivity. (High wages without accompanying productivity mean either losses or inflation.) The concept of *mass consumption* is the key to our society. When his employees received low wages, they were not able to purchase cars; cars were for rich people. As long as cars were only for rich people, the total output of cars was small. And, looking at the problem as a businessman, Ford saw that even though his profit/car was satisfactory, his total profit was limited due to the small number of cars he could sell because of their high price.

To maximize his total profits (not because he was a social visionary), Ford set up his assembly line, made a standardized product ("any color you want as long

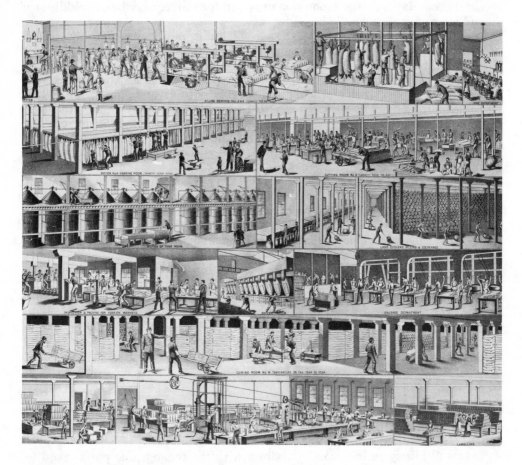

Figure 2.1 *Assembly lines* did not begin with the automobile, for the lithograph "Interior View of a Modern First Class Pork Packing & Canning Establishment of the United States of America" (Courtesy of the Chicago Historical Society) was published in 1880. Titles under the respective pictures are: The Office; Killing Benches Nos. 3 & 4 (capacity 700 pr. hour); Fire Department (at dinner); Section No. 4 Hanging Room (capacity 10,000 hogs); Cutting Room No. 3 (capacity 5000 pr. day); Section of Tank Room; Lard Coolers, Filling & Cooperage; Inspection & Packing for Foreign Markets; Boiler Room; Sausage Department; Curing Room No. 4 Temperature 38 Fah. Year to Year; Polishing; Canning Department Filling by Machinery, Meat Untouched by Hand, Soldering; Labelling. A question might be why it took so long to apply assembly line technology to other industries.

as it's black'') at high volume (and thus low cost), and paid his workers well (permitted by the high volume, which was permitted by the high sales, which were permitted by the high wages and low costs, which were permitted by the high volume which was permitted by the high sales which were permitted . . .). The natural result, as Horatio Alger would say, was that Henry Ford became the richest man in the world.

Yet when all is said and done, the "main effect" of mass consumption of automobiles may really be less important than the "side effect" of personal mobility obtained as a result of the automobile. In the USA it has increased our radius of daily travel from the limits of the streetcar to approximately 50 miles; this in turn has changed population distributions in both cities and small towns. Although factories were located in central cities at one time in order to obtain employees, effectively no factories have been built in central cities in the USA since 1940. Trade and distribution jobs began moving out of the central cities in the 1950s; during the 1960s 9 of the 10 largest American cities lost population. It is difficult to determine in advance the main effects of developments; it is exceedingly difficult to anticipate the side effects and interactions occurring after a time lag.

Table 2.1   Prices of Fords Vs. Wage/Hour

Year	Type of Model T	Price ($)	Wage/hr of unskilled labor in Ford factory	Normal hr/ day	Hours of work required to buy car
1908	Roadster	825	.19	10	4340
1913	Runabout	500	.26	10	1923
1918	Runabout	434	.50	8	870
1923	Runabout	265	.75-.85	8	331

Source: Courtesy of Ford Archives.

Frank and Lillian Gilbreth: Key Concept—Motion Study

Frank was a super-organized man. An example of his organizing (publicized in *Cheaper by the Dozen*) was his posting of French grammar on the wall of his bathroom. While sitting on the toilet, his children were expected to learn French! He also had the tonsils of all his children taken out at one time; he recorded the surgeon's hand motions with a movie camera. (Unfortunately his assistant forgot to put film in the camera.) But now you know what kind of man he was. His business was construction. A key construction job was laying brick. Frank, who had learned the bricklayer's trade as a teenager, applied his analytical skills to the study of bricklaying. He did this by studying the job in great detail—micromotion study.

In the existing method, bricks were dumped in a heap on the scaffold; the bricklayer bent over and picked up a brick. He then inspected it on all sides to select the best side for the wall face.

In Gilbreth's method, the bricks were inspected when they were unloaded from the freight car. They were placed with the best side up in packets of 90 lb. The packets of oriented bricks were then placed on the scaffold. The scaffold was redesigned so that it could be raised or lowered easily, thereby reducing the distance the bricklayer had to reach. The mortar box and the packets were placed on the scaffold so that while the bricklayer picked up a brick with one hand, he scooped his trowel full of mortar with the other hand. The mortar was made of a standardized consistency so that the motion of tapping the brick into place with the trowel was no longer required. The number of motions/brick was reduced from 18 to 4.5. In a test of the new method, bricklayers laid 350 bricks/hr while the previous record for this type of construction had been 120 bricks/hr.

The reason Gilbreth's results attracted such attention was the task studied—bricklaying. Men had laid brick since antiquity. If there was any skill in which no more change could be anticipated it seems that bricklaying, with its 3,000 years of experience, certainly would be it. Yet one man, by systematic study, had made a 300% improvement over the experience of 3,000 years! A powerful example.

When Frank died his wife Lillian continued his work on micromotion study. The many small elements of jobs have been named *therbligs*—Gilbreth spelled backward (almost).

Lee de Forest: Key Concept—Three-Electrode Tube (triode)

The telegraph was a great boon to speedier communication, but it required a wire connecting the sender and receiver. Marconi developed the *wireless telegraph*, the sending of dots and dashes without intervening wires. Edison, who had been a telegrapher himself, had discovered the *Edison effect*—electrons flowed across a gap between *two* electrodes inside a vacuum tube. Fleming developed the concept into a device—the two-electrode tube or rectifier. In 1906, de Forest added the *third* electrode (the grid). A varying weak signal on the grid could be converted into a varying strong signal between the cathode and anode. Signals could now be continuous (rather than just discrete) and could be amplified. In 1912, he added the concept of a feedback circuit. Radio and the age of electronics began.

Lee, who received a Ph.D. from Yale in 1899 on electromagnetic waves, never received much popular acclaim. Seniors at Yale voted him "nerviest and homeliest man in the class." Although he eventually had 300 patents, even his development of the feedback circuit was challenged by Armstrong. In 1934, after 14 years of litigation, the score was 6 court verdicts for de Forest and 6 for Armstrong. The US Supreme Court made it 7-6 with its decision that de Forest had invented the feedback circuit 2 months before Armstrong.

SUMMARY

The "democracy" of ancient Greece was based on slaves to do the work. Now we have machine slaves rather than human slaves. Machine power comes from internal and external combustion, often distributed and applied locally by electricity. We compute and communicate with machines. We have organizational concepts such as interchangeable parts, assembly lines, scientific study of work, and high pay for workers.

Our lives have been profoundly influenced by the technical culture although most historians write only of the other culture—that of poets and politicians, the nonproducing verbalizers rather than the nonverbalizing producers.

SHORT-ANSWER REVIEW QUESTIONS

2.1 What are the "two cultures"?

2.2 How are accurate machine tools and interchangeable parts related?

2.3   Prime movers are essential to an industrial society. Why was the steam engine important? Why was the electric motor important? Why was the internal combustion engine important?

2.4   Describe Taylor's shovel experiment using the words *task, subjects, controlled variable, criterion, results,* and *application.*

2.5   Who were the Gilbreths and why were they important?

THOUGHT-DISCUSSION QUESTIONS

1. Discuss the concept that engineers and scientists are really the critical "change agents" who affect our lives.

2. Discuss the concept that "technology" includes public libraries and technical journals as well as physical devices.

REFERENCES

1. Asimov, I. *Biographical Encyclopedia of Science and Technology.* Garden City, NY: Doubleday and Co., 1964.

2. Camp, W. Recollections of H.L. Gantt. *Journal of Industrial Engineering,* Vol. 12, No. 3 (May-June, 1961): 217-223.

3. Finch, J. *The Story of Engineering.* Garden City, NY: Doubleday Anchor, 1960.

4. Klaw, S. Frederick Winslow Taylor. *American Heritage* (Aug-Sept, 1979): 26-39.

5. Klemm, F. *A History of Western Technology.* New York: Charles Scribner, 1959.

6. Kranzberg, M. and Pursell, C., eds. *Technology in Western Civilization.* Toronto: Oxford University Press, 1967.

7. Morison, S. *The Oxford History of the American People.* New York: Oxford University Press, 1965.

8. Muller, E. and Karrasch, K. Die grosste Dauerleistung beim Schaufeln. *International Z. Physiol. einschl. Arbeitsphysiol.,* Vol. 16 (1956): 318-24.

9. Reynolds, N. Dr. Lillian Moller Gilbreth. *Industrial Engineering,* Vol. 4, No. 2 (Feb. 1972): 28-33.

10. Tichauer, E. Gilbreth revisited. New York: American Society of Mechanical Engineering, paper 66-WA/BHF-7, 1966.

11. Tuchman, B. *The Proud Tower.* New York: Bantam Books, 1967.

12. _____. *Those Inventive Americans.* Washington, DC: National Geographic Society, 1971.

13. Wyndham, C.; Morrison, K.; Williams, C.; Heyens, A.; Margo, E.; Brown, A.; and Astrup, J. The relationship between energy expenditure and performance index in the task of shovelling sand. *Ergonomics,* Vol. 9, No. 5 (1966): 371-378.

14. Wyndham, C.; Morrison, J.; Viljoen, J.; Strydom, N.; and Heyns, R. Factors affecting the mechanical efficiency of men shovelling rock in stopes. *Journal of the South African Institute of Mining and Metallurgy,* (October 1969): 53-59.

3

WORK SMART NOT HARD

PRODUCTIVITY AND LIVING STANDARDS

People have worked hard throughout history. People who extol the good old days as they protest technology usually picture themselves at Versailles rather than wielding a hoe. Millet's painting, "The Man with the Hoe," depicting agriculture labor in France during the 1830s, inspired Edwin Markham to write "The Man with the Hoe."

> Bowed by the weight of centuries he leans
> Upon his hoe and gazes on the ground
> The emptinesses of ages in his face
> And on his back the burden of the world.
>
> Is this the Thing the Lord God made and gave
> To have dominion over sea and land,
> To trace the stars and search the heavens for power,
> To feel the passion of Eternity?
>
> What gulfs between him and the seraphim!
> Slaves of the wheel of labor, what to him
> Are Plato and the swing of Pleiades?
> What the long reaches of the peaks of song,
> The rift of dawn, the reddening of the rose?

They left agriculture for the factory to improve their standard of living. And yet even this improved standard of living was low compared to today. Most people today consider Bob Cratchit (Scrooge's clerk in *A Christmas Carol* by Dickens) to be poor. Yet in his England, the country with the highest standard of living in the world in the 1840s, he was a well-paid worker. Cratchit could read, write, and calculate—certainly valuable traits since compulsory schooling did not begin in England until 1870. His wage was 15 shillings/week (39 pounds/yr); the average wage in the United Kingdom was 30 pounds (roughly twice the mean income/capita in the USA). Indoor laborers made 5 shillings/week, and weavers made 13 shillings/60-hour week. Note that Cratchit lived in a four-room house and that his wife and 5 of 6 children did not work.

The move to the city had drawbacks, as in 1900 the British government reduced the minimum height for soldiers to 60 inches from the 63 inches set in 1883 due to the lack of nourishment received by the ordinary citizen during this time. Yet progress continued. The height of Dutch Army inductees was 168 cm in 1850, 170 in 1900, and 175 in 1950.[7]

Table 3.1 gives a concise view of the change, at 70-year intervals, in the standard of living in the USA since the first census in 1790. To help predict "2001," data are given for the "halfway" point between 1930 and 2000. Figure 3.1 gives the output/work-hour from 1890 to 1970.

Man has climbed laboriously over hundreds of obstacles to successively higher and higher ledges of income. The rich nation is the exception; the techniques that made entire nations rich are the pivotal events of history. Was it done by working harder? No. By working more efficiently? Yes. We now are more productive—producing more output for the same input.

The benefits from increased productivity are not always taken in increased material goods; they can be spent on health care, wine, weapons, or increased leisure. Of course, with more productivity, there is more to share; "a rising tide lifts all boats." The choices have interactions. Education may be considered a goal in itself; yet the correlation coefficient between educational level and gross national product/capita in 75 countries is .89.[6] Is education chicken or egg? The same question can be asked about the high positive correlation between freedom of the press and GNP/capita.

Before 1914 the 60-hr week was typical for most industrial workers in France, Germany, England, Canada, and the USA. By 1922 the 48-hr week was in general practice in industry throughout Europe, Australia, New Zealand, and Latin America. In 1926 Ford introduced the 40-hr week in the USA. By 1948 the 40-hr week was the norm in the USA, Australia, New Zealand, and the USSR.[4]

Table 3.2 shows the decline in working hr/week in various countries; Table 3.3 shows how some countries put more emphasis on leisure than others.

Table 3.3 is based on the nominal hours worked by a "male manufacturing worker" assuming full employment, zero illness, and zero absenteeism. They are representative values rather than official statistics. (In the USA in 1972 average hours were 40.8 in manufacturing but 34.5 in wholesale trade, 34.1 in services, and 40.4 in transportation.) Using the USA as an example, a "typical" male industrial worker would enter the labor force at age 18 and retire at 65—giving 47 years of work. A "typical" work week is 5 days but he receives 15 working days vacation (less when younger, more when older) and 8 paid holidays, so he works 237 days/year. The "typical" work day of 8 hours (not including lunch) minus two "coffee" breaks of 10 minutes each gives 7.67 hours of work/day. The total of 85,000 hours is affected by absenteeism, strikes, illness, unemployment (voluntary or involuntary), overtime, second jobs, more or less schooling, and early or late retirement. The 85,000 hours must support not only the worker's needs for food, furniture, and frivolity but also support the children, aged, blind, and other nonproducers (either directly or through taxes). In this book we will discuss "how to increase the size of the cake" rather than "how to cut the cake"; that is, how to multiply not how to divide. (It seems the job of engineers is to increase the pie size and the job of politicians is to divide it.)

The output from the 85,000 hours can be increased by working efficiently (WORK SMART) or by working with more effort (WORK HARD). Work Smart is the desirable alternative because (1) there is more potential for improvement through reducing the excess work than through making the worker work harder,

Table 3.1   A Concise View of the Change in the Standard of Living of the Common Man in the USA since 1790. The Change Was Due to Improved Productivity.

Index	1790	1860	1930	1970-72
Population	4,000,000	31,000,000	123,000,000	213,000,000
Housing	Single-family log cabin	Frame houses. 93% of dwellings single family	Frame houses. 67% of dwelling single family. Multistory of brick with steel frame. 3.8 population/dwelling unit; 4.8 rooms/dwelling unit (median).	Frame houses. Apartments, condominiums. 63% own their own home. 3.1 population/dwelling unit; 5.0 rooms/dwelling unit.
House furnishing and equipment	Homemade furniture. No running water. Privy. Fireplace heat. Illumination by candle.	Factory-made furniture. Hot & cold running water in homes of rich. Stove (wood, coal). Illumination by kerosene and gas lamps. Matches.	Factory-made furniture. 　　　　　　Urban Rural Electricity　96%　31% Indoor toilet　92　11 Running water　92　18 Mech. refrig.　56　15 Central heat　58　10 6,000,000 pianos	% of Dwelling Units Flush toilet + H & C water + bath or shower　94% Running water　98 Telephone　94 TV　99+ Air conditioning　45 Home freezer　32
Food & drink	Local supply, little variety. Preserve by salt, pickling, & smoking.	Regional supply. Preserve also with ice. .2 cans of food/yr/capita. 178 lb red meat/yr/capita (carcass weight).	Regional & world supply. Mech. refrig. transport and home. 54 cans of food/yr/capita. 141 lb red meat/yr/capita.	Regional & world supply. Fresh at all seasons, frozen, freeze dried, convenience foods. 101 cans of food/yr/capita. 192 lb red meat/yr/capita (49 fowl, 11 fish). Food cost = 16% of disposable income.
Clothing	Made at home. Linen, wool, leather.	Some factory clothing, especially men's. Cotton & wool.	Factory-made for men & women. Rayon & silk plus cotton & wool.	Factory-made for all. Synthetic fabrics. Fabric treatments (permapress, soil and water resistant).

(continued)

Index	1790	1860	1930	1970-72
Health	Life expectancy at birth = 36. Doctors trained as apprentices. First hospital (1750). First medical school (1765). No public responsibility for health or sanitation.	Life expectancy at birth = 40. 149 hospitals; .0009 beds/capita (1873). 40 medical schools. 175 doctors, 18 dentists, 26 nurses/100,000 pop. First comprehensive water and sewer system (Chicago).	Life expectancy at birth = 60. 6150 hospitals; .009 beds/capita. 125 doctors/100,000 pop.	Life expectancy at birth = 71. 7100 hospitals; 0.0075 beds/capita. 174 doctors, 57 dentists, 353 nurses/100,000 pop. 36% of pop. have major medical insurance. Medicare.
Transportation & communication	Postal expense = $.01 yr/capita. No public roads.	Roads by local government. 35,000 miles of RR track.	227 pieces of mail/yr/capita. 200,000 miles of RR track. 53% of families own a car (1937). Some commercial air transport.	409 pieces of mail/yr/capita. .6 telephones/capita. 204,000 miles of RR track. 3,700,000 miles of highway. 83% of families own 1 or more cars (.47 cars and .1 truck/capita). 150,000,000 air passengers/yr. Man on moon.
Work & leisure	Child labor; work until death; dawn to dark; 6-day week; no vacations.	65-70 hr/wk for city & factory workers; no vacations. Circus, vaudeville, nonprofessional baseball. Newspaper circulation = .05/day/capita.	40 hr/wk or less for 50% of wage earners. Begin work at 16-18; retire without pay before death; some vacations. Unions 15% of workers; 30% white collar (1940). 90,000,000/wk movie attendance. Radio. 6000 golf courses. Baseball and football pro sports.	56% of people over 16 hold at least one job. 40 hr/wk or less average for entire work force. Retire at 65 (Social Security & company benefits). 6 or more paid holidays for 97% of work force; 15 days paid vacation after 15 years for 93%. Unions 23% of workers; 50% white collar. Newspaper circulation = .29/day/capita. 6600 radio, 707 TV stations; 1463 symphony orchestras, 713 opera companies, 763 museums, 11,000 golf courses. 100% of homes have TV. 3.7 visits/yr/capita to federal parks or recreation areas.

(continued)

Index	1790	1860	1930	1970-72
Education	No public education	Some tax-supported libraries. 80% literacy. Average schooling = 434 days. First land grant university (Kansas State University).	Education (including secondary & college) government responsibility. Free education for 12 grades. Median school completed by pop. over 25 = 8.0 yr. 96% literacy. School expense = 3.1% of GNP. 6200 public libraries.	Median school completed by population over 25 = 12.2 years. Literacy = 99%. School expense = 8.0% of GNP. 7109 public libraries.
Government	New York City government expense/yr/capita = $1.87.	150 public and private water sypply systems. New York City government expense/yr/capita = $10.52.	New York City government expense/yr/capita = $189 (1935). Taxes = 10% of GNP. Government (all levels) workers = 9.7% of all workers.	Federal $729, state & local $731 expense/yr/capita (nationwide). New York City government expense/ yr/capita = $1207. Taxes = 31% of GNP. Public responsibility for unemployment and recreation. Government (all levels) workers = 19.3% of all workers (1975).

Table 3.2   Actual Weekly Hours of Work (Including Overtime) in Manufacturing[5]

Country	1953	1958	1963	1968	1973
Australia[b]			42.8	43.7	43.2*
Austria[a]			38.6	38.6	34.6
Belgium			41.2	39.5	36.8*
Canada[b]	41.3	40.2	40.8	40.3	39.6
Czechoslovakia[a]			47.8	44.5	43.4
Denmark[a]			39.8	37.8	35.5*
Finland[a]	44.1	42.6	44.1	39.1	38.2*
France[a]	44.5	45.3	46.3	45.3	43.5
Germany (West)[b]	48.0	45.5	44.3	43.0	42.9
Ireland[a]	45.1	45.1	44.6	43.3	42.4
Japan[a]	48.4	46.5	45.5	44.6	42.0
Netherlands[a]	48.9	48.6	46.6	45.3	43.3*
New Zealand[a]		39.8	40.5	40.2	40.5
Norway[a]		40.8	38.4	36.7	32.7
Spain[a]			44.8	44.1	43.3
Switzerland[b]	47.7	46.8	45.5	44.6	44.3
USSR[a]			40.2	40.5	
USA[b]	40.5	39.2	40.5	40.7	40.7

*1972 value

NOTE: Due to different definitions in each country, hours between countries are not strictly comparable: [a]Hours actually worked, [b]Hours paid for.

and (2) people don't like to work hard and therefore resist efforts made to make them work hard.

COMPONENTS OF PRODUCTIVITY

As mentioned before, productivity is the ratio between input and output. One definition of output gives four essentials: labor, materials, energy, information. The special character of a technical society is that the materials, energy, and information replace labor.[1] There are four classic factors: land, materials, machines, and labor. (Sometimes materials and machines are called capital.) Overriding these four factors is a fifth factor, technology (the combination of scientific, engineering, and managerial techniques).

Improved productivity from *land* might be using better seed to grow 10% more corn/acre, or better trees which will mature in 20 years instead of 25, or better scheduling so that a factory produces 10% more radios. Fertilizer or insecticides may increase crop yield. Output/unit of land increases.

Improved productivity from *materials* might be use of a collector container to catch the drips from barrels of viscous chemicals so that 99.7% of a container's contents are used instead of 98.5% or use of a noncorrosive material to extend the life of a bridge or truck. Insulation reduces fuel oil need. Output/unit of material increases.

Improved productivity from *machines* might be scheduling a truck to haul materials both going and coming rather than returning empty, or using a ceramic cutting tool in a lathe so a higher speed can be used. A machine's physical life might be 50,000 hr. Use of 1 shift (2000 hr/yr) would let the machine be used 25 years. Two shifts (4000 hr/yr) would improve machine utilization and decrease the risk of obsolescence. (Because the problems of shift work are social rather than physiological, shift work may not be utilized if public transport or recreation or shopping are closed during part of the day.) Output/unit of machine time increases.

Table 3.3 Nominal Working Hours/Lifetime For a "Typical" Male Manufacturing Worker in Various Countries in 1975. Full Employment, Zero Illness, Zero Absenteeism, And Zero Overtime Are Assumed as Well as No "Early" Retirement.

	USA	Germany	France	UK	Netherlands	USSR	S. Africa	India	Hong Kong	Korea
1. Working years	47	48	49	49	49	43	42	39	42	37
Leave labor force	65	65	65	65	65	60	60	55	60	55
Enter into labor force	18	17*	16	16	16	17	18	16	18	18
2. Working days/year	237	227	230	235	231	243	225	284	321.5	263.0
Days/week	5	5	5	5	5	5	5	6	6.5	5.5
Days/year	260	260	260	260	260	266**	260	312	338.0	286.0
Vacation	15	22	22	18	22	15	22	18	6.5	11.0
Holidays	8	11	8	7	7	8	13	10	10.0	12.0
3. Hours/day	7.67	7.5	8.27	7.5	7.5	7.5	7.5	7.17	8	8
Hours (excluding lunch)	8.00	8.0	8.6	8.0	8.0	7.5	7.5	7.5	8	8
Breaks	0.33	0.5	0.33	0.5	0.5	0.0	0.5	0.33	0	0
Hours/lifetime	85,000	82,000	93,000	86,000	85,000	78,000	71,000	79,000	108,000	78,000

*Part time vocational school from 16-18
**4 hours on one Saturday/month

Improved productivity from *labor* might be improvement in the work methods of a nurse to permit attending to more patients, or a simplified form so that a clerk could calculate more vouchers/hr. Use of a fixture to hold parts can permit assembly with two hands instead of one. Output/unit of time increases.

Although the examples assumed the same input with an increase in output, improved productivity also can occur by decreasing input for the same output, by increasing output faster than input, or by decreasing input more than the output decrease. Productivity should be recorded in nonmonetary units so comparisons are not distorted by inflation. Example indices are vouchers/week from the accounts payable office, number of student credit hours/teacher, and area cleaned/day by the janitors.

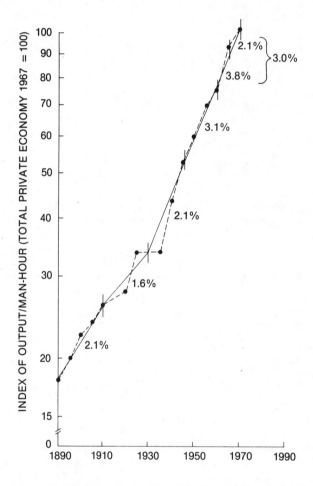

Figure 3.1 *Output/work-hour* (productivity) has increased 2% to 3% per year in the USA. Will past be prologue?

It also will be emphasized that productivity is a mixture of the factors of land, materials, machines, and labor. Later-developing nations have an advantage in that they can selectively accept ideas from an ever-larger store of transnational knowledge. This knowledge is not just physical hardware but also social knowledge (Quality Circles, interlibrary loans, double entry bookkeeping, agricultural extension agents). The popular press often writes as if only reduction in labor costs is meaningful and ignores improved productivity for the other factors. Historically, the developed countries have substituted cheap energy and materials for labor (see Figure 3.2). The rise in the price of oil since 1973 has resulted in substitution of labor and capital for energy. In the USA, labor costs now dominate while in the lesser-developed countries a shortage of foreign exchange or surplus of labor make material cost or machine cost dominant. In addition, some industries are labor-intensive, some skill-intensive, and some knowledge-intensive. Thus engineers in different industries have different objectives.

TOTAL TIME FOR A JOB OR OPERATION

Any program to raise productivity by reducing time/unit must consider the workers' fears (1) that they will work themselves out of a job, and (2) that employers will get all the benefits of the higher productivity. Policies to ensure adequate employment and satisfactory distribution of the benefits of the productivity are not merely desirable parts of productivity programs—they are the foundation. (The good of the whole conceals the cost to the few.)

Reducing time/unit gives costs: (1) erosion of individual skills and experience, (2) the need for some workers to change their jobs and perhaps their place of residence, and (3) some individuals may never be able to make the changes required.[3] Thus workers, individually and collectively, need to have the costs of improved productivity not fall too heavily on any person. One of the key aspects of Japan's high annual productivity increases is that the male Japanese worker is hired until age 55 and thus does not resist technological change. In the USA, employers attempt to use normal turnover and expansion of sales to cushion layoffs; severance pay and unemployment pay are a backup system.

People also differ on who should get the benefits of higher productivity: the workers through higher wages, the society through lower prices, or the person who risks his capital through greater profits. Wars have been fought over this issue. When political rhetoric is brushed aside, the answer is that benefits must be split among the three. Naturally there is always discussion on the amount of the split. Samuel Gompers expressed labor's opinion concisely when asked "What did labor want?" "More."

Time/unit can be considered to be composed of[3]:

Basic work content
Extra work content

1. Due to poor product design

2. Due to poor manufacturing methods

3. Due to poor management

4. Due to poor workers

For simplicity the following examples will emphasize manufacturing but the same concept also applies to banking, retailing, transportation, health services, etc. Also for simplicity we will consider productivity for a product (e.g., making a shirt) or an operation (e.g., selling the shirt) to depend only on the time content (labor or machine).

Extra Work Content Due to Poor Product Design

Improper design. For example, don't design a product to be an assembly of weldments when a casting is more economical. (In some situations, weldments are better than castings.) Design for easy maintenance. Don't omit a protective

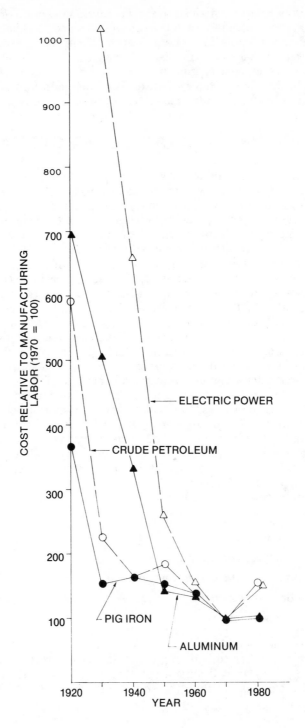

Figure 3.2 *Relative costs* of energy, materials, and labor change over time. Manufacturing labor wage in 1970 ($3.36/hr) is the base. As an example calculation, the price of crude petroleum was $3.18/barrel in 1970, so the ratio was 1.06. In 1960 the price of oil/labor was 2.88/2.26 = 1.27. Thus 1.06 × 1.27 = 1.35. Until 1970 (actually until 1973) the relative cost of energy vs. labor declined rapidly and the relative cost of materials vs. labor declined slowly. Since 1973 the cost of energy vs. labor has risen fairly rapidly while the cost of materials vs. labor has risen slowly.

coating for steel used in a corrosive atmosphere. Don't use steel for truck trailers when payload is limited by highway weight limitations; use aluminum so more payload can be moved per trip. Design containers to stack to reduce cost of shipping empty.

Nonstandardization. Don't use a special thickness washer instead of a standard washer. Use standard materials. Lack of standardization splits production volume between two sizes, increases paper work, and makes supply of spare parts more difficult.

Incorrect quality standards. Don't use plastic which breaks instead of steel. The converse problem is "gold-plating" in which an engineer overdesigns. An example would be use of precision threads when standard threads are sufficient. (We all tend to design to be safe so the cumulative result of multiple levels of designers being safe is super super super safe.)

Material Wastage. Don't use a large size of paper for a form when a small size will do, or design a shaft with both large and small diameters so excess material is lost when it is machined, or select an engine with poor fuel economy so that excess fuel is used.

Extra Work Content Due to Poor Work Methods

Incorrect size or type of machine or tool. For example, don't use a large truck when a small truck is sufficient, or the wrong type of screwdriver, or a nonadjustable chair so the typist becomes tired quickly.

Incorrect use of the proper tool. Don't have poor maintenance of a truck, or a cutting tool running at too slow a speed, or fail to instruct a worker on how to use a tool.

Poor layout. Don't locate supplies so that distance traveled by people or material is excessive. Failure to use the correct size bin can cause excess trips.

Poor work methods. Use a telephone instead of writing a letter or making a physical visit. Don't furnish parts in only one bin so that the worker uses one hand instead of two or have the bin 20 cm away when it could be 15. Don't have the waitress' order-taking form without a printed listing of the food so that unnecessary time is spent in writing instead of making a checkmark. Don't cut one piece of cloth at a time—cut 60.

Extra Work Content Due to Poor Management

Too wide a variety of products. Standardization is inadequate and productions runs are low. If a market of 100,000 units/year must be divided among 10 models instead of 4, use standard components (motors, brackets, switches) in all 10 models so that they, at least, still have the benefit of higher volumes.

Inadequately designed product. Resources are wasted due to costly rework and repair as well as interrupting the flow of operations in the factory. The customer has wasted money.

Poor production scheduling. Setup time is increased, production runs are too short, and workers work overtime one month and then are surplus the following month. A "unit train" hauling coal shows how scheduling can improve productivity.

Inadequate maintenance of equipment. Workers are plagued with equipment which breaks down. Quality of items manufactured may be defective.

Poor quality assurance policies. When machines are not set at the proper setting, scrap results. Then the product either has to be reworked or, worse, a poor-quality product is sent to the customer. Schedules might not be met.

Inadequate concern for working conditions and safety. Excessive time and product are lost due to accidents and near-accidents. Pain and suffering occur unnecessarily.

Extra Work Content Due to Poor Workers

Not spending full time on the work. Workers fail to start on time, quit early, and stretch breaktimes.

Being absent without cause. Since the replacement worker usually is not as efficient as the regular worker, time/unit increases. Replacements may not be available.

Causing scrap through sheer carelessness. Most quality problems are due to poor management but a small minority are from worker carelessness.

Causing an accident by carelessness. Again, my personal opinion is that management must make a job not just "fool proof" but "damn fool proof."

From the foregoing items, in most operations or jobs the effect of the worker in causing extra work is relatively minor compared to the effect of poor product design, manufacturing methods, and management. Obtaining better productivity (the benefits to be split among the workers, the society, and the providers of capital) is thus a shared responsibility. The importance of working smart not hard can be demonstrated to students or managers with the pegboard demonstration.

PEGBOARD DEMONSTRATION

Figure 3.3 is a photograph of the pegboard task: "work smart" in front, "work hard" at the right. The task is to put 30 pegs into the 30 holes. Figure 3.4 gives a left-hand/right-hand chart of the method used by the instructor in demonstrating Condition A: blunt end of peg into nonchamfered hole. The students should time the instructor doing one assembly to the instructions "Work at a pace you can

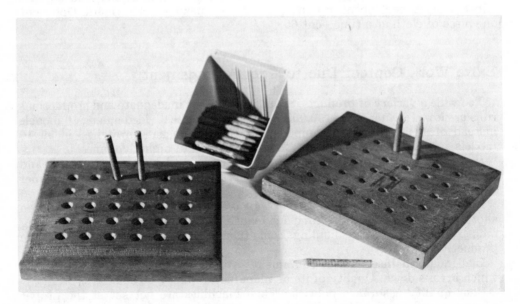

Figure 3.3 *Work Smart Not Work Hard* can be demonstrated with assembly of pegs into a pegboard.

maintain for 8 hours; assume you are paid by the hour." For reference, when 51 students assembled 10 boards, their mean time for 10 boards for Condition A was 1.02 min.

Condition B demonstrates WORK FAST. Use the method for Condition A but with instructions "Work at a pace you can maintain for 8 hours; assume you are paid by the piece." Mean time/assembly for 10 assemblies for the 51 students was .81 min—a reduction of 21%.

Condition C demonstrates WORK SMART—that is, reduce the excess work content. For *product design* changes, use chamfered holes instead of non-cham-

ACT BREAKDOWN

SKETCH

Study File No. _____ Date _____
Oper. Name—Equip. Description _____
   Pegboard Assembly—Method A & B

Tools Used_____ 1 bin for pegs _____

Part Description pegboard with non-beveled holes (30)-3/8 D   5/16 D peg

Part No. _____
Routing Hrly. Cap. _____

Study Hrly. Cap. _____
Analysis By _____ SK _____

Step No.	LEFT HAND DESCRIPTION	OBJECT	ACT	Process	ACT	OBJECT	RIGHT HAND DESCRIPTION	
1			W		G	peg	in bin	} Repeat
2			W		P	peg	in hole—column 6	} 5 times
3			W		G	peg	in bin	} Repeat
4			W		P	peg	in hole—column 5	} 5 times
5			W		G	peg	in bin	} Repeat
6			W		P	peg	in hole—column 4	} 5 times
7			W		G	peg	in bin	} Repeat
8			W		P	peg	in hole—column 3	} 5 times
9			W		G	peg	in bin	} Repeat
10			W		P	peg	in hole—column 2	} 5 times
11			W		G	peg	in bin	} Repeat
12			W		G	peg	in hole—column 1	} 5 times
13								
22								
23								
24								
25								
26								

03080-219 - 12-66 -- IE-74-1   (OVER)

Figure 3.4 *Detailed analysis* of the pegboard assembly shows that the right-hand activity includes only the acts of get and place while the left hand has only inactivities of wait. (Acts are get, place, and dispose; inactivities are hold, wait, process, and drift.) The analyst "worked smart" by using G for get, P for place, and W for wait. The analysis is typed for clarity—normally it would be handwritten.

fered holes (turn board over) and insert the pointed end of the peg instead of the blunt end. For *manufacturing methods* changes, use two parts bins and two hands instead of one and fill the center holes first to avoid moving the hands over a barrier. Preorient the pegs in the bins. (Preorientation involves a cost but cost would be low if it could be done by the previous operator or by a vibratory feeder.) Figure 3.5 gives a left-hand/right-hand chart for Condition C. Working at the pace of Condition A, the mean time/assembly for 10 assemblies was .47 min—a reduction of 54%.

The effect of better management can be demonstrated by long production runs. To save demonstration time, use the learning curve rate of 94% calculated by Youde, who timed 300 consecutive assemblies of the pegboard.[8] The 94% means that every time output is doubled, the new time is 94% of the previous time. Thus if time for 10 is .47, then time for 20 would be .94 (.47) = .4418; time for 40 would be .94 (.4418) = .4153; time for 80 would be .94 (.4153) = .3908, etc.

The potential improvement for better product design, manufacturing methods, and management is unlimited. Convincing workers to work harder is difficult. Sweat, as they know, is inversely related to wealth.

The British posted a guard on the cliffs of Dover in 1812 to watch for an invasion by Napoleon. The job was abolished in 1935. They should have worked smart not hard.

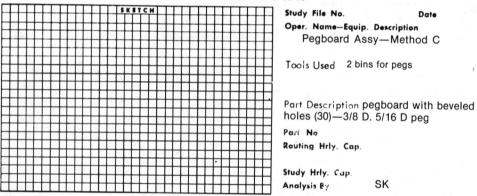

Figure 3.5 *Reduction of excess work* content by "working smart" demonstrates the benefits of using your brain instead of your muscles.

SUMMARY

The rich nation is the exception; techniques which make not only individuals but entire nations rich are the pivots of history. Our standard of living, which can be taken in leisure, health care, wine, or weapons as well as consumer goods, depends on our productivity. Productivity comes from efficient technology (combination of scientific, engineering, and managerial techniques) applied to land, materials, machines, and labor. The key to productivity is to WORK SMART, not to WORK HARD.

SHORT-ANSWER REVIEW QUESTIONS

3.1 Output/work hour (productivity) in the USA has increased at what approximate rate since 1900?

3.2 Graph the ratio of labor cost versus the cost of a material, of oil, and of electrical power in the USA since 1900.

3.3 How many hours of work/lifetime is typical in many different cultures? Show calculations.

3.4 There are four classic factors of production: land, materials, machines, and labor. Give an example of how productivity can be increased by each of the four.

3.5 Why is it more important to work smart than work hard?

THOUGHT-DISCUSSION QUESTIONS

1. Discuss the relation of living standards and productivity.

2. Discuss the approach of "increasing the size of the cake" versus the approach of "how to cut the cake."

REFERENCES

1. Arnold, J., The frontier in space. *American Scientist*, Vol. 68, No. 3 (1980): 299-301.

2. International Labour Office. *Higher Productivity in Manufacturing Industries.* Geneva, Switzerland, 1954.

3. _____, *Introduction to Work Study.* Geneva, Switzerland, 1969.

4. Evans, A. Work and Leisure, 1919-1969. *International Labour Review*, Vol. 99, No. 1 (1969): 35-59.

5. Evans, A. *Hours of Work in Industrialized Countries.* Geneva, Switzerland: International Labour Office, 1975.

6. Harbison, F. Education for development. *Scientific American*, (September 1963): 140-47, 209.

7. Tanner, J. Earlier maturation in man. *Scientific American*, 218, (January 1968): 20-27.

8. Youde, L. A study of the training time for two repetitive operations. Master's thesis, State University of Iowa, 1947.

PART II
THE DESIGN
PROCESS

4

ENGINEERING DESIGN

SCIENTIFIC METHOD

If asked to list important inventions, the ordinary citizen tends to list devices such as the wheel, transistors, and the electric motor. Just as important, however, and perhaps even more important are *concepts.* Examples of important concepts are technical societies, public libraries, and the scientific method.

Mowrer stated the importance of research[8]:

In plants change occurs almost entirely by means of the evolutionary mechanism; but in animals there is the capacity for another kind of "evolution" of change; namely learning. Learning by actual doing involves hazards. A still higher level of advantage accrues to organisms which can explore their environments not only in terms of actual performance but also perceptually.

The scientific method is an efficient method of doing research (learning by doing). It becomes even more useful if combined with another concept, education (perceptual exploration of the environment), so that students need not "reinvent the wheel."

We can use a formula to predict when a beam will break; we do not need to build the beam. We can use a model to predict the best production schedule; we do not need actually to use all possible schedules. We can predict the amount of material handling required for a proposed plant; we do not need to build all the possible plants. But, if we are going to trust the output of the formula (model) (theory) and not physically evaluate alternatives, it is essential that the formula (model) (theory) is valid for the conditions of use.

Table 4.1 gives the five steps of scientific method as well as an example application to the optimum height for a typewriter keyboard. The essence of the scientific method is the feedback of *data* vs. *theory.* The ancient Greeks were good at the first three steps, but they considered it beneath their dignity to see if their theories checked with the reality. Mere data gathering without an underlying theory also is not very useful. The powerful combination is theory plus data with a comparison of the data versus theory until the error is acceptable: a negative-feedback control circuit.

Table 4.1   The Scientific Method Has Five Steps.
Step Five, The Critical Step, Compares Data Versus the Prediction

Step	Example
1. Clearly state the problem you are trying to solve.	What is the optimum height at which a typewriter should be placed?
2. Construct a hypothesis or model.	Speed and accuracy of touch typing depend upon fatigue.
3. Apply analysis to the model. From the analysis, predict what will happen in various conditions.	Fatigue is caused primarily by supporting the arm weight; the higher the hand is held, the greater the torque which must be resisted and the greater the fatigue.
	Since the typist is assumed to be seated, the optimum keyboard position will be in the typist's lap (i.e., at the lowest feasible position).
4. Design and perform an experiment with the real situation. Compare vs. the model.	Have a number of individuals type the same material at a number of heights. Record their speed, accuracy, and preferred work heights.
5. Compare data from the real situation vs. the model's predictions:	See Principle 2 of chapter 14.
a. If the difference between the prediction and the data (called error) is acceptable, accept the model.	
b. If the error is too large, revise experiment or model until error is acceptable.	

ENGINEERING DESIGN

Engineering design, although related to the scientific method, differs from it. Remember the five steps of engineering design with the acronym DAMES where D = Define the problem, A = Analyze, M = Make search, E = Evaluate alternatives, and S = Specify solution. See Table 4.2.

Readers interested in more detail on the design process should read Middendorf, Asimow, and especially Krick.[7,1,6]

Define the Problem Broadly

Usually the designer is not given the problem; instead the designer is confronted with the existing solution. The current solution is not the problem but just one solution among many possible solutions. The broad, detail-free statement of the problem should include the number of replications, the criteria, and the schedule. Using the example of Table 4.2, the replications are "10,000/yr," the criteria are "reasonable quality and low manufacturing cost," and the schedule is "within 5 days." Putting in too much detail makes you start with defending your concept rather than opening minds (yours and your clients') to new possibilities. At this stage, the number of replications should be quite approximate (within ±500%). The importance of giving criteria is that there usually are multiple criteria (cost, quality, simplicity, etc.) rather than just one criterion. (Nadler,

Table 4.2   The Five Steps of the Engineering Design Procedure Can Be Remembered by the Acronym DAMES—Define, Analyze, Make search, Evaluate, Specify

Step	Comments	Example
Define the problem broadly.	Make statement broad and detail-free. Give criteria, number of replications, schedule.	Design, within 5 days, a workstation for assembly of 10,000/yr of unit Y with reasonable quality and low mfg. cost.
Analyze in detail.	Identify limits (constraints, restrictions). Include variability in components and users. Make machine adjust to man, not converse.	Obtain specifications of components and assembly. Obtain skills and availability of personnel. Get restrictions in fabrication and assembly techniques and sequence. Obtain more details on cost accounting, scheduling, and trade-offs of criteria.
Make search of solution space.	Don't be limited by imagined constraints. Try for optimum solution, not feasible solution. Have more than one solution.	Seek a variety of assembly sequences, layouts, fixtures, units/hr, hand-tools, etc.
Evaluate alternatives.	Trade-off multiple criteria. Calculate benefit/cost.	Alt. A: installed cost $1000; cost/unit = $1.10 Alt. B: installed cost $1200; cost/unit = $1.03
Specify solution.	Specify solution in detail. Sell solution. Accept a partial solution rather than nothing. Follow up to see that design is implemented and that design reduces the problem.	Recommend Alt. B. Install Alt. B1, a modification of B suggested by the supervisor.

on the other hand, recommends starting with an ideal (extreme) solution and then "back off.") The schedule identifies priorities and allocation of resources that can be used both in the design process and the replication of the products from the design.

An important distinction between science and engineering is that the scientist wants a precise answer while the engineer is willing to settle for a practical answer. Consider the following challenge. A girl sits on one end of a bench; a boy sits on the other end; the distance between them is X. In the first minute, they decrease the distance between them by 50%; in the second minute by a further 50%, in the third minute by a further 50%, etc. Will they ever meet? The scientist ponders and says "Never!" while the engineer smiles and says "Close enough for practical purposes!"

Analyze in Detail

Amplify step 1 with more detail on replications, criteria, and schedule.

- What are the needs of the design users (productivity, style, comfort, accuracy, esthetics, etc.)?
- What should the design achieve?

- What are the design limits (also called constraints and restrictions)?
- What are the characteristics of the population using the design? For example, if designing an office typing workstation, the users would be adults within certain ranges (age from 16 to 65, weight of 50 to 100 kg, etc.).

Since people vary, designers can follow two alternatives. (1) Personal selection. Make the design and fixed characteristics and make the people adjust to the device. One example would be a fixed dimension chair and selecting people to fit the chair; another would be a machine-paced assembly line and forcing each worker to work at the speed of the master cam. (2) Fit the task to the worker. The design adjusts to varying characteristics of the users. One example would be a chair which adjusts to individuals of different dimensions: another would be a human-paced assembly line with buffers so that all workers work at their own pace.

Make Search of Solution Space

At this stage the engineer designs a number of alternatives; one of the key distinctions between science and engineering is that in science there is only one solution while in engineering there are a number of solutions. Although the solution space is cut down by economic, political, esthetic, etc. constraints, there is more than one feasible solution; see Figure 27.2. However, of the many feasible solutions (solutions which work), the engineer should try to get the best solution—the *optimum*. The best will be a trade-off of the various criteria, which also change from time to time, so the designer must be careful not to eliminate alternative designs too early. Another problem is the tendency of designers to be satisfiers rather than optimizers. That is, designers tend to stop designing as soon as they have one feasible solution—they have satisfied the problem. For example, when designing an assembly line, the engineer may stop as soon as there is a solution; when laying out the factory, the designer may stop as soon as one satisfactory layout is made. In order to get an optimum design, there must be a number of alternatives to select from; alternatives also suggest further alternatives, so stopping too soon can severely limit the solution quality and acceptance.

Evaluate Alternatives

A scientist tends to look for the single formula which describes one criterion of a situation; the engineer must trade off multiple criteria, usually without any satisfactory trade-off values. For example, one design of an assembly line may need .11 min/unit while another design may need .10 min/unit; however, the first design may give more job satisfaction to the workers. Which assembly line design should be used? How do you quantify job satisfaction? Even if you can put a numerical value on it, how many "satisfaction units" equal a 10% increase in assembly labor cost?

A simple ranking of alternatives is sufficient in some cases (good, better, best). More precise is a numerical ranking, using a single criterion with an equal interval scale (method A requires 1.1 min/unit while method B requires 1.0 min/unit: design A requires 50 m² of floor space while Design B requires 40 m² of floor space). Most managers, however, prefer a comparison combining all the various costs and benefits in terms of money. Even if the various costs and benefits can be put in terms of money, they may be "different kinds" of money. For example, you must add operating costs (such as labor cost/unit), capital costs (machine purchase costs), maintenance costs (machine lubrication costs), product quality costs (reduced product failures in the field after three years of use), environmental costs (CO concentration in the work area reduced from 40 ppm to 30 ppm), etc. See Table 7.6 for one approach.

Specify Solution

Scientists can state their conclusions in terms that only Ph.D.'s can understand but engineers must communicate with the ordinary individual, so abstract theories must be translated into "nuts and bolts." Then this individual can have the audacity to say "No!" to your beautiful design! A humbling experience. The improvement occurring in a situation is a function of the quality of the design times the acceptance of the design. If "they" don't "buy" it, nothing happens. Engineers therefore accept modifications in their beautiful designs to gain acceptance; striving for the "whole loaf or nothing" usually gets "nothing." Then, after the design has been accepted, follow up to see that it is put into practice. (Many a good idea has been accepted with beaming smiles by people who have a firm resolve to let the implementation die due to apathy.) Then, when the design has been implemented, see whether it has reduced the original problem—are you part of the solution or part of the problem?

Although I have presented some of the difficulties of engineering design, engineering is a very satisfying profession. Herbert Hoover expressed it well:

It is a great profession. There is the fascination of watching a figment of the imagination emerge through the aid of science to a plan on paper. Then it moves to realization in stone or metal or energy. Then it brings jobs and homes to men. Then it elevates the standards of living and adds to the comforts of life. That is the engineer's high privilege.

The great liability of the engineer compared to men of other professions is that his works are out in the open where all can see them. His acts, step by step, are in hard substance. He cannot bury his mistakes in the grave like the doctors. He cannot argue them into thin air or blame the judge like the lawyers. He cannot, like the architects, cover his failures with vines and trees. He cannot, like the politicians, screen his shortcomings by blaming his opponents and hope the people will forget. The engineer simply cannot deny he did it. If his works do not work, he is damned . . .

On the other hand, unlike the doctor, his is not a life among the weak. Unlike the soldier, destruction is not his purpose. Unlike the lawyer, quarrels are not his daily bread. To the engineer falls the job of clothing the bare bones of science with life, comfort, and hope. No doubt as years go by the people forget which engineer did it, even if they ever knew. Or some politician puts his name on it. Or they credit it to some promoter who used other people's money . . . But the engineer himself looks back at the unending stream of goodness which flows from his successes with satisfactions that few professions may know. And the verdict of his fellow professionals is all the accolade he wants.

WHAT TO STUDY *(PARETO DISTRIBUTION)*

Engineering time is a valuable resource; don't waste time on unimportant problems. Allocate design time to the important problems; neglect the minor problems. To check quickly whether a project is worth considering, calculate: (1) savings/yr if material cost is cut 10%, and (2) savings/yr if labor cost is cut 10%.

The concept of the *Pareto distribution* may help you find the important problems (see Figures 4.1 and 4.2). (Lorenz also used curves to demonstrate the concentration of wealth but Vilfredo Pareto's name is now associated with the concept of the "insignificant many and the mighty few.") Herron says the Pareto curve (also called ABC curve) can be approximated by a log-normal distribution.[4]

The key concept is that the bulk of the problem (opportunity) is concentrated in a few items. For example:

A few time standards	cover	most of the direct labor hours
products	produce	most of the sales dollars
products	have	most of cubic storage required/item
products	produce	most of the profit
operations	have	most of the quality problems
delivery routes	have	most of the stops/route
days	have	most of the outgoing or incoming orders
half-hours/day	have	number of truck arrivals/half-hour
salespersons	sell	most of the product
farmers	grow	most of the food
cities	have	most of the population
engineers	have	most of the patents
professors	have	most of the publications
individuals	commit	most of the crimes
individuals	have	most of the money
individuals	eat	most of the food at a party
individuals	drink	most of the beer

Table 4.3   List Categories with the Most Important One First and the Least Important Last. Then Plot the Cumulative Percent as in Figure 4.1

Item Number	Operation	Work-Hours per year	Labor Cost/Hr	Labor Cost/Yr	Percent of Total	Cumulative Percent
1	Hand polish	65,000	8.75	568,750	15.7	15.7
2	Form grind	52,000	9.30	483,600	13.4	29.1
3	Drill and bore	56,000	8.25	462,000	12.8	41.9
906	Insert fittings	120	8.00	960	.0	99.9
907	Apply nameplates	60	8.50	510	.0	100.0
				3,600,000	100.0%	100.0%

Pareto diagrams are just a special form of histograms (frequency counts); the key difference is that the categories are put in sequence of largest first (instead of a random order) and the cumulative curve is plotted. Table 4.3 shows an example table of operations arranged in order of largest annual cost first. This then is translated into a cumulative figure such as Figure 4.1. For other examples, see Figures 27.4, 27.5, and 27.6.

Farmers have always known that most of the butterfat is in the cream, not in the skim milk. Therefore, using the Pareto concept, if your design concerns crime, it should concentrate on the few individuals who cause most of the crimes; if your design is to affect food consumption at a party, it should concentrate on the few individuals who eat the most; if it is to improve product quality, it should concentrate on the few components which cause most of the problems, etc.

In order to maximize your design productivity, work on several projects at the same time rather than spending all your time on one project and then going on to the second project (i.e., parallel projects not series projects). Not only will your time be spent more effectively, but the idea quality also will be better.[9]

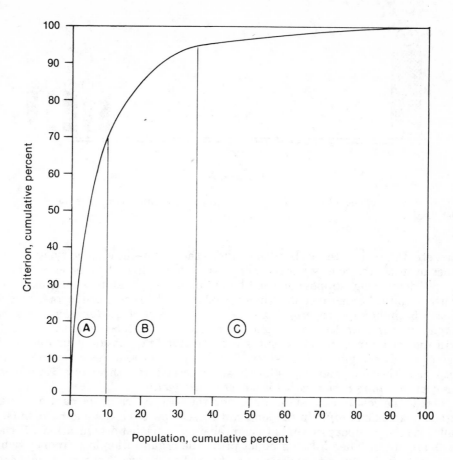

Figure 4.1 *Fight giants.* Many populations have a *Pareto distribution* in which a small proportion of the population has a large proportion of the criterion. Inventories, for example, can be classified by the ABC system. "A" items may compose 10% of the part numbers but 70% of the inventory cost. "B" items may compose 25% of the part numbers and 25% of the cost so the total of A + B items composes 35% of the part numbers but 95% of the cost. "C" items thus compose the remaining 65% of the part numbers but only 5% of the cost. The concept is to concentrate your effort on "A" items.

COST ALLOCATION

Organizations allocate costs into various categories to aid decision making. Table 4.4 gives a hypothetical cost breakdown of an electric fan. Different organizations, industries, and countries will have different categories and different ratios of costs in the categories depending on the product, industry, amount of competition, etc. A monopoly, for example, may have a higher profit/unit.

Direct materials and direct labor (also called "touch" labor as they touch the product) are the most easily allocated costs. Of course, there may be difficulties, such as when multiple products use the same material or the same worker is assigned to work on many different products. For managerial control, firms usually keep raw material costs separate from purchased components costs; their total is the direct material cost. *Direct material + direct labor equals prime costs.*

Burden or overhead is the next level of costs to be allocated. It can be divided into indirect labor, indirect materials, and capital costs. Indirect labor includes the salaries and wages of the clerks, engineers, technicians, supervisors, inspec-

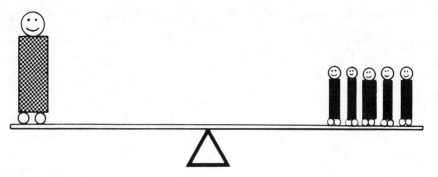

Figure 4.2 *Insignificant many and the mighty few* is another name for the Pareto distribution.

tors, etc. Indirect materials burden includes the cost of the electricity, water, and other utilities; the cost of various supplies such as paper clips, pencils, lubricating oil, degreasing compounds used before painting, grinding wheels, and toilet paper. Capital burden includes the cost of the fork trucks, conveyors, machine tools, the building, property taxes, etc. *Prime cost + factory burden equals manufacturing cost,* the cost of getting an item to the factory door.

In most circumstances customers are not standing outside your door in the rain clamoring to buy your product. To sell a product requires catalogues, product warranties, etc., and wages for those concerned with this work. *Manufacturing cost + selling cost equals total cost at the factory.*

Next is the cost of capital (interest) and the cost of risk of the capital; usually totaled and called *profit.* In the long run these expected costs and risks must be built into the product price in both capitalist and socialist organizations. If these costs are not included in the product price, the organization loses money, which must come either from the owner's assets or the taxpayer's assets. *Total cost + profit equals selling price at the factory.*

Since in normal circumstances the customer does not come to the door, there are additional costs for distribution of the product. They include shipping and expenses of the retailing organization. *Factory price + distribution cost equals consumer price.*

Within the factory, the engineer probably will deal mostly with the manufacturing cost of $11.02. Actually, fans probably will not cost $11.02 each day because on some days more scrap is made, on some days more or less labor is used, and so forth. The $11.02 is not the real cost but is a *standard* cost. Standard costs make assumptions about such things as the standard amount of material, a standard price/unit of material, a standard amount of labor, a standard price/unit of labor, a standard amount of scrap, a standard cost of each type of defect, a standard use of energy/unit, a standard amount of overhead/unit. Unfortunately it is difficult to keep standard costs accurate over time as conditions change. For example, standard labor cost may be based on prorating the setup cost over 10,000 units/yr. However, over the years production may have changed to 5,000/yr or 15,000/yr—causing an error in the standard setup hours/unit and thus an error in the standard cost. Another common problem is that the standard labor hours/unit (say painting time of 10 s/unit) may not be the actual time (painting time may be 9 s or 12 s). When making a cost analysis, try to use actual costs instead of standard costs.

When jobs are designed, a common criterion is the cost/unit, that is, we judge workstation A to be better than workstation B because labor cost/unit is lower for workstation A. Labor cost/unit should never be the sole criterion because material and overhead costs also are important. The following section discusses

Table 4.4   Hypothetical Cost Accounting Breakdown of the Price of An Electric Fan. The Percent of Each Cost of the Total Varies Widely by Product And Industry. For Automobiles, Manufacturing Cost Might Be Split 48% For Material (Raw and Purchased), 20% For The Direct Labor, And 32% For Burden.

Category	Example	Cost/fan, $	
Raw materials (need further work)	Steel for fan blades	.80	
Purchased components and assemblies (used as is)	Motor, bearings, knobs	1.50	
Direct labor	Time to stamp out blades, paint blades, assemble knob to shaft	2.25	
		PRIME COST	4.55
Indirect labor burden	Fork truck drivers, industrial engineers	5.05	
Indirect materials burden	Toilet paper, light bulbs	.45	
Capital burden	Depreciation for conveyor, building	.97	
		MANUFACTURING COST	11.02
Selling cost	Warranties, brochures, sales salaries	2.00	
		TOTAL COST AT FACTORY	13.02
Profit	Interest, risk	.98	
		FACTORY SELLING PRICE	14.00
Distribution cost	Shipping, distributor expenses, profit	6.00	
		CONSUMER PRICE	20.00

some of the important money costs to consider; chapter 10 discusses the problem of criteria in more detail.

RETURN ON INVESTMENT

Do the benefits of a design outweigh the costs? Making the benefit/cost comparison requires three steps:

1. Determining what is changed due to the design (e.g., product quality is better).
2. Putting the changes into monetary units (e.g., improved quality is worth $0.02/unit).
3. Calculating the total benefits vs. costs (e.g., all the changes totaled give benefits of $4,700/yr; all the costs total $1,400/yr).

Most errors in decision making are due to poor data for steps one and two; engineers and accountants tend to "overkill" step three with exotic formulas and complex calculations to four decimal points when they have errors of ±50% in steps 1 and 2. Spend approximately 90% of your time in steps 1 and 2 and 10% in step 3; you will make better decisions than the person who reverses the ratio. The secret is valid data. Leave out judgments and opinions unless supported by data and quantified into dollars.

Figure 4.3 gives a cost analysis which emphasizes steps 1 and 2; it is oriented to metal-working manufacturing, but you can adapt it to your own industry. The example will evaluate use of a special purpose screwdriver for use in making automobile tune-ups in a garage.

Key information required is: (1) project life, (2) savings/yr, and (3) one-time cost.

Project __SCREWDRIVER__

Part name _____ Part number _____ Used in Dept. _Tuneup_

Volume __800__ pcs/yr _____ pcs/day _____ pcs/hr __1.0__ hrs/pc

Labor cost/hr __$8.00__ Engineer __SK__

Life of application __5__ yrs Date __22 Feb__

A. Annual cost on the controlling operation	Existing Method	Best Manual Proposal	Best Mechanized Proposal
Direct labor, $/unit	.044	.033	
Relief labor, $/unit	-	-	
Downtime, $/unit	-	-	
Maintenance, $/unit	-	-	
Direct material, $/unit	Reference	0	
Indirect material, $/unit	.007	.009	
Perishable tools, $/unit	-	-	
Tool regrind (repair), $/unit	-	-	
Utilities, $/unit	-	-	
Inspection, $/unit	Ref.	0	
Product quality, $/unit	Ref	-.010	
Rework & scrap, $/unit	.002	.001	
Other (specify) *Pain and Suffering* $/unit	Ref.	-.010	
Total, $/unit	.053	.023	
Total, $/yr			
Line 1. Savings/yr		$24	

B. One-time cost, $			
Equipment, $	Ref.	3	
Jigs, fixtures for equipment, $	-	-	
Installation, $	-	-	
Operator retraining, $	Ref.	.67	
Engineering, $	Ref.	48.00	
Line 2. Total one-time costs, $	Ref.	51.67	

C. Benefit/cost calculations

Line 3.	Total savings during application life = $	120	(line 1 x yrs)
Line 4.	One-time cost = $	51.67	(line 2)
Line 5.	Net savings = $	68.33	(line 3-4)
Line 6.	Return on investment, %	26%	line 5 x 100
			line 4 x appl. yrs

Figure 4.3 *Cost analysis forms* reduce omitting relevant data. The data shown are for a proposed specialized screwdriver to be used for automobile tune-ups.

Life of the Application

The top of the form starts with basic information such as *project, part name, part number,* and *where used.* The first key item is *life of the application.* Life can be limited either by the life of the equipment (say a lathe would be worn out in 5 years) or by life of the product (say a fixture made obsolete by an anticipated change in product design in 3 years). Assume the screwdrivers have an application life of 5 years.

The *volume/yr* is important. Be careful to calculate pcs/hr and pcs/day by dividing by the proper number of days the product is made/yr; that is, most products are *not* made continuously so the rate/hr usually cannot be multiplied by 2,000 hours/yr to obtain annual output. In many cases volume/hr may change over the life of the application. Estimate volume/yr for each year; for a simple level of analysis (such as with this form) use the average; for a more precise analysis, use each year's estimate to calculate the benefits/cost for each year (see an engineering economics text such as Smith's for the solution techniques).[10] In our example, assume 800 tune-ups/yr with an average time of 1.0 hr/tune-up.

An important question is the *labor cost/hr.* Most cost reductions are justified on labor savings; the question is what the proper labor-cost rate to use is. A worker may be paid $6/hr. Because of fringe benefits (vacations, holidays, pensions, etc.), the cost to the organization may be $9/hr. The burden in the factory may be allocated in proportion to direct labor cost (say 300% of direct labor), so cost/hr may be given as $9 + 3 ($9) = $36/hr. My recommendation is to use the direct labor cost value ($9/hr) rather than direct labor plus burden ($36). If labor cost is reduced there is no reason to believe burden cost will be reduced; in fact, often burden expense increases (say for more electrical power). Even for a constant burden cost, a lower amount of direct labor will mean that the burden rate/direct labor-hour will increase. Be very suspicious of cost reduction proposals which use cost rates including burden. Burden changes can be listed separately. These imprecise cost estimates may be important in finely balanced decisions. In our case, assume labor cost of $8/hr.

Next record information for *three* alternatives: the *existing solution,* the *best manual proposal,* and the *best mechanized proposal.* The reason for requiring the best manual proposal is that engineers love machines and devices and thus often have a bias toward solutions which involve machines and devices. In our example, only simple hand tools seem feasible so the mechanized alternative will not be considered.

Information is needed on two types of costs: annual costs and one-time costs.

Annual Costs

Annual costs, constituting the second key item, are determined by calculating the cost/unit and then multiplying by annual volume. In some situations it might be easier to calculate annual costs for each subcategory directly.

- Direct labor is the cost of labor exerted specifically on this specific operation. An existing screwdriver for auto tune-ups may require 20 s/tune-up vs. 15 s/tune-up for a proposed screwdriver. In the example, existing cost would be (20/3,600) (8) = $.044/tune-up while proposed cost is .033.

- Relief cost is the cost for substitute labor on an assembly line (e.g., 7 workers may work at 6 stations). In this auto tune-up example, there are no relief costs.

- Downtime is the cost of idle equipment or workers at this or other workstations. On tightly linked jobs, downtime becomes very important since downtimes add. In the screwdriver example we will assume that other workers are not tightly linked to this job and downtime is zero for both alternatives.

- Maintenance is the cost of equipment maintenance. Assume zero maintenance for both screwdrivers.
- Direct material is material used in the product. Assume $12 for materials with either tool so there is no cost difference.
- Indirect materials covers miscellaneous supplies and materials. Assume that the existing tool causes one stripped setscrew/50 tune-ups but the proposed tool, because it permits more torque, will probably have one stripped setscrew/40 tune-ups. Assume a setscrew costs $.10 plus $.25 for the labor required to go get the replacement screw from the stock room. Thus the existing tool cost is .35/50 = $.007/tune-up while the proposed cost is .009.
- Perishable tools (tool bits, grinding wheels, etc.) are used up by the process. Assume zero for both screwdriver alternatives.
- Tool regrind (repair) is the repairing, resharpening, reworking of tools. Assume neither screwdriver will need repair.
- Utility costs include electricity, water, heat, and light. Assume no change in utility costs occurs with either screwdriver.
- Inspection costs include inspection both by the worker and by separate inspectors. Assume inspection cost of $1/tune-up regardless of the screwdriver used.
- Product quality is the improvement (degradation) in the product as expressed in warranties, lost customers, good will, etc. Assume that the proposed tool will give a very slightly better tune-up; estimated value is $.01/tune-up.
- Rework and scrap is the cost of the product quality before it leaves the department. Assume the existing tool requires rework in 1/70 while the proposed tool will require rework for 1/140; the rework time is estimated as 60 s. Then existing cost is (60/3,600) (1/70) (8) = $.002/tune-up while the proposed is .001.
- Other costs are any additional costs you find relevant. Assume, in this case, that the proposed screwdriver will cause less fatigue, less muscle pain, and fewer cracked knuckles; you value this as $.01/tune-up.

The total cost considered for the existing screwdriver then is .044 + .007 + .002 = $.053/tune-up while for the proposed tool it is .033 + .009 − .01 + .001 − .01 = $.023/tune-up. Thus savings/tune-up is $.03/tune-up; annual savings are .03 × 800 = $24/yr.

One-Time Costs

Next consider the *one-time costs*, the third key item.

- Equipment cost is the purchased cost (including tax and delivery) of the equipment to your receiving dock. In this example, assume the existing tool costs $1 while the proposed tool costs $3. However, the existing tool will last the application life and its cost already has been paid so consider its cost as zero. For a more expensive tool it may be worth considering its decline in value over the application life; that is, the difference between what you could sell it for now and what you could sell it for at the end of the application life.
- Jigs and fixtures often are required for equipment use. Assume as zero in the screwdriver example.

- Installation cost refers to the costs of getting the equipment from your dock to installed and working. Typical expenses are for millwrights, electricians, and plumbers. Assume as zero in this example.

- Operator retraining refers to the loss in output while the operators adjust to the new procedure, tool, or device. Assume retraining for the proposed tool will require about 5 min; thus its one-time costs would be (5/60) (8) = $.67.

- Engineering costs (often forgotten in cost analyses) include time to investigate the new method, determine alternatives, calculate costs, sell recommendations, and install the new method, if accepted. For our example, assume it took the engineer 4 hr to read about the proposed tool, talk to people, make this estimate, etc. Assume the engineer's wages are 150% of the mechanic's wages/hr so engineering cost is 4 × $12/hr = $48.

Thus total one-time costs for the existing alternative are $0; one-time costs for the proposal are $3 + .67 + 48 = $51.67.

Benefit/Cost Calculations

Benefit/cost calculations then give total savings, during the application life, of $24 × 5 yr = $120. Subtracting the one-time expenses of $51.67 gives total benefit of the project of $68.33. Benefits/yr = $68.33/5 = $13.67. Thus the annual return on investment = $13.67/51.67 = .26 × 100 = 26%.

In some cases it is difficult to estimate the amount of saving from a project. A useful technique is to calculate the savings for several alternatives, such as a 1%, a 2%, and a 5% improvement in productivity. The question then becomes not the exact improvement but whether the improvement is likely to be greater than the minimum required. For example, for a person costing $10/hr (including fringes), the annual cost is $20,000 (assuming a 2000 hr work year). A 1% change would be $200, a 2% = $400, and a 5% = $1000. Then a new chair costing $100 and improving output 1% (5 min/day) would pay for itself in .5 years. If it improved output .1% then it would pay for itself in 5 years (i.e., a 20% return on investment). The question management then needs to answer (assuming 20% is satisfactory) is "Do you think an improved chair will improve output at least 30 s/day"?

Organizations usually require new projects to have a proposed return on investment greater than they presently are making. Table 4.5 gives some example return on investments.

EVOLUTIONARY OPERATION
OF PROCESSES (EVOP)

When designing, there are major decisions and minor decisions to be made—major and minor in their effect on the outcome. Through trial and error, experimentation, luck, prototypes, pilot projects, and so forth, the engineer designs a process. Eventually there comes a stage in which the job "goes into production." There are still possibilities for improvement, but each individual change has only a small improvement potential. Thus they are not considered worthy of spending additional engineering resources investigating them and delaying production.

For example, consider painting part 123 in Department A. Considering that we could vary paint-thinner ratio, distance of part 123 from the paint nozzle, the nozzle diameter, the air pressure, the paint temperature, the temperature of part

Table 4.5 Return on Sales and Investment of Selected Organizations in 1980. Each Year *Fortune* Reports the Results for the 500 Largest U.S. Industrial Firms (May); Industrial Firms From 501 to 1000 (June); the Largest Nonindustrials (July); And the 500 Largest Industrials Outside the U.S. (August)

Name	Type	Sales Rank	Sales	Assets	Net Income, $	Percent Return on	
			(000,000 omitted)			Sales	Assets
Exxon	Ind.	1	103,142	56,576	5,650	5.5	10.0
National Steel	Ind.	100	3,706	3,446	83	2.2	2.4
Emhart	Ind.	200	1,802	977	67	3.7	6.8
McGraw-Hill	Ind.	300	1,000	785	86	8.6	10.9
Hart, S. & Marx	Ind.	400	674	402	22	3.3	5.6
Fiat-Allis	Ind.	500	447	285	54	12.3	19.3
Bandag	Ind.	600	330	226	27	8.2	12.2
Burndy	Ind.	700	253	195	24	9.5	12.3
Continental Steel	Ind.	800	191	101	4	2.3	4.4
Essex Chemical	Ind.	900	152	90	5	3.6	6.1
Reading Industries	Ind.	1000	124	37	1	1.5	5.1
Sears & Roebuck	Retail	1	25,194	28,053	606	2.4	2.2
UAL	Transp.	1	5,041	4,041	21	.4	.5
AT&T	Utility	1	125,450	50,791	6,079	4.8	12.0
Royal Dutch/Shell (Neth-Britain)	F.Ind.	1	77,114	68,518	5,174	6.7	7.5
Guest Keen (Britain)	F.Ind.	100	4,471	3,518	209	4.7	5.9
Asahi Glass (Japan)	F.Ind.	200	2,303	2,462	103	4.5	4.2
Hanson Trust (Britain)	F.Ind.	300	1,552	749	57	3.7	7.7
Morinaga Milk (Japan)	F.Ind.	400	1,098	475	2	.3	.6
Olida & Caby (France)	F.Ind.	500	826	238	4	.5	1.8

123, etc., what is the optimum value of each of these variables? Or consider drilling holes in part 345 in Department C. Considering that we could vary drill rake angle, drill material, drill rpm, drill feed rate, coolant type, coolant volume, etc., what is the optimum value of each of these variables? To experiment would seem to require too much engineering time and expense.

Box and Hunter put forth the simple yet powerful concept that a process produces two things: (1) items for sale and (2) information about the process.[3] That is, while we are painting part 123, we also are generating information about the effects of the various variables affecting the painting. That is, we have been running an experiment—we just didn't realize it and collect and analyze the data!

Box and Hunter proposed that, to minimize the engineering expense and to study the process in its "production" version, the "experiment" be run by the plant operation personnel (the painter or drill press operator) on the production machines—thus eliminating the cost of the experimenter and the cost of the lab.

In most experiments, the experimental design attempts to maximize the amount of information obtained from the experiment, usually by minimizing the effects of "noise" (variability due to the process or the measurement of the process). This can be visualized as "cutting the weeds" (see Figure 4.4). The normal experiment tries to "cut the weeds" so the "crop" can be observed. However, there is an alternate strategy: repeat the experiment over and over until the signal appears through the noise. Using the crop analogy, that means that, if you watch long enough, eventually the crop rises above the weeds. Evolutionary

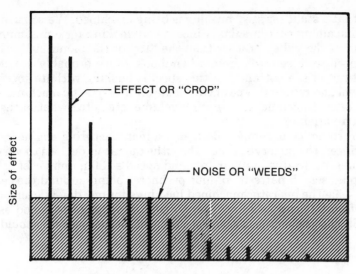

Effects arranged in descending importance

Figure 4.4 *Traditional experimental design* reduces the noise level to detect the effect—"cutting the weeds to detect the crop." EVOP uses another approach, continued experimentation until the signal shows through—"letting the crop grow above the weeds."

Operation of Processes (EVOP) follows this second strategy, using the assumption that experimentation costs nothing since the process is producing product for sale.

Box and Hunter said that if the experiment is to be run by "shop" personnel (i.e., no college, no statistical training) and the primary objective of producing products for sale is not to be endangered, then the experiments must be simple and cautious so as to cause no scrap. The technique is called Evolutionary Operation of Processes (EVOP) since it follows the two essentials of evolution: (1) small variations and (2) selection of favorable variants.

Table 4.6 gives some possible applications of EVOP. In most applications, the criterion seems to be the *yield* of the process. EVOP usually is used to maximize the use of material or energy or quality rather than to minimize the cost of labor. The very simple example in Table 4.6 is from Konz (see Box and Draper for an excellent text on EVOP which gives not only administrative procedures for EVOP but also simple and elegant statistical techniques).[5,2] EVOP is really another version of the operations research problem of "hill climbing"; EVOP, however, deals with a "noisy" signal.

Table 4.6   Example Applications of EVOP

Application	Criterion	Some Variables
Turning	Machining cost Surface finish	Feed, speed, tool geometry
Welding	Weld strength	Cooling rate, amps, rod type
Painting	Scrap rate	Paint-thinner ratio, gun distance
Casting	Yield	Pouring temperature, additive percentages, sand additive percentages
Chemical processes	Yield	Time, temperature, percentages of constituents and catalysts

Assume that a stainless steel bushing is being machined. We will use as the criterion the machining cost/bushing. Since we are looking for a minimum, look for the "bottom of the valley" rather than the "top of the mountain." The present feed and speed were selected from a handbook so we probably are close to the bottom but not at the bottom. For this specific bushing with its *specific* type of stainless steel and required type of cut, made on this *specific* machine, with a *specific* brand of cutting fluid at a *specific* volume, etc., etc., what is the optimum value of each variable?

An EVOP committee (composed of, say, a manufacturing engineer, a quality-control engineer, the supervisor, and the lathe operator) decides to vary feed and speed. At present, feed = .140 mm/rev and speed = 30 m/min; cost = $.25/unit. They set up a "search pattern" of four perimeter points around the center point (see Figure 4.5). The operator machined bushings for one day at each point, keeping track of output and tool life. The .16 and 31 point looked good as cost was $.22/unit. Should they shift to this point and run a new search; should they con-

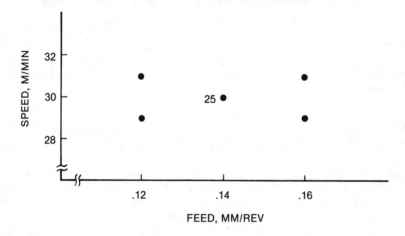

Figure 4.5 *Focus of a search pattern* is an initial center point (at .14 mm/rev and 30 m/min) with four perimeter points. Initial cost = $.25/unit.

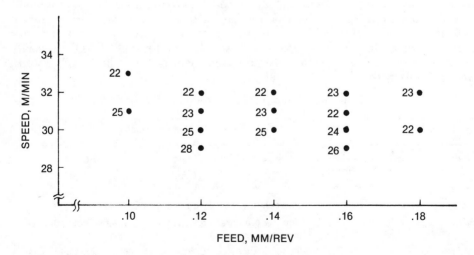

Figure 4.6 *Optimum value* of feed and speed is selected as feed of .16 mm/rev and speed of 31 m/min for an estimated savings of $.03/unit.

tinue the search using the existing pattern; or should they try a different variable, such as volume of cutting fluid? (In poker terms, should they bet, stay, or throw in the hand?) In our example they decided to continue at the same five points for another week. After three weeks, they shifted to .14 and 32 as a center point and started a new search pattern. In general, it is good practice to overlap search areas as it helps to prevent "falling off a cliff." Eventually, after some months, the data shown in Figure 4.6 were available. They decided, as the new standard for feed and speed on this bushing, to machine the bushing with a feed of .14 mm/rev and a speed of 31 m/min for an anticipated savings of $.03/piece. EVOP savings, though worthwhile, rarely are spectacular.

SUMMARY

Engineering design differs from the scientific method. Remember the five steps of engineering design with the acronym DAMES: (1) Define the problem broadly, (2) Analyze in detail, (3) Make search, (4) Evaluate alternatives, and (5) Specify solution.

Concentrate your valuable design time on the "mighty few" rather than the "insignificant many."

Determine the value of your proposal through benefit/cost analysis. A quick analysis using good data will be more valid than a highly mathematical analysis using crude and incomplete data.

Use EVOP to improve processes even after production begins.

SHORT-ANSWER REVIEW QUESTIONS

4.1   List the five steps of engineering design (DAMES).

4.2   Give five examples of Pareto distributions.

4.3   What is a typical return on investment after taxes for US industry? What is a typical return on sales?

4.4   Why should cost reductions include an analysis for "best manual proposal" as well as "best mechanized proposal"?

4.5   The EVOP concept is that a process produces what two things?

THOUGHT-DISCUSSION QUESTIONS

1. "Learning by doing involves hazards. A still higher level of advantage accrues to organisms which can explore their environments not only in terms of actual performance but also perceptually." Discuss.

2. Using the acronym DAMES, give examples for each step for the problem of "What is the best cutting speed for a lathe when cutting stainless steel?"

REFERENCES

1. Asimow, M. *Introduction to Design.* New York: Prentice-Hall, 1962.

2. Box, G., and Draper, N. *Evolutionary Operation: A Method of Increasing Industrial Productivity.* New York: Wiley, 1969.

3. Box, G., and Hunter, J. Condensed calculations for evolutionary operations programs. *Technometrics,* Vol. 1 (1959): 77-95.

4. Herron, D. Industrial engineering applications of ABC curves. *American Institute of Industrial Engineers Transactions,* Vol. 8, No. 2 (1976): 210-18.

5. Konz, S. Selecting feed and speed under factory conditions. *Tool and Manufacturing Engineer,* Vol. 55, No. 7 (1965): 31-33.

6. Krick, E. *Methods Engineering.* New York: Wiley, 1962.

7. Middendorf, W. *Engineering Design.* Boston: Allyn and Bacon, 1971.

8. Mowrer, O. *Learning Theory and the Symbolic Process.* New York: Wiley & Sons, 1960.

9. Pelz, D., and Andrews, F. *Scientists in Organizations: Productive Climates for Research and Development.* New York: Wiley & Sons, 1966.

10. Smith, F. *Engineering Economy: Analysis of Capital Expenditures.* Ames: Iowa State University Press, 1973.

PART III
AIDS IN JOB DESIGN

5

SEARCH

CREATIVITY

As discussed in chapter 4, engineering design requires a definition of the problem, analysis, a search for a solution, an evaluation of alternatives, and a specification of the solution. This chapter discusses searching for a solution. Edison, who had more inventions than anyone else in the history of the world, expressed it well: "Genius is 1% inspiration and 99% perspiration." What are some of the techniques to make the "perspiration" effective?

UNSTRUCTURED SEARCH: BRAINSTORMING

When trying to solve a problem, we often focus too narrowly. That is, possible solutions are rejected in our own minds due to assumptions, which, if considered carefully, turn out not to be limiting. What is desirable is to obtain many possible solutions; reducing them later is easy compared to the problem of getting a "large solution space."

In brainstorming, creativity is considered an "orchid" which will be "blighted" by the "frost" of criticism; therefore no criticism is allowed.

Brainstorming will be illustrated by the following example.

Consider the problem depicted in the process chart and the flow diagram of Figures 7.7 and 7.8. Joe College was quite inefficient in supplying himself with beer. What are some other ways of obtaining beer?

The discussion leader asks for suggestions from the group.

The first rule is no criticism (even if it is called evaluation). The leader publicly displays the idea (say, on a blackboard) using different words if possible. (Paraphrasing the idea is a useful technique to clarify the meaning.) The public display encourages others to combine and modify ideas. No individual is associated with an idea—this reduces social pressure to accept or reject an idea because it came from someone in the group with high or low status. The leader, knowing the tendency of some individuals to be extroverts, should discourage one or two individuals from dominating the session by recognizing others in the group first or even, if necessary, saying "let's hear what others have to say." Ten minutes is sufficient for most brainstorming sessions.

Some suggestions to this beer problem given by my college classes are in Table 5.1. From brainstorming ideas, it usually is possible to select two or three for detailed evaluation.

Table 5.1   Brainstorming Suggestions to Improve Beer Drinking
(see Figures 7.7, 7.8)

Idea
Knock hole in wall to shorten distance
Switch position of TV and sofa
Move refrigerator
Drink from can
Move wastebasket
Put magnetic opener on refrigerator
Use pull-top can so no opener is needed
Have wife bring the beer
Move refrigerator next to sofa
Don't bother to inspect glass
Get two beers at a time
Keep beers in a cooler next to sofa

There is evidence that group participation inhibits creative thinking; the "best bet" is the pooled individual efforts of people physically apart rather than getting them into a group.[3,5]

STRUCTURED SEARCH: CHECKLISTS

Some individuals feel that the best approach is a systematic approach rather than merely hoping someone has a "brainstorm."

Perhaps the most literary checklist is by Rudyard Kipling:

> I have six honest serving men
> They taught me all I know
> Their names are who and what and when
> And why and where and how.

Victor Morales expanded this into the "critical examination sheet" shown in Figure 5.1. The boxes ask suggestive questions about purpose, place, sequence, person, and means. The first column describes the existing situation; the second and third columns try to bring out disadvantages of the present method and advantages of alternatives; the fourth column asks for a decision on a portion of the problem. The questions try to force the users to list alternatives.

Another technique using the "5 W's and an H" is to write each on the top of a sheet of paper.[1] Then, on the sheet with the "WHO," write down all questions concerning who and the problem; on the sheet with "WHAT," all questions concerning what and the problem; etc. Who questions might be: who should do it, who should bring the material, who should supervise, who should inspect. Don't worry about the sequence of questions—at this stage you are trying to generate ideas. The second stage is to answer one of the questions on one of the sheets—such as Joe Roberts should do it. Now write Joe Roberts on the top of a sheet of paper and make up six questions concerning Joe starting with who, why, what, where, when, and how. Examples might be: Who is Joe Roberts? Why did I select Joe? What exactly should Joe do? Where should Joe do it? When should Joe do it? How should Joe do it? Then go on to the next question on the stage 1 sheet, answer it, put it on a sheet of paper and ask who, why, what, where, when, and how. Stage 3 is to review all the questions from stages 1 and 2 for possible ideas to solve the problem.

There are also more specific checklists for specific situations. Paul van Wely and the Philips Ergonomics group developed the checklist shown in Table 5.2. While at Kansas State as a visiting professor, he developed his "Eleven Commandments" (see Table 5.3). Boer and Burger modified an extensive ergonomics

Situation Studied _____   Date _____

Why Studied _____   Study by _____

Where _____   _____ minutes/occurrence

How Often Repeated per Year _____   _____ minutes/year

	1	2	3	4
Purpose	What is achieved?	What would happen if it weren't done?	What could be done and still meet requirements?	What should be done?
Place	Where is it done?	Disadvantages of doing it there: Advantages of doing it elsewhere:	Where else could it be done?	Where should it be done?
Sequence	When is it done? After:	Disadvantages of doing it then:	Advantages of doing it sooner: Advantages of doing it later:	When should it be done?
Person	Who does it?	Why that person?	List two others who could do it.	Who should do it?
Means	What equipment and methods are used? Equipment: Method:	Disadvantages of equipment: Method:	How else could it be done? Advantages:	How should it be done?

Figure 5.1 *Kipling's "six honest men"* are amplified on a critical examination form.

checklist developed by Burger, Frieberger, Hultgren, De Jong, Lehmann, Murrell, and Grandjean.[2] It uses a hierarchical question system to allow nonrelevant questions to be omitted and thus reduce analysis time.

Table 5.2 Ergonomics Checklist

A. DIMENSIONS
1. Has a tall man enough room?
2. Can a petite woman reach everything?
3. Is the work within normal reach of arms and legs?
4. Can the worker sit on a good chair? (height, seat, back)
5. Is an armrest necessary, and (if so) is it a good one? (Location, shape, position, material)
6. Is a footrest required, and (if so) is it a good one? (Height dimensions, shape, slope)
7. Is it possible to vary the working posture?
8. Is there sufficient space for knees and feet?
9. Is the distance between eyes and work correct?
10. Is the work plane correct for standing work?

B. FORCES
1. Is static work avoided as far as possible?
2. Are vises, jigs, conveyor belts, etc., used wherever possible?
3. Where protracted loading of a muscle is unavoidable, is the muscular strength required less than 10% of the maximum?
4. Are technical sources of power employed where necessary?
5. Has the number of groups of muscles employed been reduced to the minimum with the aid of countersupport?
6. Are torques around the axis of the body avoided as far as possible?
7. Is the direction of motion as correct as possible in relation to the amount of force required?
8. Are loads lifted and carried correctly, and are they not too heavy?

C. NOISE
1. Is noise reduced to the minimum by means of technical measures at the source?
2. Are the sources of noise so insulated they hinder as few people as possible?
3. Is reflected noise reduced to the minimum by provisions on walls and ceilings?
4. Are the noisiest apparatus located as far as possible from the ears?
5. Can sound signals and verbal instructions be readily distinguished from the ambient noise?

D. LIGHTING
1. Is the lighting adequate for the nature of the work?
2. Has the fact that old lamps emit less light than new ones been taken into account?
3. Has glare on account of naked light sources, windows, or their reflection in shiny surfaces been avoided?
4. Has too great a brightness contrast between work position and surroundings been avoided?
5. Is reading of meters not impeded by reflection of light sources?

E. CLIMATE
1. Are the surroundings where the work is done not too warm or too cold for the nature of the work?
2. Have effective measures to prevent a high radiation temperature been taken?
3. Can the humidity of the air be kept within acceptable limits?
4. Are processes calling for a high degree of humidity insulated to the greatest possible extent?
5. Are draughts avoided and, at the same time, is there adequate ventilation?

F. INFORMATION
1. Is the quantity of information received by the worker adequate and yet confined to essentials?
2. Does the information arrive in good time and is it perceived through the right sense?
3. Is the information clear and unambiguous?
4. Is urgent information given via the ears?
5. Particularly in the case of lengthy inspection tasks, is seeing replaced by hearing?
6. Can sound signals having different meanings be readily distinguished from each other?
7. Can prealignment, assembly, and setting be carried out rapidly and efficiently by touch?
8. Can the positions of components, control knobs, control buttons, and tools be perceived by touch? *(continued)*

G. DISPLAYS
1. Can the meter be read quickly and correctly, according to a measure immediately suitable for use and with the required degree of accuracy?
2. Is the desired degree of accuracy really necessary?
3. Is the type of meter selected efficient?
4. Is the scale graduated properly and as simply as possible?
5. Are the letters, figures, and graduate marks clearly visible at the required range?
6. Is the pointer simple and distinct, and does it pass close to the scale without concealing the figures?
7. Are reading errors caused by parallax avoided?
8. Is a warning given, if the meter fails?

H. CONTROLS
1. Have the knobs, wheels, grips, and pedals been adapted to the specific requirements of fingers, hands, and feet? (Location, dimensions, shape, direction of motion, counterpressure)
2. Have the controls been located logically, in the sequence of operation, and can they be easily recognized by shape, dimensions, marking, and color?
3. Are pedals avoided for standing work and confined to two for sedentary tasks?
4. Are all the selected controls efficient?

I. PANELS
1. Is positioning of the controls in relation to the meters logical?
2. Is the relation between the direction of motion of the control and that of the deflection of the pointer or the reaction of the apparatus (e.g., On-Off) logical?
3. Has the panel the correct shape and dimensions in connection with sitting posture, grasping range, and direction of vision?
4. Are the meters, knobs, and buttons that are most important and most frequently employed appropriately located?
5. Are the meters so grouped and positioned in relation to each other that they can be read quickly and faultlessly?
6. Do the scale divisions and subdivisions of different meters correspond as far as possible?
7. Have larger panels been made easier to scan through separation of groups of meters, knobs, and buttons—spatially, with the aid of colors or by means of different planes?
8. If possible, is the process represented on the panel in the form of a diagram?
9. Are warning lamps clear and have they been located in the central part of the field of vision?

J. PACED WORK
1. Are at least three vacant positions available at any moment?
2. Is the component to be mounted supplied ready-aligned?
3. Are the assembly positions equipped with ample guides for the feed movements? (Holes, stops, etc.)
4. Are faulty components removed on the basis of preliminary inspection?
5. Should the answers to questions 2, 3, and 4 be negative, are there at least six vacant positions?

NOTES: a. Have you considered whether it might be useful to make a dummy of the machine or work setup?
b. Have you considered whether it might be useful to discuss the work setup with the operator concerned, the supervisor, or the departmental manager?
c. Have you considered whether it might be useful to discuss the work setup with your liaison for ergonomic matters?

Table 5.3   The "Eleven Commandments" by van Wely[6]

I AIM AT MOVEMENT
Use muscles, but *not* for holding or fixation.
Movement reduces monotony.
II USE OPTIMUM MOVEMENT SPEEDS
Too quick or too slow movements are fatiguing and inefficient; try to find the specific optimum speed of each movement.
III USE MOVEMENTS AROUND THE MIDDLE POSITIONS OF THE JOINTS
Long duration or frequent use of extreme positions of a joint, especially under load, are harmful and have poor mechanical advantage; yet occasional extreme positions, while not loaded, are desirable.
IV AVOID OVERLOADING OF MUSCLES
Keep dynamic forces to less than 30% of the maximum force that the muscle can exert; up to 50% OK for up to five minutes.   *(continued)*

Keep static muscular load to less than 15% of the maximum force that the muscle can exert.

V AVOID TWISTED OR CONTORTED POSTURES

Do not use pedals in standing work.

Use arm supports only when the upper arms cannot relax and be approximately perpendicular to the floor.

Bending the head backward causes glare and sore necks.

VI VARY THE POSTURE

Any fixation causes problems in muscles, joints, skin, and blood circulation. Do not use pedals in microscope work.

VII ALTERNATE SITTING, STANDING, AND WALKING

Continuous sitting for more than one hour and continuous standing for more than ½ hour are in the long run too fatiguing.

Standing for more than 1 hour a day is fatiguing and causes physical abnormalities.
Concrete floors are very fatiguing; elastic supports such as wood, rubber, or carpet are better.

VIII USE ADJUSTABLE CHAIRS

For sitting longer than ½ hour continuously or longer than 3 hours a day, use adjustable chairs and footrests (if necessary). When adjusting remember:

a. Seat so that elbows are at about the height of the working plane

b. Footrest so that no pressure occurs in the back of the knees

c. Backrest so that the lower back is supported (bottom of backrest 100 to 200 mm above seat)

IX MAKE THE LARGE MAN FIT AND GIVE HIM ENOUGH SPACE: LET THE SMALL WOMAN EASILY REACH

For standing it is essential that the working plane be adjustable; a platform for the short person is a simple solution.

Keep the working plane within 50 mm of elbow height.
Note: The working plane height is usually above the table height.

X TRAIN IN CORRECT USE OF EQUIPMENT: FOLLOW UP

People must be instructed and trained in good working postures: sitting, standing, and especially lifting are often done in the wrong way.

XI LOAD PEOPLE OPTIMALLY

Neither maximum nor minimum are optimum.

An optimum load (physical and mental) gives better performance, more comfort, less absenteeism, and less harm.

STRUCTURED SEARCH: SEARCH

Since others could make a checklist, so can I (and so can you). My six-item list can be remembered by the acronym, SEARCH, where:

S = Simplify individual operations
E = Eliminate unnecessary work and material
A = Alter sequence
R = Requirements
C = Combine operations, elements, and equipment
H = How often

Simplify Individual Operations

One example is found in the design of a routing memo in an office (see Figure 5.2 for one example that reduces communication work). Consider also the problem of a motel desk clerk in describing to a customer where his room is. A poor technique would be an oral description; a good technique would be a preprinted map upon which the relevant room and route could be indicated and the map given to the customer. Routes to two or three different locations can be painted or taped on the floor to reduce time of both customers and service personnel. Or, a poor technique for disposing of a part from a punch press would require the operator to reach into the die and remove it; a better technique would be air ejection.

OFFICE TRANSMITTAL SLIP

$\mathcal{K}ansas\ \mathcal{S}tate\ \mathcal{U}niversity$

Department of Industrial Engineering
Manhattan, Kansas

To:

_____

_____

_____

_____

_____

_____

( ) See Me
( ) Type
( ) File
( ) Signature
( ) Refer
( ) Necessary Action
( ) Comment
( ) Information
( ) Note, Sign, Return
( ) Per Conversation
( ) _____

Comment:

Date: _____ Signature: _____

Figure 5.2 *Routing memo forms* substitute a quick checkmark for writing phrases such as "please see me" or "for your signature."

Eliminate Unnecessary Work and Material

This item really should be listed first, but I list it second to fit in my acronym. For a first example, consider the business letter shown in Figure 5.3. Ignoring the content (which in itself probably could be shortened with no loss), how much extra work does the format require for the typist?

Consider the date. Does it require more work to type or write Nov. 1, 1976 or 1 Nov. 1976? The first way requires one additional character out of 10 or 10% more work. Consider using 1 Nov 76 and save 4 characters or 40% of the work in writing every date.

Consider the salutation "Gentlemen" and the closing "Very truly yours." Do they add anything to a business letter except cost? If your boss won't agree to omitting them, perhaps substituting "Sirs" for "Gentlemen" and "Sincerely" or "Yours" for "Very truly yours" would be accepted.

Consider the initials in the lower left corner. What is their purpose? The JPJ identifies that the letter was by Jones, who signed it at the right. In the vast

Nov. 1, 19——

Atlas Shoe Stores, Inc.
232 Fifth Avenue
Brooklyn, New York 10014

Gentlemen:

Thank you for your prompt notification of damage to your shipment
of Sept. 24th.
We sincerely regret any inconvenience the loss of some of the sizes
may have caused you.
We will be pleased to take care of all the necessary inquiries to the
freight company. In the meantime, we are shipping you the missing
numbers by prepaid express.

Very truly yours,

John P. Jones
Sales Manager

JPJ:mcp

Figure 5.3 *Business letter formats* often require extra typing work. How could it be modified?

majority of situations, this redundant information can be eliminated. The only situation for which I can conceive as requiring initials at the left is as an internal code to indicate situations in which individuals other than Jones dictated the letter; thus in most situations there should be no capitalized initials at the left. The lower case initials identify the typist. They are needed only in multitypist offices. In addition, a single letter would be sufficient as identification; three are not needed.

Alignment of the various items along the left margin also can reduce the number of excess keystrokes. Figure 5.4 shows the modified letter; it requires 434 strokes (including spaces and indexing) instead of 486.

1 Nov ——

Atlas Shoes Stores, Inc.
232 Fifth Avenue
Brooklyn, New York 10014

Thank you for your prompt notification of damage to your shipment of
Sept. 24th.

We sincerely regret any inconvenience the loss of some of the sizes may
have caused you.

We will be pleased to take care of all the necessary inquiries to the
freight company. In the meantime, we are shipping you the missing numbers
by prepaid express.

John P. Jones
Sales Manager

m

Figure 5.4 *Modified format* of Figure 5.3 is shown. What format is used for letters from your organization?

As a second example, I personally was responsible for some unnecessary work when I worked for a midwestern firm. A department making springs from coil stock found one batch defective due to faulty material. I instituted a policy of cutting a segment from each coil received and testing its quality. It seemed to be a reasonable policy at the time. With the benefits of hindsight, I now realize that the persistent cost of the inspection far outweighs the cost of a very rare failure due to poor material. The best policy would have been no inspection along with a notification to purchasing to watch that vendor more closely. Although "doing something" (such as starting a new inspection procedure) seemed a good idea, I had added a new overhead expense that probably never will be questioned.

The third example of eliminating unnecessary work is the book purchasing procedure now followed by the Kansas State University library. The original policy was to purchase all books requested by faculty members plus books judged important by the head librarian. During the middle 1960s the head librarian realized that the library was processing thousands of individual orders for books. More important was the realization that, within approximately three years after a technical book was issued by a "major" publisher, there was over a 95% chance that the KSU library had ordered it. The present policy is to have a standing order with major publishers for 100% of books (in certain categories). The publisher sends the book when published (thus getting it to the library 1 to 36 months sooner) and sends a monthly bill for all the books sent that month. The library gets more books, sooner, and at a very substantial savings in paperwork cost.

A fourth example is pruning of mailing lists and internal distribution lists. Once/yr send a letter saying that unless you complete the form and return it, you will be taken off the list. Another technique is to make the continuation form page two of the material—this will detect who reads the material. Presorting mail by the user can speed mail delivery and reduce costs. At each department's mail drop, have three bins: internal, local, and out-of-town. Or perhaps internal this building, other internal, and US mail. The Post Office does the same with local and out-of-town boxes.

A fifth example is machining coolant. Formerly concentrate was added whenever the level became low. Then it was realized that it was the water that was evaporating, not the concentrate. Now the concentration is measured and water is added as necessary; this not only gives the proper concentration but saves concentrate.

Alter Sequence

Altering the sequence may help by (1) reducing the need for or difficulty of other operations, (2) reducing idle or delay time, or (3) reducing material handling costs.

Changing the sequence often eliminates the need of cleaning or deburring operations. Machining after hardening instead of before is an example of reduced difficulty. Consider the sequence of ordering a pitcher of beer, having it "drawn," and then paying your money. If the bartender takes your money while the pitcher is being filled, you can get back to your table sooner! The same principle applies to paying your bill while waiting for a meal to be assembled or cooked. Another food example is a waitress bringing coffee to a table on her initial trip instead of the second trip; this reduces delay time as well as reduces travel distance. In factories, group similar components into "parts families" and schedule "relatives" together to reduce delays and material handling costs.

Requirements

Two aspects of requirements will be mentioned: *quality* (or capability) and *initial vs. continuing* costs.

Quality costs. The general shape of the cost vs. quality (capability) curve is shown in Figure 5.5. The specific numbers vary, but the important point is the *shape* of the curve; it's a *curve,* not a straight line. That is, beyond a certain point additional precision (capability) has a very high cost. For a specific component, is the specified quality too high? What are the costs and the benefits of reducing a tolerance by 10%? Remember the old joke that a tolerance is the smallest number that a design engineer can think of. For example, do you need to buy a unit which can operate to "military" specifications (e.g., operate from 0 to 20,000 m) or will your unit always be on the ground?

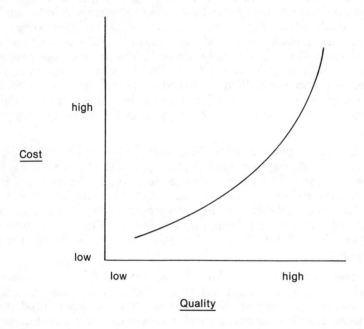

Figure 5.5 *Cost rises* disproportionately as quality (or product capability) improves. Do you really need all that quality?

Another quality cost consideration is *fabrication costs vs. assembly costs.* How much money should be spent in fabrication to make assembly easy? Engineers tend to design using "worst case" analyses so they put many tolerances on components to prevent possible assembly problems. It may be desirable to question a fully interchangeable component policy and use some selective assembly. In *selective assembly,* if bushing *A* doesn't fit on shaft *B*, the assembler may occasionally have to try a second bushing. Another technique is to sort *bushings* into small, medium, and large and *shafts* into small, medium, and large. The additional sorting and assembly cost is less than the savings in fabrication cost. Note, however, that if parts are not interchangeable then spare components cannot be sold, only spare assemblies.

Initial vs. continuing costs. The second aspect of requirements is initial vs. continuing costs. A common problem is reluctance to spend now to save later. The savings later can come from either reduced operating or maintenance costs.

A personal example was the bracket on my car's air cleaner. It was too thin—saving the company perhaps .1 cent/bracket. It failed every 15,000 km—forcing me to buy a new bracket until, in desperation, I welded a washer to it to make it strong enough. Yet the auto companies have found that people are reluctant to pay attention to any cost of a car except first cost. For many complex military items (such as jet aircraft), the maintenance cost (over its life) is 30 times the initial price. As the oil filter advertisement says, you can pay me now (for the filter) or pay me later (for the repairs to the engine). In fish farming, when cupro-nickel cages replace nylon nets, which must be cleaned often due to algae fouling, there are large maintenance savings.

Combine Operations, Elements, and Equipment

In a more general sense, this is the argument for special-purpose equipment vs. general-purpose equipment. A simple example is a car wax combined with a car cleaner. By combining both materials into a single compound, you can clean and wax your car in one step instead of two. Another example is making a duplicate typed copy by using a special paper (carbon paper)—one typing makes multiple copies. A machining example is the combination drill and countersink shown in Figure 5.6. Instead of drilling the hole in one operation and countersinking in the second, the hole is drilled and countersunk all in one operation.

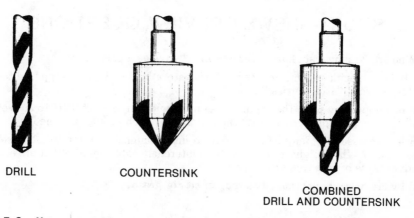

DRILL   COUNTERSINK

COMBINED
DRILL AND COUNTERSINK

Figure 5.6 *Using special purpose equipment* (a combined drill and countersink) makes it possible to drill and countersink a hole in one operation instead of two; the engineer must trade off the extra capital cost vs. the reduced operating costs.

How Often?

A way of considering how often is to consider it an economic lot size problem; see chapter 7 for calculations. The trade-off is setup cost vs. cost of inventory (capital, spoilage). For example, when preparing a meal should a large portion be prepared and part frozen for use later? Technologies change the answers over time. Freezers and microwave ovens have decreased the storage problem for food and encouraged larger batches for cooking. In manufacturing, numerical controlled machine tools have lower setup costs and encourage smaller batches.

The how often question is especially relevant to maintenance, inspection, and service activities.

Consider the problem of the janitor replacing toilet and towel rolls. How often need he inspect? If each toilet holds two rolls of paper instead of one, then less frequent service is required. In addition, if a one-roll holder is used, the janitor usually removes the "stub roll" and throws it away—thus wasting material. Or consider the replacement of failed light bulbs. Should each failed lamp be replaced or should lamps be replaced in groups? Should inspectors inspect an operation once an hour, once a day, or once a week? One useful technique is to classify operators or operations as "red," "amber," or "green." The red operators (say a new worker in the department) are inspected once/hr, while the amber operators are inspected twice a day, and the green operators are inspected only when the inspector has nothing else to do. This tends to put the resources (the inspectors) where the problems are.

SUMMARY

Before the evaluation of alternatives can begin, there must be a number of alternatives to select from. This chapter deals with generating alternatives.

Although some like to "freewheel" and generate many ideas with brainstorming, most prefer the staid, systematic series of questions developed by people who have worked on the same type of problem before.

Which checklist to use depends upon the situation. Why not make one of your own which is relevant to your specific industry and plant and even your department?

SHORT-ANSWER REVIEW QUESTIONS

5.1 Who are Kipling's six honest serving men? Give as a poem.

5.2 Give the specific techniques used in "brainstorming." Why is it supposed to work? Does it work "as advertised"?

5.3 When improving jobs, the engineer can use the acronym SEARCH to suggest possible areas of improvement. List the six areas. Which should be done first?

5.4 Selective assembly may be preferable to interchangeable parts. What is selective assembly? When might it be better than interchangeable parts? What are some disadvantages of selective assembly?

5.5 Why are operators classified as red, amber, or green?

THOUGHT-DISCUSSION QUESTIONS

1. Discuss, with its manager, the format of the business letter of a specific firm. What does the manager think of your ideas?

2. Discuss unstructured vs. structured search. If you were a supervisor, which would you pay your employees to use? Why?

REFERENCES

1. Beardsley, J. Ingredients of successful quality circles. *Transactions of 2nd Annual International Association of Quality Circles*, (1980): 139-45.

2. Boer, K., and Burger, G. Ergonomics checklist. *Value Engineering*, (September 1968): 153-64.

3. Dunnette, M.; Cambell, J.; and Jaastad, K. The effect of group participation on brainstorming effectiveness for two industrial samples. *Journal of Applied Psychology*, Vol. 47, No. 1 (1963): 30-37.

4. Kellerman, F.; van Wely, P.; and Willems, P. *Vademecum: Ergonomics in Industry.* Eindhoven, Netherlands: Philips Technical Library, 1963.

5. Taylor, D.; Berry, P.; and Block, C. Does group participation when using brain-storming facilitate or inhibit creative thinking? *Administrative Science Quarterly,* Vol. 3 (1958): 23-47.

6. van Wely, P. Design and Disease, Engineering Experiment Station Report 86, Manhattan: Kansas State University, 1969.

VALUE ENGINEERING

CONCEPT

Excess work can be caused by poor product design (see chapter 4). The goal of value engineering, which is applied to procedures (*software*) as well as products (*hardware*), is to reduce the excess cost in a design. The fundamentals of the "systematic approach" of value engineering were originated by Miles of General Electric in 1947. The basic concept is that many existing designs can be improved substantially since the original design has excess costs. One common reason is that no design engineer can be expert in all phases of engineering, manufacturing, and purchasing; thus, the design may have potential for improvement. The excess cost may be due to a change in material or labor prices since the original design (either absolute or relative; see Figure 3.2), changing technology, different applications than originally envisioned, and lack of time to make a good original design. In contrast to this "second guessing" of existing designs, many firms now use value engineering techniques during the original design process. After all, why not do it right the first time? That's when change costs are smaller and benefits the greatest.

TECHNIQUE

Table 6.1 gives the six steps.

First, select which product or procedure to study. The Pareto distribution (mighty few and insignificant many) helps focus the problem. If a product costs $5/unit and is used 3 times/yr, then its annual cost is $15.00. If you spend $500 of engineering time to reduce unit cost to $4, then you have spent $500 to save $3/yr. However, if you save 10% on another product which had annual costs of $50,000, then you spent $500 to save $5000—a better buy! Note that $5000/yr can come from different combinations: a $5,000 unit made once/yr, a $1000 unit made 5 times/yr, a $10 unit made 500 times/yr, or even a $.10 unit made 50,000 times/yr.

Table 6.1　Six Steps of Value Engineering.[2]
Note Similarities to Table 4.1.

---

1. Select the problem.
2. Get information.
3. Define and evaluate functions.
4. Create solutions.
5. Evaluate solutions.
6. Recommend a solution.

Within this range of potentially profitable projects, select products which are similar to other products (cousins); they are good to study as the "cousins" may suggest alternate approaches and also may be additional application areas. Some projects come from noticing something on a stroll through the plant; chance favors the prepared mind.

Second, get information. Get from marketing, engineering, manufacturing, and purchasing. The diverse challenges facing the value engineering project usually cause projects to be done by teams as no one individual has such broad expertise. Because replacing special company components with standard purchased components is a common solution, the team usually has a purchasing member.

In addition to marketing and engineering data, get cost and specification information. Cost information includes labor costs (setup time, time/unit, cost/hr), material costs, packaging costs, tooling costs, and equipment costs. Specifications can be divided into customer and internal; yet another division is real vs. imaginary. Specifications often are figments of the imagination (e.g., based on a historical situation no longer relevant); that is, the engineer has been self-constrained.

Third, define and evaluate the function. Distinguish between value and function. Consider two tie clips: one a paper clip and one a diamond pin. Both accomplish the same function but one has a value of $.01 and the other a value of $1000. The functional value is the same but the personal (prestige, esteem) value is different. The engineer must consider not only function characteristics but also prestige characteristics. That is, not only what a product or service can do but also what will make it sell.

Mudge gives four rules for defining function[2]:

1. Express functions in one noun and one verb.

2. Give work functions in action verbs and measurable nouns.

3. Give sell functions in passive verbs and nonmeasurable nouns.

4. Divide functions for each component into the primary (basic) one and secondary functions.

Table 6.2 gives examples of "good" nouns and verbs as well as some poor ones. Words such as component, parts, nut, bolt, and bracket are poor because they are too vague and don't identify what the function is. In units with 4 to 6 components, each component should have all its functions listed (*bushing:* provide support, provide adjustment, provide location, transmit force, provide connection; *key:* transmit force, provide location; *lock washer:* induce friction, transmit force, provide location). In units with more components, use the Pareto principle and just analyze the "giants."

Table 6.2   Mudge Recommends Describing Work Functions With Action Verbs and Measurable Nouns.[2] Sell Functions Should Be Described in Passive Verbs and Nonmeasurable Nouns.

WORK FUNCTIONS

Action verbs		Measurable nouns		
amplify	interrupt	access	fluid	message
apply	make	circuit	force	oxidation
change	modulate	contamination	friction	protection
collect	prevent	current	heat	radiation
conduct	protect	damage	insulation	repair
control	provide	density	light	voltage
create	rectify	energy	liquid	weight
emit	reduce	flow		
enclose	remove			
establish	repel		Undesirable nouns	
filter	retain	article	device	table
hold	shield	component	part	wire
impede	support			
induce	transmit			
insulate				

SELL FUNCTIONS

Passive verbs		Nonmeasurable nouns		
decrease		appearance	exchange	prestige
improve		beauty	features	style
increase		convenience	form	symmetry
optimize		effect		

Then, for each component, identify which of the many functions is the basic function and which the secondary ones.

Then consolidate and evaluate the functions on a form such as Figure 6.1. List all of the functions which appear for the components on the top portion of the form. Then using the "key letter" from the top portion of the form, compare function A (provide connection) vs. function B (transmit force). A *major* difference in importance ("obvious" decision) is assigned a 3; a *medium* difference in importance ("short" time to make decision) is assigned a 2; and a *minor* difference ("considerable" thought to decide which is more important) is assigned a 1. If A is more important and the difference is medium, write A2 at the intersection of the A row and B column. Now compare function A vs. function C. If A is more important than C and the difference is medium, write A2 at the intersection of the A row and C column. Continue until all the comparisons are made. Then total the "points" for each letter on the top portion of the form. You now have established the basic function (the function with highest weight). The remainder are secondary functions. The purpose of defining and evaluating the design's functions is to focus your thinking for the next stages—create solutions and evaluate solutions. Perhaps it will "provoke you to the point of thought." Your objective is to compare cost of a function vs. value of a function.

Fourth, create solutions. Solutions can be created either through unstructured search (such as brainstorming) or structured search (such as checklists). See chapter 4 for examples.

Fifth, evaluate solutions. In practice, the fourth and fifth steps form a loop which is repeated a number of times until a number of solutions are considered. Perhaps the most important thing is to have multiple solutions from which to pick; do not stop with the first solution which works. Don't forget the selling functions.

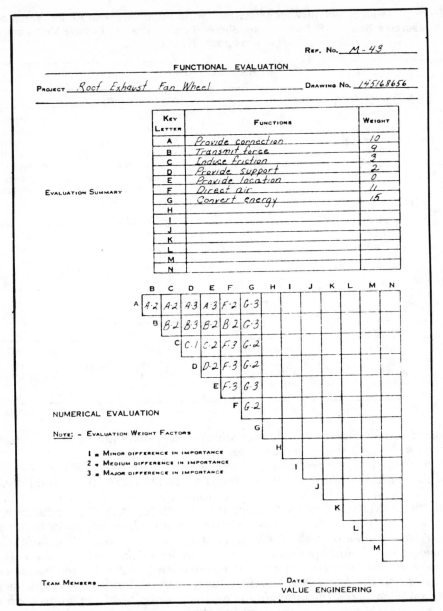

Figure 6.1 *Functional evaluation* is done in three steps: (1) List functions (without weights) on top form. (2) Tabulate differences in importance of the various functions (using key letter from upper form to determine row on lower form). (3) Total weights for each function letter in lower form and enter as "weight" in upper form.[2]

Sixth, recommend a solution. Get all the information on the project together for the presentation to the decision maker. Determine exactly what needs to be done to implement the change (such as stop existing production in 43 days; change drawings 821, 822, and 917; rewrite service spares manual on pages 18 and 19; change standard cost for item, etc). Make the proposal to the decision maker. Then, if accepted, get the paperwork started. (If the proposal is not accepted, can you settle for "half-a-loaf" rather than none?) Follow up to make sure the change is implemented. Audit the actual savings vs. your proposal so future calculations can be made more accurately.

EXAMPLE APPLICATIONS

Men trust rather to their eyes than to their ears.
The effect of precepts is, therefore, slow and
tedious, while that of examples is summary and effectual.

Seneca

The examples primarily come from Mudge and ASTME; the construction examples come from *Value Engineering*.[1,2,3]

- **"O" Rings.** A chemical company used a special "O" ring at the rate of 2,000/month. It saved $17,000/yr by ordering standard rings in lots of 50,000 (i.e., a two-year supply).

- **Nameplate.** The company nameplate used a nonstandard aluminum alloy. A standard alloy saved $9,500/yr.

- **Bushing and gasket material.** The specified material was red although the standard material is black. Switching to standard black saved $10,000/yr.

- **Molded name.** The company name was molded into a porcelain lamp receptacle for "advertising." This made it a special item for the vendor. However, the only time the name was seen was when the lamp was changed by the maintenance man. Removing the company name gained a 10% price reduction from the vendor.

- **Brass oil-level gauge.** A special brass gauge was used on some compressors. A standard gauge, used on other company products, saved $500/yr.

- **Special nut.** A "special" nut was made "in house" at $13.30 per hundred at the rate of 375,000/yr because suppliers did not make "specials." When purchasing asked for a quote anyway, the vendor said they needed a minimum order of 25,000 for specials. When told of the 375,000/yr, the vendor said "It's just become a standard!"; cost savings was $7/100.

- **Motor guard.** A guard for a totally enclosed electric motor on a candy machine cost $20. It turned out its real function was to prevent people touching the motor; complaints had been received on the original machines that the "motor ran hot." The guard was replaced with a $.25 adhesive nameplate which read "This motor normally runs hot."

- **Test paperwork.** An extensive testing program (350,000 pages of test data with 4,500,000 entries) was challenged. It was discovered that the contract specified testing only during the research and development phase but it was not necessary during production. Savings on the contract were $720,000.

- **Rotary switch.** The original design had 167 parts and required 21 hr/assembly. Value analysis generated 36 useful ideas. The revised unit had 10 parts and required .1 hr/assembly.

- **Carton.** A side-opening carton replaced a top-opening carton; savings were 8%. Then the supplier suggested changing the carton material from gloss white to buff white; this saved 40%. Both changes were applied then to a variety of other cartons.

- **Milled contacts.** Copper contacts were hand-filed to remove the burrs from milling; cost was $1.58/unit. A forged contact at $.43/unit required no filing and also carried 20% more current.

- **Magnolia duct spacers.** Magnolia wood (which does not contaminate transformer oils) was being used for spacers. A value analyst took 3 years to experiment, test, and modify the material to chipboard; savings were $50,000/yr.

- **Packaging tape.** An airframe manufacturer used costly acetate fiber tape in packaging. Gummed paper provided the same function and saved $3000/yr.
- **Shipping box.** An electric company packaged reels of punched tape in $1 plastic boxes. The new box (for $.15) is a plastic sandwich box from a dime store.
- **"Dead sharp" gaskets.** "Dead sharp" was considered to require lathe cutting by the vendor. When the high price was challenged, punching was considered satisfactory by the buyer; the price was reduced 20%.
- **Metal cabinet.** The cabinet was stiffened by spotwelding stiffeners. The spot-welding indentations were expensive to conceal. The stiffeners were deleted by making the cover thicker and bonding two attachment brackets. The saving on the contract was $10,000.
- **Dam culvert.** A culvert on Smithland Dam on the Ohio river was narrowed to 14 feet from 16 feet; savings were $695,000.
- **Two-stage contract.** The Bankhead lock needed replacement. When blasting for the new lock, contractors were afraid of damaging the existing lock; bids were high. The value engineering proposal was to divide the job into two stages. Stage 1 (purpose of get information) used highly skilled blasting crews and experimentation and data recording in blasting a small hole. Results were given to contractors for bid on stage 2—the big hole. The savings of $3,300,000 received the Presidential Management Improvement Award for 1971.
- **Underground cables.** Contractor suggested modifying material enclosing cables for the Air Force. The Government Services Administration (GSA) and the contractor split the $100,000 savings. (The GSA, to encourage value engineering, splits savings in initial construction costs 50-50 with the contractor.)

SUMMARY

Value engineering emphasizes reduction of excess work caused by poor design of products or procedures. Standardization is used. Perhaps the key concept is the emphasis on value and function and the stripping away of nonessentials.

SHORT-ANSWER REVIEW QUESTIONS

6.1 What is the goal of value engineering?

6.2 Functions are divided into which two categories?

6.3 List Mudge's four rules for defining function.

6.4 Discuss the distinction between functional value and prestige (esteem, sales) value. Give examples.

THOUGHT-DISCUSSION QUESTIONS

1. Why is value engineering necessary? Why don't engineers "design it right" the first time?

2. Why does value engineering encourage "fighting giants"?

REFERENCES

1. ASTME. *Value Engineering in Manufacturing.* Englewood Cliffs, NJ: Prentice Hall, 1967.

2. Mudge, A. *Value Engineering.* New York: McGraw-Hill, 1971.

3. *Value Engineering* 1973. Hearings before Senate Committee on Public Works, Serial 93-H15. Washington: Government Printing Office, 1973.

7

OPERATIONS ANALYSIS

This chapter presents analysis techniques for the "big picture" (analysis of several operations or tasks) and then the "little picture" (analysis of an individual job). For the "miniature picture" (analysis of specific motions), see chapters 8 and 14.

BETWEEN OPERATIONS

Location of One Item

Locating just one item in an existing network of customers is a more common problem than it appears; Table 7.1 gives some examples of this broad class of problems. The *item* can be a person, a machine, or even a building; the *network of customers* can be people, machines, or buildings; and the *criterion minimized* can be movement of people, product, energy, or even service time. In the example, consider the new item as a machine tool and the customer network (circles in Figure 7.1) as other machine tools with which the new item will exchange product.[7] The criterion is to minimize movement distance of the product. In most real problems there are only a few possible places to put the item; the remaining places already are filled with other machines, building columns, aisles, etc. Two feasible locations for the item are A and B (rectangles in Figure 7.1).

Travel between a customer and A or B can be a straight line (e.g., conveyors) or rectangular (e.g., fork trucks down an aisle). In some problems the travel may be neither (e.g., one-way aisles); in these cases measure the distance on a map.

Quantify the importance of each customer to the item. Common indices are shipments/month or pieces/day.

The operating cost of locating the item at a specific feasible location is:

$$MVCOST = WGHTK \, (DIST) \qquad (7.1)$$

where $MVCOST$ = index of movement cost for a feasible location
$WGHTK$ = weight (importance) of the Kth customer of N customers
$DIST$ = distance moved

$$MVCOST = \sum_{K=1}^{N} (|X_{i,j} - X_k| + |Y_{i,j} - Y_k|) \qquad \text{(for rectangular)} \qquad (7.2a)$$

$$MVCOST = \sum_{K=1}^{N} \sqrt{(X_{i,j} - X_k)^2 + (Y_{i,j} - Y_k)^2} \qquad \text{(for straight line)} \qquad (7.2b)$$

For the two locations given in Table 7.2, Table 7.3 shows the $MVCOST$. Movement cost at B is about 67,954/53,581 = 126% of A.

Table 7.1 Examples of Locating an Item in an Existing Network of Customers With Various Criteria to be Minimized.

New Item	Network of Customers	Criterion Minimized
Machine tool	Machine shop	Movement of product
Tool crib	Machine shop	Walking of operators
Time clock	Factory	Walking of operators
Inspection bench	Factory	Movement of product or inspectors
Xerox machine	Office	Movement of secretaries
Warehouse or store	Market	Distribution cost
Factory	Warehouses	Distribution cost
Electric substation	Motors	Power loss
Civil defense siren	City	Distance to population
AIIE meeting place	Locations of AIEE members	Distance traveled
Fire station	City	Time to fire

Assume you wish to know the cost at other locations than A and B for the assumption of the above problem. By calculating costs at a number of points, a contour map can be drawn. This indicates that the best location is $X = 42$ and $Y = 40$ with a value of 32,000. Thus Site A is 6,000 from the minimum and Site B is 13,000 from the minimum.

The example, however, made the gross simplification that movement cost per unit distance is constant. Figure 7.2 shows the more realistic assumption where most of the cost is loading and unloading (starting and stopping) or paperwork and cost of moving, when "acceleration and deceleration" are omitted, is very low. More realistically:

$$DIST = L_k + C_k \, (|X_{i,j} - X_j| + |Y_{i,j} - Y_k|) \qquad \text{(for rectangular)} \qquad (7.3a)$$

$$DIST = L_k + C_k\sqrt{(X_{i,j} - X_k)^2 + (Y_{i,j} - Y_k)^2} \qquad \text{(for straight line)} \qquad (7.3b)$$

where L_k = load + unload cost (including paperwork) per trip between the Kth customer and the feasible location

C_k = cost/unit distance (excluding L_k)

Assume for customers 1, 2, and 3 that L_k = \$.50/trip and C_k = \$.001/m; for customers 4, 5, and 6, L_k = \$1./trip and C_k = \$.002/m. Then the cost for alternative A = \$854 + 79.07 = \$933.07 while the cost of B = \$854 + 117.89 = \$971.89. Thus B has a movement cost of 104% of A. When making the decision where to locate the new item, use not only the movement cost but also installation cost, capital cost, and maintenance cost. Note that the product ($WGHTK$) ($DIST$) (that is, the \$854) is independent of the feasible location; it just adds a constant value to each alternative.

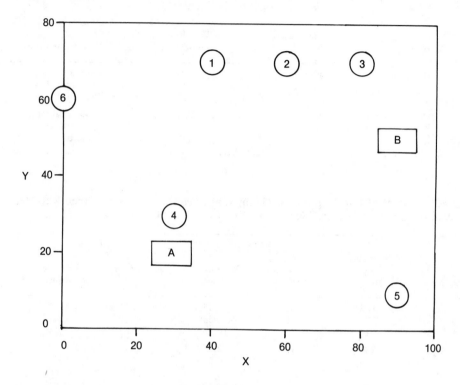

Figure 7.1 *Should the customers* (in circles) be served from location A or location B? Table 7.2 is shown on the coordinates.

Cost need not be expressed in terms of money. Consider locating a fire station where the *customers* are parts of the town and the *weights* are expected number of trips in a 10-yr period. Then *load* might be 1 min to respond to a call, *travel* is 1.5 min/km, and *unload* might be 1 min; the criterion is to minimize mean time/call.

Note also that the distance cost may rise by a power of 2—the *inverse square law*—for such problems as location of a siren or a light.

Systematic Layout of Multiple Items

In contrast to the previous section on location of one item in an existing network of customers, this section discusses arrangement of the entire facility. This layout can be a "block" (e.g., department within factory) layout or a "detailed" (e.g., machine within a department) layout. A variety of computer programs have been written. They are variations on two popular techniques: travel charting and Muther's relationship charting—part of the Systematic Layout Planning Technique.[12,13] Systematic Layout Planning (SLP) will be drawn upon since travel charting uses product material handling movement (i.e., equation 7.2) as its *only* criterion. Systematic Layout can be used at different levels: departments within a plant, machines within a department, or even displays within a display panel.

Table 7.2   Customers 1 to 6 Can Be Served Either From Location A (X = 30, Y = 20) or From Location B (X = 90, Y = 50). Which Location is Better?

Customer	Coordinate X	Y	Weight or importance	Movement type
1	40	70	156	Straight line
2	60	70	179	Straight line
3	80	70	143	Straight line
4	30	30	296	Rectangular
5	90	10	94	Rectangular
6	0	60	225	Rectangular

Table 7.3   Cost of Locating a New Machine at Locations A and B. Since WGHTK Was Pallets/Month and DIST Was in Meters, MVCOST = M – Pallets/Month.

Customer	Weight, Pallets/ Month	SITE A		SITE B	
		Distance, Meters	Cost, M—Pallets/ Month	Distance	Cost, M—Pallets/ Month
1	156	51	7,956	54	8,424
2	179	58	10,382	36	6,444
3	143	71	10,153	22	3,146
4	296	10	2,960	80	23,680
5	94	70	6,580	40	3,760
6	225	70	15,750	100	22,500
			53,781		67,954

Step 1. The relationship chart (see Table 7.4) is the first step. Divide the facility into convenient activity areas (office, lathes, drill press, etc.). For more than 15 areas, analyze in 2 sections (e.g., layout of assembly departments and layout of component departments). For "Closeness Desired Between Areas," assign a letter grade: A = absolutely necessary, B = important, C = average, D = unimportant, and E = not desirable to be close. (Muther's technique uses 6 levels instead of 5.) Letters are used rather than numbers as numbers imply more precision to the judgment than is available. Avoid too many A relationships. About 10% As, 15% Bs, 25% Cs, and 50% Ds is a good goal. Support A, B, and E relationships with a "Reason for Closeness." Reasons will depend upon the problem but common reasons are: 1 = product movement, 2 = supervisory closeness, 3 = personnel movement, 4 = tool or equipment movement, 5 = noise and vibration.

Step 2. Assign floor space to each activity area, along with physical features and restrictions (see Table 7.5). Remember Moore's corollary to Parkinson's Law "Inventories expand into whatever space is available, regardless of the need to maintain the inventory."[11] If the layout is a group of machines within an area, add space to the space for the machine alone. Consider space for the operator, for maintenance access, for movement of parts of the machine, and for local storage of product and supplies.

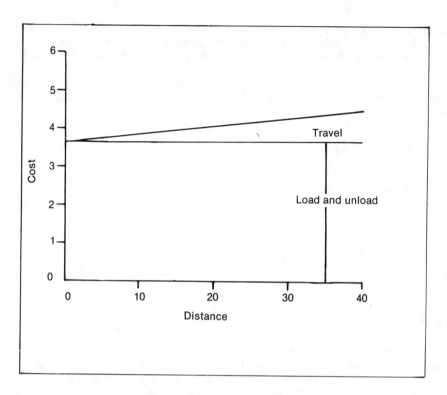

Figure 7.2 *Material handling cost* is almost independent of distance. See Principle 2 of chapter 13 for additional comments.

Step 3. Make an "activity relationship diagram" (see Figure 7.3). First, *list* all the A relationships from the relationship chart, then the Bs, Cs, Ds, and Es. Then make a diagram with just the As. Then add the Bs, keeping in mind the E restrictions. (For Es, walls and other barriers permit physical closeness while reducing visual and auditory distraction.) Then add the Cs; then the Ds.

Step 4. Make a *scaled* layout of at least two trials from step 3 using the areas and restrictions from step 2 (see Figure 7.4). First, use scaled pieces of stiff paper for each department; sketches tend to "get set in concrete" too soon. Then sketch in some alternatives with a pencil. Don't forget possible modifications of area shape and even size. The reason for at least two layouts is that engineers are satisfiers rather than optimizers—they tend to stop designing as soon as they have a solution which "works."

Step 5. Evaluate the alternatives (see Table 7.6). The relevant criteria and their weights will change from situation to situation. Grade each layout (A = Excellent = 4; B = Good = 3; C = Average = 2; D = Fair = 1; and E = Bad = 0) and calculate the layout's "grade point" (grade x weight). If there is an existing layout, include it as one alternative. Defining the best as 100%, calculate the percent for the alternatives. Have the affected people "sign off" on the evaluation form. Then go back and select features from the alternatives to get an improved set of designs.

Step 6. Detail the layout (make a working drawing) and present these improved alternatives to management personnel. After their modifications, install the final design.

Table 7.4 Step 1 of Systematic Layout is Identify the Desired Closeness Between Areas With Letter Grades. Give Reasons For Letters A, B, and E. Thus B/2 For Drill Press-Lathes Indicates a B Importance For Closeness for Reason 2.

Area Number	Area Name	Office 1	Lathes 2	Drill Press 3	Punch Press 4	Plating 5	Shipping 6	Die Storage 7
1	Office							
2	Lathes	D						
3	Drill Press	D	B/2					
4	Punch Press	E/5	D	B/2				
5	Plating	D	C	D	D			
6	Shipping	C	D	D	C	B/2		
7	Die Storage	D	D	D	A/4	D	B/4	

Assembly Line Balancing

A number of workers may be coupled together in an assembly line. See Principle 3 of chapter 13 for decoupling techniques. There remains, however, the problem of how to divide the total task when the operations are not decoupled; the problem is known as *the line balancing problem.*

Table 7.7 gives 10 elements and their work times; Figure 7.5, a *precedence diagram,* shows the element sequences. The problem, given the element times, the element precedences, and the required units/min from the assembly line, is to determine the number of workstations, the number of workers at each station, and the elements to be done at each station. The problem is to *balance the line* so total idle time is minimum.

First, what is the total number to be assembled and in how long a time? Thus, 20,000 units could be produced in 1000 hr at the rate of 20/hr, 500 hr at 40/hr, 250 hr at 80/hr, or many other combinations. For a first draft, assume we wish to produce 20,000 units at the rate of 20/hr (that is, 1 unit/.05 hr of "clock time"). Since we are dealing with rigidly coupled workstations, each workstation thus will take 1000 hr/20,000 units = .05 hr/unit; "cycle time" = .05 hr.

Second, guess an approximate number of workstations by taking total work time divided by cycle time; .1818 hr/ (.05 hr/workstation) = 3.63 workstations. Use 4 workstations.

Table 7.5 Step 2 of Simplified SLP is Identify Amount Desired for Each Area and Give Physical Features and Restrictions

Area Name	Desired m²	Restrictions
1 Office	50	Air conditioning
2 Lathes	40	Minimum of 10 m long
3 Drill press	40	
4 Punch press	50	Foundation
5 Plating	30	Water supply, fumes, wastes
6 Shipping	20	Outside wall
7 Die Storage	50	Crane
	280	

Third, make a trial solution as in Table 7.8 and Figure 7.6. Identify each workstation with a cross-hatched area. Remember not to violate precedence; for example, elements 1 and 5 can't be done at one workstation and 2 at another. Then calculate idle percentage; 9.1% in our example.

With larger assembly lines the problem complexity grows rapidly; grouping the elements into *zones* (which either prevent or require certain elements from being done at the same workstation) is one attempt at simplification. Most actual lines also have changes in product volume; i.e., 20/hr in May, 24/hr in June, 26/hr in July, etc., so the line must be rebalanced often. Another complication is that actual lines often are multiple product lines: first we assemble a 4-door Pontiac station wagon with a V6 engine, then a 2-door Oldsmobile hardtop with a V6 engine, then a Pontiac 4-door sedan with a V6 engine and air conditioning, etc. In addition, the mix changes (i.e., in March we produce 70% with air conditioning and in April 80%).

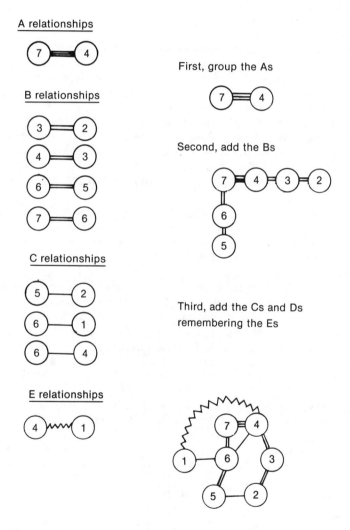

Figure 7.3 *Step three,* the activity relationship diagram, first groups the As, Bs, Cs, Ds, and Es (left side of figure). (It may help to think of the lines as "rubber bands" pulling areas together. The wavy line for the Es is a "spring" keeping them apart). Then group all the As (right side of figure), add the Bs to the As (remembering the Es), and then add the remainder.

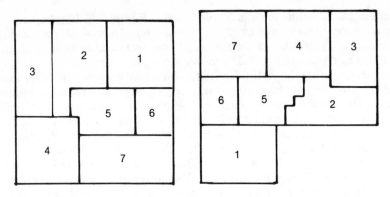

Figure 7.4 *Step four,* scaled layout, makes several alternative arrangements. It may be desirable (to keep the overall building shape rectangular) to slightly modify some of the desired areas from step 2.

The problem complexity and the need for repeated solutions has led to considerable effort to use computer simulation to solve the problem. At present, computer solutions seem only to be good first drafts which should be modified by an engineer. The basic problem is that the computer believes what is given to it is a fact, which cannot be "bent" or "modified."

Mariotti discusses some useful modifications.[10]

One, consider element sharing. That is, operators/station need not equal 1.0; you can have more than one operator/station (this permits a cycle time less than an element time as an operator does every 2nd or 3rd assembly) or you can have less than one operator/station (operators move between stations or work can be done off-line).

Two, elements really are not indivisible. For example, element 9 might be "Pick up screwdriver, drive 20 screws, release screwdriver" with a time of .0167 hr. For balancing purposes it might be desirable to have element 9a "Pick up screwdriver, drive 15 screws, release screwdriver" and 9b "Pick up screwdriver, drive 5 screws, release screwdriver." You have added extra work to the task (an extra pickup and release of a screwdriver) but may be able to cut the total cycle time. (Be careful about this technique since, when conditions change, the reason for the extra work may be forgotten and the unnecessary work retained without reason.)

Third, times are not fixed. This refers to cycle time and to element time. At the start we assumed a cycle time of .05 hr which meant the line would run 1,000 hr or 1,000/8=125 days. It may be more efficient to have a cycle time of .048 hr so the line would run for 20,000 × .048=960 hr or 960/8=120 days. In addition,

Table 7.6 Step 5 of Simplified Systematic Layout Planning is Evaluate the Alternatives. Criteria and Weights Depend Upon the Specific Management's Goals.

Criterion	Weight	Present	Alternative 1	Alternative 2
Minimum investment	6	A 24	B 18	A 24
Ease of supervision	10	C 20	C 20	B 20
Ease of operation	8	C 16	C 16	C 16
East of expansion and contraction	2	C 4	C 4	B 6
Total points		64	58	66
Relative merit		97%	88%	100%

standard time for an element of, say, .0167 hr will not be met exactly for each and every cycle by each and every operator; times will vary from unit to unit and operator to operator. Thus it may be possible to have "tight" stations for better than average operators and "loose" stations for the boss' nephew (or a new or old worker). If buffers are inadequate, put the best operator or shortest cycle in the middle of the line. The worst design is the worst operator or longest cycle at the end of the line.

Table 7.7   Elements and Work Times for Assembly Line Balancing Problem. Each Element Time is Assumed Constant. In Practice, Each Element Time is a Distribution.

Element	Work time/ unit, hr
1	.0333
2	.0167
3	.0117
4	.0167
5	.0250
6	.0167
7	.0200
8	.0067
9	.0333
10	.0017
	.1818

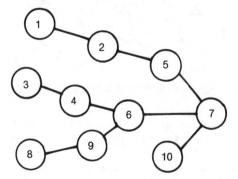

Figure 7.5 *Precedence diagrams* show the sequence required for assembly. The lines between the circles are not drawn to scale; that is, elements 4 and 9 both must be completed before element 6 but 9 could be done before or after 4. Precedence must be observed; thus elements 3, 4, and 9 could not be assigned to one station and elements 8 and 6 to another. However, 8, 9, and 10 could be done at one station.

Flow Diagrams and Process Charts

In contrast to the Systematic Layout and the location of one new item procedures, flow diagrams, and their associated process charts, are a technique for organizing and structuring a problem rather than a solution technique. Although a good engineer should be able to notice what is important in a problem and make corrections without the aid of a flow diagram and process chart, somehow solutions become more obvious with them.

Table 7.8 Trial Solution for Assembly Line Balance Problem

Station	Element	Element time, hr/unit	Station element time, hr/unit	Station idle time, hr/unit	Line idle time, hr/unit
1	1	.0333			
	2	.0167	.0500	0	0
2	8	.0067			
	9	.0333	.0400	.0100	.0100
3	3	.0117			
	4	.0167			
	6	.0167	.0451	.0049	.0149
4	5	.0250			
	7	.0200			
	10	.0017	.0467	.0033	.0182

Idle percent = .0182/(4 × .05) = 9.1%

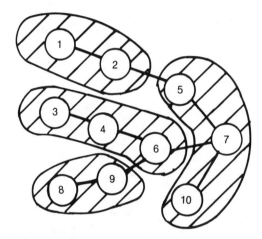

Figure 7.6 *Solution* of Table 7.8 is shown graphically.

Consider the problem of Joe College, who wants to improve his beer-drinking efficiency. The situation is shown in Figure 7.7, the process chart, and Figure 7.8, the flow diagram.

Figure 7.9 gives the five standard symbols for process charts: operations are circles, transportations or moves are arrows, inspections are squares, storages are triangles (upside down piles?), and delays (unplanned storage) are capital Ds. A circle inside a square is a combined operation and inspection. Some people put a number inside each symbol (identifying operation 1, 2, 3, or inspection 1, 2, 3) while others don't. Some people darken the circle for "do" operations but not for "get ready" and "put away" operations. Since inspection without change was a square, inspection with change (an operation) is a circle inside a square. Since a process chart is primarily a concise communication tool to yourself, take your choice.

At the end of the analysis, summarize the number of operations, moves, inspections, storages, delays, and the total distance moved. Estimate times for storages and delays. Usually it is not worthwhile to determine or record operation or inspection times since this between-operations-analysis usually is not concerned with within-operations methods. Process charts, because they give a "bird's-eye view" of the operation, often serve as methods documentation for others; you may wish to make a polished copy for them.

PROCESS CHART

SUBJECT CHARTED _____ Joe College _____   DATE __1 Oct__
                                                             CHART BY __SK__
OPERATION _____ Get beer _____   CHART No. _____
DEPARTMENT _____ Apartmant _____   SHEET No. __1__ OF __1__

DIST. IN FEET	TIME IN MINS.	CHART SYM-BOLS	PROCESS DESCRIPTION
		▽	Sit on sofa watching TV
19		⇒	Go for beer
	.1	①	Get beer
8		⇒	Go for glass and opener
	.1	②	Get glass and opener
	.1	⊔	Inspect glass
5		⇒	Go to sink
	.2	③	Wash glass
	.2	④	Open bottle and pour into glass
5		⇒	Return opener
	.1	⑤	Replace opener
7		⇒	Go to wastebasket
	–	⑥	Dispose of bottle
19		⇒	Go to sofa
			Summary: Operations 6 .7 min
			Moves 6 63 feet
			Inspect 1 .1 min
			Storage 1

Form L-1

Figure 7.7 *Single-object* process charts follow a single object. The single object in this example is a person.

Expense account paperwork was simplified at Intel by cutting 25 steps to 14 with the aid of a process chart.[7] Steps 5 and 7 were eliminated by having the accounts-payable clerk, instead of the cash-receipts clerk, collect refunds or unused traveler's checks. Steps 8, 10, and 11 were eliminated as another department also did these steps. Step 14 was eliminated as costing more than it saved. Step 19 proved unnecessary. The delays (steps 2, 6, 9, and 18) were cut. Expense accounts now are processed in days instead of weeks. Expense accounts less than $100 just require a petty cash voucher and a visit to the cashier. When the job was analyzed, the process chart was arranged horizontally and displayed in large horizontal strips of paper in a conference room. Each step then was questioned in detail by a group.

FLOW DIAGRAM

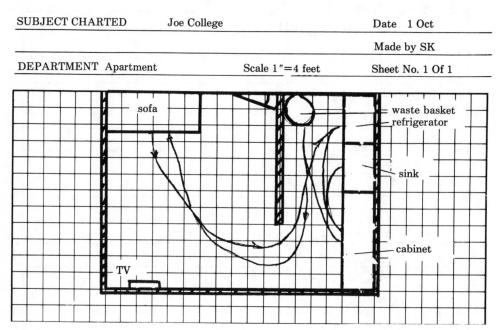

Figure 7.8 *Flow diagrams* are a graphical communication device used with process charts.

Figure 7.9 *Five standard symbols* for process charts are: a circle (operation), arrow (move), square (inspection), triangle (storage), and D (delay). A circle in a square is a combined operation and inspection.

Draw—not to drafting standards—the flow diagram to scale on cross-section paper; it serves as a communication aid and shows overall relationships.

There are two types of flow diagrams: single object and assembly. Figures 7.7 and 7.8 showed the single-object type. Either an operator or an object is followed (i.e., follow the beer drinker or the beer container). Examples of following a person are: vacuuming office, making bed, stocking shelves, changing tire, unloading semitrailer, handling material in the sandblast room, waiting on table, and loading/unloading dishwasher. Examples of following an object are: purchase order preparation, check processing, and machining casting. Figure 7.10 shows how to handle rework and scrap.

Figure 7.11 shows the second type, an assembly process chart. Assembly flow diagrams tend to point out the problems of disorganized storage. Other examples are: making pizza, switch assembly, potting plants, making a whiskey sour, blood sample analysis, and reloading shotgun shells. There also are disassembly charts (beef packing-house, portioning pie).

The purpose of flow diagrams and process charts is to focus your thinking; use them with critical examination forms, Kipling's six honest men, and checklists (see chapter 5).

Multi-Activity Charts

Figure 7.12 shows a multi-activity chart. In different forms it has different names. If the columns represent people, it may be called a gang chart; if some columns are people and some machines, it may be called a man-machine chart. If one column represents the left hand and one the right, it may be called a left-hand right-hand chart; see Figure 7.13.

There can be two or more columns. The time axis (drawn to a convenient scale) can be seconds, minutes, or hours. The purpose of a multi-activity chart is improved utilization of a *column*. Improved utilization can mean less idle time, rebalanced idle time, or less idle time of an expensive component; see Principle 4 of chapter 16 for additional comments on utilization. One common approach to improve equipment utilization is double tooling or double fixtures. Putting two sets of tools on an indexing fixture reduces idle fixture rotation time. Double fixtures permit loading and unloading the fixture while the machine is processing the next unit. The typical situation is one person for one machine. One person for two machines (.5 person/machine) also might be possible. Another possibility is .67 person or .75 person/machine. Do this with 2 people/3 machines or 3 per 4. The workers work as a team rather than individuals. Since all work on all machines, absenteeism is not so severe a problem. For task analysis, the chart shows time conflicts.

For each column, give cycles/yr, cost/min, and percent idle. Make the idle time distinctive by cross-hatching, shading, or red color.

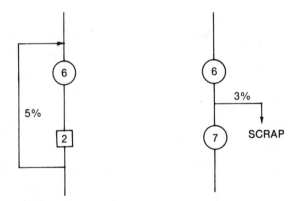

Figure 7.10 *Flow diagrams* are very useful for material handling analysis. Rework (left figure) and scrap (right figure) are important considerations.

A disadvantage of the multi-activity chart is that it requires a standardized situation. Nonstandardized situations are very difficult to show; for them, use computer simulation.

Example charts and columns might be: make lead molds (operator, machine, cooler); milling casting (operator, machine); cash check (cashier, customer 1, customer 2); and serve meal (customer, waitress, cook).

94

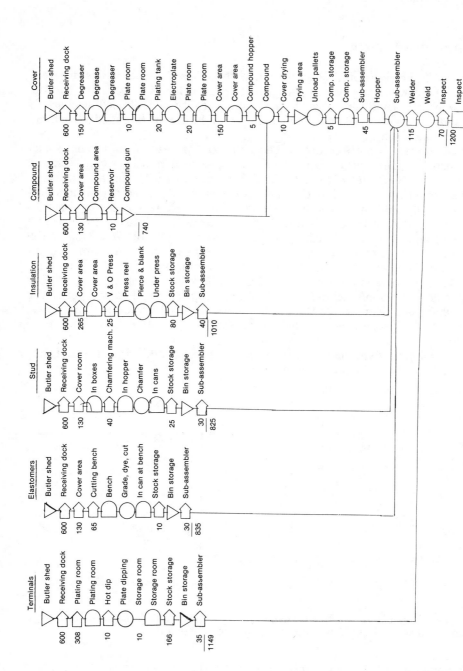

Figure 7.11 Assembly process charts are useful for methods analysis and as a graphical communication tool (for others as well as yourself).

WITHIN AN OPERATION

Fish Diagrams

Fish diagrams (see Figure 7.14), the cause side of cause-and-effect diagrams, also are known as Ishikawa diagrams due to their development in 1953 by Professor Ishikawa while on a quality control project for Kawasaki Steel.[6] They can be

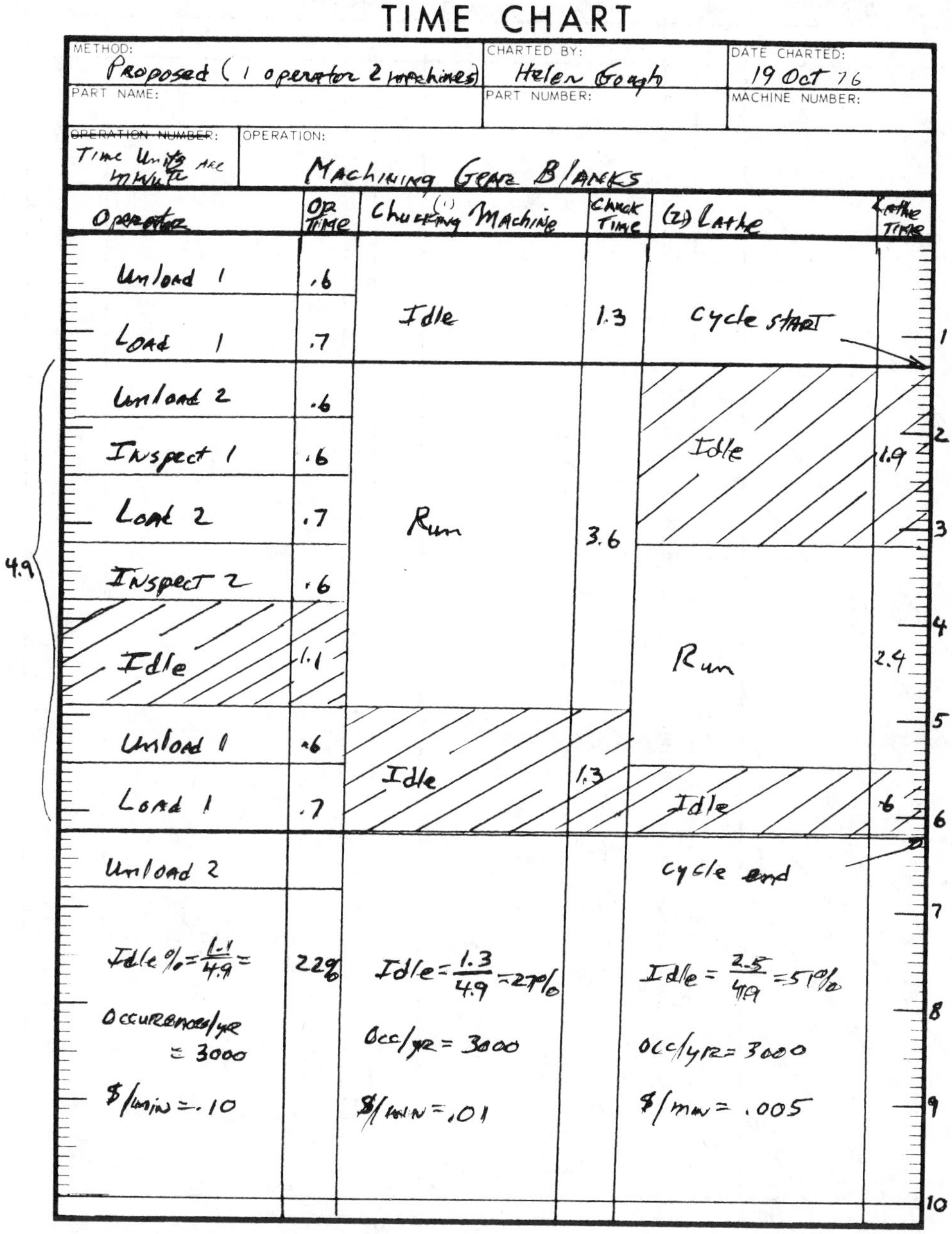

Figure 7.12 *Multi-activity charts* have many forms. The basic concept is to have two or more columns using a common time axis. Can one operator run two machines (i.e., .5 operator/machine)? Can you have teams so that two operators run three machines (.67 operator/machine) or three operators run four machines (.75 operator/machine)?

MULTIPLE ACTIVITY CHART

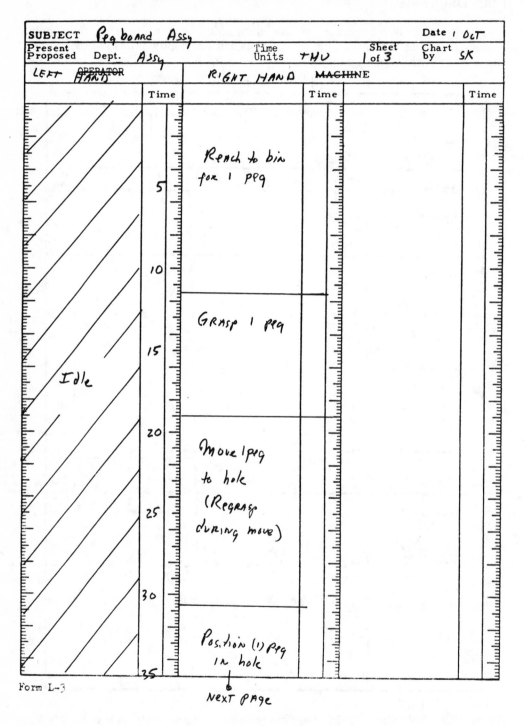

Form L-3

Figure 7.13 *Left-hand right-hand charts* are a type of multi-activity charts. Figure 8.1 shows a more common approach for a detailed hand analysis.

considered a multi-dimensional list. Fish diagrams are widely used in Japanese *Quality Circle* meetings—meetings of about 10 production workers who meet on their own time to try to improve their jobs; in 1979 there were 6,000,000 people in 600,000 groups in Japan.[3]

Start with a "fishhead"—a specific definition of a specific problem. Then put in a "backbone" and other bones. A good diagram will have three or more "levels" of bones (backbone, major bones, minor bones on the major bones). The diagram gives an easily understood overview of a problem and tends to trigger suggestions.

Figure 7.14 has improved grinding efficiency for the "head" and excessive grinding, working environment, method, machine, and man as the "major bones." Figure 7.15 was used at Bridgestone to reduce the variability of the viscosity of the splicing cement used in radial tires. Figure 7.16 shows the variability of the four operators before the *Quality Circle* studied the problem and after. For other examples, consider defective berry boxes as the head, and equipment, material, workers, box design, and delays as the major bones; making a better pie shell (flour, fat, shaping shell, tools used, water, mixing ingredients, and baking); drilling a well (drill bit grinding, working environment, method, machine, operator); bartending serving efficiency (method, materials, operator, environment, and cash register); improved physiology lab teaching (skeletons, manuals, rats, cadavers, exams, instructor, assistants, visual aids, and handouts); barber shop utilization (sales promotion, environment, work policies, method, and man); and stabilizing the rice price in Indonesia (market monitoring, harvest forecasting, stock reporting, procurement operation, storage system, sales operation, and stock movement).

Decision Structure Tables

Decision structure tables are a version of *if statements* in computer programs; they also are known as protocols or *contingency* tables (see Tables 26.3, 26.4, and 26.5 for examples).

The discussion in chapter 26 emphasizes making better-quality decisions due to (1) better decision analysis (higher quality personnel make the decision, using complex analysis techniques if necessary), and (2) less time pressure at the time the decision is made. However, in addition, decision structure tables are a good tool for methods analysis since they make the analyst consider all possibilities and force thoroughness.

Some example applications are: spot welding, bowling prices, firewood prices, library checkout procedure, refund policy, spare parts prices, hiring policy, food scoop equivalents, bowling position for spares, CO_2 level for furnace, and flight scheduling.

Breakeven Charts

A basic design problem is whether to use general-purpose equipment (low capital costs and high operating costs) or special-purpose equipment (high capital costs and low operating costs). At some production quantity (pieces, kg, etc.), the costs of the two methods are equal—the *breakeven point*.

The following example concerns making a part on an engine lathe vs. a turret lathe. Table 7.9 gives the information. The fixed cost, in this example, is setup cost + tooling cost. For the engine lathe, fixed cost is $11.05; for the turret lathe, it is $111.00. Initially, the engine lathe has an advantage of $111.00 − 11.05 = $99.95. For variable cost, consider both labor cost/unit and machine cost. The engine lathe variable cost is .2 hr ($10.50/hr) = $2.10 while the turret lathe is .1(11) = $1.10. The turret lathe picks up an advantage of $2.10 − 1.10 = $1.00 for every unit.

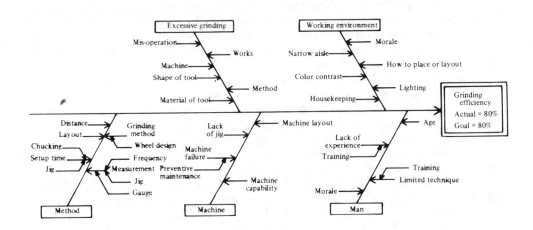

Figure 7.14 *Fish diagrams* serve as a useful communication tool (yourself and others). They show the many facets of a problem as well as levels of importance.

Figure 7.17 shows the cost equations graphically. The breakeven point, the point at which costs of both alternatives are equal, is:

$$111 + 1.10 \, N = 11.05 + 2.10 \, N$$
$$N = 100 \text{ pieces}$$

That is, the engine lathe's initial advantage of $99.95 is overcome by unit 100. Below 100 units, theory says use the engine lathe; above 100 units use the turret lathe.

In practice the problem is not so simple since errors or incorrect assumptions often are made.

First, consider tooling costs. If you already have the tooling, then the cost is zero for any additional project. On the other hand, if proposed tooling can be used by more than one project, spread the cost over all the projects.

Second, remember operating efficiency. That is, the standard time/unit from cost accounting may not be the same as the actual time. If, for example, the lathe department has been operating at 120% of standard, then both setup and operating times may be 1/1.20=83% of standard. Accurate operating times especially are important as a small error in labor costs has a large effect on the breakeven point.

Third, remember the effect of learning curves on labor time. See chapter 24 for more on learning curves. The basic point is that time/unit varies with production quantity; time is not constant. That is, instead of saying the engine lathe operator could make one unit every .2 hr regardless of the quantity made, time/unit may be .20 hr at 50 units; .18 at 100 units; and .16 at 200 units. Conversely, at less than 50 units, time might be .22 at 25 units and .24 at 13 units. The alternative with a lower labor content probably will have less learning. Figure 7.17 shows that the breakeven point may be shifted very much if learning is considered.

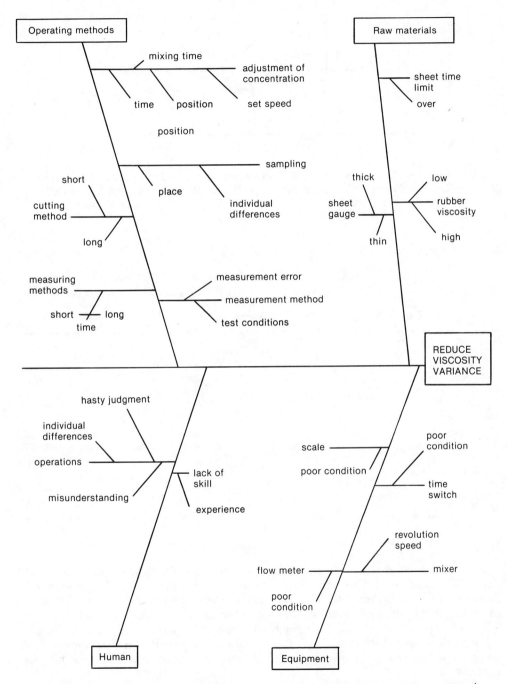

Figure 7.15 *Fish diagram* used at Bridgestone Tire to reduce cement viscosity variability.[4]
The four main categories were raw materials, operating methods, equipment, and human. The
problem turned out to be an improper method standard as variability was greater for those who
followed the standard procedure.

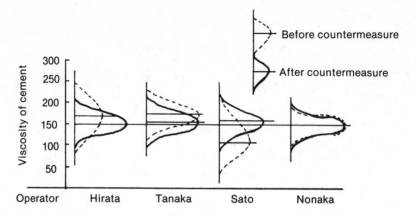

Figure 7.16 *Distribution curves* of four operators before and after the *Quality Circle* project.[4] Reducing variability is an excellent technique for improving quality as the effects of various changes are seen much more clearly for processes with little variability.

Table 7.9   Information for Breakeven Example

Item	Engine Lathe		Turret Lathe	
Machine cost, $/hr				
Capital (depreciation)	.50		1.00	
Other burden	4.00		4.00	
Labor cost, $/hr		4.50		5.00
		6.00		6.00
Machine + labor, $/hr		10.50		11.00
Fixed costs				
Setup cost, hr	.1		1	
Setup cost, $		1.05		11.00
Tooling costs, $		10.00		100.00
Total, $		11.05		111.00
Variable costs				
Manufacturing time, hr	.2		.1	
Mfg. variable cost, $/piece		2.10		1.10

Fourth, gamble if the odds are with you. For example, for an order of 80, the "best" alternative is the engine lathe. However, the advantage (using straight line assumptions) is only $20. Thus, use the turret lathe if there is the possibility of repeat orders or other jobs on which the turret lathe tooling could be used.

It also is assumed that (1) both machines are available for work, and (2) there is other work to keep both machines busy. In summary, breakeven charts are useful analysis tools, but be sure to check all the assumptions before making the decision.

Economic Lot Size

When designing a job, a key question is how many identical units will be made at a time before shifting to a different model or product. (Most jobs involve working on *batches*—first batch A, then B, then C, etc. A job is unusual if the

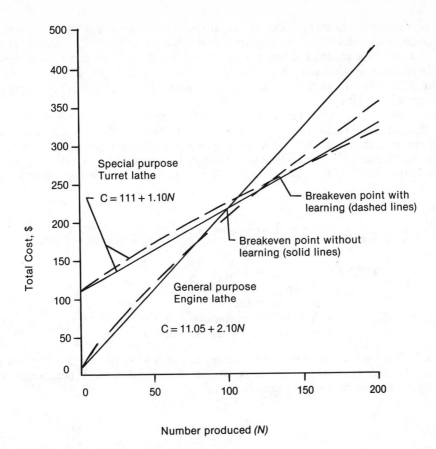

Figure 7.17 *Breakeven charts* show the effect of quantity on cost A vs. cost B or cost vs. income. This example shows the cost of using the engine lathe is lower than the cost of using the turret lathe—up to 100 units. The dotted lines show the effect of learning. Assume a 90% curve and standard at 50 units for the engine lathe; a 95% curve and standard at 100 units for the turret lathe (see chapter 24 for more on learning). Using learning increases the breakeven quantity from 100 units to 135.

Table 7.10   Information For Economic Lot Size Example

Item			Value
Annual sales, units			10,000
One-time costs, $/lot			
Paperwork	20		
Tool setup	30		50
Continuing costs, $/unit			
Material cost, $/unit		.10	
Labor cost, $/unit		.20	
Manufacturing overhead, $/unit		.50	
Direct cost, $/unit		.80	
Inventory cost, % of inventory value/yr			
Obsolescence, deterioration	1		
Storage	1		
Capital	18	20	
Inventory cost, $/yr/unit			
.20 ($.80/unit)			.16

same product is made throughout the entire year.) The economic lot size problem is to balance one-time costs (paperwork and setup) vs. continuing costs (inventory); it can be formulated either within an organization (how many to make) or outside an organization (i.e., purchasing or how many to buy).

Table 7.10 presents data for an example problem. The solution will give the size of the *equal-size* lots.

Setup costs. First, consider setup cost. If the annual requirement of 10,000 is made in 1 lot, then total setup cost/yr = (1 lot) ($50/lot) = $50. If 2 lots of 5,000 each are made, then total cost = $100; if 4 lots of 2,500 each are made, then total cost = $200, etc. These annual costs also can be expressed in cost/unit; that is $50/10,000 = $.005 $/unit; (50 × 2)/10,000 = .010 $/unit; (50 × 4)/10,000 = .020 $/unit.

Using symbols:

$$STCTPU = \frac{\text{Annual setup cost, \$/yr}}{\text{Total units/yr}} = \frac{S\ (N)}{A} = \frac{S\ (N)}{Q\ (N)} = S/Q \qquad (7.4)$$

where $STCTPU$ = Setup cost, $/unit
S = Setup cost, $/lot
N = Number of lots/yr
A = Annual requirements, units/yr
Q = Quantity (number) of units/lot

Figure 7.18 shows setup costs/unit getting larger and larger as the number/lot gets smaller. Note the steeply rising curve for "small" lot sizes. If setup cost was the only cost, then we would make large lots.

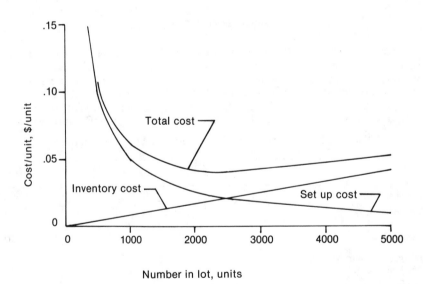

Figure 7.18 *Total costs* are composed of setup costs + inventory costs. We want to have a lot size such that total costs are as close as possible to the minimum.

Inventory costs. There is, however, another cost—inventory. Inventory cost, a continuing cost, depends on (1) the number in inventory and (2) cost of keeping a unit in inventory.

The number in the inventory depends on the quantity we put in inventory each time we put some in, on timing on when we put units in, and the sales (withdrawal) pattern. Figure 7.19 shows a simple version of reality. Starting with an "empty bin," we put Q items into stock. The stock level becomes $0 + Q = Q$. Then sales are absolutely equal every day so the stock level declines in a straight line toward zero. Just at the second that the last item is sold, a new shipment arrives and the stock level goes back up to Q. The average amount in inventory is the beginning amount plus the ending amount divided by 2 or $(Q + 0)/2 = Q/2$. Geometrically we substitute a rectangle with height $Q/2$ for a triangle with height Q—both shapes have the same area. For a single lot of 10,000, the average inventory is $10,000/2 = 5,000$; for 2 lots of 5,000, the average inventory is $5,000/2 = 2500$, for 4 lots of 2500, the average inventory is $2,500/2 = 1250$.

Cost of keeping an item in inventory depends upon the cost (value) of the item, the loss in value over time (spoilage, deterioration, obsolescence), the cost of capital, and the physical cost of the storage (building, stockracks, heat, clerk's salaries). The cost of the item is the cost to that stage—usually materials, labor, and overhead—but not sales expense or profit. The relevant cost of capital is not the interest on a bond; use the cost of alternate uses of capital in your organization—your organization's expected return on investment. That is, if money is tied up in inventory, it cannot be used to buy equipment, build buildings, etc. For computational simplicity, the physical cost of storage (say $100,000/yr) is rephrased into annual percentage cost ($100,000 to store $10,000,000 of inventory for 1 year = 1% of value/yr).

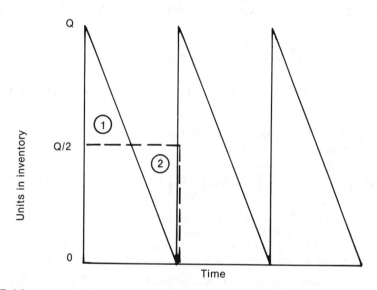

Figure 7.19 *Inventory,* in the most simple economic lot size formula, is represented by a triangle with height, Q. Constant sales eventually result in zero inventory at which instant a new shipment of Q arrives. The idealized triangle is replaced with an equivalent area rectangle (dotted lines) as triangle 1 = triangle 2.

Thus, inventory cost is (Amount in inventory) (Inventory cost/unit) or, using symbols:

$$INCTPL = (Q/2)\ (Cp) \tag{7.5}$$

where $INCTPL$ = Inventory cost per lot, $
          $Q/2$ = Average Inventory, units
          C = Cost per unit, $/unit
          p = Percent inventory cost, %

Converting this to inventory cost/unit:

$$INCTPU = INCTPL/A = QCp/2A \qquad (7.6)$$

where $INCTPU =$ Inventory cost/unit, \$/unit
      A = Annual requirements, units

See Figure 7.18; inventory cost/unit versus Q rises as a straight line. Then total cost = Setup cost + Inventory cost.

$$TOCTPU = S/Q + QCp/2A \qquad (7.7)$$
where $TOCTPU =$ Total cost per unit, \$/unit

Generally we are interested in the minimum of this total curve, $QMIN$, or, more formally, where the slope equals zero. From calculus, this means setting the first derivative equal to zero and solving for Q:

$$Q/d \; cost = 0 = -SQ^{-2} + Cp/2A$$

$$QMIN = \sqrt{\frac{2AS}{Cp}} \qquad (7.8)$$

Graphically, for a two-curve situation *where one curve is a straight line,* the minimum is at the point where the two component curves are equal. Or, from equation 7.8:

$$QMIN = \sqrt{\frac{(2)\,(10,000)\,(50)}{(.2) \qquad (.8)}} = 2,500 \text{ units}$$

In practice the problem is not so simple as errors or incorrect assumptions often are made.

1. If, due to varying sales or production capability requirements, you will not produce *equal-size* lots, then equation 7.8 is not applicable; it might give a $QMIN$ too high or too low.

2. Remember to base tooling decisions over the total units produced, not just Q, the quantity released for this order.

3. Obsolescence often is overestimated. For example, if a product has a 5-year life, obsolescence is not 20%—it is zero until at least the end of the model production and probably longer if spare parts are considered. Overestimating obsolescence tends to make predicted $QMIN$ smaller than actual $QMIN$.

4. Consider your sales pattern. If you ship most of your production shortly after you produce it, your average inventory is not $Q/2$ but is $Q/3$, $Q/4$, or even $Q/10$. Overestimating the number in storage or the time in storage tends to make predicted $QMIN$ smaller than actual $QMIN$.

5. Equation 7.8 does not consider at least 4 different costs: overage, quality, learning, and inflation.

Overage costs consider the extra units needed in a lot; to get 20 completed units, you start with 25 at the first operation to allow for setup units and scrap. But to get 100 completed units, you also start with 5 extra so 5/100 is a smaller percentage than 5/20. Since the cost of overage is not in equation 7.8, the predicted $QMIN$ is smaller than actual $QMIN$.

Quality is assumed to be the same for both small and large lots. In practice, once a process is set up correctly, errors are relatively few. Many setups mean many opportunities for errors. Thus predicted $QMIN$ is lower than actual $QMIN$.

Learning is assumed to be zero; that is time/unit was the same for lot sizes of 10,100 or 1,000. In practice, time/unit is lower for larger lots. Thus predicted *QMIN* is smaller than actual *QMIN*.

Inflation does not exist in equation 7.8. That is, future costs are assumed to equal present costs. Unlikely! Thus the predicted *QMIN* is lower than actual *QMIN*.

Finally, note the shape of the total cost curve in Figure 7.18. It is "saucer shaped," not "cup shaped." That is, departure from predicted *QMIN* becomes expensive only with large departures from *QMIN*, and especially departures to small lot sizes.

Table 7.11   A Summary Of The Analysis Techniques Given in This Chapter

Analysis Technique	Purpose of technique	Typical criterion
BETWEEN OPERATIONS		
Location of one new item	Locate one new item in an existing facility.	Minimize material handling costs.
Arrangement of entire facility	Rearrange all items in a facility.	Minimize material handling cost.
Assembly line balancing	Assign tasks to stations on an assembly line.	Minimize balance delay time.
Flow diagrams and process charts	Get overview of a task.	Eliminate unnecessary operations, movements, and inspections.
Multi-activity charts	Study coordination problems: between hand, people, and machines.	Reduce delay time.
WITHIN AN OPERATION		
Fish diagrams	Organize cost reduction analysis.	Reduce costs; improve quality.
Decision structure tables	Have better-quality decisions by workers; better communication of procedures.	Improve quality of procedure implementation.
Breakeven charts	Decide which process to use.	Minimize manufacturing cost.
Economic lot size	Decide how many to make per order.	Minimize total of setup cost + inventory cost.

SUMMARY

Man can explore physically (i.e., trial and error in a physical situation). However, he also can explore perceptually (use formulas and mathematical models). Chapter 7 presents some "tools of the trade" for designing jobs (see Table 7.11 for a summary). Between operations, techniques can focus on arranging equipment to minimize material handling cost; on decoupling assembly lines to reduce balance delay; on eliminating operations, moves, and inspections altogether; and on improving coordination. Within-operation techniques can focus on getting an overview to better understand relationships, on improving decision quality, on deciding between alternative processes, and on determining how many to make per order. However, the engineer is cautioned not to make the final decision based solely on the calculations. The techniques are just aids to make decisions, not the decision themselves. The techniques will give you a "first draft" of the decision. Only rarely is the first draft answer the final answer.

SHORT-ANSWER REVIEW QUESTIONS

7.1 Sketch movement cost versus distance.

7.2 What are the two types of flow diagrams?

7.3 In a flow diagram and process chart, what are the standard symbols for an operation, transportation, inspection, delay, and storage?

7.4 What is the purpose of a multi-activity chart?

7.5 The economic lot size problem is to balance which two types of costs?

THOUGHT-DISCUSSION QUESTIONS

1. Make a fish diagram for some problem that some other person has. Discuss the problem with the person while constructing the diagram. Does the resulting diagram help in resolving the problem?

2. Engineers should use formulas (such as for economic lot size) as aids to decision making but should not "let the formula make the decision." Discuss.

REFERENCES

1. Amsden, D., and Amsden, R., Ed. *QC Circles: Applications, Tools and Theory*, Milwaukee, Wisc.: American Society for Quality Control, 1976.

2. Anderson, D. New plant layout information system. *Industrial Engineering*, Vol. 5, No. 2 (1973): 32-37.

3. Aoki, J. Japanese productivity: What's behind it? *Modern Machine Shop*, Vol. 52, No. 4, (1979): 117-125.

4. Cole, R. *Work Mobility and Participation*. Berkeley: University of California Press, 1979.

5. Hicks, P., and Cowan, T. Craft-M for layout arrangement. *Industrial Engineering*, Vol. 8, No. 5 (1976): 30-35.

6. Inoue, M., and Riggs, J. Describe your system with cause and effect diagrams. *Industrial Engineering*, Vol. 3, No. 4 (1971): 26-31.

7. Konz, S. Where does one more machine go? *Industrial Engineering*, Vol. 2, No. 5 (1970): 18-21.

8. Lee, R., and Moore, J. CORELAP—computerized relationship layout planning. *Journal of Industrial Engineering*, Vol. 18, No. 3 (1967): 195-200.

9. Main, J. How to battle your own bureaucracy. *Fortune*, Vol. 103, No. 13, (June 29, 1981): 54-58.

10. Mariotti, J. Four approaches to manual assembly line balancing. *Industrial Engineering*, Vol. 2, No. 6 (1970): 35-40.

11. Moore, J. Computer methods in facilities layout. *Industrial Engineering*, Vol. 12, No. 9 (Sept. 1980): 82-93.

12. Muther, R. *Practical Plant Layout*. New York: McGraw-Hill, 1955.

13. Muther, R., and McPherson, K. Four approaches to computerized layout planning. *Industrial Engineering*, Vol. 2, No. 2 (1970): 39-42.

14. Muther, R., and Wheeler, J. Simplified Systematic Layout Planning. *Factory* (1962) (August): 68-77; (September): 111-119; (October): 101-109.

15. Tompkins, J. and Moore, J. *Computer Aided Layout: A User's Guide*, Monograph 1. Norcross, Ga.: American Institute of Industrial Engineers, 1978.

8

PREDETERMINED TIME SYSTEMS

Predetermined time systems can be used both as a methods tool and to determine time/unit. I chose to insert the topic in the methods section of the book so it is in chapter 8. Others may wish to emphasize the time aspect; they would label it chapter 25.

HISTORY AND DEVELOPMENT

Frederic Taylor, developer of "scientific management," first applied the scientific method to the mundane world of work rather than just to laboratory experiments. Frank and Lillian Gilbreth took a more detailed look at work than Taylor and broke work into 17 micro elements. Through micro motion analysis of bricklaying, a task which had been done for 2,000 years before Gilbreth, Frank improved bricklaying productivity 300%! Step 1 is determine an efficient work method. Step 2 is determine the time/unit for the efficient work method. Predetermined Time Systems (PTS) can be used for step 2 as well as for step 1.

Table 8.1 gives the 17 therbligs, as refined by Barnes.[3]

Therbligs were the first step in that they broke work down into elements. The second step was to assign time values to each of the elements. Then, to determine the total time for a task, the times for each of the elements are totaled.

In ordinary words the concept is similar to constructing a building. A building is composed of elements—doors, walls, beams, bricks, plumbing. The structure is the sum of the elements. Likewise, a job is also considered to consist of elements, the total of which is the sum of the elements. In formal words, the assumption is that each job element is independent and additive; that is, each element does not affect what happens before or after it.

A number of critics of Predetermined Time Systems (PTS) have shown these independence and additivity assumptions are not perfectly valid for all situations. That is, time for an element is affected by the preceding and following ele-

Table 8.1   Therbligs And Their Present Descendants

Therblig	Code	Definition
Transport empty	TE	Reach for an object with an empty hand. Became REACH or TURN in MTM (Methods-Time Measurement) and ARM MOTION in Work-Factor.
Search	Sh	Search begins when the eyes or hands begin to hunt for an object and ends when the object has been found. Incorporated into the various cases of REACH and GRASP of MTM and into the "work factors" of ARM MOTIONS and GRASP.
Select	Se	Select one object from several. Often combined with search. Incorporated into the cases of GRASP in both MTM and Work-Factor.
Grasp	G	Grasp begins when the hand or fingers first make contact with an object and ends when control is obtained. Grasp is a key element in both MTM and Work-Factor.
Transport loaded	TL	Transport loaded is moving an object (or an empty hand vs. resistance). Became MOVE or TURN in MTM and an ARM MOTION with work factors in Work-Factor.
Pre-position	PP	Pre-position for the next operation is locating an object in a predetermined place or position. Becomes either a REGRASP or POSITION in MTM but is called PRE-POSITION in Work-Factor.
Position	P	Position begins when the hand begins to turn or orient an object during a move and ends when the object has been placed in the desired location. Became POSITION in MTM and part of ASSEMBLE in Work-Factor.
Assemble	A	Assemble is placing one object on or in another object. Became MOVES, REGRASPS, and POSITION in MTM and ASSEMBLE in Work-Factor.
Disassemble	DA	Disassemble is separating one object from another object of which it is an integral part. Not an element in MTM except when resistance to separation is present (DISENGAGE). Became DISASSEMBLE in Work-Factor.
Release load	RL	Release load is letting go of an object. Became RELEASE in MTM and RELEASE in Work-Factor.
Use	U	Use begins when a hand begins to manipulate a tool or device and ends when the hand ceases the application. Described by individual elements in MTM. Became USE in Work-Factor.
Hold	H	Hold is retention of an object after grasp with no movement taking place. Not an element in MTM. One of two types of Balance Delay in Work-Factor (the other type is wait).
Inspect	I	Inspect is examining an object (with sight, hearing, touch, odor, or taste) to compare vs. a standard. Covered very briefly by MTM with VISUAL ACTION; covered in much more detail by Work-Factor with MENTAL PROCESS.
Avoidable delay	AD	Avoidable delay is a delay which operators may avoid if they wish. Not an element in either MTM or Work-Factor.
Unavoidable delay	UD	Unavoidable delay is a delay beyond the operator's control. If it occurs due to an interruption in the process the time should be considered by adding an allowance or by having the operator "clock out." A second possibility is a delay of one part of the body while another part is busy. This is considered by the SIMO Table in MTM; it is called Balance Delay in Work-Factor.
Plan	Pn	Plan is a mental reaction preceding the physical movement. Covered by learning allowances in both MTM and Work-Factor as both systems assume an "experienced" operator.
Rest to overcome fatigue	R	Rest is a fatigue or delay factor to permit the operator to recover from fatigue incurred by the work. Not given in either the MTM or Work-Factor systems as both systems assume a percentage will be added to the times for "personal, fatigue, and delay time."

ment. However the PTS do compensate for these assumptions—at least to some extent—since most elements occur in standard sequences. However, at the present state of knowledge, there is no realistic alternative to the PTS. They should be used with caution since their time values, especially in the hands of inexperienced users, often can depart from "actual" times by 25%—50%. Trained users do better. However, actual times also vary depending on worker and work pace on a specific day.

The concept of basic, universal units of work with accompanying "standard" amounts of time is an attractive idea. Many felt called to develop a system; few were chosen. The most popular systems (as of the present) are *Methods-Time Measurement* and *Work-Factor;* they are described in this chapter. Some other systems (which may or may not be more accurate, easier to apply, more consistent, more relevant to a specific industry, etc.) are: Motion-Time Analysis by Segur, Dimensional Motion Times by Geppinger, the Minneapolis-Honeywell System, the General Motors System, the Springfield Armory System by Olsen, the Elemental Time Standards of Western Electric, and Basic Motion Timestudy by Presgrave and Bailey.[11,21]

METHODS-TIME MEASUREMENT

Basic Concept

Methods-Time Measurement (MTM), descending from therbligs as did all the PTS, was developed by Maynard, Stegemerten and Schwab from motion pictures of sensitive drill press operations at Westinghouse.[19] The MTM Association, which publishes news of research and applications in the MTM Journal, has emphasized improving ease of application and development of simplified versions of MTM. MTM is probably the most widely used PTS in the world.

This section describes the basic MTM system (MTM-1) as well as two simplified systems (MTM-2 and MTM-3). The MTM material was reviewed by Karl Eady, Director of Training Development. The MTM tables are reprinted with the permission of MTM, 9 Saddle River Road, Fair Lawn, NJ 07410. People who wish to actually use MTM (in contrast to students) should take a training course approved by the MTM Association (approximately 24 to 80 classroom hours) so they can use the procedures accurately and consistently.

MTM-1

Motions are broken down into 10 categories: Reach, Move, Turn, Apply Pressure, Grasp, Position, Release, Disengage, Body (leg-foot, horizontal, and vertical) Motions, and Eye Motions. Times for each of these are given in Time Measurement Units (TMU), which is a fancy name for .000 010 hr. Thus 1 TMU = .000 010 hr = .000 600 min = .036 s. Conversely, 1 s = 27.78 TMU; 1 min = 1,667 TMU; and 1 hr = 100,000 TMU.

The times are for an experienced operator working at a "normal" pace (100%). (See chapter 22 for a definition of normal.) No allowances are included in the times. (See chapter 23 for allowances.)

Rivett developed learning curves for four tasks; MTM standard was achieved at 900, 135, 3,100, and 3,300 cycles.[23] Chapter 20 of Karger and Bayha gives a complicated technique to apply learning curves to MTM values.[14] In their two examples, the worker achieved MTM standard after 1900 cycles for method 1 and

2300 cycles for method 2. Chapter 24 of this text also discusses learning curves. For Rivett's four examples, at 200 cycles a worker would be expected to produce at 82%, 108%, 71%, and 74% of standard. Thus, complaints about inaccurate MTM times may be primarily due to failure to apply learning allowances. An important point is that the MTM standard will not be met for occasional work; for anything less than, say, 2,000 cycles additional time will have to be given.

Reach. Reach is usually movement with an empty hand or finger while Move is generally movement with an object in the hand. Reach is subdivided into five cases (see Table 8.2), which can be modified in some instances by removing the effect of acceleration and/or deceleration.

The five cases of Reach are:

A. Reach to an object in a fixed location, or to an object in the other hand, or on which the other hand rests. (The concept is of minimum eye control and emphasis on proprioceptive feedback.)

B. Reach to a single object whose general location is known. Location may vary slightly from cycle to cycle. (The concept is that some visual control is necessary and that the following grasping motion will be simple and not slow down the reach.)

C. Reach to object jumbled with other objects in a group so that search and select occur. (The concept is that considerable visual, muscular, and mental control is necessary and that the following grasp motion will be complex so the hand will slow down during the terminal portion of the reach to prepare for the complex grasp; it is the most difficult reach.)

D. Reach to a very small object or where accurate grasp is required. (The concept is that considerable visual control is necessary and that the following location grasp motion will be precise so the hand will slow down during the terminal portion of the reach to prepare for the careful grasp.)

E. Reach to an indefinite location to get the hand in position for body balance or next motion or out of the way. (The concept is of minimum mental control. The movement often is "limited out" as other motions are done simultaneously.)

Table 8.2 gives the TMU for each case of Reach for various distances. The distances are for the motion path of the hand knuckle or fingertip rather than the straight line distance between two points although this fine point is often overlooked by users not intensively trained in MTM. Novices also tend to overlook the reduction in Reach distance due to shoulder movement and pivoting—called *body assist.*

If a movement is not given in the table, the user can interpolate (time for R15A is average of time for R14A and R16A) or just use the next higher value. The MTM Association recommends interpolation when the tables are given in inches but use of the next higher value when the tables are given in even cm or multiples of 5 cm.

In addition to breaking Reach down into five cases and a number of different distances, the effect of acceleration and deceleration can also be considered. The normal motion has the hand stopped at the beginning and end of the motion. Hands also could be in motion either at the beginning or end of the cycle so either an acceleration or deceleration time can be omitted. The shortest time would occur in a situation in which the hand was in motion both at the beginning and end of the cycle and so both acceleration and deceleration can be omitted. When one acceleration or deceleration can be omitted, time is decreased as is shown by the hand-in-motion columns; when both acceleration and deceleration can be omitted, double this decrease is subtracted.

Table 8.2 MTM-1 reach times (TMU) are for five cases at various distances. Case A and B reaches also can omit an acceleration or deceleration. A typical code is R6B.

Distance Moved Inches	Time TMU				Hand in Motion		Case and Description
	A	B	C or D	E	A	B	
¾ or less	2.0	2.0	2.0	2.0	1.6	1.6	A Reach to object in fixed location, or to object in other hand or on which other hand rests.
1	2.5	2.5	3.6	2.4	2.3	2.3	
2	4.0	4.0	5.9	3.8	3.5	2.7	
3	5.3	5.3	7.3	5.3	4.5	3.6	
4	6.1	6.4	8.4	6.8	4.9	4.3	B Reach to single object in location which may vary slightly from cycle to cycle.
5	6.5	7.8	9.4	7.4	5.3	5.0	
6	7.0	8.6	10.1	8.0	5.7	5.7	
7	7.4	9.3	10.8	8.7	6.1	6.5	C Reach to object jumbled with other objects in a group so that search and select occur.
8	7.9	10.1	11.5	9.3	6.5	7.2	
9	8.3	10.8	12.2	9.9	6.9	7.9	
10	8.7	11.5	12.9	10.5	7.3	8.6	
12	9.6	12.9	14.2	11.8	8.1	10.1	D Reach to a very small object or where accurate grasp is required.
14	10.5	14.4	15.6	13.0	8.9	11.5	
16	11.4	15.8	17.0	14.2	9.7	12.9	
18	12.3	17.2	18.4	15.5	10.5	14.4	
20	13.1	18.6	19.8	16.7	11.3	15.8	
22	14.0	20.1	21.2	18.0	12.1	17.3	E Reach to indefinite location to get hand in position for body balance or next motion or out of way.
24	14.9	21.5	22.5	19.2	12.9	18.8	
26	15.8	22.9	23.9	20.4	13.7	20.2	
28	16.7	24.4	25.3	21.7	14.5	21.7	
30	17.5	25.8	26.7	22.9	15.3	23.2	
Additional	0.4	0.7	0.7	0.6			TMU per inch over 30 inches.

Reach motions are described in a "shorthand" of R (for Reach), the distance moved, the case of Reach (A, B, C, D, or E) and whether acceleration or deceleration is considered (an "m" before or after the code). For example, a 14-inch case A Reach is coded "R14A;" an 8-inch case B Reach is coded "R8B." If the hand is in motion at the beginning of the motion, then these motions would be "mR14A" and "mR8B." If the hand is in motion at the end of the motion, write "R8Bm."

Move. Move differs from Reach in that in Move the hand is usually holding something; occasionally the hand is pushing or dragging an object.

Move is subdivided into three cases (see Table 8.3). As with Reach, each Move can be refined to consider the effect of acceleration or deceleration or both. In addition there is an additional refinement for object weight or resistance to movement.

The three cases of Move are:

A. Move object to the other hand or against stop. (The concept is that there is little need to control the last portion of the Move, except perhaps to prevent damage to the object.)

B. Move object to an approximate or indefinite location. (The concept is a Move in which some control is needed at the end of the Move but not a great deal of control.)

C. Move object to exact location. (The concept is a Move with considerable control needed at the end of the Move. A case C Move very often is followed by a position.)

The three cases of Move thus differ (as did the five cases of Reach) by the nature of their destination. Thus MTM, in effect, says that all elements are not purely independent but that both Moves and Reaches are influenced by the motions preceding or following them.

The shorthand for a 5-inch case C Move is M5C.

As with Reach, the hand acceleration or deceleration may not be needed for some Moves. If omitted, it is indicated by a "m" before or after the code; for example, mM6B, or M6Bm.

The last adjustment to Move is the effect of the weight of the object moved. Weight up to 2.5 lb/hand is included in the time shown in the Move table. Thus, if

Table 8.3   MTM-1 Move times (TMU) are for three cases at various distances. Case B moves can omit acceleration and/or deceleration. The effect of weight is calculated by multiplying Effective Net Weight/hand by the Dynamic Factor and then adding the Static Constant. A typical code without weight is M8A; a typical code with a weight of 7 lb is M8A7.

Distance Moved Inches	Time TMU			Hand In Motion	Weight Allowance			Case and Description
	A	B	C	B	Wt. (lb) Up to	Dynamic Factor	Static Constant TMU	
¾ or less	2.0	2.0	2.0	1.7				
1	2.5	2.9	3.4	2.3	2.5	1.00	0	
2	3.6	4.6	5.2	2.9				
3	4.9	5.7	6.7	3.6	7.5	1.06	2.2	A Move object to other
4	6.1	6.9	8.0	4.3				hand or against stop
5	7.3	8.0	9.2	5.0	12.5	1.11	3.9	
6	8.1	8.9	10.3	5.7				
7	8.9	9.7	11.1	6.5	17.5	1.17	5.6	
8	9.7	10.6	11.8	7.2				
9	10.5	11.5	12.7	7.9	22.5	1.22	7.4	
10	11.3	12.2	13.5	8.6				B Move object to
12	12.9	13.4	15.2	10.0	27.5	1.28	9.1	approximate or indefinite location
14	14.4	14.6	16.9	11.4				
16	16.0	15.8	18.7	12.8	32.5	1.33	10.8	
18	17.6	17.0	20.4	14.2				
20	19.2	18.2	22.1	15.6	37.5	1.39	12.5	
22	20.8	19.4	23.8	17.0				
24	22.4	20.6	25.5	18.4	42.5	1.44	14.3	C Move object to exact
26	24.0	21.8	27.3	19.8				location
28	25.5	23.1	29.0	21.2	47.5	1.50	16.0	
30	27.1	24.3	30.7	22.7				
Add'l.	0.8	0.6	0.85		TMU per inch over 30 inches			

EFFECTIVE NET WEIGHT

Effective Net Weight (ENW)	No. of Hands	Spatial	Sliding
	1	W	$W \times F_c$
	2	W/2	$W/2 \times F_c$

W = Weight in pounds
F_c = Coefficient of Fiction

both hands move a 5-lb object, no extra time is allocated. If an object is slid rather than lifted, then take the object weight times the coefficient of friction (0.4 for wood-wood and wood-metal, 0.3 for metal-metal).

Next, all weights between 2.5 and 7.5 lb are considered alike. First, the TMU value is multiplied by a factor and then a constant is added. For example, a M6B = 8.9. For a 5-lb weight, time would be M6B5 = 8.9 (1.06) + 2.2 = 11.8. For an 18-lb weight moved 12 inches with both hands, time for an M12C would be M12C9 = 15.2 (1.11) + 3.9 = 20.8.

MTM also has a more precise method of allowing for weights. Additional time for weight has a "static component" for obtaining control and a "dynamic component" for additional travel time. The static time component in TMU = .975 + .345 (weight, lbs). The dynamic time component is 1.1%/lb, at any given distance.

Table 8.4 gives the times for a variation of moving: *cranking*. For continuous cranking for 5 revolutions against a 10-lb load with a 6-inch diameter crank, the time would be 5 × 12.7 = 63.5 TMU plus 5.2 for start and stop = 68.7 TMU. Then the resistance is considered by multiplying by 1.11 (from Move table) and adding 3.9 giving a total of 68.7 × 1.11 = 76.3 + 3.9 = 80.2 TMU. The code is 5C6-10.

Table 8.4   MTM-1 cranking motion times (TMU) are for light resistance. A typical code is 6C5. With 10-lb resistance: 6C5-10.

Diameter of Cranking (INCHES)	TMU (T) Per Revolution	Diameter of Cranking (INCHES)	TMU (T) Per Revolution
1	8.5	9	14.0
2	9.7	10	14.4
3	10.6	11	14.7
4	11.4	12	15.0
5	12.1	14	15.5
6	12.7	16	16.0
7	13.2	18	16.4
8	13.6	20	16.7

FORMULAS:

A. Continuous Cranking (Start at beginning and stop at end of cycle only)
$$TMU = [(N \times T) + 5.2] \cdot F + C$$
B. Intermittent Cranking (Start at beginning and stop at end of each revolution)
$$TMU = [(T + 5.2) F + C] \cdot N$$

C = Static component TMU weight allowance constant from move table
F = Dynamic component weight allowance factor from move table
N = Number of revolutions
T = TMU per revolution (Type III Motion)
5.2 = TMU for start and stop

Table 8.5   MTM-1 turn (forearm swivel) times (TMU) vary with angle turned and weight in the hand. Typical codes are T30S and T45M.

Weight	Time TMU for Degrees Turned										
	30°	45°	60°	75°	90°	105°	120°	135°	150°	165°	180°
Small—0 to 2 Pounds	2.8	3.5	4.1	4.8	5.4	6.1	6.8	7.4	8.1	8.7	9.4
Medium—2.1 to 10 Pounds	4.4	5.5	6.5	7.5	8.5	9.6	10.6	11.6	12.7	13.7	14.8
Large—10.1 to 35 Pounds	8.4	10.5	12.3	14.4	16.2	18.3	20.4	22.2	24.3	26.1	28.2

Turn. The third type of motion with the hand is Turn; see Table 8.5. Turn is a movement that rotates the hand, wrist, and forearm about the long axis of the forearm. The amount of time depends on the number of degrees turned as well as on the weight of the object or the resistance against which the turn is made.

Table 8.6 MTM-1 apply pressure times (TMU) have two categories. An APB is an APA plus a regrasp.

Symbol	TMU	Full Cycle Description	Symbol	TMU	Components Description
APA	10.6	AF + DM + RLF	AF	3.4	Apply Force
			DM	4.2	Dwell, Minimum
APB	16.2	APA + G2	RLF	3.0	Release Force

Apply Pressure. Apply Pressure (see Table 8.6) is the application of force without resultant movement. An APA is the basic element; an APB is an APA plus a Regrasp (G2).

Grasp. The next four motions—Grasp, Position, Disengage, and Release—are the skill motions. Improvement in performance times usually is a result of reductions in times for these motions rather than increased speed of movement for Move, Reach, or Turn.

Grasp is the motion used when the purpose is to gain control of an object or objects; it almost always is followed by Move. Grasp is divided into five categories; pickup and select are subdivided further (see Table 8.7).

Type 1 Pickup Grasp. Usually follows an A or B Reach.
  Case 1A Small, medium, or large object by itself, easily grasped
  Case 1B Very small object or object lying close against a flat surface

Table 8.7 MTM-1 grasp times (TMU) are given for five types of grasp; pickup and select are subdivided further. A typical code is G1A.

Type of Grasp	Case	Time TMU	Description
PICK-UP	1A	2.0	Any size object by itself, easily grasped
	1B	3.5	Object very small or lying close against a flat surface
	1C1	7.3	Diameter larger than ½″ — Interference with Grasp
	1C2	8.7	Diameter ¼″ to ½″ — on bottom and one side of
	1C3	10.8	Diameter less than ¼″ — nearly cylindrical object.
REGRASP	2	5.6	Change grasp without relinquishing control
TRANSFER	3	5.6	Control transferred from one hand to the other
SELECT	4A	7.3	Larger than 1″ × 1″ × 1″ — Object jumbled with other
	4B	9.1	¼″ × ¼″ × ⅛″ to 1″ × 1″ × 1″ — objects so that search
	4C	12.9	Smaller than ¼″ × ¼″ × ⅛″ — and select occur.
CONTACT	5	0	Contact, Sliding, or Hook Grasp.

Case 1C1	Interference with Grasp on bottom and one side of nearly cylindrical object; diameter greater than .50 inch
Case 1C2	Interference with Grasp on bottom and one side of nearly cylindrical object; diameter .25 to .50 inch
Case 1C3	Interference with Grasp on bottom and one side of nearly cylindrical object; diameter less than .25 inch
Type 2	Regrasp. This Grasp is used to change or improve control of an object which had previously been grasped. It often is performed during a Move and is "limited out."
Type 3	Transfer grasp. This Grasp is used to transfer control of an object from one hand to the other.
Type 4	Jumbled Grasps. Follows a C Reach.
Case 4A	Object jumbled with other objects so search and select occur; larger than 1 inch × 1 inch × 1 inch
Case 4B	Object jumbled with other objects so search and select occur; .25 inch × .25 inch × .12 inch to 1 × 1 × 1 inch
Case 4C	Object jumbled with other objects so search and select occur; smaller than .25 inch × .25 inch × .12 inch
Type 5	Contact, sliding, or hook Grasp. This Grasp usually occurs between a Reach and Move; no time is required to make contact with an object.

Position. "Original" position is the collection of minor hand movements (distance moved to engage no more than 1 inch) for aligning, orienting, and engaging one object with another object. It usually follows a C Move.

Position times (see Tables 8.8) vary with amount of pressure needed to fit, with symmetry of the object, and with ease of handling.

There are three classes of fit:

1. *Loose*   "No" pressure required (gravity sufficient). Code = 1.
2. *Close*   "Light" pressure required (1 APA). Most common fit. Code = 2.
3. *Exact*   "Heavy" pressure required (3 APA + G2). Code = 3.

There also are three classes of symmetry:

1. *Symmetrical* (Code = S)	This class is demonstrated by a round peg in a round hole. The concept is that, no matter in which orientation the part might happen to be, no rotation is necessary for assembly.
2. *Nonsymmetrical* (Code = NS)	This class is demonstrated by a cylinder with a key. There is one and only one orientation in which the two parts will mate. A turn of 180° (either clockwise or counterclockwise) is the maximum rotation required. Because some preorientation is done during most moves, only 75° is considered to be typical (i.e., an extra 4.8 TMU is allowed).
3. *Semisymmetrical*	All Positions that are neither symmetrical nor nonsymmetrical are considered semisymmetrical. An average turn of 45° is considered typical.

There are two classes of ease of handling: easy and difficult. They differ by a Regrasp, that is, 5.6 TMU. All flexible materials are considered difficult. A complete Position code includes all three variables; e.g., P1SE, P2SSE, P1NSD,

Table 8.8   MTM-1 Position (align, orient, and engage) times depend on the pressure required, the symmetry and the ease of handling. Using the top table (Original Position), a typical code is P2SSE, denoting close fit, semisymmetrical orientation, and easy handling. Insertion up to one inch is implied in all original Position codes.

The middle table (Supplementary Position) classifies fit in terms of radial clearance between the mating objects as 21, 22, or 23. In addition to align, orient, and primary engage time, a secondary engage time is added to account for varying insertion depth. Select the Position time from one of the four Depth of Insertion columns. For example, to align, orient, and engage a semisymmetrical object with a radial clearance of 0.25 inch between the mating parts to a depth of one inch, coded P21SS4, requires 14.6 TMU. The Align Only column is used for surface alignments having no engagement depths as when sliding a ruler to a point.

Class of Fit		Symmetry	Easy To Handle	Difficult To Handle
1—Loose	No pressure required	S	5.6	11.2
		SS	9.1	14.7
		NS	10.4	16.0
2—Close	Light pressure required	S	16.2	21.8
		SS	19.7	25.3
		NS	21.0	26.6
3—Exact	Heavy pressure required	S	43.0	48.6
		SS	46.5	52.1
		NS	47.8	53.4

SUPPLEMENTARY RULE FOR SURFACE ALIGNMENT

P1SE per alignment: $> \frac{1}{16} < \frac{1}{4}''$ P2SE per alignment: $< \frac{1}{16}''$

*Distance moved to engage—1″ or less.

Class of Fit And Clearance	Case of Symmetry†	Align Only	Depth of Insertion (per ¼″)			
			0 $> 0 \leq \frac{1}{8}''$	2 $> \frac{1}{8} \leq \frac{3}{4}$	4 $> \frac{3}{4} \leq \frac{5}{4}$	6 $> \frac{5}{4} \leq \frac{7}{4}$
21 .150″ to .350″	S	3.0	3.4	6.6	7.7	8.8
	SS	3.0	10.3	13.5	14.6	15.7
	NS	4.8	15.5	18.7	19.8	20.9
22 .025″ to .149″	S	7.2	7.2	11.9	13.0	14.2
	SS	8.0	14.9	19.6	20.7	21.9
	NS	9.5	20.2	24.9	26.0	27.2
23* .005″ to .024″	S	9.5	9.5	16.3	18.7	21.0
	SS	10.4	17.3	24.1	26.5	28.8
	NS	12.2	22.9	29.7	32.1	34.4

*BINDING—Add observed number of Apply Pressures.
DIFFICULT HANDLING—Add observed number of G2's.

†Determine symmetry by geometric properties, except use S case when object is oriented prior to preceding Move.

Table 8.8   (continued). The lowest table gives Secondary Engage times (TMU) as a function of class of fit and depth of insertion. It is used as a separate code when the radial clearance changes after initial engagement and before completing the insertion. A typical code is E22-2 following a P21 Position.

Class of Fit	Depth of Insertion (per ¼″)		
	2	4	6
21	3.2	4.3	5.4
22	4.7	5.8	7.0
23	6.8	9.2	11.5

P3SD. Alignment to a point or line (without engagement) within .25 to .5 inch requires only an M__C Move, within .06 to .25, an M__C Move plus a P1SE or P1SD, and less than .06, an M__C Move plus a P2SE or P2SD. The middle table of Table 8.8 gives an alternative description for Position while the lower table gives the times for Secondary Engage.

Secondary Engage is a component of Supplementary Position data which is not a variable in original Position (top table of Table 8.8). It is the time required to insert an object into another for varying depths of insertion. It is used in analyses as a separate code when the radial clearance of two mating objects changes during insertion as may be the case in chamfering or countersinking.

Disengage. Disengage is the breaking of contact between one object and another. It includes the involuntary movement resulting from the sudden end of resistance (see Table 8.9). The three factors of class of fit, ease of handling, and care in handling are combined in a two-factor table. The three classes of fit are given as:

1. *Loose*   Very slight effort, blends with subsequent Move (recoil up to 1 inch)

Table 8.9   MTM-1 disengage times (TMU) depend on the class of fit and the ease of handling. The lower table aids in selecting the proper category. A typical code is D2E.

Class of Fit	Supplementary	
	Care in Handling	Binding
1—LOOSE	Allow Class 2	
2—CLOSE	Allow Class 3	One G2 per Bind
3—TIGHT	Change Method	One APB per Bind

Class of Fit	Height of Recoil	Easy to Handle	Difficult to Handle
1—Loose—Very slight effort, blends with subsequent move	Up to 1″	4.0	5.7
2—CLOSE—Normal effort, slight recoil	Over 1″ to 5″	7.5	11.8
3—TIGHT—Considerable effort, hand recoils markedly	Over 5″ to 12″	22.9	34.7

2. *Close*   Normal effort, slight recoil (over 1 inch to 5 inches)

3. *Tight*   Considerable effort, hand recoils markedly (over 5 inches)

Again there are only two classes of ease of handling: easy and difficult. A complete Disengage code includes both fit and ease of handling; for example, D1E, D1D, D2D.

Table 8.10   MTM-1 Release times (TMU) have only two subdivisions; the normal Release is coded RL1.

	SUPPLEMENTARY	
Case	Time TMU	Description
1	2.0	Normal release performed by opening fingers as independent motion.
2	0	Contact Release.

Release. Release is the relinquishing of control of an object by the hand or fingers. Table 8.10 shows that there are only two categories. The most common Release, a simple opening of the fingers, is given 2 TMU. In the contact Release, for which no time is allowed, the Release begins and is completed at the instant the following Reach motions begins.

The two codes are RL1 and RL2.

Body, Leg, and Foot Motions. The previous tables gave motions of the hand and arm. Table 8.11 gives motions of the leg-foot, horizontal torso motions, and vertical torso motions. The description of the motions in the table are reasonably self-explanatory. A *Foot Motion* is hinged at the ankle; a *Leg Motion* is hinged at the knee or hip or both, and the body centerline doesn't move appreciably. In a *Sidestep* the time depends on the distance the centerline moves. A pace is considered to cover 2.8 feet or 34 inches at the relatively fast pace of 3.57 miles/hr. For walking through obstructed areas, use 17 TMU/pace. Use 15 TMU/pace for stairs. For carrying a load up to 35 lb, use a 30-inch pace and 15 TMU/pace. For vertical torso motions, going up takes more time than going down.

Eye Times. In addition to motions of the limbs and torso, time is allowed, in certain cases, for activities of the eyes. Table 8.12 gives the two basic elements: *Eye Focus* and *Eye Travel*, and the formula for *Read*. *Eye Focus* is the focusing of the eye once it has an object in its line of sight. *Eye Travel* is the movement of the eyes from one point to another. From geometry, when T/D = 1, the angle swept by the eyes is 45°; for T/D = 2, the angle is 90°; for T/D = 3, the angle is 135°. Thus, 15.2 TMU is allowed per 45° of sweep (.33 TMU/degree) with the limitation, however, of a maximum allowable for Eye Travel of 20 TMU, because the eyes are generally restricted to a maximum movement of 70°. Reading 100 words takes 505 TMU (18.2 s).

Combined Motions and Limited Motions. So far the motions have been described as if the person performed one at a time. Often we have combined motions or simultaneous motions.

Combined motions are those which occur when two more motions are performed by the same body member at the same time (*Turn* a part in the hand while moving it; *Regrasp* during a *Move*). The time to allocate is the greater of the two times. Thus a Regrasp during an M3A would be given 5.6 for time with a slash line through the M3A. A *Regrasp* during an M6A would be given 7.0 for time with a slash line through the G2. See Figure 8.1 for a combined motion.

Table 8.11 MTM-1 times (TMU) for motions of the leg and torso have three major subdivisions. Typical codes are W5P and SS15C1.

Type	Symbol	TMU	Distance	Description
LEG–FOOT	FM	8.5	To 4″	Hinged at ankle
	FMP	19.1	To 4″	With heavy pressure
MOTION	LM—	7.1	To 6″	Hinged at knee or hip in any direction
		1.2	Ea. add'l inch	
SIDE	SS–C1	17.0	<12″	Use Reach or Move time when less than 12″; complete when leading leg contacts floor
		0.6	12″	
			Ea. add'l inch	
STEP	SS–C2	34.1	12″	Lagging leg must contact floor before next motion can be made
		1.1	Ea. add'l inch	
TURN	TBC1	18.6	—	Complete when leading leg contacts floor
BODY	TBC2	37.2	—	Lagging leg must contact floor before motion can be made
WALK	W–FT	5.3	Per Foot	Unobstructed
	W–P	15.0	Per Pace	Unobstructed
	W–PO	17.0	Per Pace	When obstructed or with weight
VERTICAL	SIT	34.7	—	From standing position
	STD	43.4	—	From sitting position
	B,S,KOK	29.0	—	Bend, Stoop, Kneel on one knee
	AB,AS,AKOK	31.9	—	Arise from Bend, Stoop, Kneel on one knee
MOTION	KBK	69.4	—	Kneel on both knees
	AKBK	76.7	—	Arise from Kneel on both knees

HORIZONTAL MOTION

Table 8.12   MTM-1 times (TMU) for the eye are Eye Focus, Eye Travel, and Read. Eye Focus assumes Eye Travel is completed; Eye Travel is the movement from one point to another at .33 TMU/degree of motion. Reading time, TMU, = 5.05 N, where N = the number of words.

Eye Travel Time = 15.2 × T/D, with a maximum value of 20 TMU.
where T = the distance between points from and to which the eye travels.
D = the perpendicular distance from the eye to the line of travel T.

Eye Focus Time = 7.3 TMU.

SUPPLEMENTARY INFORMATION

—Area of Normal Vision = Circle 4″ in Diameter 16″ from Eyes
—Reading Formula = 5.05 N where N = The Number of Words.

METHODS ANALYSIS CHART.   REFERENCE NO. _____

PART __30 Wooden pegs & (1) pegboard__ DATE _21 July 76_ STUDY NO. _1A_
OPERATION __Assemble 30 pegs to pegboard__ ANALYST __SK__ SHEET NO. _1_ OF _1_ SHEETS

DESCRIPTION — LEFT HAND	No.	L H	Time	R H	No.	DESCRIPTION — RIGHT HAND
① Reach to (1) peg in bin w/RH. Pick up and place nose FIRST in hole. See sketch on back for hole sequence. Repeat 30 times.						
			13.6	R11C	1	Reach to bin for (1) peg from table edge
			333.5	R8C	29	Reach to bin for (1) peg from pegboard center
			219.0	G4A	30	Grasp (1) peg
			354.0	M8C	30	Move (1) peg to hole
			—	G2	30	Regrasp during Move
			168.0	P1SE	30	Position (1) peg in hole
			60.0	RL1	30	Release peg
			11.2	R11E	1	Return hand to table edge
			1159.3			

D.	ELEMENT DESCRIPTION	ELEMENT TIME TMU	CONVERSION FACTOR					
			LEVELED TIME	% ALLOWANCE	ELEMENT TIME ALLOWED	OCCURRENCES PER PIECE OR CYCLE	TOTAL TIME ALLOWED	
1	Get 1 peg at a time. Place in board. Repeat 30 times	1159.3						

Figure 8.1 *Combined motion* (Regrasp during the Move to the hole) is shown with a slash through it. Note how repeated steps can be described with the number column. Time to assemble 30 pegs in the pegboard = 1159.3 TMU = .70 min.

Motions also occur "simultaneously," such as right and left hand, hands and feet, or eyes and hands. If they are *truly* simultaneous, allow only the longer time. If they are only *apparently* simultaneous, allow both times. Use Table 8.13 to decide between truly and apparently simultaneous. Truly or apparently depend on which combinations are considered, on whether they are in the area of normal vision (objects within 4 inches of each other at a distance of 16 inches from the eye), and the amount of practice (one common definition is 500 cycles although others use 1,000 or 2,000).

Thus for truly simultaneous motions such as an M8A with the left hand and an M10A with the right hand, allow only the 11.3 of the M10A. If apparently simultaneous but according to Table 8.13 not actually simultaneous (such as an M8C with the left hand and an M8C with the right hand), allow 11.8 plus 11.8—a total of 23.6.

In many cases where truly simultaneous motions are not allowed by Table 8.13, one hand can drift toward the target while the other does the motion. For example, while the right hand does an M8C, the left hand does an M8B. The M8B can be done simultaneously with the M8C and is "limited out"; it is conventional to circle limited motions (see Figure 8.2 for an example). Then, when the M8C is completed, the left hand has only a small distance (an MfC where f stands for fractional) remaining. The time allowed is 11.8 for the right hand plus a 2.0 for the left—a total of 13.8.

MTM-2 and MTM-3

How much error can be tolerated in the time determined for a specific task: 1%, 5%, 10%, 30%? How much are you willing to pay for accuracy? The answers are not the same for all situations so the International MTM Directorate (an associa-

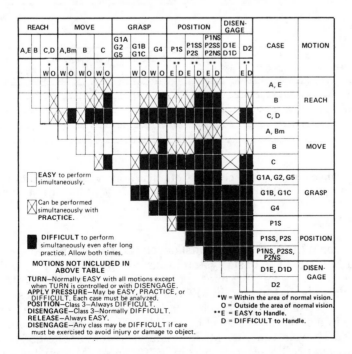

Table 8.13 If motions are easy to perform simultaneously, allow just the longer of the two times. If they can be done with practice, use judgment on whether to allow both times. If difficult, allow both times.

122

METHODS ANALYSIS CHART

REFERENCE No. _____

PART _30 Wooden pegs & (1) pegboard_ DATE _21 July 76_ STUDY No. _1B_

OPERATION _Assemble 30 peg to pegboard_ ANALYST _JK_ SHEET No. _1_ OF _1_ SHEETS

DESCRIPTION — LEFT HAND	NO.	LH	Time	RH	NO.	DESCRIPTION — RIGHT HAND
① Reach into bin w/ BH simo. Get (1) peg in each hand. Move peg to board and beveled end first. See sketch on back for hole sequence.						
To bin from table edge	1	R11C	13.6	R11C	1	to bin from table edge
To bin from board center	14	R8C	161.0	R8C	14	to bin from board center
Turn hand 60°	14	T60		T60	14	Turn hand 60°
Grasp (1) peg	15	G4A	109.5			
			109.5	G4A	15	Grasp (1) peg
Move (1) peg toward board*	15	M8B	172.0	M8C	15	Move (1) peg to board
Regrasp during Move	15	G2		G2	15	Regrasp during Move
Turn hand 60°	15	T60S		T60S	15	Turn hand 60°
			84.0	P1SE	15	Position peg in hole
Move (1) peg to hole	15	M4C	70.0			
Position peg in hole	15	P1SE	84.0			
Release peg	15	RL1	30.0	RL1	15	Release peg
Hand to table edge	1	R11E	11.2	R11E	1	Hand to table edge
			809.8			

*Although two rows are within area of normal vision, lack of practice precludes simo motion.

	ELEMENT DESCRIPTION	ELEMENT TIME TMU	CONVERSION FACTOR	% ALLOWANCE	ELEMENT TIME ALLOWED	OCCURRENCES PER PIECE OR CYCLE	TOTAL TIME ALLOWED
			LEVELED TIME				
2.	Get (1) peg in each hand, Place in board. Repeat 15 times	809.8					

Figure 8.2 *Limited motions* are circled. Note that working smart with two hands gives an assembly time of 809.8 TMU = .49 min or 70% of the one-hand method.

tion of 12 national MTM organizations) developed two simplified systems called MTM-2 and MTM-3. (The basic system described previously is called MTM-1.)

In MTM-1, it may take 350 times the cycle time to analyze the task while in MTM-2 it takes about 150, and in MTM-3 it takes about 50 times.[17] Figure 8.3 shows expected error limits (95% confidence) as a function of the nonrepetitive manual time in a task and the level of analysis. For example, for a specific task with 1,000-TMU cycle time, an MTM-1 analysis would take about 3.5 hr and would be expected to be accurate within ±7%; an MTM-3 analysis would take about .5 hr and would be accurate within ±20%. Magnusson found that the use of MTM-2 gave times 0.1% higher than MTM-1 while MTM-3 gave values 0.9% higher than MTM-1. (This "bias" is considered to be insignificant.) The remaining random error (after this systematic error) will decrease with increasing "sample size" or, in this case, with longer cycle times. Thus a 10,000 TMU cycle with MTM-1 would be expected to be accurate within ±2%. MTM-2 and MTM-3 require considerable analyst judgment.

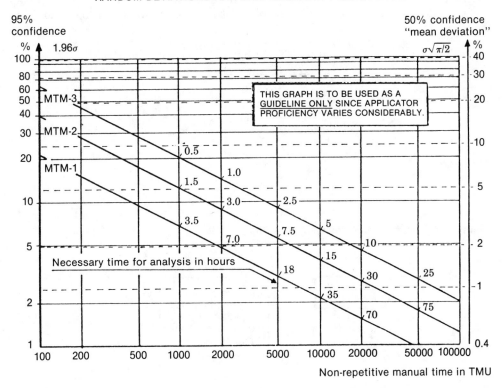

RANDOM DEVIATIONS AND TIME REQUIRED FOR ANALYSIS

SOURCE: Copyrighted by the MTM Association for Standards and Research. No reprint permission without written consent from the MTM Association 16-01 Broadway, Fairlawn, NJ 07410.

Figure 8.3 *Analysis time and accuracy* for MTM-1, MTM-2, and MTM-3 can be predicted given the nonrepetitive cycle time.[17] A task with 2000 TMU (.02 hr) would take 350 (.02) = 7 hr to analyze and would have an accuracy of ±4.9% (with 95% confidence) if analyzed by MTM-1.

Table 8.14 MTM-2 is a very simplified version of MTM-1; task analysis time should take about 150 times the cycle time. Times are in TMU.

			MTM-2			
RANGE	Code	GA	GB	GC	PA	PB	PC
Up to 2"	−2	3	7	14	3	10	21
Over 2" to 6"	−6	6	10	19	6	15	26
Over 6" to 12"	−12	9	14	23	11	19	36
Over 12" to 18"	−18	13	18	27	15	24	36
Over 18"	−32	17	23	32	20	30	41

GW 1-per 2 lb				PW 1-per 10 lb.	
A	R	E	C	S	F	B
14	6	7	15	18	9	61

MTM-2. Table 8.14 shows the MTM-2 card; there are only 39 times in all.

The two key motion categories are GET (combining Reach, Grasp, and Release) and PUT (combining Move and Position). Which of the 15 values of GET or PUT to use depends on the case of GET or PUT, the distance, and the weight or resistance to motion.

Figure 8.4 shows the decision tree for GET to determine case. Then the user estimates (not measures) the distance and uses one of the five rows. Third, if necessary, add 1 TMU/2 lb moved if the object to be moved weighs 4 lb or more per hand.

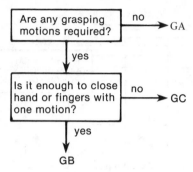

Figure 8.4 *Decision tree* for GET in MTM-2.

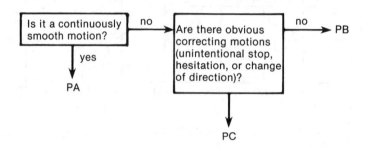

Figure 8.5 *Decision tree* for PUT in MTM-2.

Figure 8.5 shows the decision tree for PUT to determine case. Then the user estimates (not measures) the distance and uses one of the five rows. Third, if necessary, add 1 TMU/10 lb moved if the object weighs 4 lb or more per hand.

There remain seven motions:

A—Apply Pressure	An action with the purpose of exerting muscular force on an object.
R—Regrasp	A hand action performed with the purpose of changing the grasp on an object.
E—Eye Action	An action with the purpose of either recognizing a readily distinguishable characteristic of an object or shifting the aim of the axis of vision to a new viewing area.
C—Crank	A motion with the purpose of moving an object in a circular path of more than half a revolution with the hand or fingers.

S—Step	Either a leg motion with the purpose of moving the body or a leg motion longer than 12 inches.
F—Foot Motion	A short foot or leg motion when the purpose is not to move the body.
B—Bend and Arise	A bend, stoop, or kneel on one knee, and the subsequent rise.

See Figure 8.6 for an MTM-2 analysis of the two-hand assembly of the pegboard. MTM-3. Table 8.15 shows the MTM-3 card. Now there are only 10 times. The two key motion categories have been reduced to HANDLE (getting

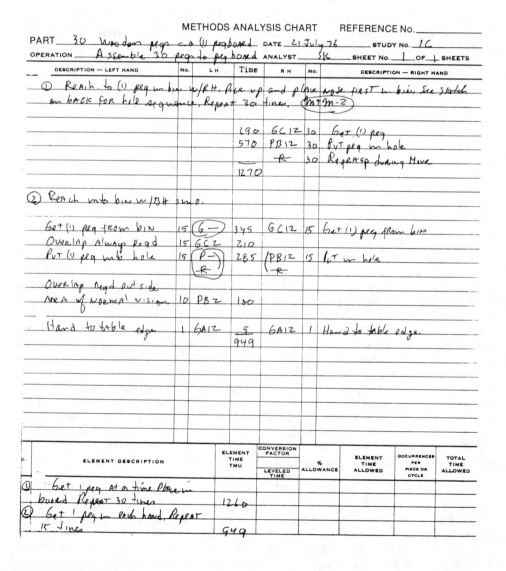

Figure 8.6 *MTM-2 analysis* of the one-hand and two-hand assembly of the pegboard previously analyzed in Figures 8.1 and 8.2 with MTM-1 showing total time = 1,270 TMU = .76 min and 949 TMU = .57 min.

control over an object with the hand or fingers and placing the object in a new location) and TRANSPORT (placing an object in a new location with the hand or fingers). Which of the 4 values of HANDLE or TRANSPORT to use depends on the case and the distance.

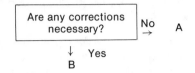

Figure 8.7 *Decision tree* for both HANDLE and TRANSPORT in MTM-3.

Figure 8.7 shows the decision tree for both HANDLE and TRANSPORT. Distances are estimated as either equal to or less than 6 inches or over 6 inches. The 7 additional elements of MTM-2 have been reduced to two: SF and B. SF combines the S and F categories of MTM-2 while B is the same as in MTM-2. Crank would now be a TRANSPORT while Apply Pressure, Regrasp, and Eye Motion are included in the HANDLE and TRANSPORT.

See Figure 8.8 for a MTM-3 analysis of the pegboard task assembly.

WORK-FACTOR

Basic Concept

The idea for the system was conceived at the Philco Radio Corp. in Philadelphia, Pennsylvania, between 1930 and 1934 because rate setting with stopwatch time study was unacceptable to union employees working under tightly controlled incentive pay systems. The original motion time data were developed between 1934 and 1938 by a group of experienced time study engineers directed by Joseph Quick, assisted by Samuel Benner, William Shea, and Robert Koehler. The purpose of the 1934 research was to develop an objective work measurement technique to eliminate the stopwatch and the observer's performance rating judgments in establishing work standards and setting production output rates.

The Motion Times were applied worldwide for work measurement in actual factory operations at Radio Corporation of America (RCA) between 1938 and 1946. They first were published in 1945. James Duncan and James Malcolm, working with Quick, developed techniques to simplify its application. Quick, Duncan, and Malcolm described the Detailed, Ready, and Abbreviated Systems in detail.[22] In 1949 they began their research on human mental processes which led to the compilation of the Mento-Factor Manual of Detailed Work-Factor Mental Process Times in 1965.

Work-Factor time is based on the output level of the "average experienced operator working with good skill and good effort." Popularly, it is identified as an "incentive work pace." If the MTM pace = 100%, then Work-Factor and British Standard 3138 = 120%.[4] That is, if the time for a specific task was 1.0 min/unit according to MTM, then Work-Factor time would be $1/1.2 = .833$ min/unit. Naturally this ratio is averaged over many tasks because, for specific tasks, details of the application rules might lead to slightly larger or smaller differences. No allowances are included in the Work-Factor time tables. Workers are expected to achieve standard time in 200 to 400 cycles[31]; see Table 24.6.

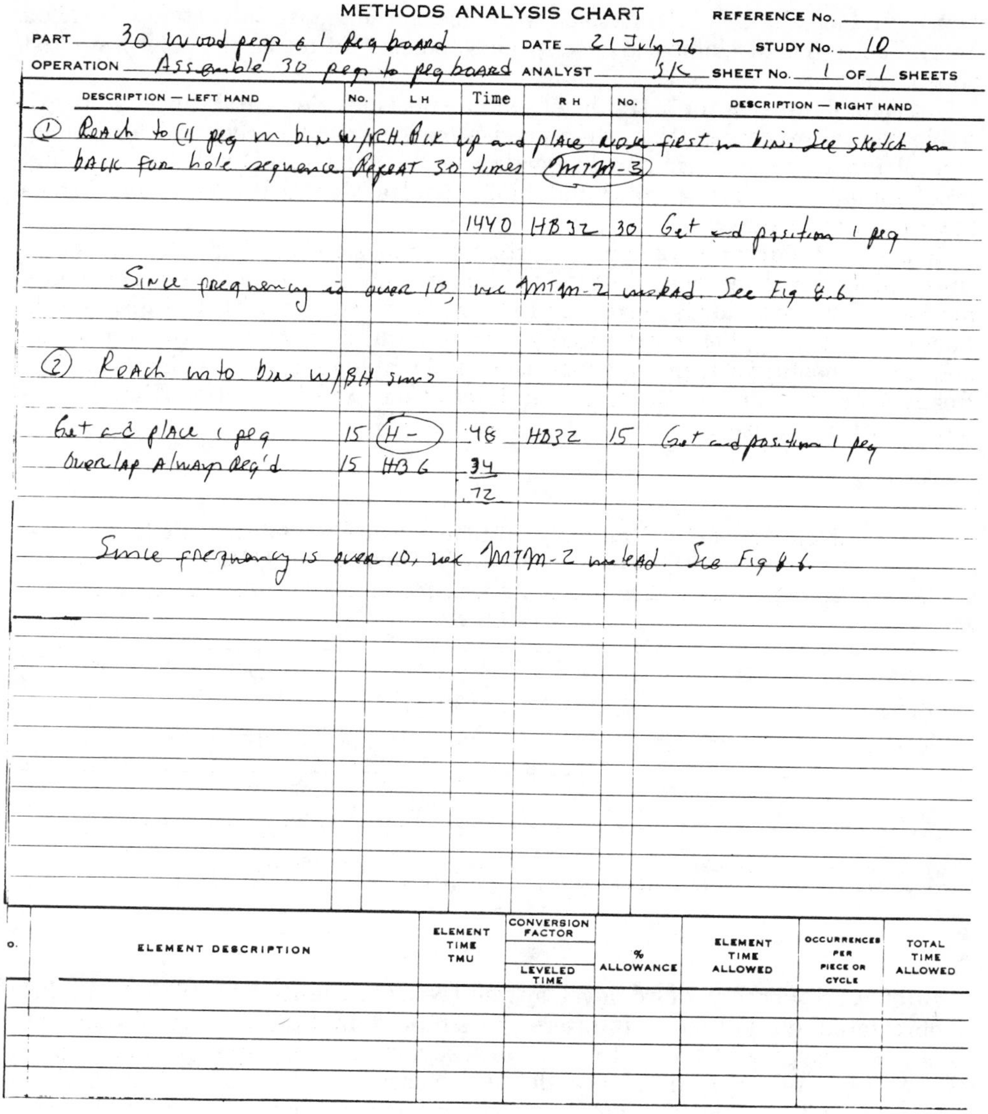

METHODS ANALYSIS CHART   REFERENCE No. _____

PART ___30 wood pegs & 1 peg board___ DATE __21 July 76__ STUDY No. __10__
OPERATION __Assemble 30 pegs to peg board__ ANALYST ___SK___ SHEET No. __1__ OF __1__ SHEETS

DESCRIPTION — LEFT HAND	No.	L H	Time	R H	No.	DESCRIPTION — RIGHT HAND
① Reach to (4 peg in bin w/RH. Pick up and place hole first in bin. See sketch on back for hole sequence. Repeat 30 times (MTM-3)						
			1440	HB32	30	Get and position 1 peg
Since frequency is over 10, use MTM-2 method. See Fig 8.6.						
② Reach into bin w/BH sim2						
Get and place 1 peg	15	H –	48	HB32	15	Get and position 1 peg
Overlap always req'd	15	H36	34			
			.72			
Since frequency is over 10, use MTM-2 method. See Fig 8.6.						

o.	ELEMENT DESCRIPTION	ELEMENT TIME TMU	CONVERSION FACTOR		ELEMENT TIME ALLOWED	OCCURRENCES PER PIECE OR CYCLE	TOTAL TIME ALLOWED
			LEVELED TIME	% ALLOWANCE			

Figure 8.8   *MTM-3 analysis* of the one-hand and two-hand assembly of the pegboard previously analyzed in Figures 8.2 and 8.6 shows how you revert to MTM-2 if the frequency is over 10 (except for SF).

The Work-Factor text was reviewed by Emerson Boepple, Coordinator of Research and Development of Wofac Company. The figures and tables are reprinted with the permission of Wofac Company, Fellowship Road, Moorestown, NJ 08057.

There are three system levels—Detailed, Ready, and Brief—each of which is based on the original Detailed time data. Each system is self-sufficient, yet each is completely consistent and interchangeable if a mixture is desired.

Detailed Work-Factor® (1st level system) provides elemental times for motion study and measurement of mass production (i.e., fingers, hands, arms, legs, feet, trunk, and head). Use when exact work measurement is important. Detailed Work-Factor has 31 elemental descriptions and its Motion Time Tables have 764 time classes.

Detailed Mento-Factor® (1st level system) provides elemental times for identifiable mental processes. Use when precise measurement is required for human mental functions occurring in inspection (audio, visual, kinesthetic), reading, proofreading, calculating, color matching, etc. Its Mental Process Time Tables have 710 time classes.

Ready Work-Factor ® (2nd level system) provides a simplified table of motion times which are averages of times for finger, hand, arm, foot, leg, and trunk motions. Time standards correspond closely to those set with Detailed Work-Factor but it is applied more rapidly. Ready requires a shorter training period and thus is useful for training supervisors and employees in work simplification. Ready Work-Factor has 10 elemental descriptions and its Motion Time Tables have 154 time classes.

Table 8.15   MTM-3 has only 10 times; analysis time should take about 50 times the cycle time. Times are in TMU.

		MTM-3			
RANGE	CODE	HA	HB	TA	TB
Up to 6″	−6	18	34	7	21
Over 6″	−32	34	48	16	29
		SF	18	B	61

Brief Work-Factor® (3rd level system) was introduced in 1975 to replace Abbreviated Work-Factor. It offers the simplest motion timetable combining Reaches, Grasps, and Moves to form work segments called Pickups. It is applied to situations requiring much less detailed measurement, such as nonrepetitive operations with long cycle times occurring in maintenance, clerical, and indirect labor functions. Brief Work-Factor has 8 elemental descriptions, and its Time Tables have 32 classes.

WOCOM® is a computerized version of all the Work-Factor systems (except Brief which is so simple to apply that the computer is not useful). *WOCOM's* modules include the Ready Work-Factor System, the MTM-1 System, Multiple Regression Analysis, Learning Time Allowances, Production Line Balancing, Standard Time Data Development, and Standard Time Data Revision. The concept is to eliminate routine calculations and thus substantially reduce engineering time and improve accuracy.

If you wish to apply the Work-Factor Systems (in contrast to the cursory knowledge of a student who receives an appreciation course), take a training course ranging from 10 class hours for Brief to 80 for Ready to 120 for Detailed Work-Factor.

The remainder of this chapter will deal only with the second level, Ready Work-Factor. Times for Ready Work-Factor are in Ready Work-Factor Time Units (RU); 1 RU = .060 s = .001 min = .000 017 hr. Conversely, 1 hr = 60,000 RU, 1 min = 1000 RU, and 1 s = 16.7 RU.

Ready Work-Factor

There are eight standard elements: Transport, Grasp, Release, Pre-position, Assemble, Use, Disassemble, and Mental Process. Walk is classed as a Transport.

Transport. Table 8.16 gives Transport (Reach and Move) times.

For the degree of difficulty, use 1 Work-Factor, if it occurs, for each of the following manual control factors involved in the Transport Motion (Definite Stop, Steer, Precaution, or Change Direction).

Table 8.16   Transport (Reach and Move) times (RU) for Ready Work-Factor. All distance and weight ranges are over the smaller value and up to and including the larger value. To determine Work-Factors due to weight or resistance (lb), enter lower table on the appropriate row, go right to appropriate weight/limb, and then up to read the Work-Factor. −X means over the previous distance, up to and including X; e.g., 20 in the upper table means "more than 10 inches, up to and including 20 inches."

			WORK-FACTORS				
			0	1	2	3	4
		Distance,	Very				Very
Code	Description	Inches*	Easy	Easy	Average	Difficult	Difficult
4	Very short**	− 4	2	3	4	5	6
10	Short	−10	4	5	6	7	8
20	Medium	−20	5	7	9	11	13
30	Long	−30	7	9	11	13	15
40	Very long	−40	9	11	13	15	17

*For trunk motions, multiply distance by 2.
**Use this row also for Forearm Swivel motions.

	WORK-FACTORS				
Body member	0	1	2	3	4
Finger-hand	−1	− 2	−3	− 5	Over 5
Arm	−2	− 4	−6	−10	Over 10
Foot	−3	− 8	Over 8	−	−
Leg	−5	−16	Over 16	−	−
Trunk	−7	−32	Over 32	−	−

Definite Stop Work-Factor is used for motions which are stopped voluntarily by the operator at or near a specific location (> 2 inch diameter sphere or circle). All reach motions followed by grasp have at least a Definite Stop. All move motions, other than tosses, motions to indefinite locations, or motions suddenly arrested by rigid objects, have at least a Definite Stop. Examples of Definite Stop are: reach to a pack of cigarettes, move a bolt to a place on a die surface, and move the hand to touch a large button.

Steer Work-Factor is used for motions which steer a body member or object to a Target. For tolerances ≤ 2 inches, use Steer plus Definite Stop. For tolerances ≤ 5/8 inches, use Move followed by an Assemble. Examples are move bolt to hole, move to touch small button, and move washer to end of bolt.

Precaution Work-Factor is used for motions which must be made carefully. Examples are move razor to face, reach near moving saw blade, and move open container full of liquid.

Change Direction Work-Factor is used for motions which follow a curved path sharper than a circle. Examples are reach to 15-inch distant object over 10-inch

Table 8.17 Example Transport Motions With their Work-Factors and Times

Description	Distance Code	Manual Control WF	Weight WF	Code	Time, RU
Push small switch button with finger.	4	0	0	4-0	2
Toss 5 lb bracket 18 inches to truck.	20	0	0	20-0	5
Move arm 15 inches aside to indefinite location after placing part in machine.	20	0	0	20-0	5
Return trunk 11 inches to normal relaxed position. (double distance for trunk)	30	0	0	30-0	4
Reach 24 inches to grasp handle of screwdriver.	30	Definite Stop	0	30-1	9
Reach 15 inches to grasp bolt from bin.	20	Definite Stop	0	20-1	7
Move .25 inch bolt 6 inches to hole in bracket.	10	Definite Stop; Steer	0	10-2	6
Move eraser 6 inches to specific location on typed letter.	10	Definite Stop; Steer	0	10-2	6
Reach 18 inches to grasp piece of wood near revolving circular saw blade.	20	Definite Stop; Precaution	0	20-2	9
Reach 20 inches over side of box 12 inches high to grasp casting.	40	Definite Stop; Change Direction	0	40-2	13
Move 3 lb bolt 36 inches to hole in I-beam.	40	Definite Stop; Steer; Weight	1	40-3	15
Pull machine lever 2 inches against stop (3 lb resistance)	4	0	1	4-1	3
Toss 3 lb object 24 inches into truck.	30	0	1	30-1	9
Toss 8 lb piece of scrap iron 24 inches to scrap pile.	30	0	3	30-3	13
Push heavy carton 15 inches along conveyor with one arm (15 lb resistance).	20	0	4	20-4	13
Using leg, depress pedal 5 inches against 15 lb resistance.	10	0	1	10-1	5

high barrier and move casting 20 inches to fixture over the side of a 12-inch high box side.

In addition to a maximum of 1 WF each for Definite Stop, Steer, Precaution, or Change Direction, Work-Factors can be given for weight. The lower portion of Table 8.16 indicates that a 1 lb weight moved by the hand would be 0 WF but a 1.5 lb weight moved by the hand would be 1 WF. A 5 lb weight moved by the arm would be 2 WF.

Table 8.17 has example Transport Motions and their codes. Typical codes would be 20-0, 30-1, 10-2, and 40-3; that is, the distance, a dash, and the number of Work-Factors.

Now that the degree of difficulty for the motion has been determined (0 to 4 WF), estimate the distance moved or reached. Measurement is unnecessary since estimations are within the accuracy of the Ready Work-Factor System. Then read the time from Table 8.16. An "average" Move for a "medium" distance takes 9 RUs or, if you wish, a 2 WF Transport for over 10 to including 20 inches takes an average of 9 RUs.

Grasp. Contact Grasps are allowed as part of the preceding Reach and have no time added from the Grasp Table. Table 8.18 gives the time for the other Grasps as well as Release and Pre-position. In a Visual Grasp, the operators see their fingers while grasping; in a Blind Grasp, they don't. Single Motion Grasps require only 1 Finger Motion; Pinch Grasps require 0 WF and Wrap-Around Grasps require 1 WF. Multiple Motion Grasps require more than 1 Finger Motion; that is, 2, 3, or 4 WF depending on their complexity. Table 8.19 has example Grasps and their codes; the Grasp code is the number of Work-Factors (with × 2 if weight grasped is more than 3 lb), a dash, and symbols for the add-on times, if any.

Table 8.18   Grasp, Release, and Pre-Position Times (RU) for Ready Work-Factor

	WORK FACTORS				
No. of finger motions	Single Motion (Simple)		Multiple Motion (manipulative)		
	0	1	2	3	4
DESCRIPTION					
Element[a]	Very Easy (Pinch)	Easy Wrap-Around Transfer	Average	Difficult	Very Difficult
Simple and Manipulative Grasp—Visual[c]	1[b]	2[b]	3	5	8
Simple and Manipulative Grasp—Blind[c]	1[b]	2[b]	4	6	8
Major Dimension, inches			−.25	−.25	−.25
Diameter, inches			.25	−.25	All
Thickness, inches			.05	−.05	All
Complex Grasp—Visual[d]	1	2	3	5	—
Complex Grasp—Blind[d]	1	2	4	6	8
Release[a]	1	2	—	—	—
Pre-Position[e]	4	5	6	7	8

[a] Contact grasp and release have zero time.
[b] Multiply time by 2 for weight >3 lb.
[c] Add 2 RU if SIMO manipulative; SIMO does not apply to simple grasp.
[d] Add 1 RU if entangled (e) or nested (n); add 1 RU if slippery (slp); add 2 RU if SIMO.
[e] Multiply RU by percent occurrence to closest 25%; round result to nearest RU.

Release. (Code = R1) Contact Release (the opposite of Contact Grasp) requires no time. Single-motion releases require only 1 Finger Motion; Gravity Releases (the opposite of Pinch Grasp) get 1 WF but Unwrap Releases (the opposite of Wrap-Around Grasp) get 2 WF. Multiple motion releases require more than 1 Finger Motion; they occur only when the object clings or sticks to the fingers and are analyzed using Transport rules.

Pre-position. (Code = PP) Pre-positions (turning or orienting an object to a correct position for a subsequent element) that can be completed by a single arm motion or forearm swivel are allowed for in the move and have no time added from the pre-position row in Table 8.18. One hand pre-positions are 2 WF if the object's longest dimension is ≥3/8 inch, 1 WF if >10, and 0 WF if in between. Two-Hand pre-positions are 3 WF if the longest dimension is ≤10 inches and 4 if it is >10. A very easy pre-position (0 WF) would be a one-hand manipulation of a

Table 8.19   Example Grasp Motions With Categories, Work-Factors, and Times

Description	Category	WF	Code	Time RU
Place hand against carton to push along conveyor.	Contact grasp	—	Gr-C	0
Place fingers on piece of paper to hold it steady during writing.	Contact grasp	—	Gr-C	0
Grasp pencil from desk.	Single motion—pinch grasp	0	0	1
Transfer ruler from one hand to the other.	Single motion—transfer grasp	1	1	2
Grasp handle of suitcase.	Single motion—wrap-around grasp	1	1	2
Grasp 1 inch diameter piece of pipe.	Single motion—visual	1	1	2
Grasp 8 lb piece of pipe.	Single motion—wrap-around grasp	1	1-x2	4
Grasp bracket (major dimension .75 inch) from box of brackets.	Multiple motion—blind	2	2-B	4
Grasp flat punched card from table.	Multiple motion—visual	3	3	5
Grasp 2 bolts (.37 diameter, 2 long) SIMO from pile.	Complex blind—SIMO	2	2-Bs	6
Grasp paper clip (1 long, .03 thick) from pile.	Complex visual—entangled	3	3-B	6
Grasp ends of tape stuck to surface.	Multiple motion—visual	4	4	8
Grasp washer (thickness .015 inch; diameter of .22 inch).	Complex visual—oily	4	4-slp	9
Grasp dished, oily washer (thickness .12; diameter .25) from pile.	Complex visual—nested-oily	4	4-nslp	10

cylinder between 3/8 and 4 inches long. Easy (1 WF) is for over 4 to 10 inches long. Average (2 WF) would be one-hand manipulation of an object $\leq 3/8$ inch long. Difficult (3 WF) would be two-hand manipulation of a cylinder 4 inches long. Very difficult (4 WF) would be two-hand manipulation of a solid over 10 inches long.

If the pre-position does not occur every cycle, multiply the RU value by percent occurrence (rounded to nearest 25%); round resulting time to nearest RU.

For example, pre-position (turn end for end) a 1-inch-long bolt following grasp in 0 WF. RU = 4. Since this pre-position occurs only 50% of the time, allow 4 RUs. Pre-position a 1/4 inch long screw in the fingers following grasp in 2 WF. RU = 6. Since this pre-position occurs only 50% of the time, allow 3 RUs. Pre-position (turn over) a flat board $9 \times 9 \times 1/4$ inch using both hands. RU = 7. If it occurs every cycle, allow 7 RUs.

The pre-position code is the number of Work-Factors and the percent occurrence (nearest 25%).

Assemble. (Code = Asy) Assemble follows Move and has the subelements of

Align (bring plug and target opposite each other), Upright (put plug and target on the same axis), Index (rotate plug and target), Insert (move plug into target or vice versa), and Seat. Assemble has two classes: *mechanical assemble* and *surface assemble*. In *surface assemble* no mechanical features aid correct orientation. Examples are put a book on a desk top, a stamp on a letter, or a nail point on a mark.

Assemble times and adjustments are given in Table 8.20. The move immediately preceding assemble always contains both Definite Stop and Steering Work-Factors.

Assemble is based on the plug (male-type object) vs. target (female-type object) concept. The plug can be put into the target or the target onto the plug.

For Mechanical Assemble, first determine whether the target is closed or open. Closed targets permit the plug or target to move along only one axis (such as a washer on a bolt or a hammer head onto a hammer handle). Open targets reduce the number of aligns necessary as movement is possible in two axes (such as a bolt into vise jaws, a cylinder into a slot). Second, determine target size, plug size, and plug/target ratio. If the target size varies (as with a beveled hole), use the outside diameter of the bevel. If the plug diameter varies, the mating dimension is whichever of the two following is greater: (1) actual diameter across the flat, or (2) .33 times the plug shank diameter, if the plug shaft is rounded or pointed.

Adjustments are given as a multiplier of the align time. The basic value combines Align, Upright if required, and Insert.

Gripping distance adjustment occurs when the fingertips are more than 2 inches from the insert point. Between targets adjustment occurs when two or more plugs and targets are being assembled simultaneously. A temporary blind adjustment occurs when the target or plug or both are in the operator's view during the move preceding assemble but not during the aligns that are a part of assemble. In a permanent blind assemble, the view is blocked during both the move and the assemble.

The assemble code gives, on one line, closed target (CT) or open target (OT), the row used, and the column used (e.g., C.T. − <3/8 − >.9). Align, Add-ons, Index, SIMO, and Seat are each on separate lines. Table 8.21 shows example calculations.

Use. (Code = Use) The act of using machines, equipment, instruments, tools, or devices is called Use. In fact, a body member itself may be a tool (mark with a finger).

Operator-controlled motions (tighten screw with screwdriver, drive nail, cut with scissors, paint with brush) have their times determined from the appropriate tables of Transport, Grasp, etc.

Process-controlled motions (pour liquid, pull caulking-gun trigger) are determined from time study or mathematical calculation.

Machine or process time (tool cutting time) also is determined from time study or mathematical calculation.

Disassemble. (Code = Dsy) Analyze Disassemble from the Transport table.

SIMO-Assemble. Add SIMO time when assemble, pre-position, or multiple-motion grasp occurs entirely or partially simultaneously—one in each hand. All other elements can be done without SIMO time. See Table 8.22.

For assemble, add 50% of align time. For pre-position, add 50% of total pre-position time. For multiple-motion grasp, add 2 RU.

Mental Process. (Code = MP) Mental process is very simplified in Ready Work-Factor and includes only Focus, Inspect, React, and Mento. Time is allowed only when they do not occur simultaneously with manual motions.

Eye Focus (2 RU) (Code = Fo) is given each time the eyes shift to a new location. At distances of 12-18 inches, 1 Focus is given for each "inspection unit" (a 3-inch square) inspected.

Table 8.20 Assemble time (RU) and adjustments for Ready Work-Factor. The table gives base time (total assemble time) followed by the align time in parentheses. If there is an align time, align time may require adjustments. If appropriate, multiply align time by factor in adjustment table. If SIMO, add .5 (align time + align adjustment time). Then, if appropriate, add times for Index. Then, if appropriate, multiply total time by factor in weight table. Rounding is done at each step of the calculation as shown in Table 8.21.

	MECHANICAL ASSEMBLE						SURFACE ASSEMBLE	
							Open Target	Closed Target
	Open Target			Closed Target				
Plug/Target Ratio	$-.4$	$-.9$	$>.9$	$-.4$	$-.9$	$>.9$		
Target dimension, inches — up to 1/8	6(4)	6(4)	10(4)	9(7)	9(7)	13(7)	8(5)	12(9)
Target dimension, inches — 1/8 to 3/8	3(1)	4(2)	8(2)	5(3)	6(4)	10(4)	5(2)	6(3)
Target dimension, inches — Over 3/8	2(0)	3(1)	7(1)	2(0)	3(1)	7(1)	3(0)	3(0)

Align adjustment multiplier:

| | Distance, inches | | | | | | |
	-1	-2	-3	-5	-7	-15	>15
Gripping distance	—	—	.1	.2	.3	.3	.7
Distance between	—	.2	.3	.5	.7	*	*
Temporary blind	—	.2	.3	.5	.7	1.5	—
Permanent blind	.3	.5	.7	1.5	2.5	5.0	—

*Consider as 2 assemblies; >15 inches requires MP of 5 RU between the 2 assemblies.

SIMO adjustment to align: .5 (Align time + Align adjustment time)

Index   Mechanical 3 RU   Surface 4 RU

Weight adjustment:

| | Weight, lb | | | | |
	-2	-4	-6	-10	>10
Weight multiplier	1.0	1.3	1.5	1.5	2.0

Inspect Interval (3 RU/Interval) (Code = I) is given following Focus. One Inspect Interval occurs, at a specified viewing distance and under specified lighting, for:

1. Determining the presence and identity of or the absence of one or any group of visible and distinguishable characters, objects, symbols, or characters (referred to as Work-Factor Inspection Characters). A visible and distinguishable Character subtends a visual angle of not less than 1 minute. (At "normal fix distance," a character dimension subtending 1 min = .004 400 in.) Under perfect visual conditions, 1 min of arc can be seen with 20/20 vision. For practical applications, however, minimum dimension characters are not sufficient (see chapter 17 for additional discussion).

2. Determining the presence or absence of exactly 1, 2, 3, or 4 Inspection Characters.

3. Recognizing ≤ 3 digits of a number.

4. Recognizing words ≤ 6 letters, provided the word is common to the reader's language. For familiar words of 7 or 8 letters, give 2 Inspect Intervals. For familiar words of ≥ 9 letters or for unfamiliar words, give 1 Inspect Interval for each 3 letters.

Table 8.21   Examples of Mechanical Assemble and Surface Assemble in Ready Work-Factor

DESCRIPTION Mechanical Assemble	Target Plug Ratio	Target	Base	Times, RU Align Adjustment	Index	Total
Assemble .25 inch diameter wooden dowels to .5 inch hole in board, one at a time.	.5	Closed	3	—	—	3
Assemble .49 inch diameter steel shaft to bearing with .50 inch diameter.	.98	Closed	7	—	—	7
Assemble end of .06 inch diameter shaft into .25 inch diameter hole.	.24	Closed	5	—	—	5
Assemble washer with .5 inch diameter hole over end of .25 inch diameter bolt.	.50	Closed	3	—	—	3
Same as previous example except 2 assemblies, one in each hand, are made simultaneously; assemblies 6 inches apart.	.50	Closed	3	2	—	5*
Assemble 1/4 × 1/4 × 1/2 inch piece of steel with square cross section in open jaws (.31 open target) of vise.	.80	Open	4	—	3	7
Surface Assemble						
Move .5 inch diameter metal disc and place on flat surface within a .56 inch diameter circle (target = .06 inch).	—	Closed	12	—	—	12
Place 15 inch ruler at 2 points 6 inches apart, for drawing straight line.	—	Open	8	4	4	16**

*Distance between targets = .7 × 1 = .7 = 1. SIMO factor (see section on SIMO-Process "Times") = .5 × 2 = 1.
**For 6 inch distance between targets the factor is .7; 5 × .7 = 3.5 = 4.

5. Even if there are <3 items in a group, give 1 Inspect Interval for each group of numbers, letters, or symbols.

6. When an Inspection Unit is inspected for the presence or absence of a specified Inspection Character, give 1 Inspect Interval for the initial inspection and 1 Inspect Interval for each Character (such as a scratch, spot, stain, or scar) that could be mistaken for the specified Character.

React is divided into *Anticipated Simple React* (2 choice decision in which the signal is anticipated) for which 0 time is allowed; *Unanticipated Simple React* (2 choice decision in which the signal is unanticipated) for which 2 RU is allowed: *Anticipated Choice React* for which 2 RU is allowed if there are 3 to 5 choices and 4 RU is allowed if there are more than 5 alternative choices and *Unanticipated Choice React* (select from multiple alternatives) for which 2 RU is allowed (8 RU maximum) for each 2 alternate choices, from which the operator must select the correct one.

Mento (1 RU) (Code = Mt) is subdivided into Memorize, Compute, and Recall. *Memorize* (2 RU) is the Work-Factor term for the act of fixing a number (word, idea) in the mind so it is available for use by the operator a few seconds or minutes later. It is not comparable to memorizing a poem, speech, or table of numbers. Memorize occurs (1) immediately following Compute, or (2) when memorized information must be retained in instant readiness to be used after an intervening Manual Motion or a Mental Process (other than an Inspect Interval). *Compute* (2 RU) is the Work-Factor term for adding, subtracting, multiplying, or dividing one-digit numbers. For numbers with >2 digits, Compute is performed in steps using 1 digit of each number per step. *Recall* (2 RU) is the Work-Factor term for recalling previously memorized information. Recall is given if a motion or Mental Process other than 1 Inspect Interval intervenes between Memorize and the use of the memorized information.

Table 8.22   SIMO-Factor, Mental Process, and Walk Times (RU)
for Ready Work-Factor

Element	Times, RU
SIMO-Factor:	
Multiple-motion grasp	2
Assembly	.5 (Align time)
Pre-position	.5 (Pre-position time)
Mental Process:	
Focus	2
Inspect	3/Inspection
React	2
Mento	1
Walk:	
Normal	12 + 8 (Number of paces)
Restricted	12 + 10 (Number of paces)
Up and down steps:	
General	10 (Number of steps)
Restricted	13 (Number of steps)
Stand up	13
Sit down	9

Walk. See Table 8.22. A pace is 30 inches. Multiply the number of paces by 8 RU/pace and add 12 RU for start plus stop; speed is 3.7 mph. For restricted travel (restricted space, slippery floors, heavy loads) give 10 RU/pace; speed is 2.96 mph.

Figure 8.9 gives a Ready Work-Factor analysis of the one-hand pegboard assembly; Figure 8.10 gives a Ready Work-Factor analysis of the two-hand peg-board assembly.

Note that the analyst worked smart instead of hard by writing "R to bin from table edge" instead of "Reach to bin from table edge" or "Reach to the bin from the edge of the table."

COMMENTS ON PREDETERMINED TIME SYSTEMS

In theory, Predetermined Time Systems (PTS), when applied by a "trained" analyst, can "accurately" predict the amount of time for a task.

There is extensive evidence that the theory and reality don't agree. If the reader is interested in the extensive literature on the subject, see Wehrkamp and Smith; Nadler and Wilkes; Gomberg; Geppinger; Abruzzi; Heckler, Green, and Smith; Buffa; Buffa and Lyman; Burns and Simerson; Simon and Simon; Frederick; Schmidtke and Stier; Bailey; Sellie; Taggart; Davidson; Schmidtke and

Part Name Pegboard - 30 Pegs	Sheet No. 1 of 1	Company	Department	Part No.	Sub.	Oper. No.

Operation Name & Description: Asy pegs to board - 1 hand

LEFT HAND Elemental Description	Analysis	Time Units	Cumulative Time	Time Units	Analysis	RIGHT HAND Elemental Description	No.
			7	7	20-1	R to 1st peg from edge of table	1
			10	3	2-	Gr 1st peg	2
			12	2	0-50%	P P peg	3
			18	6	10-2	M peg to hole	4
			23	5	CT-.4-3/8	Asy beveled end of peg to hole	5
			24	1	0-	Rl peg	6
			29	5	10-1	R to 2nd peg	7
			46	17	El 3-8	Gr and Asy 2nd peg	8
			442	396	18 x EL 9-10	PU and Asy 18 more pegs	9
			447	5	10-1	R to 21st peg	10
			448	1	0-	Gr peg (isolated)	11
			454	6	10-2	M peg to board - PP internal	12
			459	5	CT-.4-3/8	Asy peg to hole	13
			460	1	0-	Rl peg	14
			622	162	9 x EL 12 - 17	PU and Asy 9 more pegs	15
							16
							17
Hold Board	BD	622	622				18
							19
							20
							21
							22
							23
							24

	Total 622		Time in Minutes .622	Multiplier

Date 1 July 76   Analyst E. Boepple

13.5/2/ (69/1)

Figure 8.9 *Ready Work-Factor* analysis of one-hand pegboard assembly which was analyzed by MTM-1 in Figure 8.1. Total time = 622 RU = .62 min.

Stier; and Sanfleber.[1,2,6,7,8,9,10,11,12,13,20,24,25,26,27,28,29,30] I also have written on this subject in Konz, Jeans, and Rathore with comments on the effect of angle of movement (a variable neglected by the PTS) and in Konz and Rode with comments on weight allowances.[15,16] An impartial observer must conclude that the PTS cannot "always accurately" predict the time a worker will take for a task.

So? Do you throw away your bowling ball because it gives strikes only 7 times in 10?

In this imperfect world we must use the available tools—even though not perfect—until better tools come along.

One problem with the Predetermined Times Systems is that they are not "automatic," that is, analyst judgment is required. Different analysts get different times for the same job because of different interpretations of the various rules. Very detailed rules have been found to be feasible if applied by a computer, but the programming is very complex. At least two computer systems for applying MTM-1 and one for applying MTM-2 are in worldwide use.[18]

The popularity of the "quick and dirty" systems such as MTM-2 and Ready Work-Factor suggest most managements do not need a great deal of accuracy and are quite concerned with analysis cost. The point here is: "What is the purpose of making a PTS study?" The purpose is (1) make a methods analysis of the job to determine an efficient work method, and (2) determine the amount of time necessary to do the job.

PTS force the analyst to consider use of both hands, whether a grasp is simple or complex, whether the distance moved is 6 or 9 inches, etc. If a PTS system analysis results in an efficient work method, then it has accomplished its most important task.

The second and less important task is to assign a time to the efficient work method. Time per unit is necessary for:

1. Cost accounting. How much should we charge for the unit?

2. Scheduling. If we want to be done by the 15th, when should we start?

3. Evaluation of alternatives. Should the job be done with one hand or two?

4. Acceptable days work. How many should Joe do in a week?

5. Pay-by-results. What should Betty Jo's pay be today?

In most of these applications a 10% deviation from the estimated time is not critical. This especially is true if the error is a consistent error. That is, if a PTS time system consistently estimates times that are 7-12% shorter than an organization achieves, the supervisor can adjust without much difficulty by just adding 10% to the "standard" time.

SUMMARY

PTS developed from the therbligs of Gilbreth and Barnes; they offer detailed methods analysis as well as times for each element.

The two most popular systems are Methods-Time Measurement (MTM) and Work-Factor (WF). Each system has several levels of detail, allowing a trade-off of analysis accuracy versus analysis time.

The text in the chapter describes MTM-1, MTM-2, and MTM-3 as well as Ready Work-Factor.

Although there are accuracy problems with the PTS, there doesn't seem to be, at the present state of development, any serious hindrance to their use—if the user is trained in their use and realizes their limitations.

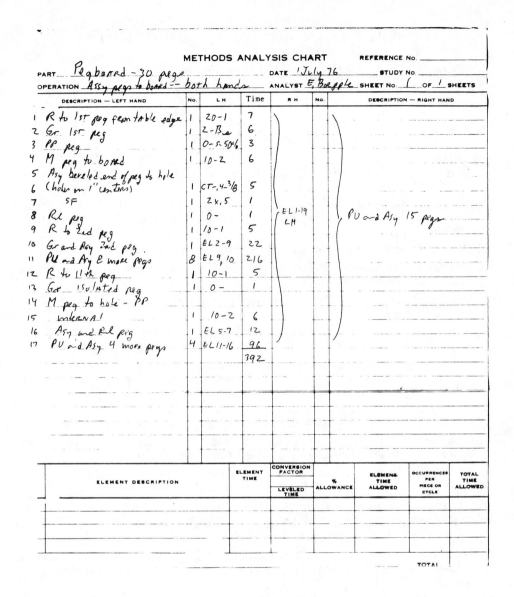

METHODS ANALYSIS CHART   REFERENCE No. _____

PART _Pegboard - 30 pegs_____   DATE _1 July 76_   STUDY No. _____
OPERATION _Assy pegs to board -- both hands_   ANALYST _E. Boepple_ SHEET No. _1_ OF _1_ SHEETS

	DESCRIPTION — LEFT HAND	No.	L H	Time	R H	No.	DESCRIPTION — RIGHT HAND
1	R to 1st peg from table edge	1	20-1	7			
2	Gr. 1st peg	1	2-B₄	6			
3	PP peg	1	0-S 50%	3			
4	M peg to board	1	10-2	6			
5	Asy beveled end of peg to hole						
6	(holes on 1" centers)	1	CT-,4-³⁄₈	5			
7	SF	1	2x,5	1			
8	Rl peg	1	0-	1	EL 1-19		PU and Asy 15 pegs
9	R to 2nd peg	1	10-1	5	LH		
10	Gr and Asy 2nd peg	1	EL 2-9	22			
11	PU and Asy 8 more pegs	8	EL 9, 10	216			
12	R to 11th peg	1	10-1	5			
13	Gr isolated peg	1	0-	1			
14	M peg to hole - PP						
15	internal	1	10-2	6			
16	Asy and Rl peg	1	EL 5-7	12			
17	PU and Asy 4 more pegs	4	EL 11-16	96			
				392			

ELEMENT DESCRIPTION	ELEMENT TIME	CONVERSION FACTOR		% ALLOWANCE	ELEMENT TIME ALLOWED	OCCURRENCES PER PIECE OR CYCLE	TOTAL TIME ALLOWED
		LEVELED TIME	ALLOWANCE				
							TOTAL

Figure 8.10 *Ready Work-Factor* analysis of the two-hand assembly analyzed with MTM-1 in Figure 8.2, with MTM-2 in Figure 8.6, and with MTM-3 in Figure 8.8. Total time = 392 RU = .39 min.

SHORT-ANSWER REVIEW QUESTIONS

8.1 What are the two purposes of making a PTS analysis?

8.2 What is the pace for MTM and for Work-Factor?

8.3 At which number of cycles is a worker predicted to achieve MTM or Work-Factor time?

8.4 In MTM-1, what is the difference between a Reach and a Move?

8.5 How does MTM handle the times when both motions may be simultaneous?

THOUGHT-DISCUSSION QUESTIONS

1. Why are there several levels of MTM and Work-Factor?
2. Why should you use a PTS if it doesn't always give the correct answer?

REFERENCES

1. Abruzzi, A. *Work, Workers and Work Measurement,* New York: Columbia University Press, 1956.
2. Bailey, G. Comments on an experimental evaluation of the validity of predetermined time systems. *Journal of Industrial Engineering,* 12 (September-October 1961): 328-330.
3. Barnes, R. *Motion and Time Study.* New York: Wiley & Sons, 1968.
4. Boepple, E. Coordinator of Research and Development of Work-Factor, personal communication, May 17, 1977.
5. Boepple, E., and Kelly, L. How to measure thinking. *Industrial Engineering,* Vol. 3, No. 7 (1971): 8-15.
6. Buffa, E. The additivity of universal standard data elements. *Journal of Industrial Engineering,* 8 (1957): 327-334.
7. Buffa, E., and Lyman, J. The additivity of the times for human motor response elements in a simulated industrial assembly task. *Journal of Applied Psychology,* 42 (1958): 379-383.
8. Burns and Simerson. Fundamentals of predetermined time standards and a comparison of 5 systems. *Industrial Management,* Vol. 11-16 (February 1959).
9. Davidson, H. On balance—the validity of predetermined elemental time systems. *Journal of Industrial Engineering,* 13 (May-June 1962): 162-165.
10. Frederick, C. On obtaining consistency in application of predetermined time systems. *Journal of Industrial Engineering,* Vol. 11, No. 1 (January-February 1960): 18-19.
11. Geppinger, H. *Dimensional Motion Times.* New York: Wiley & Sons, 1955; pp.2-6.
12. Gomberg, W. *A Trade Union Analysis of Time Study* (especially chapter 15). Englewood Cliffs, New Jersey: Prentice-Hall, 1955.
13. Heckler, D.; Green, D.; and Smith, D. Dimensional analysis of motion: X. Experimental evaluation of a time-study problem. *Journal of Applied Psychology,* 40 (August 1956): 220-227.
14. Karger, D. and Bayha, F. *Engineered Work Measurement.* New York: Industrial Press, 1965.
15. Konz, S.; Jeans, C.; and Rathore, R. Arm motions in the horizontal plane. *AIIE Transactions,* 1 (December 1969): 359-370.
16. Konz, S. and Rode, V. The control effect of small weights on hand-arm movements in the horizontal plane. *AIIE Transactions,* 4 (September 1972): 228-233.
17. Magnusson, K., The development of MTM-2, MTM-V and MTM-3. *Journal of Methods-Time Measurement,* 17 (February 1972): 11-23.
18. Martin, J. The 4M data system. *Industrial Engineering,* 6 (March 1974): 32-38.
19. Maynard, H.; Stegemerten, G.; and Schwab, J. *Methods Time Measurement.* New York: McGraw Hill, 1948. (See also Maynard, H., ed. *Industrial Engineering Handbook.* New York: McGraw Hill.)
20. Nadler, G., and Wilkes, J. Studies in relationships of therbligs. *Advanced Management,* Vol. 2 (1953): 20-22.
21. Presgrave, R., and Bailey, G. *Basic Motion Timestudy.* New York: McGraw-Hill, 1958.

22. Quick, J.; Duncan, J.; and Malcolm, J. *Work Factor Time Standards.* New York: McGraw Hill, 1962. (See also Maynard, H., ed. *Industrial Engineering Handbook.* New York: McGraw Hill.)

23. Rivett, H. Learning curve prediction development using MTM and the computer. *MTM Journal,* 1 (February 1972): 32-42.

24. Sanfleber, H. An investigation into some aspects of the accuracy of predetermined motion time systems. *International Journal of Production Research,* Vol. 6, No. 1 (1967): 25-45.

25. Schmidtke, H., and Stier, F. An experimental evaluation of the validity of predetermined elemental time systems. *Journal of Industrial Engineering,* 12 (May-June 1961): 192-204.

26. Schmidtke, H., and Stier, F. Response to the comments. *Journal of Industrial Engineering,* 14 (May-June 1963): 119-124.

27. Sellie, C. Comments on an experimental evaluation of the validity of predetermined elemental time systems. *Journal of Industrial Engineering,* 12 (September-October 1961): 330-333.

28. Simon, J., and Simon, B. Duration of movements in a dial setting task as a function of the precision of manipulation. *Journal of Applied Psychology,* 43 (December 1959): 389-394.

29. Taggart, J. Comments on an experimental evaluation. *Journal of Industrial Engineering,* 12 (November-December 1961): 422-427.

30. Wehrkamp, R., and Smith, K. Dimensional analysis of motion: II. Travel-distance effects. *Journal of Applied Psychology,* Vol. 36, No. 3 (1952): 201-206.

31. Work-Factor, *Work-Factor Learning Time Allowances,* Ref. 1.1.2, January, 1969.

9

OCCURRENCE SAMPLING

PROBLEM

Assume that in your organization material is moved with a fork truck. There seem to be long delays. Upon what do you bias your opinion? The question is asked, "Do we need another truck or is the present one idle too much?" This leads to a more specific question: "What is the present utilization of the truck?" It is important to remember that the purpose of occurrence sampling is to obtain information in order to make decisions; the purpose is not to demonstrate a knowledge of statistical theory.

We might have someone follow the truck for a specific time period—say, 20 working days—and record the following type of information for each day:

0700 Went to shipping dock
0706 Park outside foreman's office
0711 Left office with orders
0712 Enter first freight car with load

This would be a continuous time study. It gives a complete picture of the situation while it is studied, assuming the past is the same as the future. The problem is the expense of the study.

In order to cut the expense of the study, the truck might be observed for only 5 days or 1 day or even half a day. The expense has been cut, but we now have the problem of a representative sample; perhaps the Monday morning on which we made the study is not representative of the entire month. We could reduce the problem by following the truck for 2 hours on Monday morning and 2 hours on Tuesday morning. We still might worry about idleness in the afternoon. However, we could observe for 1 hour on Monday morning, 1 hour on Monday afternoon, 1 hour on Tuesday morning, and 1 hour on Tuesday afternoon. We still might worry about before and after breaks; about Wednesday, Thursday, and Friday; about the first week of the month vs. the second week, etc.

The end result is a sample composed of a large number of very short intervals: discrete sampling instead of continuous sampling. All time studies are samples from a population. The conventional time study (a "movie") is a continuous sample of n cycles (assumed underlying statistical distribution is normal). Occurrence sampling (a series of "snapshots") is a technique in which there are gaps of occurrences between the sample readings (assumed underlying statistical distribution is binomial). Although the statistical calculations are valid even if there are no gaps between the events sampled, continuous recording of occurrences does present questions as to whether the sample represents the population.

Occurrence sampling of times was first used by Tippett in the British textile industry in the early 1930s; it was introduced to the USA about 1940 under the name "ratio delay" since it is often used to study the ratios of various delays.[14] It also is known as *work sampling* since the times sampled often are times of people working. More correct, however, is *occurrence sampling* since what are sampled are occurrences of various types of events. These events can be delays (such as in equipment utilization), work task ratios (such as the proportion of material handling labor that is direct labor or the proportion of time spent on the telephone), or other ratios (such as the proportion of loads that are damaged). Work and ratio-delay are poor adjectives since no connotation of work or delay is needed for the use of the technique.

Let us return to the fork truck problem.

If we observe the fork truck many times (say, 1,000), then we will be quite confident in the information. If we observe the fork truck a few times (say, 10), our confidence will be much less. But, there is a large difference in the cost of obtaining the information—the historic conflict between information and cost of obtaining information. You must make a *trade-off*. A small sample gives a low cost of information but a high risk that the sample is not representative of the population. A large sample gives a high cost of information but a low risk that the sample is not representative of the population. (See Figure 9.1.)

The sampling problem then becomes:

1. Obtaining a sample whose size gives the desired trade-off between cost and risk

2. Obtaining a sample representative of the population

The sample size problem will be discussed first.

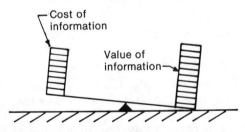

Figure 9.1 *Trading off information* (which increases with $\sqrt{n}$) vs. cost of information (which increases with n) is a common engineering problem.

REQUIRED NUMBER OF OBSERVATIONS

There is a general principle of statistics that the information obtained from a sample is a function of the square root of sample size, n. (It comes from the standard deviation of the error having that square root.) That is, if we get 2 units of information from a sample size of 4, then to get a total of 4 units of information, we need a sample size of 16; to get 8 units of information we need a sample size of 64.

Cost of obtaining information, however, generally increases directly with the sample size, n, rather than $\sqrt{n}$.

In ordinary words, it is the law of diminishing returns. As n increases there comes a point beyond which additional information is not worth the additional cost. Assume we can decide whether or not to buy a new fork truck if we know that the present truck is idle between 6% and 8% of the time. Then additional samples telling us that the idle percent is between 7.06% and 7.07% are not worth the additional cost.

How many observations should you make? It depends. It depends on:

A = Desired absolute accuracy
p = Percent of occurrence
c = Confidence level desired

First, the standard deviation of a percentage, σ_p, is:

$$\sigma_p = \sqrt{\frac{p\,(1-p)}{n}}$$

where p = percentage (decimal) of occurrence
n = number of observations

Second, since sample sizes generally will be "large" (over 30), the normal distribution is assumed. See Figure 9.2.

Third, make the distinction between relative and absolute accuracy. This simple distinction, if not made, will cause much grief. If the mean percent of occurrence, $\bar{p} = 40\%$, then 10% relative accuracy is $(.1)\,(.4) = 4\%$ absolute accuracy; 20% relative accuracy is 8% absolute accuracy; and 30% relative accuracy is 12% absolute accuracy. We need to use absolute accuracy in the calculations, but most people think in terms of relative accuracy. Be sure management understands the distinction when they make a statement such as "Make the study accurate to within 10%." The confusion arises because both the criterion and accuracy are in percent. If 10% accuracy on 25 lb is requested, then the 10% relative accuracy is not confused with the 2.5 lb absolute accuracy.

Fourth, how confident do you wish to be in your conclusions? Quite confident, very confident, absolutely confident, etc., are not precise enough so confidence must be expressed in numbers—70% confident, 90% confident, 99.9% confident. What confidence expresses is the *long run* probability that the sample mean is within the accuracy limits.

The probability calculations are similar to flipping a "true" coin; 8 of 10 is quite likely; 80 of 100 is very unlikely. It is the *number* of occurrences, not the percent, that is critical.

The formula to show the relationship between p, n, desired accuracy, and desired confidence level is:

$$A = z\sigma_p$$

or $$A = z\sqrt{\frac{p\,(1-p)}{n}} \tag{9.2}$$

or $$n = \frac{z^2\,p\,(1-p)}{A^2}$$

$$n = \left(\frac{z}{s}\right)^2 \frac{(1-p)}{p}$$

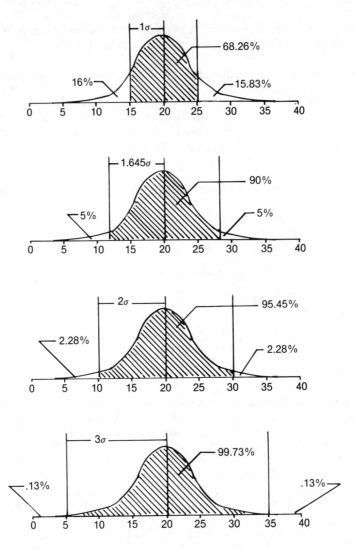

Figure 9.2 *Normal distributions* are symmetrical. The proportion of the total distribution included by a number of standard deviations does not depend on the value of the mean (see Table 9.1). Table 26.7 (the normal distribution) gives the area included for numbers of standard deviations other than above (Note: it gives area from —∞ to z, not area within ± z).

where
p = mean percent occurrence, decimal
s = relative accuracy desired, decimal
$A = sp$ = absolute accuracy desired, decimal
z = number of standard deviations for confidence level desired (see Table 9.2)
n = number of observations

To determine the number of observations required:

1. Make a sketch, giving mean and A (see Figure 9.3).

2. Determine value of z (see Table 9.1).

3. Solve equation 9.2.

Table 9.1   Confidence Level for z Values of Equation 9.2.

z, Number of Standard Deviations	Corresponding Confidence Level, %
±1.0	68
±1.64	90
±1.96	95
±2.0	95.45
±3.0	99.73

This may be clearer with some examples.

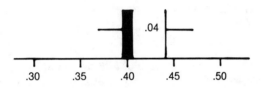

Figure 9.3   *Examples 1 and 2 are shown in a "one-side" sketch.*

Example 1
  Given   Estimated idle percent of fork truck = 40%
                  Relative accuracy desired = ±10%
                  Confidence level desired = 68%

  To Find   How many observations, *n*, are required?
  Solution   (.1) (.4) = .04 = 4% (see Figure 9.3)
                  68% confidence equals ±1 standard deviation
                  (from Table 9.1)
                  Therefore:

$$.04 = 1\sigma_p$$

$$.04 = 1\sqrt{\frac{(.4)\,(.6)}{n}}$$

$$.04 = \sqrt{\frac{(.4)\,(.6)}{n}}$$

$$.0016 = \frac{.2400}{n}$$

$$n = .2400/.0016 = 150$$

  Statement   If the sample average of 150 observations is 40% then I can
                  say with 68% confidence that the long run idle percentage is
                  between 36% and 44% with 40% being the most likely esti-
                  mate if the situation does not change.

148

Example 2
Given

Estimated idle percent of fork truck = 40%
Relative accuracy desired = ±10%
Confidence level desired = 95%

To Find
Solution

How many observations, n, are required?
(.1) (.4) = .04 = 4% (see Figure 9.3)
95% confidence equals ±1.96 standard deviations
Therefore:

$$.04 = 1.96\sigma_p$$

$$.04 = 1.96\sqrt{\frac{(.4)(.6)}{n}}$$

$$.0016 = 3.8416\frac{.24}{n}$$

$$n = .922/.0016 = 576$$

Statement

If the sample average of 576 observations is 40% then I can say with 95% confidence that the long run idle percentage is between 36% and 44% with 40% being the most likely estimate if the situation does not change.

Example 3
Given

Estimated percent of girls wearing sweaters = 40%
Relative accuracy desired = ±20%
Confidence level desired = 68%

To Find
Solution

How many observations, n, are required?
(.2) (.4) = .08 (see Figure 9.4)

$$.08 = 1\sigma_p$$

$$.08 = \sqrt{\frac{(.4)(.6)}{n}}$$

$$.0064 = \frac{.2400}{n}$$

$$n = .2400/.0064 = 37.5$$

Statement

If the sample average of 38 observations is 40% then I can say with 68% confidence that the long run percentage of girls wearing sweaters is between 32% and 48% with 40% being the most likely estimate if the situation does not change.

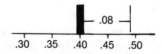

Figure 9.4 *Example 3* is shown in a "one-side" sketch.

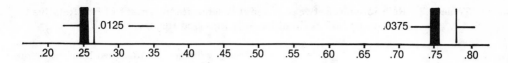

Figure 9.5 *Examples 4 and 5 are shown in a "one-side" sketch.*

Example 4

Given   Estimate percent of overcast days = 25%
              Relative accuracy desired = ±5%
              Confidence level desired = 95%

To Find   How many observations, n, are required?
Solution $(.05)(.25) = .0125 = 1.25\%$ (see Figure 9.5)

$$.0125 = 1.96\sqrt{\frac{(.25)(.75)}{n}}$$

$$n = 4618$$

In Example 4, we required "considerable" accuracy (5%) for a "small" target (25%). As proof that it is easier to aim at a big target, consider Example 5.

Example 5

Given   Estimated percent of days that are not overcast = 75%
              Relative accuracy desired = ±5%
              Confidence level desired = 95%

To Find   How many observations, n, are required?
Solution $(.05)(.75) = .0375$ (see Figure 9.5)

$$.0375 = 1.96\sqrt{\frac{(.25)(.75)}{n}}$$

$$n = (1.96)^2 (.25)(.75)/(.0375)^2 = 512$$

Table 9.2 shows the relation between p and n for a 5% value of *relative* accuracy and 95.45% confidence level. The table displays digitally the importance of not shooting at a small target. If we require ±5% relative accuracy about an occurrence with a mean of 1%, then we need 158,400 "shots" to hit the target (which has a width from .95% to 1.05%).

Examples 4 and 5 pointed out the importance of precise statement of the problem. Contrast, from the table, the number of observations required for an occurrence with $p = .1$ ($n = 14,400$) vs. a nonoccurrence of $p = .9$ ($n = 175$).

Example 6 works the problem after gathering the data with accuracy being the unknown and z being specified since p and n are by then known. You could specify A and calculate z if you wish.

Example 6

Given   In a completed study: Machines idle   1400
                                     Machines working   2600
                                     Total observations   4000

To Find   If I must be 95.45% confident of the answer, what is

Table 9.2 Relationship Between *p* and *n* When a Requirement of 5% Relative Accuracy and 95% Confidence is Imposed Regardless of *p*.

p (Decimal)	5% of *p*	Resulting Range about *p* (Target)	Required Number of Observations
.01	.0005	.0095 to .0105	158,400
.02	.0010	.0190 to .0210	78,400
.03	.0015	.0285 to .0315	51,700
.04	.0020	.0380 to .0420	38,400
.05	.0025	.0475 to .0525	30,400
.10	.0050	.0950 to .1050	14,400
.15	.0075	.1425 to .1575	9,070
.20	.0100	.1900 to .2100	6,400
.30	.0150	.2850 to .3150	3,730
.40	.0200	.3800 to .4200	2,400
.50	.0250	.4750 to .5250	1,600
.60	.0300	.5700 to .6300	1,070
.70	.0350	.6650 to .7350	685
.80	.0400	.7600 to .8400	400
.90	.0450	.8850 to .9450	175
.95	.0475	.9025 to .9975	85

a. The absolute accuracy level?
b. The relative accuracy level of percent idle?

Solution $p = 1400/4000 = .35 = 35\%$

$$A = 2\sqrt{\frac{(.35)(.65)}{4000}} = \pm.015 = \pm1.5\% \text{ absolute accuracy}$$

Relative accuracy $= .015/.35 = \pm4.3\%$ on machines idle

$= .015/.65 = \pm2.3\%$ on machines working

Statement From the sample of 4000 observations I am 95.45% confident that the long run idle percent is between 33.5% and 36.5% with 35% being the most likely estimate if the situation does not change.

The above procedure is based on the sequence of plan study, do study, analyze data. If you are willing to analyze the data as you gather data (i.e., sequential sampling), the sample size can be reduced in some situations.[3] Now let's turn to the second part of the problem, a representative sample.

REPRESENTATIVE SAMPLE

There are three problems in obtaining a representative sample: stratification, influence, and periodicity.

Stratification

If you are taking a sample on political opinion, it is desirable to keep data from different strata (Kansas vs. New York, old vs. young, rich vs. poor, male vs.

female, city vs. rural) separate. In the same way, when sampling for work design purposes, it is desirable to stratify your sample. Sample from morning and afternoon, first and second shift, Monday and Thursday, etc. Data can always be combined.

The same item of information can belong simultaneously to several strata. A political poll could identify the same person as belonging to the male stratum, the Kansas stratum, the college graduate stratum, etc. By proper planning and by identification of the data, it is possible to "reuse" the data; that is, predict votes for males, for Kansas, for college graduates, and so forth. Data on the fork truck could belong to several strata: the morning stratum, the first shift stratum, etc.

If the sample is not properly stratified, then you will fail to obtain potentially important information; more important, your sample will not faithfully represent the universe and the risk of making incorrect assumptions about the population is greatly increased. The sampled population should be a good representation of the target population—the population to which your decision will apply.

An example might help. In 1936 the *Literary Digest* took a poll to see whether Landon or Roosevelt would win. They took a random sample of the people listed in the phone book. The poll predicted a Landon victory! Roosevelt won big, because people who weren't able to afford a phone voted heavily for Roosevelt. The sample had been unbiased but it also had not been representative.

An anecdote of George Bernard Shaw points up the importance of assumptions when you have mixtures. Beautiful woman to GBS: "Let's get married and have children with my beauty and your brains." "What if they had my beauty and your brains?"

Stratified samples have a higher efficiency (a smaller number of observations is required for a given risk) than nonstratified samples. More precisely, if the percent of occurrence is constant in all strata, then efficiency is the same for stratified and nonstratified sampling.[5] The higher efficiency is due to the lower variance (error) when strata are used. Stratification might not help, but it can't hurt.

Using the fork truck example, assume you took samples of 100 on Monday, Tuesday, Wednesday, and Thursday. Assume further that the truck was idle 10% of the time on Monday, 30% on Tuesday, 50% on Wednesday, and 30% on Thursday. Overall, $\bar{p} = (10 + 30 + 50 + 30)/400 = .3$. The variance is $(.3)(.7)/400 = .000\ 525$ and standard deviation $= .023$. Using a confidence band of ± 2 standard deviations you would estimate the idle time as $30\% \pm 5\%$.

If you had kept your data stratified, you could predict idle times of .1, .3, .5, and .3. Variances would be .000 900, .002 100, .002 500, and .002 100; standard deviations would be .030, .046, .050, and .046. Your estimate of idle then would be $10\% \pm 6\%$ on Monday, $30\% \pm 9.2\%$ on Tuesday, $50\% \pm 10\%$ on Wednesday, and $30\% \pm 9.2\%$ on Thursday.

Influence

The second problem of a representative sample is influence. The event being sampled must not change its behavior because it is being sampled. In many situations, changing behavior is not a problem. If you are observing the color of passing automobiles, the auto will not change from green to red because you looked at it. When recording free-throw percentage in a basketball game, the player will not make or miss the shot because you are keeping score in the stands.

But, if you always sample only the top of the barrel, bad apples seem to avoid the top. If Harry Nardowell can see you coming on your sampling round, he may start working just before you record your observation. Even if Harry can't see you coming but can anticipate your presence since you always come at a specific time—you are periodic—the occurrence is not likely to be representative of the population.

Periodicity

Both the situation studied and the sample of the situation can be periodic or nonperiodic. For convenience we will call nonperiodic situations random even though they don't satisfy a mathematical definition of randomness. For engineering purposes they are sufficiently random.

The worst combination is a periodic situation and periodic sample with the same period length and initial point. For example, assume that Bill Kukenburger, the fork truck driver, every hour on the hour drives by the desk on the shipping dock where Lisa Nimtz works and checks the time on his watch vs. the clock above her desk. If Sam Helal, your time study technician, observed Bill only every hour on the hour, his study would show that Bill did nothing but set his watch. If Sam observed Bill only at 14 minutes after the hour, he would never know the importance Bill attaches to having his watch set correctly.

Fortunately, although some machine cycles such as cam-controlled machines are completely periodic, most situations are not. But the possibility of a periodic situation, although rare, should be considered.

The most common technique of avoiding problems of influence and periodicity of the situation is to make the sample have no pattern. The easiest way to make a sample without a pattern is to make the sample random.

Random Samples

What is desired is a random selection of times to do the sampling. Assume that it has been calculated (see the section on the required number of observations) that we need 10 observations on the first shift on Monday. That shift runs from 7:00 a.m. to 3:30 p.m. with 10-minute coffee breaks at 9:00 and 1:30 and a 30-minute lunch break at 11:30. Label the minutes of the shift from 1 to 530. We need to have a representative sample including 1 to 119, 130 to 269, 300 to 389, and 400 to 530.

The selection of the 10 random times from the 460 alternatives (530-10-30-10 = 460) can be done by putting 460 numbers on cards and drawing 10 or it can be done by using dice. It has been demonstrated many times that nonmechanical methods of selection have patterns. A very common pattern is selection of numbers divisible by 5. The most practical way of avoiding a pattern is to use a random number table.

Our next step then is to consult a random number table such as Table 9.3. We want to get random numbers which range from 001 to 480. Although the table is *printed* in blocks of three, the sequence is without pattern so the digits can be used in blocks of two, three, four, nine, or whatever.

To be precise, close your eyes and touch the table with a pencil; this selects the first number. Then if the first digit is odd, work up in the table; if it is even, work down. If you are not a purist, start in the upper left hand corner and work down; the sequence will be random, but it also will be the same sequence every time.

Let's assume your pencil touched the ninth number from the top of column 6. The number is 413. One of the 10 times to make an observation is minute 413.

Since the first digit, 4, was even, work down in the table. The next number is 168. An observation should be made at minute 168.

The next number is 679. This number does not match one of our possible observation times so it is ignored. Next is 807; it is ignored also. Next is 660; it is

Table 9.3   Random Numbers. All Digits Have Equal Probabilities. The Sequence of Digits Has No Pattern No Matter Which Direction You Move in the Table (Up, Down, Right, or Left).

055	946	090	448	484	262	866	709	215	965
377	581	299	769	989	571	093	274	080	345
237	314	819	383	771	826	432	461	290	888
426	456	446	502	940	674	067	984	296	147
058	314	689	338	028	326	355	013	649	130
604	693	293	677	885	237	010	607	790	854
328	936	541	717	374	919	214	734	912	564
798	775	751	834	129	780	432	725	086	256
451	370	364	974	131	413	863	702	462	622
206	720	296	942	836	168	233	219	872	571
679	552	230	488	685	679	177	806	287	646
865	692	160	848	614	807	929	802	832	944
667	018	105	282	789	660	445	003	735	862
551	514	984	310	208	101	432	620	094	792
235	587	038	871	121	942	074	328	623	632
414	337	184	222	776	380	271	105	779	582
093	586	647	215	391	907	499	906	809	678
902	721	537	183	856	687	118	632	834	231
989	222	232	477	170	171	712	650	011	654
742	979	974	710	082	326	884	474	392	281
118	501	436	502	856	956	883	429	643	548

ignored. The next number, 101, becomes the third time for an observation. The next numbers to be used are 380, 171, 326, 093, 432, 067, and 355.

The observation times selected are 413, 168, 101, 380, 171, 326, 093, 432, 067, and 355. Putting them in order gives:

067 = 8:07	168 = 9:43	326 = 12:26	380 = 1:20
093 = 8:33	171 = 9:46	355 = 12:56	413 = 1:53
101 = 8:51			432 = 2:12

Sometimes the same number occurs twice in the random selection. For example, the second number selected was 168. The ninth number also might have been 168 instead of 067. We can "double count" the 168 observation and make only 9 trips to observe the fork truck instead of 10; we use 10 observations in the calculations. This double counting does save one trip to the shop. On the other hand, it may confuse the technicians. It may even lead them to double count a few extra times to same themselves some trips when their feet hurt. Computers can be used to generate the random sampling times. Programs have been published by Weingast and by Whitehouse and Washburn.[15,16]

Randomization with restrictions. Stratified random sampling has three advantages over nonstratified random. (1) It is more efficient. (2) It is easier to calculate the observation schedule than with a purely random sample. Using our fork truck example, if we needed 200 observations in all, it is easier to calculate a schedule of 20 shifts with 10 observations each than when the observation schedule is randomized over the entire 20 shifts. (3) Additional observations can be added once the study has begun. If, after the study began, we decided to take 250 observations instead of 200, with stratification it is easy to add 5 more shifts. Without stratification, it is impossible to randomize over the entire period since some of it has already passed.

It is also possible to use other restrictions to help ensure a representative sample. These restrictions become important for small sample sizes (say n less than 30). For example, you might require that within the strata of each day at least one-third of the observations occur before and after the midshift break. Ideally one-half would occur before and one-half after but, due to the lack of pattern of a random number table, it is possible to get 8 of the 10 observations in the morning.

Maximum and minimum intervals between observations to ensure a representative sample usually are not desirable. If you are dissatisfied with your sampling plan because time between observations exceeds some maximum interval (say there is no observation for a 4 hr period), then your problem most likely is that your sample number is too small. Requiring a minimum interval is not good either. At the extreme, the minimum interval is zero; that is, one observation is counted twice. This is valid. Remember that what is being determined is the percent of occurrences rather than the number of occurrences. That is, you are trying to determine the percent of time the fork truck is idle rather than the number of times it is idle so observing one delay interval twice is satisfactory.

There may be restrictions in the schedule for ease of observation. For example, your observer may have a department meeting every Thursday morning from 9:00 to 10:00. You may schedule no observations during this time; the problem is that the sample has a systematic difference from the population. Engineering judgment must be used in these situations. The goal is not a statistically perfect sample; it is obtaining reasonably valid information with reasonable cost of information gathering so that valid decisions are made.

Periodic samples. We use the random sample when there are problems of periodicity or influence. Yet random samples are a lot of bother to calculate and tend to disrupt the observer's day since the observer's day is broken up into segments of various lengths. Observers would like to use periodic samples.[9]

The typical assumption is that the people being observed get tired of modifying their behavior if observed for a long enough time, so periodic samples can be used. An assumption! The periodicity of the situation and its phase relationship with the periodicity of the sample are also matters for engineering judgment. If you know the periods and their phases accurately, there is little need for the occurrence sampling study.

Nonetheless, if there are a very large number of observations, say 1,000, then periodic sampling may be satisfactory. My personal recommendation is to avoid periodic sampling since it adds an additional possible error to the data. Your basic desire is to get clean data rather than "probably" clean data.

DATA GATHERING

To further reduce the problem of influence, some organizations use a random device such as a coin or random number table to randomize the direction of approach to the situation. That is, sometimes you come from the left, sometimes right, etc. Several different observers may be used at different times. For these complicated situations, a tour log (Figure 9.6) may be useful.

For many situations a simple data form, such as that in Figure 9.7, is very satisfactory. The observer simply puts a tally mark in the appropriate space. At the completion of the study the total and percentages are calculated. By planning ahead, the raw data and calculations can be on the same sheet. This reduces errors and keeps everything together.

Thilgen and Procopio feel that data from most occurrence sampling is too general to accurately answer the questions "What are we doing?," "Who's doing it?," and "How is it being done?"[13] They recommend recording, for each observation on each individual, the function (responsibility) and method (means of accomplishment). For an engineering department, functions might be reports, parts analysis, expedite, suggestions, personal, etc., while methods might be file, order, telephone, out of department, etc.

Date	Time	Entry point	Observer
8/16	8:07	A	J
8/16	8:33	C	J
8/16	8:51	C	J
8/16	9:43	A	J
8/16	9:46	B	J
8/16	12:26	A	M
8/16	12:56	A	M

Figure 9.6   *Tour logs* help keep all the times, entry points, who will do the observing, etc. organized.

It may be that you have heard about the virtues of computers and the use of mark sense cards (Figure 9.8). The tally mark is put on an IBM card with a special pencil. A mark on a card, however, is not sufficient; the card must be identified with other information such as day of the week, morning vs. afternoon, machine studied, etc. This additional information also can be mark sensed by the observer; now the job is quite a bit more complicated than before when just one mark was needed. Another possibility is to prepunch each card with specific information such as day and morning vs. afternoon. Then the observer just takes the Monday morning card and marks on it *working* or *idle* in the appropriate location.

What's the benefit of all these cards? It's in the data processing. The computer now can calculate all the totals and percentages in one or two seconds. More important, however, is that since each observation is tied to various strata by being on the same card, a very detailed analysis of the data can be made.

The problem is the cost of setting up the card system, prepunching, getting the computer program working, waiting for your output, correcting mispunched cards, etc. An ordinary clerk using a form such as Figure 9.7 can tally 1000 observations and calculate totals and percentages in a couple of hours. Setting up the procedure to be done on a computer, even if the program is written, probably requires a day's time of an engineer. Thus, unless you have time and money to spare or the situation studied is very complex, let the clerk do the work.

As you will discover as soon as you make your first study, the real expense of a sampling study is the time required by the observer. Nothing else seems to get done during this time by the observer. This has led to automation—let a machine do the work.

In this case, the machine is a camera, a "spy in the sky." A camera is placed in the area to be observed. Then the camera takes a picture every so often. (One roll of 8-mm film at time lapse speed is like having 400 slides of an activity in proper succession.) Cost of TV tape/hr is so much lower than movie film that TV records can be continuous rather than time-lapse. Although it is possible to equip these cameras with devices that make the picture taking random, in general most organizations do not feel the extra expense is worth it and so they just take pictures periodically. They feel that since the operators do not know when the camera is working there is no influence problem. Since you will be making many observations (say 1,000 or more), the periodicity is not likely to be a problem. (You must be going to take a lot of observations or why bother with the camera?)

Situation Clark Fork Truck Observer SK Date 8/17/82

Period	TALLY Working	Idle		TOTAL			PERCENT		
		Driver absent	Driver present	Work	Idle (ab)	Idle (pr)	Work	Idle (ab)	Idle (pr)
1	1111	111	1	4	3	1	50	37	12
2	~~1111~~ 111			8			100		
3	~~1111~~ 111			8			100		
4	~~1111~~ 111			8			100		
5	~~1111~~ 111			8			100		
6	~~1111~~ 111			8			100		
7	~~1111~~ 111			8			100		
8	~~1111~~ 111			8			100		
9	~~1111~~ 111			8			100		
10	1111		1	4	3	1	50	37	12
TOTAL				72	6	2	90.0	7.5	2.5

Figure 9.7 *With forethought,* the data analysis can be done on the same sheet on which the data are recorded.

There are some problems. Improved films that use available light usually work although in some dark areas the camera just does not have enough light. Sometimes there are problems of keeping the object in focus. You might need many cameras (e.g., the fork truck study). The primary problem, however, is worker reaction. As much as people hate to have an engineer peer over their shoulder, they dislike a machine "spying" on them even more. If you plan on many studies and thus can amortize the cost of the cameras, and worker reaction is not a problem in your organization, consider use of cameras to gather your data.

DATA ANALYSIS

If you have stratified your sample, you can use two different data analysis techniques. If you have no strata (subgroups) then you cannot use these two techniques.

Comparison Between Two Strata

It may be desirable to compare the occurrences between two different strata to see if there is a statistically significant difference. The following procedure

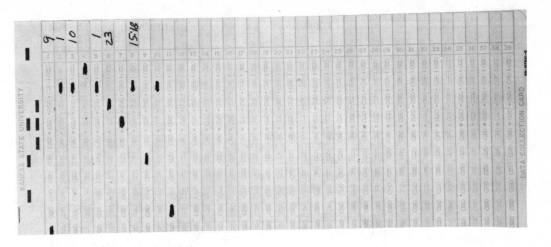

Figure 9.8 *Use a mark sense card* for occurrence sampling if you wish to get maximum use of the data. For the card above, column 1 was for the machine number and 2 for machine activity (1 = run, 2 = setup, 3 = idle, 4 = idle, unavoidable). Columns 3 and 4 were for operator activity (10 categories). Column 5 was shift, 6 and 7 were day of the month, and 8-12 were time (24-hour clock). The observer wrote the numbers down and then mark sensed later; this procedure gives more accurate data than trying to mark sense "at the observation." The machine activities and operator activities then were analyzed by time of day, machine, and shift. Note that if a tally sheet (such as Figure 9.7) is used, an observation is used only once while with the mark sensing an observation can be used multiple times to show the effect of time of day, shift, day of month, machine number, etc.

assumes that the total observations are the same for both strata; see the same reference for tables for unequal sample sizes.[11] Allen and Corn present a chi-square procedure which can be used for unequal sample sizes.[1]

Put the data in the following format:

	Class I	Class II	
Sample from strata A	r_a	s_a	r = number of occurrences
Sample from strata B	r_b	s_b	s = number of nonoccurrences
			$n = r + s$ = number of observations

For $n_a = n_b$

Example 1
Given $n_a = n_b = 16$

Strata	Times Idle	Times Working
Monday	9	7
Tuesday	1	12

To Find   Is there a difference between Monday's and Tuesday's data?

	Procedure	Number
Solution	1. Pick the smallest of the four numbers.	1
	2. Select the other number in the same column.	9
	3. Calculate "observed contrast."	$9 - 1 = 8$
	4. From Table 9.4, enter at sample size row, go across to the smallest number, or, if it is not given, the end of the row. Then go up to "minimum contrast."	16 to 1 to 6 need contrast of at least 6
	5. Compare observed vs. minimum	$8 > 6$ Therefore Tuesday's data are different from Monday's ($\alpha < .05$)

Example 2
Given

$$n_a = n_b = 75$$

Strata	Times Phone Busy	Times Phone Not Busy
1st Shift	45	30
2nd Shift	15	60

To Find

Is there a difference between the data for the first shift and second shift?

	Procedure	Number
Solution	1. Pick the smallest of the four numbers.	15
	2. Select the other number in the same column.	45
	3. Calculate "observed contrast."	$45 - 15 = 30$
	4. From Table 9.4, enter at sample size row, go across to smallest number and go up to "minimum contrast."	No row for 75 so use 70 and 80 and interpolate. 70 to 15 to 12 80 to 15 to 12 Thus minimum contrast is 12.
	5. Compare observed vs. minimum	$30 > 12$ Therefore first shift differs from second shift ($\alpha < .05$)

Control Charts

This test evaluates the effect of time (sequence) for your various strata (subgroups).

A special form of analysis of variance called a control chart was developed by Walter Shewhart during the 1930s. It has been used extensively in quality assurance work. A typical application in quality assurance would be monitoring a quality characteristic (say percent of parts with scratched paint). The percent of parts with scratched paint depends on many influences (operator, machine, type of paint, etc.), so you might think that an analysis of variance is desirable. However, in practice, there is a very strong relationship to sequence or time. That is, if you can identify *when* something occurred, you can do the detective work to find out *where and why* it happened. The control chart helps identify *when* the process changes.

The same situation applies to occurrence sampling. Although attribute control charts and measurement control charts are used in quality assurance, occurrence sampling records the data in a discrete manner (yes-no, working- not working), so we use an attribute control chart. In fact we use only one of the various types of attribute control charts—the percentage chart or p chart.

The first step in the construction of a p chart is to plot the center line of the chart, $\bar{p}$. The average percentage is calculated by totaling the number of occurrences divided by the total number of observations.

The second step is to estimate the variability from the average that could occur by chance. The spread of the distribution is quantified by the standard deviation of p, symbolized σ_p. (Note that the n of equation 9.1 used for calculating the spread uses the sample size, not the total study n.) If all the individual readings were "slid along a wire parallel to the center line," then they would form a histogram (see Figure 9.9).

The histogram describes the variability that could occur by chance. Variability beyond the histogram is assumed to occur not by chance but *from a cause.*.

How far does the histogram spread? In theory, from negative infinity to positive infinity. However, the odds get quite small once we get "out a ways" from the center. If we go out $\pm 3\sigma_p$ in either direction from the mean percentage, there are "very few" chance occurrences; if we go out $\pm 2\sigma_p$ there are "few"; if we go out $\pm 1.5\sigma_p$ there are "some," etc.

First, let's define the "very few," "few," etc. Although it's not statistically correct to use the normal as an approximation to the binomial for small sample sizes, the normal is commonly used because it is simple to use and gives answers that are "close enough." Limits including $\pm 3\sigma_p$ include 99.7% of the chance occurrences; limits including $\pm 2\sigma_p$ include 95.45% (usually rounded to 95%); limits including $\pm 1.645\sigma_p$ include 90%. The trade-off is looking for trouble where none exists vs. not looking for trouble when trouble exists. Quality assurance limits commonly are $\pm 3\sigma_p$ although the chemical industry uses $\pm 2\sigma_p$ since the consequences of a change in the process are severe in the chemical industry. If a point is beyond the $3\sigma_p$ limits, the odds are 3/1000 that the point is beyond the limit due to chance and 997/1000 that the sample differs from the population.

For occurrence sampling, my recommendation is $\pm 2\sigma_p$ or even $\pm 1.645\sigma_p$ limits. If a point is beyond $2\sigma_p$ limits, then the odds are 1/20 that chance has occurred and 19/20 that the sample differs from the population. The odds are still very good for $\pm 1.645\sigma_p$ limits—1/10 for chance and 9/10 for cause.

Figure 9.10 shows an example chart. In constructing the chart use all the observations to calculate the average, $\bar{p}$. *But,* the standard deviation for the limits uses the sample size n. That is, for a plot of 10 days, each an average of 50 occurrences, use $n = 500$ for the calculation of $\bar{p}$ but $n = 50$ for the calculation of σ_p. If n varies from sample to sample, use the average number. Note also that the limits cannot exceed 0% or 100%. Thus if $\bar{p} = 60\%$ and $2\sigma_p = 70\%$, the upper limit would be 100% and the lower limit 0%.

APPLICATIONS OF OCCURRENCE SAMPLING

The most common applications are to provide information for management decisions.[2,8,12] Examples might be work vs. idle time of inspectors, nurses,

Table 9.4 Relationships of Sample Size, Smallest Number (In Table), and "Minimum Contrast" for Tests Between Two Percentages. Binomial Distribution Used. Significance Level of $\alpha = .05$ or 2 Tails: $\alpha = .025$ for 1 Tail.

Minimum Contrast for Statistical Significance

Sample Size $n_a = n_b$	4	5	6	7	8	9	10	11	12	13	14	15	16	17	18
4-5	0														
6		0													
7-9		0-1													
10-11		0	1-2												
12-13		0	1-3												
14		0	1-2	3											
15-16		0	1	2-4											
17-19		0	1	2-5											
20			1	2-5	6										
30			0	1-2	3-5	6-10									
40			0	1-2	3-4	5-7	8-15								
50			0	1	2-3	4-6	7-10	11-19							
60			0	1	2-3	4-5	6-8	9-13	14-24						
70			0	1	2-3	4-5	6-8	9-12	13-18						
80			0	1	2-3	4-5	6-7	8-11	12-15	19-28					
90			0	1	2-3	4-5	6-7	8-10	11-14	16-23	24-33				
100			0	1	2-3	4	5-7	8-10	11-13	15-20	21-31	32-37			
150			0	1	2	3-4	5-6	7-9	10-12	14-18	19-25	26-42	26-32	33-41	
200			0	1	2	3-4	5-6	7-8	9-11	13-15	16-19	20-25	23-27	28-33	42-66
300			0	1	2	3-4	5-6	7-8	9-10	12-14	15-18	19-22	21-24	25-29	34-41
400			0	1	2	3-4	5	6-8	9-10	11-13	14-17	18-20	21-24	25-28	30-35
500			0	1	2	3-4	5	6-8	9-10	11-13	14-16	17-20	20-23	24-27	29-33

Sample Size	**19**	**20**	**21**	**22**	**23**	**24**	**25**	**26**	**27**	**28**	**29**	**30**	**31**	**32**
200	42-51	52-65	66-89											
300	36-41	42-48	49-56	57-66	67-68	79-95	96-137							
400	34-38	39-44	45-51	52-58	59-67	68-76	77-87	88-100	101-117	118-141	142-185			
500	33-37	38-42	43-48	49-55	56-62	63-70	71-79	80-89	90-100	101-113	114-128	129-147	148-172	173-234

doctors, teachers, supervisors, setup operators, material handling equipment, operating equipment, etc. In many studies, the categories of activity and of the items observed are not split into two or three but ten to twenty. That is, you might observe 10 different nurses and divide their activity into 18 different categories rather than observe just 1 nurse and divide her activity into 2 categories.

A slightly different application is to set time standards.[4,6,7] It requires a count of the units produced during the time period as well as the occurrence sample.

Assume you had the data for a secretary shown in Table 9.5. From the sample, percent of time times 40 hr gives the third column, the estimated hours spent on the activity. Then the fifth column, estimated hr/unit, can be calculated by dividing by output. Note that standards need not be set on the entire list of activities during the week. Note that the worker's work method is assumed to be good; that is, we are setting time standards, not improving productivity.

An advantage of time standards from occurrence sampling is that the study can cover a large number of cycles of varied work. A disadvantage is that no performance rating was used. Performance rating of a crude nature can be done if the observer, when making the observation (allocating the activity to phone call, typing letters, typing forms, etc.), simultaneously estimates a performance rating (90%, 120%, etc.). It should be recognized that the accuracy of the resulting standard will be improved only slightly by using this type of performance rating since instantaneous rating necessarily must be crude. Kinack recommends using conventional time study with performance rating for the important activities and using occurrence sampling for the minor activities; the rating from the important activities (say 90%) is assumed to be valid also for the occurrence sampling activities.[10]

Table 9.5   Example Data Showing How Occurrence Sampling Data Can Be Used to Determine Time Standards

Activity	Percent of Time Spent on Activity	Estimated Hours/Week on Activity	Output for Activity	Estimated Hours/Unit
Phone calls	12	4.8	147	.033
Letters	28	10.2	48	.233
Form A	3	1.2	121	.010
Form B	3	1.2	12	.100
Other	54	22.6		
		40.0		

SUMMARY

Steps in making an occurrence sampling study:

Step	Example
1. State the problem:	Determine idle percent for fork trucks in Dept. 8.
a. Give application of data.	Aid decision whether to buy new fork truck.
b. Determine categories.	Truck will be considered to be: • working • idle—driver absent • idle—driver present

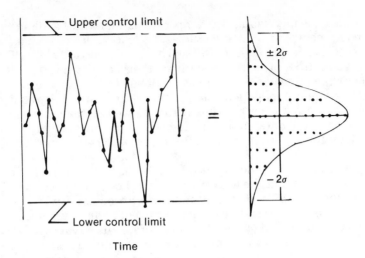

Figure 9.9 *Control charts* are a specialized form of analysis of variance which show the effect of time. Without the time scale it is a histogram (figure on right). The control limits indicate how much variability is likely to occur by chance. If a point falls outside $\pm 2\sigma$ limits, the odds are about 5% that nothing is different from the typical situation and about 95% that something has changed.

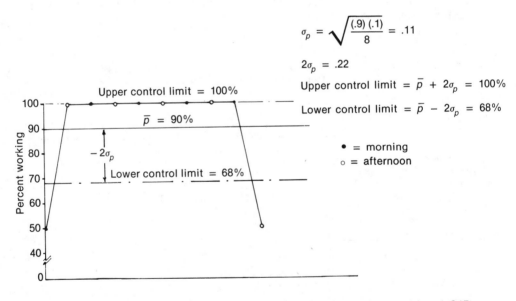

$$\sigma_p = \sqrt{\frac{(.9)\,(.1)}{8}} = .11$$

$2\sigma_p = .22$

Upper control limit $= \bar{p} + 2\sigma_p = 100\%$

Lower control limit $= \bar{p} - 2\sigma_p = 68\%$

• = morning
o = afternoon

Figure 9.10 *Control charts for occurrence sampling* should use limits either 1.645 or 2σ from the mean rather than the 3 used in statistical quality control as the cost of a type 1 error (looking for a problem when none exists) is much lower in occurrence sampling. Use all the observations in calculating the mean but only the sample size for the limits.

Step	Example
2. Calculate sample size required.	Estimate percentages are:

Estimate percentages are:
- working   75%
- idle—driver absent   10%
- idle—driver present   15%

At this stage estimates are just the best guess.

$A = 5\%$ (plus and minus) on all three
$c = $ confidence $ = 68\%$ on all three

Required n are 75, 36, and 51,
respectively, so use $n = 75$.

3. Plan to obtain a representative sample.	Take study during one entire week; half before lunch, half after.
	Divide week into 10 periods of 4 hours.
	For simplicity, take 8 observations per period for 80 total observations. Make up random times for 10 periods. Walk into observation area by one of the three entrances—specific entrance chosen randomly.
	Make up data form.
4. Make study.	See Figure 9.8.
5. Analyze results: a. Strata	a. See Figure 9.7. $\bar{p} = .90$; $A = .034$. Does not seem to be a morning vs. afternoon effect even though all idleness does occur on only 2 of the 5 days. The observed contrast is 4; required contrast is 5.
b. Time	b. See Figure 9.10. It seems there is a problem at the beginning and end of the week as points are beyond the control limits. The $2\sigma = 95\%$ assumption is not valid for small samples; our n was 8, not the 30 needed for the assumption of normality.
	A check shows production control was unorganized at beginning of the week due to late computer printout; not enough work scheduled on Friday due to lack of knowledge of what would be made the following week.
6. Make conclusion.	I am 68% confident that the long run working percent is between 87% and 93% with 90% the most likely estimate if nothing changes.
7. Make decision.	Make two decisions: a. Get production control organized. b. Don't buy new fork truck.

SHORT-ANSWER QUESTIONS

9.1   What is the purpose of occurrence sampling?

9.2   Does information from a sample increase proportionally to sample size?

9.3   The number of observations to take in occurrence sampling depends upon what three things?

9.4   Briefly describe the three problems of obtaining a representative sample.

9.5   Discuss how a time standard could be set using occurrence sampling.

THOUGHT-DISCUSSION QUESTIONS

1. Discuss the trade-off between information from an occurrence sample and cost of obtaining the information.
2. Discuss the advantages and disadvantages of using mark sense cards for occurrence sampling.

REFERENCES

1. Allen, D., and Corn, R. Comparing productive activities at different times. *Industrial Engineering,* Vol. 13, No. 3 (1978): 40-43.
2. Barnes, R. *Work Sampling.* New York: Wiley & Sons, 1957.
3. Buck, J., and Tanchoco, J. Sequential Bayesian work sampling. *American Institute of Industrial Engineers Transactions,* Vol. 6, No. 4 (1974): 318-326.
4. Connor, D. Clerks generate their own performance reports. *Industrial Engineering,* Vol. 2, No. 7 (1970): 28-34.
5. Davidson, H. Work sampling—eleven fallacies. *Journal of Industrial Engineering,* Vol. 11, No. 1 (1960): 367-371.
6. Gibson, D. Work sampling monitors job-shop productivity. *Industrial Engineering,* Vol. 2, No. 6 (1970): 12-25.
7. Frank, E. Low-cost standards for indirect labor. *Industrial Engineering,* Vol. 2, No. 8 (1970): 27-28.
8. Heiland, R., and Richardson, W. *Work Sampling.* New York: McGraw-Hill, 1957.
9. Hines, W., and Moder, J. Recent advances in systematic activity sampling. *Journal of Industrial Engineering,* Vol. 16, No. 5 (1965): 295-304.
10. Kinack, R. Activity evaluation technique—a quick and easy procedure for developing time standards. *Journal of Industrial Engineering,* 18 (May 1967): xiii-xvi.
11. Natrella, M. *Experimental Statistics,* Chapter 8, Handbook 91. Washington, DC: Superintendent of Documents, 1963.
12. Pitsch, R. Auditing incentive plans. *Industrial Engineering,* 8 (February 1976): 20-23.
13. Thilgen, G., and Procopio, J. Function-method approach to work sampling. *Journal of Industrial Engineering,* Vol. 18, No. 3 (1967): xv-xvi.
14. Tippett, L. Statistical methods in textile research. Uses of the binomial and Poisson distributions. A snap-reading method of making time studies of machines and operatives in factory surveys. *Shirley Institute Memoirs,* 13, (November 1934): 35-93. (Reprinted in *Work Sampling* by Barnes.)
15. Weingast, M. Random samples. *American Industrial Hygiene Association Journal,* Vol. 42, No. 1 (1981): A-15.
16. Whitehouse, G., and Washburn, D. Work sampling observation generator. *Industrial Engineering,* Vol. 13, No. 23 (1981): 16-18.

10

CRITERIA

ORGANIZATIONS

Goals

Survival is the first rule of any organism, whether bacterium, insect, man, corporation, or state. Carrying the analogy further, the organism must have sufficient nutrition; to an organization, money is the food. A surplus of income over outgo (called *profits* in capitalist countries and a *net favorable balance* in socialist countries) is necessary in the long run. In the short run, a "diet" will not kill the patient, if the diet is not too severe or prolonged. Normally organizations set prices at a level which more than covers costs. If the price charged is higher than the market can bear or which the government permits, then either costs must be reduced or starvation begins. One exception is transfusions of public funds; this, in effect, has the taxpayer (a nonuser of the goods or service) pay some of the cost of the goods or service.

Growth is the second rule of an organism. Biological organisms, however, mature, stop growth, and die. Social organisms, such as auto manufacturers, universities, hospitals, and governments, are composed of "replaceable parts" and attempt to grow, grow, grow. Isn't a university with 15,000 students better than one with 7,000? Isn't a firm with 20,000 employees better than one with 10,000? Isn't a hospital with 500 employees better than one with 250? Isn't a bureau with 1,000 employees better than one with 500? Although some may not feel that bigger is better, more employees do give more power, prestige, status, and income to higher managers of organization than do smaller numbers. Since the higher managers set the organization's goals, their rewards are what count. Thus, number of employees is a very important managerial criterion. Although not even admitted to exist as a criterion, it is a rare manager who would not prefer to supervise 2,000 than 1,000.

A more publicly proclaimed goal (related to survival) is a larger net income; that is, profits of $1,000,000 are considered better than profits of $200,000. A more sophisticated criterion is (net income)/(net assets); that is, return on investment of 10% is better than return on investment of 5%. See Table 4.5 for some returns on investment of major organizations.

Limits

External restraints are imposed by society (usually governments and unions) but also sometimes by "public opinion" or "moral values." For example, employees are not permitted to work in toxic environments; wages must exceed specified minimums; products must meet certain safety and health requirements; the community environment may not be degraded by air or water discharges; employees may not be discharged without a penalty, etc. Job tenure in Japan is an example of a self-imposed "moral" value limiting organizational freedom. The private firms hire some individuals for life; a mere lack of enough work to keep the employees busy is not justification, in the minds of the managers, for their discharging the workers or decreasing their pay. (They also have "temporary" workers with no security.) Many other decisions within their organizations follow from the Japanese premise that labor is a fixed cost, not a variable cost.

An *internal restraint* on the organization's goals of survival and growth is the periodic evaluation of managers. This paradox that periodic evaluation interferes with achievement of the goals is due to the short-term nature of most evaluations. Supervisors continually ask those supervised: What are you doing? Will we make a profit *this* year? Will sales meet the target *this* month? Will we make schedule *this* week? How many did we make *today*? Why isn't that machine operating *now*? The status of the organization in 10 years becomes relatively irrelevant to managers who want to survive in their jobs for another month.

INDIVIDUALS

The organization goals of survival and growth are affected also by goals of individuals: its employees and its customers. Customer goals and their relationship with the organization are beyond the scope of this book; employee goals are discussed below.

Just as organizations want their benefits from each employee to exceed the employee cost (i.e., to have income/employee of $20,000 while expense/employee is $15,000), the employees want their benefits to exceed their contribution. Employees contribute their labor. This labor contribution has two dimensions: quantity and quality. *Quantity* is hours worked; employees, to earn $50, prefer to work 6 hr than 8 hr. *Quality* of labor contribution has, from the worker's viewpoint, both physiological and psychological-social aspects.

Physiological Aspects

Physiological aspects of working are discussed in more detail in various chapters. Chapter 12, "Work Physiology," describes stress on the cardiovascular system (usually quantified by metabolic rate or heart rate) and stress on the skeleto-muscular system through manual material handling. On a more local scale, problems of individual muscles are discussed in chapter 14, "Workstations," and chapter 15, "Handtools." Chapter 19, "Toxicology," discusses the problems of chemical hazards and safety. The eyes and illumination are covered in chapter 17; the ears and noise are covered in chapter 18. Chapter 20, "Climate," describes the effect of task and environment on body temperature.

Psychological-Social Aspects

Maslow's "stairs" give perspective to the problem (see Table 10.1). Maslow said there is a hierarchy of individual wants. *Physical wants,* the lowest level, concern such basics as the want for food, shelter, and health. Once these physical wants have been satisfied, the second level of wants, *security wants,* becomes important. In job terms, security wants might be having seniority on a job or having a supervisor who doesn't threaten you. The third level, *social wants,* becomes important after the second level is satisfied. Work examples of social wants are a job with status, having a job that you enjoy, having a job with friendly co-workers, or working in a physical location in which you can talk with fellow workers while working. The fourth level, *ego wants,* concerns challenge and achievement. Does the job challenge you? Do you have a feeling of contribution or are you "just a number"? The fifth level, *self-actualization wants,* concerns personal fulfillment and realization of potential. For example, is the organization "serving mankind" or are you merely making common items such as soap, autos, or chairs? Satisfying the fifth level may call for a "missionary" type of endeavor such as "ecology," "religion," or its secular equivalent, "politics"; some find it in teaching, music, or in running their own businesses.

One question is how relevant are these wants to job design? Which of these wants, if any, are to be satisfied by the job and which outside of the job? As was pointed out in Table 3.3, males tend to work about 80,000 hr during their lifetime. In general they support a family from the earnings of these 80,000 hr as well as themselves and their wives during retirement. (In the USA, increasing participation of women in the employed labor force probably has added 20,000 to 30,000 hr/family.)

Employees may prefer income from a monotonous, boring, dirty, dead-end job, if they can own a sailboat and a Mercedes. Others may prefer less income and drive a used VW but enjoy the status of a "white collar" job.

Should an engineer be concerned with satisfying those higher wants on the job? Some points for discussion are:

1. There are dead-end jobs which must be done by someone. Idealists, perhaps with sounder hearts than heads, proclaim they would not be willing to work on an assembly line; this is true due to self-selection. Their opinions do not seem to be shared by those who actually work on assembly lines, probably because assembly line work is well paid and they are willing to trade satisfaction off the job for satisfaction on the job.

2. Societies over the world more and more guarantee the basics of life (food, shelter, clothing) whether an individual works or not. In the USA in 1975, 16% of all income was government transfer payments. In addition, net income of unemployed (due to tax policies and benefits available to those not working) often is 70% to 90% of working income after taxes. This has led to *high* seniority workers demanding that they be laid off first. Thus, if a job is dirty, boring, monotonous, dead-end, *and low paid,* finding employees is more and more difficult.

3. With increasing affluence and progressive tax structures, additional income becomes less meaningful. How motivating is additional income to a person whose primary problem is, on his Paris vacation, whether to stay at a two-

Table 10.1 Maslow's Stairs. Maslow Called Each Step a "Need" But I Prefer "Wants." A Satisfied Want Is Not a Motivator.

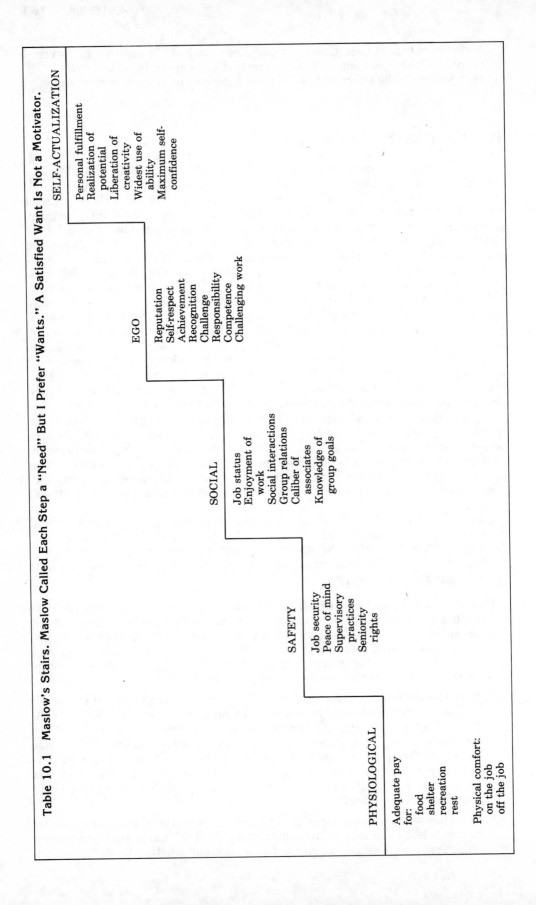

SELF-ACTUALIZATION

Personal fulfillment
Realization of
   potential
Liberation of
   creativity
Widest use of
   ability
Maximum self-
   confidence

EGO

Reputation
Self-respect
Achievement
Recognition
Challenge
Responsibility
Competence
Challenging work

SOCIAL

Job status
Enjoyment of
   work
Social interactions
Group relations
Caliber of
   associates
Knowledge of
   group goals

SAFETY

Job security
Peace of mind
Supervisory
   practices
Seniority
   rights

PHYSIOLOGICAL

Adequate pay
for:
   food
   shelter
   recreation
   rest

Physical comfort:
   on the job
   off the job

star or three-star hotel? When one auto worker was asked why he worked only four of five days, he replied "I can't make it on only three!"

4. A satisfied want does not motivate. With job enrichment it may be possible to design jobs to motivate workers; the motivated workers may work harder. However, unions, in general, oppose the concept of motivation; they favor working smart, not working hard.

5. Need there be a trade-off between fulfilling jobs and productive jobs? Can we "have our cake and eat it too"?

Examples of Satisfying Higher Wants

Security wants. In different cultures, managements have different approaches to job security. Large Japanese firms, for example, hire males at age 18 to work without layoffs until they are 55 (nenkō policy); their labor is considered a fixed cost. Since pensions don't start until 60, many hire back into their jobs at 55 as "temporary" workers (at ⅔ pay). Production fluctuations thus do not affect the permanent workers but do affect women, temporary workers, and subcontractors. The workers, in return, shift jobs freely (wages depend upon age, not the specific job), make methods and quality analyses to improve their jobs during off-job hours, often return early from vacations, etc. A side effect is that since most costs except material are fixed costs, Japanese firms continue production almost regardless of the sales volume or price/unit. Japanese managers say they cannot afford to discharge their most valuable asset in temporary downturns since they must compete vs. "cheap Asian labor" from other countries.

In the USA, Procter and Gamble, for example, guarantees 2,000 hr of work/yr for all permanent employees (those considered satisfactory after a 90-day probation period). IBM has a "no layoff" policy. The auto industry, however, guarantees income instead of work. If the worker is laid off, "supplemental employment benefits" pay up to 85% of wages for a year.

In the area of supervisory practices and seniority rights, unions have had a traditional concern. More recently US governmental agencies have exerted considerable pressure for less discrimination based on race, creed, color, sex, or age. Although organizations have complained about the extraordinary amount of paperwork now required and although some arbitrary and discriminatory practices still occur occasionally, supervisory practices have become a relatively minor problem.

Social wants. In some Western societies, employees are gaining more say in the running of the business. Most dramatic is the "Mitbestimmung" of West Germany; the concept is of joint command, of two hands on the tiller. Employees of firms with more than 2,000 employees elect almost 50% of the members of the Aufsichtsrat, the supervisory board that hires and fires the senior executives who make up the Vorstand, the executives who run the firm. Workers also elect a Betriebsrat (works council); it, rather than the union, is the focus of shop-floor grievances, negotiates with management on changes in production organization, has certain veto rights over hiring and firing. Other northern European countries have a form of Mitbestimmung with a smaller proportion of employee representatives. In England, where there is a strong "us vs. them" feeling, unions and management both reject Mitbestimmung as it requires cooperation rather than confrontation. Japanese organizations have a decentralized decision-making procedure with many different individuals "fixing their seal" on a proposal. There is a strong attempt to modify a proposal until there is 100% agreement. The Eastern bloc countries of course emphasize the idea of central authority, and workers have no voice in any decision. Yugoslavia, the maverick Communist state, has a quite successful system of "worker self-management."

In the USA there has been little formal sharing of authority between employ-ees and management but considerable informal sharing due to the changing nature of work. In 1975 the categories *professional, technical,* and *kindred* made up 15.2% of the work force and *managers* and *administrators* another 10.6%—a total of almost 26%—while *nonfarm laborers* declined to 4.9%. The changing skill levels are shifting the "shape" of the work force from a pyramid to a diamond.

At a superficial level, job titles can be changed to enhance job status. Changing janitor to building superintendent and garbage collector to sanitary engineer already may be known to you. In Texas, however, the Dallas School district calls school buses "motorized attendance modules" and their drivers "instructional facilitators." The Texas Secretary of State office advertised for an "administra-tive technician"; applicants "must have a demonstrated capacity to answer the phone when it rings without having to be told." When individuals design their own jobs, when does "individual style" become incompatible with "order" and "efficiency"?

Chapter 13 discusses many of the engineering aspects of job design; Principle 1 discusses specialization, Principle 3 assembly lines, and Principle 7 social inter-action.

Ego wants. Work can be made more challenging either by adding *variety* or by adding *responsibility*. Adding more variety has been given the name *job enlarge-ment;* examples would be to have a secretary run the duplicating machine as well as type, or to have an assembly line worker insert screws *and* tighten the screws rather than just insert *or* tighten. Adding more responsibility has been termed *job enrichment;* examples would be to have a secretary compose a letter as well as type it or to have an assembly-line worker assemble *and* inspect rather than assemble *or* inspect.

Do workers want enlargement or enrichment? Douglas Frazer, United Auto Workers president, stated: "No one has ever come up to me and asked the union to help enrich his job." Job enrichment is contrary to the traditional union approach of standard jobs, working conditions, and standard pay for all. Advo-cates of job enrichment often state that the workers will be more motivated and thus more productive. This "work hard not smart" philosophy is so irritating to unions that William Winpisinger, president of the International Association of Machinists said: "Job enrichment is a stopwatch in sheep's clothing." For example, Ford of American Telephone and Telegraph advocates enriched jobs, citing "reduced training costs."[3] Less job specialization also gives management more flexibility in job assignments and may reduce union jurisdiction problems. Salvendy commented that some people like to work at enlarged jobs; others pre-fer simplified jobs; still others do not like work of any form![4]

Bennett divided jobs into four categories: *physical* (carry, lift), *procedural* (operate, follow procedures), *social* (talk, answer), and *cognitive* (decide, answer).[1]. Job enrichment seems most interesting to social and cognitive workers. Physical and procedural workers tend to be uninterested in more challenge and respon-sibility on the job; they tend to focus their lives on their off-job hours rather than the minority of hours they spend on the job. It may be just as well because, due to extensive education (in the USA in 1975 a median of over 12 years of school), truly challenging work for everyone may be an impossible dream.

Self-actualization. In relatively few organizations do workers "throw their whole selves in." Japanese industry may be an exception. In 1974 the Matsushita Electric worker's song "Love, Light and a Dream" replaced the previous song "For the Building of a New Japan" which had been written in 1946 when rehabili-tation of the war-devastated economy was the national goal.

The song is sung at the "morning meeting," begun in 1933. Each section (10-20 people) has its own meeting, which lasts 10-15 minutes. On each day, one em-ployee, in turn, goes up to the platform and reads the company creed and seven objectives which all the others recite. Then he or she makes a 3-5 minute speech on any subject (hobbies, family, job, etc.) Then anyone else who wants to talk can do so. The meeting is closed by singing the company song. The purpose is to: (1) start the day with a refreshed recognition of the company mission, (2) to have everyone learn to speak thoughts in public, (3) to improve communication among the group, and (4) to improve the group's knowledge of each other. In 1976, the 83,000 Matsushita employees turned in 912,717 written suggestions. That's participation! American firms had company songs but they no longer are sung. Another Japanese example is "Jishu Kanri" activities.[5] In 1976 there were 170 steel plants with a total of 227,000 employees in 31,000 groups. Groups of 7-8 meet several times per month. In the steel industry, 25% of meetings were outside regular hours, 21% were during waiting time, 17% during the morning gathering time, etc. Subjects were cost 19%, safety 19%, equipment 17%, quality 16%, efficiency 12%, management 6%, errors 5%, and others 6%. Technical techniques from industrial engineering and quality control are taught by engineers to the groups which then apply the techniques to their own jobs. The results are presented (written and oral) to the others; honor and recognition are given to the group.

Love, Light and a Dream

A bright heart overflowing
With life linked together,
MATSUSHITA DENKI.
Time goes by but as it moves along
Each day brings a new spring.
Let us bind together
A world of blooming flowers
And a verdant land!
In Love, Light and a Dream.

We trust our strength together in harmony
Finding happiness,
MATSUSHITA DENKI.
Animating joy everywhere,
A world of dedication,
Let us fulfill our hopes,
Shining hopes,
of a radiant dawning.
With Love, Light and a Dream.

Lyrics by Shoji Miyazawa
Music by Kozaburo Hirai

WORK DESIGN CRITERIA

How much importance to give to organization goals and how much to give to individual goals depends upon the specific situation. The following is based on Bennett's suggestions.[2]

Safety and health are first. No job design is acceptable which endangers the workers' safety or health. However, life does not have infinite value. Decisions to mine coal, build buildings, fly airplanes, drive cars—all cause deaths. Managements must take "reasonable" precautions; of course, the definition of "reasonable" is debatable. Over the years, more emphasis has been put on safety. See Table 10.2.

Performance is second. The benefit/cost ratio must be favorable—both from the viewpoint of the organization and of the individual. For example, if expenses of a task are

$20,000/yr, then benefits must be more than $20,000 or the organization suffers a loss on the job. Both capital and operating cost must be considered; the operating cost must consider not only direct expenses but also indirect costs such as poor quality. From the individual's viewpoint, benefits from the job must exceed the physiological and psychological-social cost.

Comfort is third. Unnecessary fatigue, suffering, or pain may be eliminated by good design even though, in the short run, there is no change in output. Examples might be a well-designed chair which reduces unnecessary backaches or air conditioning of work places which reduces sweating.

Higher wants are fourth. Designing a job to encourage social contact or to make it more interesting may be feasible. Although "pleasant" jobs or "fulfilling" jobs have relatively low priority at the present time in most cultures, the importance of designing for the higher wants will increase.

Table 10.2 Historical Progress in Improving Occupational Safety
Frequency = Number of Cases Involving Days Away from Work and Deaths
Severity = Days Away from Work

Year	Frequency Severity Rate per 200,000 work hours		Deaths, Rate per 100,000 workers
1930	3.7	394	37*
1940	2.5	280	38
1950	1.9	188	27
1960	1.2	146	21
1970	1.8	133	17
1979**	2.7	50	13

*for 1933
**The recordkeeping base and reporting requirements for frequency and severity changed during the 1970s so present data are not directly comparable to data before 1970.

SUMMARY

Job design requires trade-off of multiple criteria. Not only are the criteria vague and often not even specifically defined, but generally there are no explicit trade-off equations available.

Organizations primarily are interested in growth of numbers of employees and maximizing their net income. These goals are limited by rules set by the society and unions.

Employees want "more." "More" can be income but, increasingly, "more" also describes "higher wants" such as job security, pleasant working conditions, and even "fulfilling" work.

SHORT-ANSWER QUESTIONS

10.1 List the five steps of "Maslow's stairs"; give examples for each step.

10.2 Why do unions dislike job enlargement and job enrichment?

10.3 Plot frequency rate, severity rate, and death rate for US industry since 1930.

10.4 Four job design criteria are: 1) safety and health, 2) performance, 3) comfort, and 4) higher wants. Briefly discuss the four in the order of their importance.

10.5   What is the difference between job enlargement and job enrichment? Why should any organization use either?

THOUGHT-DISCUSSION QUESTIONS

1. Who should do the "dead end" jobs?
2. Japanese firms and many European firms treat labor as fixed cost rather than a variable cost. What happens when sales increase or decrease? If sales decrease and production remains constant, where do the units produced go?

REFERENCES

1. Bennett, C. The human factors of work. *Human Factors,* Vol. 15, No. 3 (1973): 281-287.
2. Bennett, C. Designing for human what? *Bulletin of the Human Factors Society,* 3 (February 1972).
3. Ford, R. Job enrichment lessons from AT & T. *Harvard Business Review,* (January-February 1973): 96-106.
4. Salvendy, G. An industrial engineering dilemma: simplified versus enlarged jobs, *Proceedings 4th International Conference on Production Research (Tokyo).* London: Taylor and Francis, 1978.
5. Sugisawa, H., and Hiroise, K. 'Jishu Kanri' activities in the Japanese steel industry. *Proceedings of 4th International Conference on Production Research (Tokyo).* London: Taylor and Francis, 1978.

PART IV
RECOMMENDED WORK DESIGN PRINCIPLES: SCIENTIFIC BACKGROUND

11

ANTHROPOMETRY

POPULATION STATISTICS

People are different. This chapter will quantify that statement—put it in numbers. As Lord Kevin said, "When you can measure what you are speaking about and express it in numbers, you know something about it; but when you cannot express it in numbers, your knowledge is of a meager and unsatisfactory kind."

Man's prime characteristic is variability—differing in such characteristics as initiative, needs, dexterity, intelligence, visual acuity, imagination, determination, upper back strength, age, and leg length. For each characteristic, there is a distribution.

To describe the distribution of any specific characteristic, a useful number is the population mean. Another is the standard deviation, σ. Next we need a distribution; although some anthropometric characteristics are skewed, the normal distribution usually is used. Given a mean, standard deviation, and a distribution, we can calculate percentiles (Figure 11.1 has examples).

FIT THE JOB TO THE MAN

Assume a very heavy box is to be moved from point A to B. One approach (pick one worker from many) is to select a strong individual who can lift and carry the heavy load; this approach is known as *personal selection* or *fit the man to the job.* The second approach is to reduce the load so that the majority of the population can carry the load; this approach is known as *ergonomics* or *fit the job to the man.* As a general strategy, designers should follow the second approach as a key to progress has been to make the "environment" adjust to man, rather than man to the environment.

An interesting concept for selection is the "law" of similitude; Haldane called it "On Being the Right Size."[6] The basic concept is that for every mass there is an

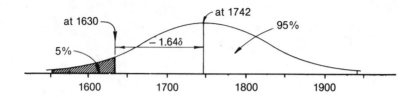

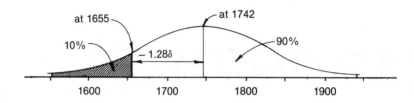

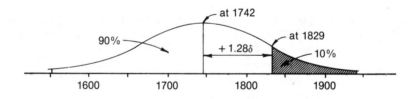

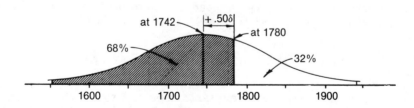

Figure 11.1 *Mean height* (no shoes) for USA males = 1742 mm with σ = 68 mm. The figure can be sketched (using Table 26.7) by: (1) Putting the mean, $\overline{X}$, at 1742. (2) Drawing the curve concave downward between $\overline{X}$ − σ (1742−68 = 1674) and $\overline{X}$ + σ (1742 + 68 = 1810) (the inflection points). (3) Drawing the curve concave upward outside this range. (4) Have the curve approach the axis at $\overline{X}$ − 3σ (1742 − 204 = 1538) and $\overline{X}$ + 3σ (1742 + 204 = 1946). For the vertical scale, if the height of the mean is 1., then the height of the curve at 1 standard deviation = .58, at 2 = .12, and at 3 = .01.

optimum surface area. Take a man 2 m tall. If you made him a giant of 20 m, then his weight would increase by 10 (for height) × 10 (for width) × 10 (for thickness) = 1,000. However, the leg bone cross section only increases by 10 × 10 = 100, so with every step he breaks his leg! The same principle applies to metabolism (increases with mass) vs. oxygen transmission through the skin (increases with area) so those giant grasshoppers you see in some movies wouldn't be able to get enough oxygen through their skins for their mass. People living in cold climates tend to be "spherical," which minimizes their surface area/volume ratio.[13] Skinny people with relatively long arms and legs, which maximize heat removal, also tend to be good basketball players! If a tool is "muscle-powered," big people not only have more massive muscles (increasing by the cube of length) but a longer moment arm (length of arm or leg).

DESIGN GUIDES

A general rule is: Design so that the small woman can reach and the large man can fit. This short rule includes several concepts.

First, don't design for the average; nobody is average—at least for more than one characteristic. Consider locating a foot pedal so it can be reached by the average worker. Then, since the average is the 50th percentile, it can't be reached by the other 50%! Let *most* of the *user population* use the device.

Most generally is defined as 90%, 95%, or 99% of a population; that is, we exclude (or inconvenience) 10%, 5%, or 1%. How many to exclude depends on the penalty for being excluded or inconvenienced and the cost of accommodating the population extremes.

For design of a door height, we might inconvenience only 0.1% since it doesn't cost much to make tall doors. In designing weights of totepans, however, we might exclude the weakest 25% of the population since they won't be moving material anyway in our factory and making all the loads small would mean more pans to be removed.

The *user population* often has different characteristics from the total population of a country. For example, most factory jobs exclude those with very high or low intelligence; jobs requiring manual dexterity have a minimum dexterity score; and inspectors can't be "blind." But industrial populations tend to be shorter, weaker, and older than military populations, so use care when selecting anthropometric values. In addition, population characteristics are changing (especially in Africa and Asia) as better nutrition allows more people to reach their full potential.

Some of the factors influencing anatomical dimensions are: age (until maturity), sex (male-female, not frequency), race, occupation (farmer, truck driver vs. accountant), clothing (especially in cold weather), and even time of day (in the morning you are about 6 mm taller since your spinal discs are uncompressed and you weigh your minimum due to respiratory and skin diffusion water loss during sleeping).

POPULATION VALUES

Dimensions

Figure 11.2 and Table 11.1 give some useful physical dimensions of USA adults.

A large part of the variation in human stature is in the length of the legs; the

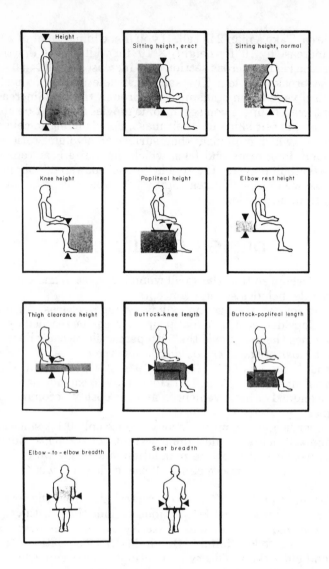

Figure 11.2 *Key human dimensions* for designers are shown in the sketches; Table 11.1 gives the values.

torso is relatively constant in length.[16] Figures 11.3 and 11.4 give the relative dimensions (proportion of height) for males and females.[4] For dimensions of the hand see Figure 15.8 and Table 15.4.

Strengths

Strengths for a specific muscle group vary greatly; expect the coefficient of variation (σ/mean) to be 50%. Strengths also are greatly influenced by the limb (arm vs. leg), by direction exerted, and, for arms, by whether it is the preferred arm. Table 12.4 gives arm strengths. Table 15.6 gives hand grip strengths and Table 15.7 gives finger strengths. Figures 14.12 and 14.14 give leg strengths. Hee-boll-Nielsen reported the average difference in muscular strength between symmetrical muscle groups was 5% to 11%.[8] Figure 12.10 shows the effect of age.

Table 11.1   Physical Dimensions (Inches) of 3,200 Subjects Selected in 1960 to Represent the US Population Between The Ages of 18 and 65.[14] Present US Industrial and US Military Populations Probably are Larger; African and Asian Populations are Smaller. See Chapanis for an Extensive Discussion of Cross-Cultural Problems.[2]

Characteristic	Population	Percentile			Standard Deviation
		5	50	95	
Height (no shoes)	US male	64.2	68.6	73.0	2.68
	US female	59.6	63.4	67.4	2.38
Sitting height, erect	US male	33.2	35.7	38.0	1.46
	US female	30.9	33.4	35.7	1.46
Sitting height, normal	US male	31.6	34.1	36.6	1.52
	US female	29.6	32.3	34.7	1.55
Knee height	US male	19.3	21.4	23.4	1.25
	US female	17.9	19.6	21.5	1.10
Popliteal height	US male	15.5	17.3	19.3	1.16
	US female	14.0	15.7	17.5	1.07
Elbow rest height	US male	7.4	9.5	11.6	1.28
	US female	7.1	9.2	11.0	1.19
Thigh clearance height	US male	4.3	5.7	6.9	.79
	US female	4.1	5.4	6.9	.85
Buttock-knee length	US male	21.3	23.3	25.2	1.19
	US female	20.4	22.4	24.6	1.28
Buttock-popliteal length	US male	17.3	19.5	21.6	1.31
	US female	17.0	18.9	21.0	1.22
Elbow-to-elbow breadth	US male	13.7	16.5	19.9	1.89
	US female	12.3	15.1	19.3	2.13
Seat breadth	US male	12.2	14.0	15.9	1.13
	US female	12.3	14.3	17.1	1.46

Summarizing: (1) The leg is approximately 3 times stronger than the arm. (2) Direction is very important with force at the nonoptimum angles being 50%-80% of force at the optimum angle. (3) The nonpreferred arm averages 60%-150% of the strength of the preferred arm (depending upon the angle and the arm). (4) There seems to be no appreciable difference between the strength of the left and right legs.

Other Characteristics

Weight and center of mass of various parts of the body can be estimated from Table 14.3.

Surface area of the total body can be estimated from equation 12.4. On the average, of the total area, the two arms and hands = 18.1% (hands 5.1% and arms 13.0%); the two legs and feet = 35.9% (legs 29.8%, feet 6.1%); the trunk 37.5%; and the head and neck 8.5%.[15]

Manual dexterity of the nonpreferred hand, if no tools are used, can be estimated as 90% to 98% of the preferred hand. With tools, time to complete a task (20 subjects) with the preferred hand was 69% of the nonpreferred hand for a screwdriver task, 87% for a wrench-turning-a-nut task, and 81% for a peg-trans-fer-with-a-pliers task[11] (see Table 15.3).

For additional anthropometric information, see Damon, Stoudt, and McFar-land for dimensions of various populations, and Roebuck, Kroehmer, and

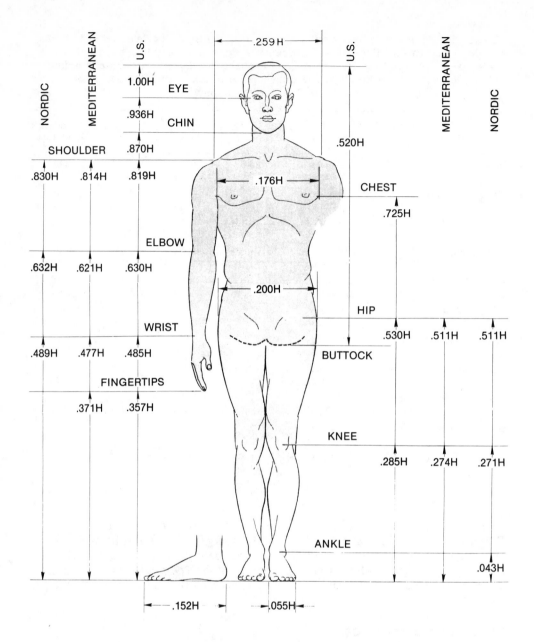

Figure 11.3 *Male proportion of height* for various body segments is given for Swedish, Mediterranean, and USA males.[4] The Swedish sample is based on 87 male industrial workers.[10] Mediterranean refers to a sample of 915 Turks, 1,034 Greeks, and 1,357 Italians.[9] The USA data came from Hansen, Cornog, and Hertzberg.[7]

Thomson for measurement techniques.[5,12] Probably the best reference is the three-volume NASA Handbook 1024.[1] Volume 1 is a 531-page, 9-chapter treatment of anthropometry; Volume 2 is a 421-page tabulation of 973 different measurements (such as waist height, sitting) with percentiles given for various population groups (such as Japanese Air Force pilots); Volume 3 is an annotated bibliography of 236 references. See chapters 14 and 15 for applications of anthropometry.

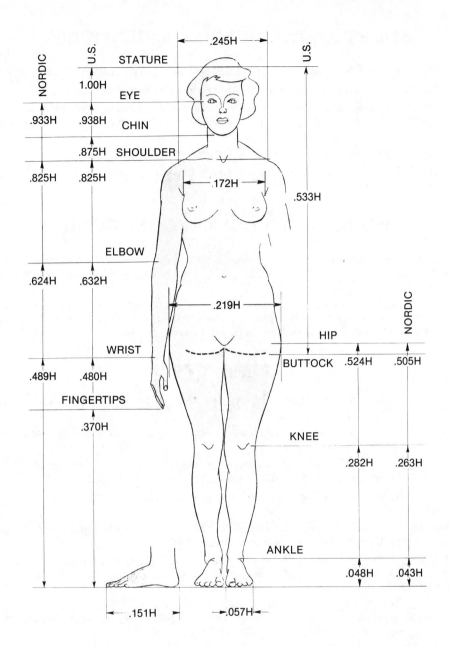

Figure 11.4   *Female proportion of height* for various body segments is given for Swedish and USA females.[4] The Swedish sample is based on 77 female industrial workers.[10] The USA data came from Churchill and Bernhardi.[3]

SUMMARY

Statistics and anthropometric data permit the designer to design to fit the job to the man. Designs should exclude only a small portion (say less than 10%, in some cases less than 1%) of the population. Design so the small woman (Nadia Comaneci) can reach and the large man (John Wayne) can fit.

184

SHORT-ANSWER REVIEW QUESTIONS

11.1 If you design a door height for the average person, what percentage hit their heads?

11.2 Contrast the two opposing strategies of ergonomics and personnel selection.

11.3 Design a totepan which 95% of the employees can lift. Assume 50% can lift 20 kg, 1 standard deviation = 5 kg, ±1.64 includes 90% of the employees and ±2 includes 95%.

11.4 Should you design for the average or for extremes? Why?

11.5 If King Kong (the giant gorilla) had really lived, why would he have had to be hollow?

THOUGHT-DISCUSSION QUESTIONS

1. What percent of people should be excluded from a manual material handling job? Discuss the justification you will give to the Equal Opportunity Commission and your boss.

2. What percent of people should be excluded from operation of an automobile brake (because they are too weak) when a power brake system fails? Discuss the justification you will give to the Product Safety Commission and your boss.

REFERENCES

1. *Anthropometric Source Book,* NASA Ref. Pub. 1024, Springfield, VA 22161: National Technical Information Service, 1978.

2. Chapanis, A., *Ethnic Variables in Human Factors Engineering.* Baltimore, Md.: The Johns Hopkins University Press, 1975.

3. Churchill, E., and Bernhardi, K. WAF trainee body dimensions: a correlation matrix. WADC Report 57-197. Wright Air Development Center, Ohio (AD-118161), 1958.

4. Contini, R. Body segment parameters, part 2. *Artificial Limbs,* Vol. 16, No. 1 (1972): 1-19.

5. Damon, A.; Stoudt, H.; and McFarland, R. *The Human Body in Equipment Design.* Cambridge, Mass.: Harvard University Press, 1966.

6. Haldane, J. *Possible Worlds.* New York: Harper and Bros., 1928. Also reprinted in Newman, J., ed. *The World of Mathematics.* New York: Simon and Schuster, 1956; pp. 952-957.

7. Hansen, R.; Cornog, D.; and Hertzberg, H. Annotated bibliography of applied physical anthropology in human engineering. WACD Report 56-30. Wright Air Development Center, Ohio (AD-155622) 1958; pp. 42-53.

8. Heeboll-Neilsen, K. Muscular assymetry in normal young men. Communication 18, Danish National Association for Infantile Paralysis, Copenhagen, Denmark, 1964.

9. Hertzberg, H. Anthropometric survey of Turkey, Greece, and Italy. Elmsford, N.Y.: Pergamon Press, 1963.

10. Lewin, T. Anthropometric studies on Swedish industrial workers when standing and sitting. *Ergonomics,* Vol. 12, No. 6 (1969): 883-902.

11. Lowden, K. Manual dexterity. Industrial engineering report. Manhattan, Kansas: Kansas State University, 1977.

12. Roebuck, J.; Kroehmer, K.; and Thomson, W. *Engineering Anthropometry Techniques.* New York: John Wiley & Sons, 1975.

13. Roberts, D. Population differences in dimensions. Chapter 2 of *Ethnic Variables in Human Factors Engineering,* A. Chapanis, ed. Baltimore, Md.: Johns Hopkins University Press, 1975.

14. Stoudt, H.; Damon, A.; McFarland, R.; and Roberts, J. Weight, Height and Selected Body Dimensions of Adults: 1960-62. Public Health Service Publication 1000-11-8, Superintendent of Documents, Washington, D.C., 1965.

15. van Graan, C. The determination of body surface area. *South African Medical Journal,* Vol. 43 (August 1969): 952-959.

16. White, R. Anthropometric measurements on selected populations of the world. Chapter 3 of *Ethnic Variables in Human Factors Engineering,* A. Chapanis, ed. Baltimore, Md.: Johns Hopkins University Press, 1975.

WORK PHYSIOLOGY AND BIOMECHANICS

METABOLISM AND THE CARDIOVASCULAR SYSTEM

Metabolism

Metabolic requirements can be subdivided into three parts: *basal metabolism*—maintains body temperature, functions, and blood circulation; *activity metabolism*—supplies energy for the muscles; and *digestive metabolism*, technically known as *specific dynamic action*—supplies energy for food digestion.

$$TOTMET = BASLMT + ACTMET + SDAMET \qquad (12.1)$$

In a laboratory we measure total metabolism. For metabolism of workers in industry, calculate metabolism from formulas.

A simple formula for basal metabolism is:

$$BASLMT = BSMET\,(WT) \qquad (12.2)$$

where $BASLMT$ = Basal metabolism, W
$\quad\quad\ BSMET$ = 1.28 W/kg for males
$\quad\quad\quad\quad\quad$ = 1.16 W/kg for females
$\quad\quad\ WT$ = body weight, kg

The reason for the difference in $BSMET$ is that females have a higher percent of body fat, and fat has a limited metabolism.

The effect of age also can be considered. Younger people have a higher basal metabolic rate/kg as: (1) children have a higher surface area/volume than adults and thus require more heat to maintain body temperature, (2) growth takes energy. The calculation takes four steps:

1. Calculate the Dubois surface area:

$$DBSA = .007\ 184\ (HT)^{.725}\ (WT)^{.425} \qquad (12.3)$$

where $DBSA$ = DuBois surface area, m²
       HT = height, cm
       WT = weight, kg

2. Correct for inaccuracies in the DuBois prediction[51]:

$$SA = .208 + .945\ (DBSA)$$
$$ = .208 + .006\ 789\ (HT)^{.725}\ (WT)^{.425} \qquad (12.4)$$

where SA = surface area, m²
       $DBSA$ = DuBois surface area, m² (equation 12.3)

3. Calculate the basal metabolism per m² of surface area:

$$BMMALE = 64.95 - .8875\ A + .0078\ A^2\ (5 < A < 70)$$
$$BMFMLE = 59.43 - .9315\ A + .0076\ A^2\ (5 < A < 70)$$

where $BMMALE$ = basal metabolism for males, W/m²
       $BMFMLE$ = basal metabolism for females, W/m²
       A = age, years

4. Calculate basal metabolism:

$$BASLMT = BMMALE\ (SA) \qquad (12.5)$$
$$ = BMFMLE\ (SA)$$

For example, assume a 20-year old male weighs 77.28 kg and is 180.3 cm tall. Then $DBSA = 1.97$ m², $SA = 2.07$ m², $BMMALE = 50.32$ W/m², and $BASLMT = 104$ W. If he were 40 years old, the $BASLMT = 87$ W. Using equation 12.2, $BASLMT = 99$ W, regardless of age.

The activity metabolism factor, $ACTFMT$, is given in Table 12.1 for various tasks. Activity metabolism is:

$$ACTMET = ACTFMT\ (WT)$$

where $ACTMET$ = activity metabolism, W $\qquad (12.6)$
       $ACTFMT$ = activity metabolism factor, W/kg
       WT = weight, kg

For example, sedentary sitting requires 28 W for a 70 kg person and 36 W for a 90 kg person.

The third portion of metabolism, $SDAMET$, specific dynamic action, is the energy required to break down and transform food within the body. Most laboratory studies of activity metabolism or basal metabolism take special precautions to exclude SDA (no food for 180 min before experimentation, no liquids for 120 min before experimentation, control of diet, etc.).

For every gram of carbohydrate burned we get 4 kcal of energy; for every gram of fat, 9; for every gram of protein, 4.3. Carbohydrate SDA, about 4% of $BASLMT + ACTMET$, lasts 2-5 hours; fat SDA, about 6%, lasts 7-9 hours; protein SDA, about 30%, starts about 1 hour after the meal and lasts about 10 hours.[31,32] For the typical mixture of the USA diet, use 10%.

$$SDAMET = .1\ (BASLMT + ACTMET) \qquad (12.7)$$

Using equation 12.2 and converting W-hr to kcal by multiplying by .86, in 24 hr a 70 kg male would have a 24-hr basal requirement of 1.28 (70) (24) (.86) = 1850 kcal. A 60 kg female would require 1.16 (60) (24) (.86) = 1435 kcal. Since there are 7700 kcal/kg of flesh, not eating at all while lying in bed would reduce the man's weight by 1850/7700 = .24 kg and the woman's by 1435/7700 = .19 kg. But in most situations, you also consider activity metabolism (Table 12.1) and specific dynamic action.

Thus, if a worker had a task requiring an activity metabolism of 1 W/kg, then, for a 70 kg man, total energy requirements = 90 W for basal + 70 W for activity = 160 W. SDA would be 16 W so a total of 176 W is expended of food value. If he worked 1 hr, 176 W-hr (.86) = 151 kcal of food are required.

If, instead of sitting, you jog at 7 km/hr for 20 min six days a week, how many extra kcal can you eat without gaining weight?

If, over time, food intake exceeds metabolism, it will affect body composition. For simplicity, divide the body into fat and lean body.

You can estimate body fat from height and weight (equations 12.8 and 12.9) from weight and waist circumference (equation 12.10) or from skinfold measurements (equations 12.11, 12.12, and 12.9).

Cowgill predicted, for adult males, from height and weight[19]:

$$DENSTY = .161 + .8 \, (HT)^{.242}/(1000 \, WT)^{.1} \qquad (12.8)$$

where $DENSTY$ = density of human body, g/mL
  = about .995 (specific gravity) since density is at body temperature of 37.5 C
  HT = height, cm
  WT = weight, kg

$DENSTY$ values of 1.02 to 1.08 (typical female value of 1.03, male of 1.06) are a mean of 0.9 for fat and 1.1 for lean body tissue (rest of body). Brozek et al. give[12]:

$$PERFAT = ((4.570/DENSTY) - 4.142) \, 100 \qquad (12.9)$$

where $PERFAT$ = percent of body that is fat

Wilmore and Behnke predicted, for adult males, from weight and waist circumference[86]:

$$PERFAT = 74 \, WACIRC/WT - 4 \, 464 \, /WT - 8.2 \qquad (12.10)$$

where $WACIRC$ = waist circumference, cm

For physically active adult males, Zuti and Golding used skinfold, a circumference and a diameter[90]:

$$PERFAT = 8.707 + .489 \, WACIRC + .449 \, PCSKIN - 6.359 \, RWDMAX \qquad (12.11)$$

where $PCSKIN$ = pectoral (front just below armpit) skinfold, mm
  $RWDMAX$ = right wrist diameter (max), cm

For adult females, Sloan et al. used skinfold to predict $DENSTY$[67]:

$$DENSTY = 1.076 - .000 \, 880 \, ARSKIN - .000 \, 810 \, ICSKIN \qquad (12.12)$$

where $ARSKIN$ = arm (back) skinfold, mm
  $ICSKIN$ = iliac crest (top of hip bone) skinfold, mm

Cowgill said equation 12.8 gives as accurate results as those using skinfold measurements due to errors in skinfold measurements made by novices.

For reference, Garn and Harper noted that lean body weight changed little for adult males between the ages of 20 and 60 but fat increased from 11.5 kg (15.7% of body weight) between 20-30 to 17.3 kg (22.4% of body weight) at age 50-60.[30] Very lean can be defined as 0-10% for men (0-12 for women), lean at 11-14% (13-17 for women), acceptable as 15-17% (18-22 for women), fat as 18-19% (23-27 for women), and obese as anything higher.

Heart, Lungs, and Circulation

Metabolism furnishes fuel and oxygen to the muscles and organs. Food and fuel are transported by the blood, which then carries off the combustion by-

Table 12.1 Reported Activity Cost for Various Activities. For Total Energy Cost, Add Cost of Basal Metabolism[a] and, if Appropriate, Specific Dynamic Action for Digestion of Foods.[b] 1.163 W = 1 kcal/hr. 1 met = 58.2 W/m^2.

W/kg	Activity
0.4	Crocheting, eating, reading aloud, sewing by hand, sewing by machine, sitting quietly, writing
0.6	Playing cards, standing relaxed, typing with electric typewriter
0.7	Paring potatoes, standing office work, sewing with foot-driven machine, standing at attention, violin playing
0.8	Dressing and undressing, knitting a sweater
0.9	Piano playing of Mendelssohn's *Song Without Words*, singing in a loud voice
1.0	Driving car, tailoring
1.2	Dishwashing, typing rapidly
1.4	Washing floors
1.5	Cello playing, light laundry
1.6	Horseback riding (walk), piano playing of Beethoven's *Appassionata*, sweeping bare floor with broom
1.7	Golf, organ playing (1/3 hand work), painting furniture
1.9	Sweeping with hand carpet sweeper
2.3	Piano playing of Liszt's *Tarantella*
2.4	Laboratory work
2.7	Heavy carpentry
3.0	Cleaning windows
3.1	Sweeping with vacuum cleaner (upright)
3.3	Walking 3.2 km/hr (2 miles/hr)[c]
3.5	Bedmaking, dancing (waltz)
4.1	Skating
4.5	Gardening (weeding)
4.9	Walking 4.8 km/hr (3 miles/hr)[c]
5.0	Horseback riding (trot)
5.1	Ping pong
5.8	Dancing (rhumba), tennis
6.6	Sawing wood
7.9	Football
8.5	Fencing
12.7	Running 9.7 km/hr (6 miles/hr) (70 kg man)[d]
14.1	Running 11.3 km/hr (7 miles/hr) (70 kg man)[d]

[a] Basal metabolism can be approximated as 1.28 W/kg for males and 1.16 for females.
[b] Digestion metabolism can be approximated as .10 (Basal metabolism + Activity metabolism)
[c] Total energy cost of walking, W/kg, = 2.031 393 + .124 122 V^2 where V = Velocity in km/hr[88]
[d] Total energy cost of running, W/kg = $-142.095/M$ + 11.045 490 + .039 678 V^2 where M = body weight, kg and V = velocity, km/hr[88]
[e] Walking stooped takes more energy; 12% more for a 10% stoop, 51% more for a 20% stoop, and 91% more for a 30% stoop.[53]

products. The lungs furnish the oxygen and eliminate the carbon dioxide; the heart is the pump (see Figure 12.1).

Usually supply of oxygen limits muscular activity rather than supply of fuel. Within the oxygen subsystem, the heart limits more than the lungs as will be shown from equation 12.13[65]:

$$1/COND = 1/ALVENT + 1/(SOLFAC)(CO) = \text{Resistance}$$

$$(12.13)$$

where

$COND$ = conductance = $\dot{U}$, liters of gas/min
$ALVENT$ = alveolar ventilation, $\dot{V}_a$, liters of gas/min
$SOLFAC$ = solubility factor, λ, liters of gas/liter of blood
CO = cardiac output, liters of blood/min

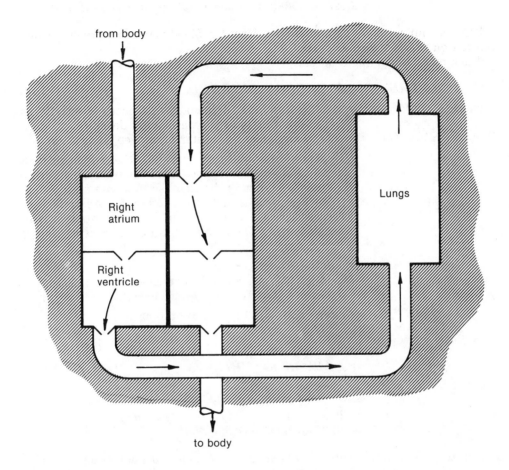

Figure 12.1 *Two pumps* form the heart. Blood from the body enters the right atrium ("little room" in Latin) where it is stored. Next, when the three-cusp (tricuspid) valve opens, blood flows into the right ventricle. When the semilunar valve opens, blood flows to the lungs so CO_2 can be eliminated and O_2 added. The first pump system is called pulmonary circulation. Blood with CO_2 reduced and O_2 increased comes back from the lungs and enters the left atrium to begin the systemic circulation. The two-leaf (bicuspid) valve then opens to permit blood into the left ventricle; then the thick walls of the ventricle contract, the aortic semilunar valve opens, and blood is pumped into the aorta and thence the body.

For oxygen, maximum $ALVENT$ is about 80, $SOLFAC$ is about 1, and maximum CO is about 25. Thus:

Oxygen resistance = $1/COND$ = $1/80 + 1/(1 \times 25)$
                  = $.0125 + .0400 = .0525$ min/liter of oxygen

That is, oxygen input to the blood depends more on cardiac output than on alveolar ventilation.

For carbon dioxide, maximum $ALVENT$ is still about 80, maximum CO is 25, but $SOLFAC$ is 5:

Carbon dioxide
resistance $= 1/COND = 1/80 + 1/(5 \times 25)$
                $= .0125 + .0080 = .0205$ min/liter of carbon dioxide

That is, elimination of carbon dioxide from the blood depends more on aveolar ventilation than on cardiac output.

When metabolism is measured:

$$TOTMET = 60\ ENERGY\ (OXUPTK) \qquad (12.14)$$

where $TOTMET$ = total metabolism, W (usually just basal + activity as SDA is eliminated by experimental design)

$ENERGY$ = energy equivalent of 1 liter of oxygen, W-hr/liter

Depends on respiratory quotient (RQ) which in turn depends upon the proportion of fat vs. carbohydrate metabolized during the exercise. Astrand and Rodahl (their Figure 14.2) indicate a RQ of .86 is satisfactory for exercise up to about 60% of maximum oxygen uptake; then RQ goes to 1.0 as uptake goes to 100%.[9] For resting, $RQ = .83$.

= 5.63 for $RQ = .83$ (rest)
= 5.66 for $RQ = .86$ (light exercise)
= 6.40 for $RQ = 1.0$ (very hard exercise)

$OXUPTK$ = oxygen uptake, V_{O_2}, liters of oxygen/min

Pulmonary ventilation is:

$$PULVNT = (LAPLOX)\ (OXUPTK) \qquad (12.15)$$

where $PULVNT$ = liters of air/min

$LAPLOX$ = liters of air/liter of oxygen
= 20-25 at rest and for work less than 15 W/min
= 30-35 during maximal work

Respiratory frequency is between 10 and 20/min at rest; most adults have a maximum of 40-45 (Astrand and Rodahl, p. 213[9]) but some athletes can go to 60.

Vital capacity (an index of an individual's lung capacity), the maximum that can be exhaled following a maximum inspiration, can be estimated with a standard deviation of 10% (Shephard, p. 112)[65] as:

$$VITALC\ \text{(men)} = 56.3\ (HT) - 17.4\ (A) - 4210 \qquad (12.16)$$
$$VITALC\ \text{(females)} = 54.5\ (HT) - 10.5\ (A) - 5120 \qquad (12.17)$$

where $VITALC$ = vital capacity (standing), mL
HT = height, cm
A = age, years

Vital capacity is reduced 5%-10% by a shift from standing to lying down.

Oxygen supply to a muscle from the blood is adjusted by heart rate, stroke volume, artery-vein $(A-V)$ differential, and blood distribution.

Heart rate. Heart rates determined by metabolic load (as opposed to emotions or vasodilation in heat) are highly correlated with incremental metabolic rates; that is, given one you can predict the other.[3] However, the linear equation has different coefficients for different individuals and types of work:

$$INCHR = K + .12\ INCMET \tag{12.18}$$

where $INCHR$ = increase in heart rate, beats/min
K = constant =   2.3 for arm work only
= −11.5 for walking or walking +
arm work
$INCMET$ = increase in metabolism, W

Due to venous pooling in the legs, arm work without leg work requires 14 more beats/min than leg work at the same metabolic rate.

Astrand and Rodahl predict maximum heart rate[9]:

$$HRMAX = 220 - A \tag{12.19}$$

where $HRMAX$ = Heart rate maximum, beats/min
A = Age, years

Arstilla, in a more recent study, gives[4]:

$$HRMAX = 200 - .66\ A \tag{12.20}$$

Thus a 50% individual at age 30 would have a maximum of 190 using Astrand and Rodahl's formula and 180 using Arstilla's. Astrand and Rodahl give 10 beats/min as the standard deviation. Thus to protect 95% of 30-year-olds, maximum for a job should be 190 − 1.645 (10) = 173 beats/min (if you like Astrand and Rodahl's formula) or 163 (if you like Arstilla's).

There are three common ways of measuring an individual's heart rate: by *palpation,* by *sound,* and by *electronics.*

- Palpation is the detection, with the fingers, of the surge of blood that follows each beat. Common locations are the wrist and the neck. Count the number of beats in 10 or 15 seconds and multiply the number by 6 or 4.

- It also is possible to detect the heart beat by laying your ear on the individual's chest. Women instinctively comfort their babies by carrying them on their left arm so the baby's head lies against the left chest and the baby can easily hear the mother's heart beat. Doctors, to be modest and to eliminate ambient sounds, use a stethoscope.

- Electronics is the most common measurement technique. Usually the entire EKG waveform is recorded but only the r wave (the spike) is counted. The most accurate, easiest, and most reliable method is to have a person look at a permanent tracing on paper and count the spikes. Automatic counting of the spikes is possible but it requires sophisticated (i.e., not trouble free) electronics.

Rather than measure the waveform's voltage, some sensors attempt to detect the wave of blood by photoelectric means. For example, a light is placed on the ear lobe and a photocell on the other side of the lobe. Then during "ebb tide" the cell detects the light and "says" "1 heart beat." A sensor also can be placed on the finger; it works by reflection. Unfortunately, slight movements of the head (finger) cause the blood to "slosh around" and the sensor becomes confused.

Electrodes, the most reliable sensor technique, usually are placed above and below the heart. They could be on the toe and forehead but, to minimize muscle noise, place them in a triangle on the chest (sternum and below and outside each nipple about 5 cm). Avoid chest hairs in placement as they give poor contact.

The signal then can be sent to the recorder. If you use telemetry, two electrodes are sufficient. If you use "hard wire," use three electrodes (one as a ground) and a shielded cable taped to the skin.

The recorder may be conventional paper but also can be magnetic tape. Small tape recorders have been developed which are carried by the worker. The heart beat can be recorded during work and the heart rate determined later.

Heart rate, a good index of task difficulty, can be estimated quite accurately simply by asking the individual his or her "perceived exertion."[5,29] The concept was first developed by Borg.[10] The new perceived exertion scale, given in Table 12.2, was specifically designed to have the scale be 10% of the heart rate:

$$HR = 10 \ (PEVOTE) \tag{12.21}$$

where

HR = heart rate, beats/min
$PEVOTE$ = perceived exertion vote (see Table 12.2)

Table 12.2   Borg's New Scale Can Be Used to Predict Heart Rate For a Variety of Tasks. Heart Rate is 10 Times the Vote. Use The Words For Guidance But Vote With a Number.

Vote	Subjective Description
6	
7	Very, very light
8	
9	Very light
10	
11	Fairly light
12	
13	Somewhat hard
14	
15	Hard
16	
17	Very hard
18	
19	Very, very hard
20	

Somewhat surprisingly, although originally developed for bicycle ergometer exercise ($r = .94$), Gamberale says it is valid also for running and walking, pushing a wheelbarrow, and lifting of weights.[29]

The variability of the interbeat interval of the EKG has been proposed as an index of "mental" load; low variability goes with high load.[41] Our research showed differences between rest and working but not among levels of work at the same task.[45] Meers and Verhaegen think sinus arrhythmia may be more an index of "emotional" load than "mental" load.[50]

Stroke volume. The second method of adjusting blood supply to the muscle is stroke volume. The amount of blood pumped by the left ventricle when sitting can be predicted as:

$$SV = STROVB + .000\ 050 \ (TOTMET - 200)$$
$$TOTMET < 500 \tag{12.22}$$

SV = stroke volume, liters/beat
$STROVB$ = male basal stroke volume (females = .9 male), liters/beat
= $SI \ (DBSA)/1000$

where SI 　= stroke index, mL/(beat-m^2)
= $53.45 - .194 \ (A)$
A 　= age, years
$TOTMET$ = total metabolism, W (see equation 12.1)

For *TOTMET* over 500 *W*, add .000 025 (*TOTMET* − 500) to equation 12.22.

Stroke volume depends upon body posture, exercise, and physical fitness. Stroke volume when lying down may be .12 liter/beat but a value of .08 is typical for sitting and standing.

Exercise with the legs improves the venous return and stroke volume may increase about .40 liter; exercise with only the arms tends to permit venous pooling in the legs and stroke volume changes little.

Maximum stroke volume is primarily a function of physical fitness. Maximum *SV* = .135 for excellent cardiovascular fitness; .120 for good; .100 for fair; .090 for poor; and .085 for very poor (see Table 12.3). World-class athletes may be as high as .200 liter/beat, giving them a sitting heart rate in the low 50s. Stroke volume peaks at about 40% of maximum oxygen consumption, *VO2MAX* (Astrand and Rodahl, p. 165)[9]: *MAXUPT = VO2MAX (WT)*

Cardiac output is the volume/min from the left ventricle:

$$CO = HR \; (SV) \tag{12.23}$$

where
$$CO = \text{cardiac output, liters/min}$$
$$HR = \text{heart rate, beats/min}$$
$$SV = \text{stroke volume, liters/beat}$$

Cardiac output can be considered to be composed of basal cardiac output plus activity cardiac output plus skin cardiac output.

Basal cardiac output can be predicted[32]:

$$COBASL \quad = CI \; (DBSA) \tag{12.25}$$

where *COBASL* = basal cardiac output, liters/min
       CI = cardiac index, liters/(min − m²)

$$= 4.2866 - .0289 \; A + .0003 \; A^2 \; (5 < A < 70) \tag{12.26}$$

       A = age, years
      DBSA = DuBois surface area of the body, m²

$$\text{(see equation 12.3)} \tag{12.27}$$

Activity cardiac output can be estimated (Astrand and Rodahl, Figure 6.17)[9] as:

$$COACT = CLMW \; (TOTMET) \tag{12.28}$$

where *COACT* = cardiac output due to activity, liters/min
    CLMW = conversion from liters of blood/hr to Watts
        = .0166 for *TOTMET* < 700 W
        = .0114 for *TOTMET* over 700 W
  TOTMET = Total metabolism, *W* (see equation 12.1)

COACT increases primarily in the muscles—up to 18 times basal flow.

Skin blood flow, *COSKIN,* [76] is:

$$COSKIN = CSKIN \; (DSKINT) + CCORE \; (DCORET) \tag{12.29}$$

where *COSKIN* = cardiac output due to vasodilation, liters/min
        = .4 for basal conditions
    CSKIN = skin coefficient, liters/C-min
  DSKINT = increase in skin temperature, C
    CCORE = core (hypothalamus) coefficient, liters/C-min
 DCORET = increase in hypothalamus temperature, C

For tasks with *TOTMET* > .2 *MAXUPT,* Rowell says the increase in *COSKIN* comes from a decrease in *COBASL* + *COACT* rather than an increase in *CO*.[64]

In a person with ordinary physical condition, vasodilation brings hot blood to the surface where it transfers heat to the environment by radiation and convection; *COSKIN* can increase up to 4 times basal. Vasodilation circulation bypasses the muscles as the blood flows from the small arteries (arterioles) to the small veins (venules) through bypasses called arteriovenous anastomoses. Heat-acclimatized persons, however, are able to sweat more and lose heat by sweat evaporation; their blood does not use the bypasses as much; therefore more blood goes to the muscles and they get a larger artery-vein oxygen differential; thus heart rate is lower (less cardiac output needed for sweat instead of vasodilation and less cardiac output due to better *A-V* differential).

Cardiac output, for a resting young man, is about 5 liters/min; maximum for a normally sedentary young man is about 25 and for a world-class athlete is about 35.

A-V differential. The arterial-venous oxygen differential is the third method of adjusting oxygen supply to the muscle.

$$OXUPTK = CO \, (AVDIF) \tag{12.30}$$

where $OXUPTK$ = oxygen uptake, $\dot{V}_{O_2}$, at std. temp. (0 C) and pressure (760 torr) dry *(STPD)*, liters of oxygen/min
CO = cardiac output, liters of blood/min
$AVDIF$ = arterial-venous oxygen differential, liters of oxygen/liter of blood

In rest circumstances the arterial oxygen content is 19 mL/100 mL of blood while the venous oxygen content is 15 mL/100 mL. That is, for every 100 mL of blood passing by the muscles, the muscles get 4 mL of oxygen. However, in an emergency (fleeing from a tiger), the muscles can get 13 mL/100 mL of blood; that is the veins drop to 6 mL. Astrand and Rodahl (their Figure 6.19) give 19 and 12 for male athletes and 16 and 11 for female athletes. In highly trained athletes, the *A-V* difference can increase to 16 mL. The coronary blood supply, even under normal circumstances, has a difference of 17; thus more oxygen for the heart must come from more blood, not an increase of the *A-V* difference.

Blood distribution. The fourth way of getting more blood to a muscle is through redistribution. During exercise, capillary density increases from 200/mm[2] at rest to 600. Muscle blood flow can change from 2 mL/100 mL of tissue to 14. Blood stored in the lungs can change from 500 mL to 1500. As exercise increases, the kidneys and intestines gradually use less blood and send their blood to skin and muscles. If food is present in the stomach, cramps may result.

If you are "underdeposited" at the bank, you go into debt. The oxygen supply from the lungs (the aerobic supply) may be too small at the start of exercise; the muscles then draw, for a short time, on the limited supply of anaerobic oxygen. Oxygen debt also can occur during a high level of exercise if total requirements exceed aerobic supply. However, debt must be repaid—with interest. Figure 12.2 shows how heart rate (primarily determined by aerobic oxygen supply) responds to constant-intensity exercise.

Heart rate cost of a task can be determined three ways.

Most simple is to subtract an individual's basal heart rate from the peak. For example, Joe's peak of 110 and basal of 70 give a task cost for him of 40 beats/min; Sam's peak of 115 and basal of 72 give 43; Mike's 125 and 85 give 40; the mean of 40 + 43 + 40 = 123/3 = 41 beats/min average increment during the task. Use the increment for individuals rather than the peaks themselves (i.e., the 41 rather than the 110 + 115 + 125 = 350/3 = 117); in statistical terms, you reduce the variance in the answer due to individuals.

Determining basal is more complicated than it might seem; people subconsciously increase their heart rate just before exercising.[48] To determine basal, measure heart rate after the end of exercise (say 3 min after light work and 10 to 15 min after heavy work).

The second method is to estimate area B (in Figure 12.2). Measured heart rate might be 100 at min 3, 106 at min 4, 108 at min 5, 110 at min 6, 110 at min 7, and 110 at min 8. Mean increment then would be 107 − 70 = 37 beats/min during the task. This method tends to underestimate the task cost.

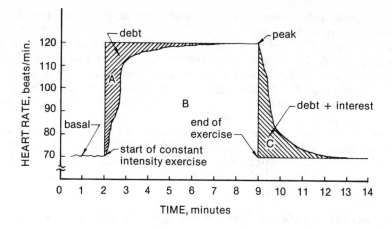

Figure 12.2 *Aerobic response* (and thus heart rate) lags the exercise onset. The debt (area A) is replaced by anaerobic oxygen; thus area A is known as "oxygen debt"; the anaerobic supply is composed of alactate debt (energy equivalent of about 1.9 liters of oxygen) and lactate debt (about 3.1 liters) (Shephard, p. 398).[65] Anaerobic oxygen is replaced during recovery (area C), but the replacement process itself uses oxygen ("interest") so area C is larger than area A. For task cost, use area B + C, not B or A + B.

The best method is to measure B + C (in Figure 12.2). Report as extra heart beats due to the task. For example, $100 + 106 + 108 + 110 + 110 + 110 + 80 + 75 + 70 = 869/9 = 97 - 70 = 27 (9) = 239$ extra beats for the task. The heart rate fluctuates from period to period, even for the same work rate. In addition, in the previous paragraph the area under the curve was estimated from a series of rectangles. Reduce these errors by plotting a smoothed curve of the heart rates; then determine area under the curve with a planimeter.[1]

Endurance work (lasting 4 to 8 hr) can be carried out at an aerobic rate of oxygen consumption which is approximately 33% to 50% of the individual's maximum ($VO2MAX$). This is about 5 kcal/min, 350 W, and 100-120 heart beats/min. The maximum depends upon the individual's sex, age, and cardiovascular training. The metabolic requirements for most jobs in the developed countries are so low that fitness depends more on leisure activity than work activity.

Table 12.3   Maximum Oxygen Consumption *VO2MAX*, mL/(kg-min), for Males and Females With Various Ages and Degrees of Cardiovascular Fitness[18]

Cardiovascular fitness	Age, years			
	Male: Under 30	Male: 30-39 Female: Under 30	Male: 40-49 Female: 30-39	Male: 50+
Very poor	<25.0	<25.0	<25.0	—
Poor	25.0−33.7	25.0−30.1	25.0−26.4	25.0
Fair	33.8−42.5	30.2−39.1	26.5−35.4	25.0−33.7
Good	42.6−51.5	39.2−48.0	35.5−45.0	33.8−43.0
Excellent	51.6+	48.1+	45.1+	43.1+

Table 12.3 gives maximum oxygen consumption in males and females of various degrees of fitness and age.

For example, a 35-year-old male with good fitness would have a maximum oxygen consumption of 39.2 − 48.0 mL/(kg-min). Assume a value of 45 mL/(kg-min) as well as a weight of 70 kg. Then his maximum oxygen consumption would be 45 × 70 = 3150 mL/min = 3.15 liters/min. Using 50% of maximum, he should not work at a rate requiring over 3.15 × .5 = 1.575 liters/min. Using equation 12.14, *TOTMET* = 60 (5.66) (1.575) = 535 *W*.

To check vs. the 110−120 heart heat criterion, use equations 12.18 and 12.2. For a 70 kg male, *BASLMT* = 90 *W*, so *INCMET* 535 − 90 = 435 *W*. For walking, *INCHR* = 42. Assuming basal heart beat rate of 70, 70 + 42 = 112.

Cardiovascular System: Sex, Age, and Training Effects

Sex. As mentioned before, women have a higher fat content than men. For minimum fat adults, blood volume is 79 mL/kg ±10%. Since fat tissue has little blood, blood volume for an adult male averages 75 mL/kg; for an adult female, 65; and for children, 60 (Astrand and Rodahl, p. 105).[9] For the same age and body weight, females also have lung volumes about 10% smaller. Women also have hemoglobin content about 10% lower than men (13.9 g/100 mL with 95% between 14.0 and 18.0); hematocrit (the relative amount of plasma and corpuscles) = 42 for women and 47 for males; arterial oxygen content = 16.7 mL/100 mL vs. 19.2 for males. Thus, for submaximal work (oxygen uptake of 1.5 liters/min), women need 9.0 liters of cardiac output to transport 1 liter of oxygen while men require only 8.0.

Although the above figures help explain the differences in athletic performance, cardiovascular differences between men and women should have little importance for industrial work since most industrial tasks should not be designed to require maximum physiological output. If very high output is required, then age and training (fitness) are probably more important than sex.

Age. Figures 12.3 and 12.4 show Shock's and Astrand and Rodahl's estimates of the effect of age on various organs. Shock gives 100% at age 30 while Astrand and Rodahl give 100% at age 25.[9,66]

Dehn and Bruce reported, from three studies, that maximum oxygen uptake declines −1.04, −.94, and −.93 mL/(kg-min)/yr for males; the decline in Dehn and Bruce's own study was −1.32 for habitually inactive males and −.65 for active males.[24] Astrand et al. reported declines, over a 21-year period, for Swedish physical education instructors of −.64 for males and −.44 for females.[8] Henschel has a good summary: "Cardiovascular capacity to perform light to moderate physical work is not grossly age-dependent up to age 65 although capacity for hard, exhausting work is strongly age-dependent, with maximum aerobic capacity between age 20 and 25."[34] Snook et al. reported older men work at a higher percent of maximum but output was the same for old and young.[73]

Training. For training of the oxygen transport system, the central circulation can be trained efficiently by engaging a large muscle mass. For peripheral circulation, train the specific muscles to be used.

Shephard states, "It is quite well established that older (40-60) sedentary men can improve their aerobic power by 10% to 20% by following suitable training for three months or more."[65]

Specific training techniques are more relevant to a book on athletics than a book on work design but it will be mentioned that less effort is needed to maintain fitness once achieved than to attain it originally. From an industrial viewpoint, if training is achieved by the work task itself, physical fitness will increase gradually over a period of weeks and months. Although cardiovascular fitness declines with lack of exercise, the engineer should not deliberately increase the metabolic requirements of a task in order to increase workers' physical fitness and thus their health.

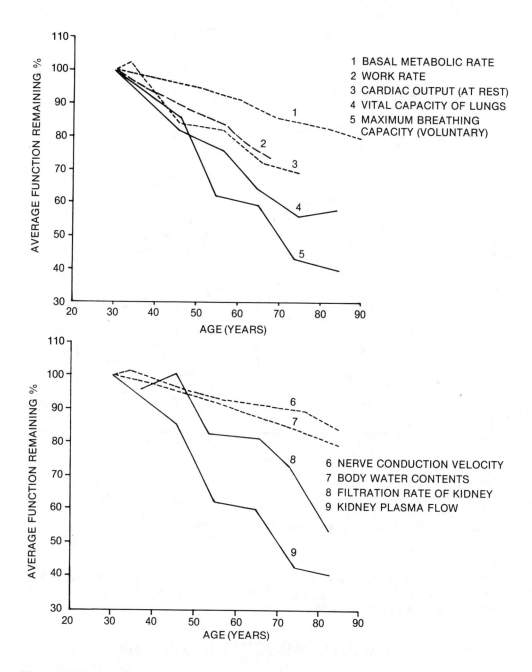

Figure 12.3 *Percent changes with age* (age 30 = 100%) were reported by Shock.[66] In early life we have surplus capacity. As age increases, capacity eventually declines below requirements.

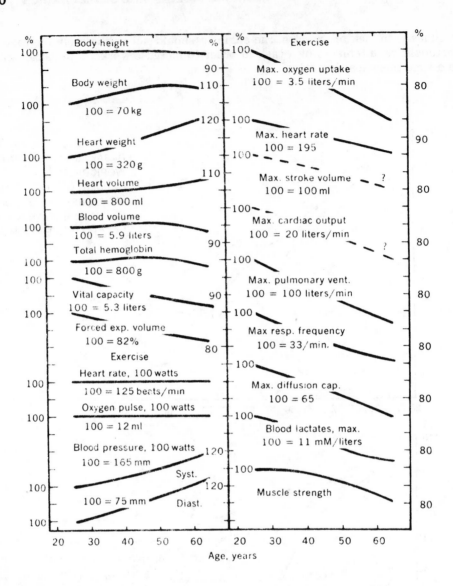

Figure 12.4 *Percent changes with age* (age 25 = 100%) were reported by Astrand and Rodahl.[9] Heart rate, oxygen pulse, and blood pressure are reported for a total metabolism of 100 *W* (oxygen uptake about 1.5 liters/min).

On the other hand, if a task has too low a metabolic requirement, it often is boring and monotonous. The engineer should strive for an optimum metabolic rate rather than a minimum or maximum.

SKELETO-MUSCULAR SYSTEM

Anatomy of Muscular Movement

Although the bones have other functions, we will discuss only the function of movement and, more specifically, movement of the limbs. The muscles have two divisions: the action muscles (protagonist) and the opposer (the antagonist).

Muscles and bones are arranged into three types of levers (see Figures 12.5, 12.6, and 12.7). First-class levers have the fulcrum in the middle; good for fine positional control. Second-class and third-class levers have the fulcrum at one end; a second-class lever has the force exerted through a longer lever arm than the resistance while a third-class lever has the force exerted through a shorter lever arm than the resistance (see Figures 12.5, 12.6, 12.7). Since individuals vary not only in the limb weight and length but also in the exact location where the muscle is attached to the limb, mechanical advantage varies from person to person.

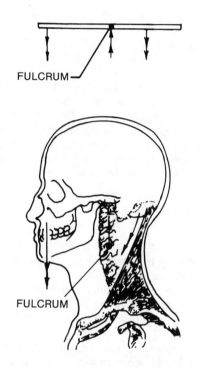

FULCRUM

FULCRUM

Figure 12.5 *First-class levers* have the fulcrum in the middle.

Mechanical advantage is strongly influenced by the included angle between the limbs. Warwick, Novak, Schultz and Berkson showed the importance of the direction of exertion.[85] For an object straight ahead, and using both hands, their male subjects could push with a force of 29.8 kg (100%), pull back with a force of 17.3 (58%), push right with 15.9 (53%), and push left with 17.0 (57%). They could lift up with 39.4 (132%) and press down with 34.7 (116%). Figure 12.8 and Table 12.4 show how muscular strength varies with the angle between the upper and lower arm. Figure 12.9 shows how sore elbow cases were influenced by the angle workers used when holding a screwdriver; that is, an improper work angle can hurt the worker's health. The various principles of work station design and hand-tool design (chapters 14 and 15) also point out how improper work angles can impair worker productivity and comfort.

Muscular Strength: Sex, Age, and Training Effects

Sex. The average female has less muscular strength than a male although individual females may be stronger than individual males. The simplest rule is that a female has 60% to 65% of the strength of a male of the same age (see Table 12.6).

Part of the difference is due to the difference in weight of these average individuals. A female of the same weight and age as a man will have about 80% of the male's strength. This 20% differential is due to physiological and cultural differences. For the same weight, females have less muscle mass than males; for the same height, female skeletons are more slender than males. In addition, there usually is a cultural difference in that female muscles are less trained than male; there also are hormonal differences. Astrand and Rodahl (p. 96) summarize: "Even with corrections for body weight, age and sex, a standard deviation of $\pm 15\% - 20\%$ is normal."[9]

Age. Figures 12.3 and 12.4 give some estimates of the decline of various functions of the body with age; Figure 12.10 and Table 12.7 give an estimate for muscular strength. Hunsicker and Greey summarize 89 references.[38] Figure 12.11 gives disc degeneration of the back.[36]

Training. With the same kind of muscle training, strength increases faster and to a greater extent among men; thus basic sex differences in muscle strength are more marked after training.[35] Mueller said, that to avoid fatigue in static work, train vs. a force double the highest static force in the job.[55] He also reported strength can be maintained either by long-interval training (1 contraction/week for 1 year) or by daily training (for 60 days) followed by weekly training. Ashton and Singh said a small amount of training (three 3 s maximal contractions 3 times/wk for 5 wk) increased isometric back strength 28% (39 kg).[6]

Carrying, Holding, and Lifting

A person may carry an object to another location, may hold it (usually after a lift and before a carry), or may lift (or lower) it from one height to another with no appreciable horizontal movement.

Carrying. Carrying for longer distances is not very common in the advanced countries as man is too expensive to use as a pack animal. Continuous carrying primarily is limited by the cardiovascular system rather than the skeleto-muscular system.

The two important principles are: (1) Minimize the moment arm of the load vs. the spine. (2) Carry large loads occasionally rather than light loads often.

For long-distance carrying (over 50 m), Datta and Ramanathan demonstrated that, to minimize oxygen consumption and increase heart rate, minimize the moment arm of the load vs. the spine (see Table 14.1).[21] Kellerman and van Wely reported minimum energy expenditure/kg-m moved was with 17 kg loads for untrained workers, but with 20-25 kg for trained workers.[43] In Asia, where manual carrying is common, teamwork is common. A loads B who carries to C who unloads B. Periodically, A, B, and C switch jobs. Drury reported students' willingness to carry a suitcase gave an exponential curve of weight vs. time (see Figure 12.12).[25]

For very short distances (2 m), such as might be used in unloading a truck, Mueller, Vetter, and Blumel emphasized the large cost for moving light (5-10 kg) loads and the importance of minimizing the vertical movement of the body (that is, don't put anything on the ground if you can help it).[56] See Figure 12.13.

Holding. Drury reported that loads on the two arms can be considered independent; that is, the subjects could hold a 25 kg suitcase in each hand just as well as they could hold a 25 kg suitcase in one hand.[25] He also reported, confirming other studies, that object weight and holding time are related with an exponential curve; see Figure 12.12. He suggested the horizontal asymptote of Figure 12.12 (that is, at time zero) could be used as an index of the maximum weight which could be lifted.

Poulsen and Jorgensen[62] looked at the opposite end of the curve—where time is

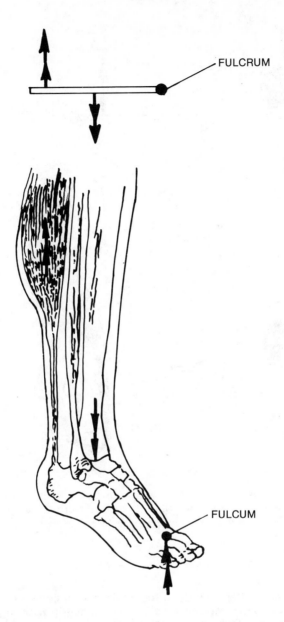

FULCRUM

FULCUM

Figure 12.6 *Second-class levers* have the fulcrum at one end, and the force has a mechanical advantage over the resistance (i.e., its moment arm in longer).

infinite and the external load is zero—as describing working while in a standing stooped posture. How strong a back do you need to work stooped without low back pain? They gave a formula:

$$MISBCK = .4\,(WT)\,(\sin \alpha)/LTSLPR \qquad (12.31)$$

where | *MISBCK* | = maximum isometric back muscle strength needed to avoid back pain while working standing stooped at angle α, kg.
| *WT* | = body weight, kg
| α | = acute angle between the back and the line of gravity
| *LTSLPR* | = long-term static load proportion of maximum isometric strength (see Principle 1 in chapter 14)

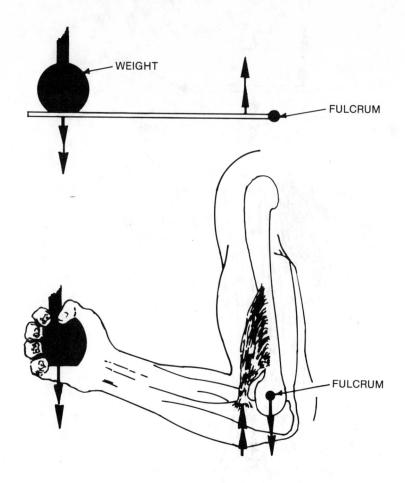

Figure 12.7 *Third-class levers* also have a fulcrum at one end but the resistance has a mechanical advantage over the force (i.e., its moment arm is longer).

Table 12.4 Strength Depends Upon the Angle, the Direction, the Arm, and the Individual. Forces (kg) Exertable on a Vertical Handgrip With Right Arm With Various Elbow Angles (Damon, Stoudt, and McFarland, Citing Hunsicker).[20] See Figure 12.8. N = 55 College Males. The Mean Coefficient of Variation Was 42%.

Movement	Elbow angle, degrees					Left arm, % of right
	60	90	120	150	180	
Pull	28.6	40.0	47.3	55.5	54.5	.94
Push	41.8	39.1	46.8	55.9	62.7	.92
Move right	19.1	16.8	15.5	15.0	15.5	1.30
Move left	23.6	22.7	24.1	24.5	22.7	.60
Up	22.3	25.5	27.3	25.5	19.5	.92
Down	23.2	24.1	26.4	21.4	18.6	.88

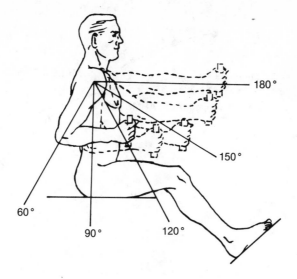

Figure 12.8 *Strength varies* greatly with the elbow angle (see Table 12.4).

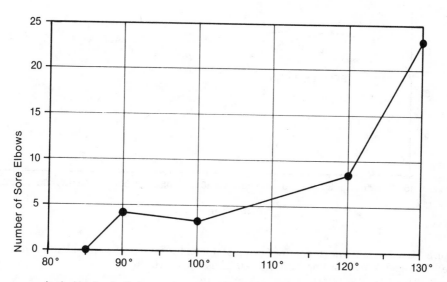

Included Angle Between Forearm and Upper Arm (In Habitual Working Posture)

Figure 12.9 *Sore elbow cases* (238 patients from workers continuously using screwdrivers) were strongly related to the habitual angle (at work) between the forearm and upper arm.[80]

They give 20° as a typical angle of working while stooped and $LTSLPR = .2$; then $MISBCK = .65\ WT$. If $LTSLPR = .15$, then $MISBCK = .87\ WT$.

For a worker a large value of alpha is probably better than a small value as, with a large value, the $MISBCK$ required greatly exceeds the actual back strength, pain quickly results, and the job is changed. With a small angle, actual strength exceeds required so no pain occurs but the worker is always under unnecessary stress. In some cases the required and actual strength are closely

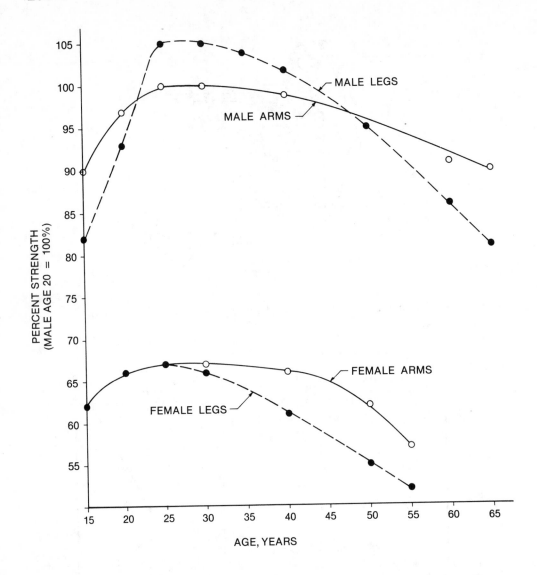

Figure 12.10 *Isometric strength* of arms and legs for men and women varies with age.[7] For each curve, strength at age 20-22 is set as 100%

matched and the worker tolerates the small pain. (Some people think work equals pain so they don't complain.) The best alternative is to redesign the job rather than trying to find a strong back.

Lifting—Repetitive. Although "occasional" lifts are the primary problem, first we will discuss the situation where an individual is lifting repetitively (say 4-6 times/min) *for an entire working day.* The definitive work by Snook and Irvine and co-workers was incorporated in the Ergonomics Guide of the American Industrial Hygiene Association.[2,70,71,72,73] They had industrial workers repetitively lift various weights and asked them to estimate what weight they would be willing to lift all day. The results can be summarized as:

$$WKLMAX = A + B \ (OBJWT) \qquad 11 < OBJWT < 25 \qquad (12.32)$$

where $WKLMAX$ = maximum workload (object weight × distance) for 50% male, kg-m/min

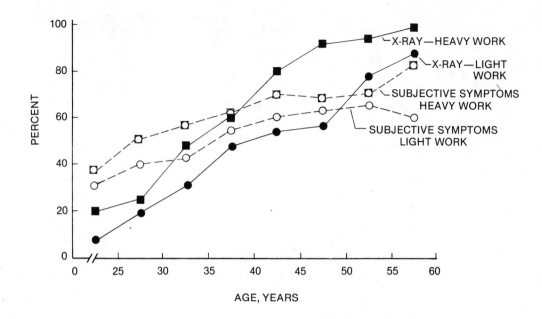

Figure 12.11 *Lower back problems* (lumbar insufficiency-lumbago-sciatica syndrome) were reported from workers in light and heavy industry (total $N = 1,200$) by Hult.[26] He used subjective symptoms of back pain and X-rays of disc degeneration.

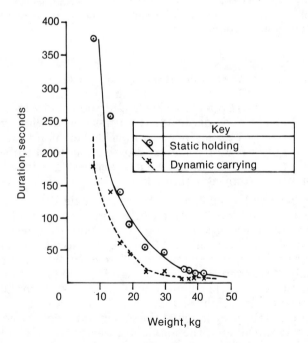

Figure 12.12 *Holding and carrying* have an exponential trade-off between time and weight.[25]

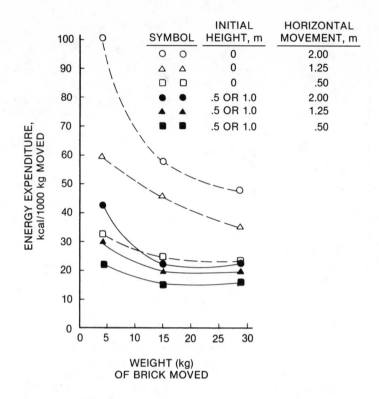

Figure 12.13 *Metabolic load* is minimized by moving large loads occasionally rather than small weights often.[56] The reason is twofold: (1) The body weight (tare weight) must be moved for each move regardless of load so fewer moves means less tare weight to be moved. (2) Occasional moves permit a rest period between moves. Note the higher cost of movements which require picking up from the ground. Note that safety may indicate small loads rather than large loads.

A	= theoretical intercept at 0 weight, kg-m
	= 15.9 for floor to knuckle-height lifts
	= 41.5 for knuckle-height to shoulder-height lifts and shoulder-height to arm-reach height lifts
B	= 1.22 for floor to knuckle-height lifts, m
	= .91 for knuckle to shoulder-height lifts
	= .46 for shoulder to arm-reach height lifts
$OBJWT$	= object weight, kg

Females are assumed to be capable of 70% of the male *WKLMAX*. Snook and Ciriello say to reduce *WKLMAX* 20% for lifting in hot environments.[69]

The adjustment for different percentiles of the population affects both *A* and *B* for floor to knuckle-height lifts but just *A* for the other two types of lift. For floor to knuckle-height lifts, subtract 3 kg-m from *A* to cover 90% of males and add 3 kg-m to cover only 10%; simultaneously change *B* to .30 for 90% and to 2.62 for 10%. For shoulder to arm-reach, change only *A* by ±11 kg-m for 90 and 10%; for knuckle-height to shoulder-height change only *A* by ±15 kg-m.

Table 12.5 expresses the proportion of the maximum weight which can be lifted at various frequencies. Defining the maximum weight which can be lifted occasionally as 1.0, Poulsen and Jorgensen estimate about .25 times the maximum weight can be lifted: 6 times/min by a 20-year-old male, 5 times/min by a 55-year-old male and a 20-year-old female, and 3 times/min by a 55-year-old female.[62]

These values should be used instead of Poulsen and Jorgensen's previous multiplier of .8 for all repetitive lifts, regardless of frequency. The Ergonomics Guide adjustments are based on equation 12.32.

With repetitive lifting, the prime problem is that lifting an object also requires lifting the weight of the body. Thus, reduce cardiovascular load by lifting heavy loads occasionally rather than light loads often. When lifting "light" (5-8 kg) loads, 70% of the total energy moves the body up and down while with "heavy" (40-60 kg) loads at the same frequency, the body uses 25% of the total energy.[40]

Lifting—Occasional. Heavy objects put an extra strain on the skeleto-muscular system—increasing the risk of injury. Approximately 25% of the "compensible industrial injuries" are due to manual material handling; the ratio seems relatively constant or increasing over the last 30 years among countries with varying degrees of mechanization.[22,61,69,82]

Back injuries and pain usually occur near the 5th lumbar disc and 1st sacral disc (L5—S1). The discs act as shock absorbers; a series of overlapping ligaments supports the vertebral column and protects the spinal cord.[58]

Skeleto-muscular ability to lift objects depends on (1) individual characteristics, (2) technique characteristics, and (3) task characteristics.

Individual characteristics include sex, age, back muscle strength, arm strength, and intra-abdominal pressure.

Technique characteristics include body posture, hand orientation, foot position and footwear, acceleration, and lifting training.

Task characteristics include object weight, ease of handling, initial and final height of object moved, angle of rotation, and lift symmetry.

A long list.

Drury developed a useful technique to show the importance of some lifting variables.[25] There is a hypothetical object weight which is the maximum which can be lifted in a specific situation. Then factors (multipliers) are developed which show the effect of the variable. For example, if weight lifting capability of females is 60% of that of males, then the factor would be 1.0 for males and .60 for females.

First, we will consider *individual characteristics*.

Table 12.6 gives the sex factor by different authors. The AIHA (Ergonomics Guide) estimate of .70 seems a little high; Drury and Pfiel's overall estimate of .60 seems better; non-selected populations may have female strength of .40 to .50 of male strength.[26] Tichauer, Miller, and Nathan point out that the force couple in the back is approximately 15% higher (for the same object) in females than males since the female hip socket is forward of L5-S1 while the male socket is below L5-S1.[81] Sonoda stated that the female spinal compression tolerance is 17% less than the male's—probably due to the smaller force-bearing area of the vertebral bodies in female spines.[75]

Table 12.7 gives the effect of age estimated by various authors. Male strength has a flat plateau between 25 to 40, then a drop. Females have their plateau from 25-35 and then drop—perhaps slightly more rapidly than men. Mueller's data show that, although over all ages women have 58% of the strength of men the same age, this ratio is maximum at 64% at age 20 and declines to 53% at age 65.[55]

Next consider the variability within a population. That is, what is the weight the 5th, 10th, or 95th percentile can lift—not just the weight the 50th percentile can lift. First we will obtain the coefficient of variation (standard deviation/mean) from a number of studies. The coefficient is larger when the study has a more varied population since age and height effects are included also. Table 12.8 gives the coefficient of variation from different authors. From these the normal distribution can be used to estimate percentiles; the estimates should be used with caution since weight lifting ability has a positive skew.[14,28]

If a coefficient of variation of .20 is used, then the 5th percentile subject is 1.645 σ below the mean and the 10th percentile is 1.28 σ. But since $\sigma = .20(X)$, the

Table 12.5 Multiplier for Repetitive Lifting by Various Authors. Single Lift Factor = 1.0. Poulsen and Jorgensen's Criterion Was Keeping Oxygen Consumption Below 50% of Maximum Capability. Ergonomics Guide's Criterion Was The Subjects' Estimate of The Weight They Would be Willing to Lift for 8 Hr/Day.

Lifts/min								VO_2			
2	3	4	5	6	7	8	9	mL/(kg-min)	Sex	Age	Author
	.50			.25			.10	27	M	20	Poulsen & Jorgensen[62]
.50			.25		.10			23	M	55	
.50			.25		.10			34	F	20	
.50	.25			.10				25	F	55	
.70	.30								M		*Ergonomics Guide* for floor to knuckle-height, knuckle to shoulder height, and shoulder to arm-reach height.[2]
1.00	.70	.50	.40								
1.00	.80	.40									

Table 12.6 Sex Multiplier For Lifting Given By Different Authors

Male	Female	N	Comment	Source
1.0	.58 to .66	510	Isometric muscle strength Range of .70 to .80 if both males and females have height of 178 cm	Asmussen and Heeboll-Nielsen[7]
1.0	.60		Expert opinion	Muchinger[54]
1.0	.65	230	Isometric push-pull forces Flexor and extensor—standing	Troup and Chapman[83]
1.0	.70		Expert opinion	Ergonomics Guide[2]
1.0	.48	46	Weight lifted. Assumes body weight of 58 kg for females and 75 for males. For equal body weights, ratios at 50, 60, 70, and 80 kg are .46, .50, .55, and .59	Poulsen and Jorgenson[82]
1.0	.65		Expert opinion	Davies[23]
1.0	.58	375	Industrial lifting data	Chaffin[14]
1.0	.52, .53, .58	11	Lab. lifting from floor to .6 m; floor to 1.2 m; and .6 to 1.2 m	Kassab and Drury[42]
1.0	.62, .49, .64		Lift from floor to knuckle height, knuckle height to shoulder height, and shoulder to arm reach	Chaffin and Ayoub[15]
1.0	.60		Expert opinion	Drury and Pfiel[26]

Table 12.7   Age Multiplier For Lifting Given By Different Authors

Population age, years										Source	Comment
20	25	30	35	40	45	50	55	60	65		
.94	1.00	1.00	1.00	.99	.95	.93	.90	.85	.82	Asmussen and Heeboll-Nielsen[7]	Male isometric strength—muscles; $N = 360$
.98	1.00	1.00	1.00	.98	.95	.89	.82			Asmussen and Heeboll-Nielsen[7]	Female isometric muscle—25 muscles; $N = 250$
.60	0.80	0.98	1.00	.85	.78	.70	.62	.55		Drury and Pfiel[26]	Males; weights lifted; $N = 45$, their Figure 2
.98	1.00	1.00	0.92	.84	.84	.84	.84	.64		Muchinger[54]	Male. Expert opinion for reasonable weight limits for occasional lifting
.97	1.00	1.00	0.93	.87	.87	.77	.67			Muchinger[54]	Female. Expert opinion for reasonable weight limits for occasional lifting
.99	1.00	0.97	0.95	.91	.90	.80	.70	.68	.65	Ufland[84]	Back muscle strength; $N = 3079$

Table 12.8   Coefficient of Variation (Between Subjects) For Lifting Given By Various Authors

Coefficient of variation	N	Comment	Source
.20	19	Male students between age 17 and 28, for lifts from floor of 1, 2, 3, 4, and 5 feet	Emanuel et al.[28]
.21			
.26			
.23			
.28			
.16			Switzer[77]
.17	9	Male industrial workers between age 25 and 37 selected for an experiment; for lifts from floor to knuckle height, from knuckle height to shoulder height, and shoulder height to arm-reach height	Snook and Irvine[70]
.44	239	Male industrial workers	Chaffin[14]
.53	136	Female industrial workers	Chaffin[14]
.24		Male industrial workers, floor to knuckle height and knuckle to shoulder	Chaffin and Ayoub[15]
.21			
.32		Female industrial workers, floor to knuckle height, knuckle to shoulder, and shoulder to arm reach	Chaffin and Ayoub[15]
.12			
.25			
.27	14	Male industrial workers between 25 and 35 for floor to knuckle height, knuckle height to shoulder height, and shoulder height to arm reach	Snook et al.[73]
.27			
.35			
.19	14	Male industrial workers between 45 and 60 for floor to knuckle height, knuckle height to shoulder height, and shoulder height to arm reach	Snook et al.[73]

multipliers are $1 - 1.645$ (.2) = .67 and $1 - 1.28$ (.2) = .74. That is, a 5th percentile person can lift 67% of what the 50% person can lift and a 10th percentile person can lift 74% of what the 50% person can lift.

What value should be used for the coefficient of variation? Drury and Pfiel recommended .18.[26] Poulsen and Jorgensen recommended a multiplier of .7 to include "practically all" the individual values; if "practically all" is the 5th percentile, this coefficient of variation also is .18.[62]

For strength, isometric back muscle strength is a good predictor of maximum load that will be lifted.[62]

$$MXLOAD \quad = 1.10 \ (ISOBCK) \quad \text{for males} \qquad (12.33)$$
$$MXLOAD \quad = \ .95 \ (ISOBCK) \ -8 \ \text{for females}$$

where $MXLOAD$ = Maximum load (single lift) for 50% individual, kg
$ISOBCK$ = Isometric back muscle strength, kg.

Konz and Coetzee found $ISOBCK$ was a good ($r = .82$) predictor of the amount males and females were willing to lift.[46] Predicted "refuse to lift" for their six males were at 35-41 kg and for their six females at 12-14 kg; actual scheduled experimental maximum weights lifted were 15 kg for males and 10 for females.

Support from the abdominal cavity muscles improves lifting ability and reduces low back pain.[27,52,57] When the individual's belly is large, the weight pulls on the L5-S1 area rather than having muscles to support the back.

From the above, is it possible to select (or eliminate) workers for lifting? Not very well.

X-rays no longer are recommended due to poor predictive ability and due to need to protect against unnecessary radiation. Due to its safety, isometric strength testing is preferred over isotonic testing. Since jobs and individual muscle group strengths vary widely, test the three to six muscle groups appropriate for a specific job.[16,89] Chaffin reported low back injuries of 92/1000 work hours when loads were greater than "strengths" vs. 32 when loads were only .5 of strength.[14]

After individual characteristics we will consider the *technique characteristics* of body posture, hand orientation, foot position and footwear, acceleration, and lifting training. At the present state of knowledge there are no numbers as there were with individual characteristics.

The straight back (squat) method is the *posture* recommended by most experts; unfortunately workers don't pay much attention to it and continue to lift with bent back and straight knees—possibly because bending the knees and thus lowering and raising the body's center of gravity requires more energy. Squat lifting also requires greater leg strength and an object which can fit between the legs. From computer modeling, Chaffin and Park calculate that stress on the back is less for the bent-back method (unless the load can be moved between the knees) so perhaps the workers know what they are doing![13] Brown says "the straight back, bent knees technique has been enthusiastically promoted for 40 years with no decrease in back injuries . . . which raises a question as to the benefit of the lifting procedures advocated."[11] Perhaps it's because "no one uses the bent knees technique."

Hand position can be summarized as (1) get a secure grip and (2) two hands are better than one. Handles should be above the center of gravity, be over 11 cm long, and have 5 cm hand clearance. Nielsen did an excellent study on tote boxes.[59] He recommended a rectangular shape of 35 cm width by 47 cm length with 51 cm as the maximum length. Handholds are especially important. A gripping block (similar to a "drawer pull") probably is best; be sure it is at least 11 cm long and has no sharp edges. A hole (3×11 cm minimum) into which the hand is inserted is satisfactory if the grip need not shift during the movement.

As for *foot position,* Konz and Bhasin reported no difference for straight-ahead lifting whether the feet were aligned or one foot was leading but, for lifting and turning, the foot opposite the direction turned should be ahead (left foot ahead when turning to the right).[44] The rationale is that a leading foot shifts the body center of gravity forward. Nonslip footwear is important as many back injuries are associated with slips and falls.[11]

Acceleration during the lift should be fast enough to get the benefit from the body's weight and momentum but not too fast. Konz and Bhasin reported that peak forces often occur during the initial lowering of the body before the object is even grasped; thus, take it "slow and easy" when lowering the body. Don't be a jerk!

Lifting training has not been effective in reducing incidence of low back pain or manual material handling injuries. The forgetfulness and careless nature of man strongly reduce the effectiveness of any teaching. A simile is the open manhole problem. The motivational approach is to put a sign on the wall "Don't fall in holes"; the training approach is to have people practice walking around open holes; the engineering approach is to put a cover on the hole. Use the engineering approach.

To reduce manual material handling injuries, the desirable approach is not personal selection nor training; the desirable technique is modify the job—that is, fit the job to the man not make the man fit himself to the job.

The primary *task characteristics* are the load itself and its locations. The load includes its weight, its compactness, and the human/container coupling; the locations include the initial and final location, lift symmetry, one vs. two hand lifts, angle of rotation, and human/floor coupling. Ridd and Davis advocate having a worker swallow a pressure-sensitive radio pill and then do their normal work; any interabdominal pressure greater than 90 mm HG is considered to lead to long-term back damage; the telemetered data indicate which portions of a task might be dangerous.[63]

In general, lowering an object is slightly easier than lifting it.

Using the multiplier effect on weight, probably the most important factor is the initial location of the load. See Tables 12.9 and 12.10. Tichauer proposed that lifting, in some cases, is limited by the spinal torque[80]:

$$SPINET = OBJWT(MOMARM) \qquad (12.34)$$

where $SPINET$ = Spinal torque which body can exert, kg-m
  $OBJWT$ = Object weight, kg
  $MOMARM$ = Moment arm of the object, m
  = $DISTSO + DISTCG$
  $DISTSO$ = Distance between spine and object (closest portion), m
  = .2 according to Drury and Tichauer, although mean male waist depth is .20 − .25 m so .3 might be a better value.[20] Note that people sometimes hold the object away from the body to avoid soiling clothes.
  $DISTCG$ = Distance from closest portion of the object to its center of gravity (hand position) = $L/2$, m

Drury experimentally confirmed equation 12.34.[25] For example, if $L = .2$, then $DISTCG = .1$ and $MOMARM = .3$ and $SPINET = .3 \ OBJWT.$ For $L = .6$, the $DISTCG = .3$ and $MOMARM = .5$ and, to keep $SPINET$ constant, then the new $OBJWT$ must be .3/.5 or .6 of the $OBJWT$ at $L = .2$.

For object compactness, Drury and Pfiel used two categories of compact = 1.0 and noncompact = .6.[26] Konz and Coetzee reported increasing a box's volume (up to a 30 cm cube) did not bother men but strongly bothered females.[46] For bags, Smith reported that people were willing to lift 10% less for a 70% full bag than a 90% full bag.[68] They were willing to lift 10% more when the bag had handles on both ends. Lovested used a Pareto analysis to determine that a minority of parts accounted for a majority of material handling strain.[49] A large amount of their problem involved picking parts out of boxes. The solution (which reduced lifting strain incidence rate 68%) was: a) to design the box so one-half of one side could be removed so the worker could get closer to the parts and b) also use "short" boxes (19″ high vs. the former 33″ high). The boxes also can be stacked. For persons working under a low ceiling, Ridd and Davis reported a 46% reduction in lifting capacity when the person stooped 10% and a 61% reduction for a 20% stoop—thus confirming the importance of posture when lifting.[63]

Table 12.9   Multiplier For Object Dimension In Fore-And-Aft Dimension

Compact	Non Compact							Comment
1.0	.60							Drury and Pfiel[26]; $N = 45$
		L, m						
.2	.3	.4	.5	.6	.7	.8		
1.0	.86	.75	.67	.60	.55	.50		Drury[25]; $N = 10$ Based on Tichauer's formula. $DISTSO = .2$
1.0	.96	.91	.86	.81	.76	.71		Ergonomics Guide, floor to knuckle height[2]
1.0	.94	.88	.82	.76	.70	.63		Ergonomics Guide, knuckle height to shoulder height and shoulder height to arm reach[2]

The final location of the load also is important. The consensus is that the higher you lift, the less you can lift—especially if the load is lifted above the shoulders (the hands need to shift location as the object goes past the chest).

As most investigators have concentrated on straight-ahead lifting with two hands on the load, there is little information on lift symmetry and angle of rotation. Konz, Dey, and Bennett showed that force platform forces differed little between 45° and 90° of turn although both differed from 0° of turn.[47] Parker, for horizontal movement at various heights, showed the important factor was the height (floor was worst, knee intermediate, hip best) rather than the degree of turn.[60] As long as the turn is relatively smooth and without contortions, turning probably presents little problem; turns combined with unbalanced body positions or slipping can present severe problems. At present the turning problem has not yet been quantified.

NIOSH lifting guidelines. The National Institute of Occupational Safety and Health (NIOSH) guidelines recognize that the approaches to improving lifting safety are through selection of people, training, and design of the lifting job.[87] The primary focus, however, is on job design for two reasons. First, job design is a more fruitful approach than selection or training. Second, selection is difficult—both from a technical viewpoint and a political viewpoint. The two easiest categories for discrimination are sex and age—as shown before, females and older people are less suitable for lifting, *on the average*. However, specific females and older people are perfectly satisfactory. For selection of specific individuals (male or female, young or old), NIOSH recommends isometric strengh testing with the specific muscle groups to be tested being the same as those used in the specific job.

The guideline concept is to give an "ideal" weight and adjust it by factors; these factors then can be used by engineers to improve job designs. Although there are a number of ways of deriving these factors, they are primarily based on

Table 12.10 Effect of Initial And Final Lift Location (No Turns) By Various Authors

INITIAL OBJECT HEIGHT	FLOOR	FLOOR	FLOOR	FLOOR	FLOOR	KNUCKLE	SHOULDER		
Final object height	Knuckle	Waist	Chest	Head	Arm-Reach	Shoulder	Arm-Reach	N	Author
Factor	1.0	1.0	.70	.55	.46			45	Drury and Pfiel[26]; weight subjects willing to lift.
	1.0					.85	.70		Snook[73]
	1.0					.94	.91	9	Snook and Irvine [70]
	1.0					.93	.97		AIHA Ergonomics Guide[2]
	1.0					1.04	.79		Males; Chaffin and Ayoub[15]
	1.0					.82	.82		Females; Chaffin and Ayoub[15]
	1.0					1.00	1.00		Males 25 to 35; Snook[73]
	1.0					.95	.81		Males 45 to 60; Snook[73]

biomechanical and psychophysical studies with metabolic (cardiovascular) approaches used only for highly repetitive lifting. The state of the art permitted giving the modifying factors only for initial horizontal location of the object, initial vertical location of the object, vertical distance moved, and lifting frequency. The guidelines apply only for:

- smooth lifting
- two-handed, symmetric lifting directly in front of the body, no twisting during lift
- compact, moderate-width load (75 cm or 30 in. maximum)
- unrestricted lifting posture
- good mechanical environment
  - •• container design includes handles, nonshifting compact load
  - •• human container/coupling includes handles (good handles are 1-1.5 in. dia with 4 in. width and 2 in. clearance)
  - •• human/floor surface coupling includes good shoes and level, unobstructed good-friction floor
- favorable visual and thermal environment

It was decided to have two limits: an Action Limit *(AL)* and a Maximum Permissible Limit *(MPL)*. The Action Limit can be done by "almost everybody." The Maximum Permissible Limit is the level that only a "few" can do. It was further decided to make the *MPL* three times the *AL*. This then further defines the "almost everybody" and the "few."

The Action Limit *(AL)* assumes:

- Musculoskeletal injury incidence and severity rates increase moderately in the exposed populations
- A 350 kg compression force on the L5/S1 disc can be tolerated by most young, healthy workers. Such forces would be created by the conditions described by the *AL*.
- Over 75% of women and over 99% of men could lift the loads described by the *AL*.

The Maximum Permissible Limit *(MPL)* assumes:

- Musculoskeletal injury rates and severity rates have been shown to increase significantly in exposed populations when work is done above the *MPL*.
- Biomechanical compression forces on the L5/S1 disc are not tolerable over 650 kg in most workers. This would result from conditions above the *MPL*.
- Only about 25% of men and less than 1% of women have the muscle strengths to perform work above the *MPL*.

The guideline formulas and adjustment multipliers are:

$$AL = (IW)(HF)(VF)(DF)(FF) \tag{12.35}$$
$$MPL = 3\,(AL) \tag{12.36}$$

where
AL = Action Limit, kg (lb)
MPL = Maximum Permissible Limit, kg (lb)
IW = Ideal conditions Weight = 40 kg (90 lb)
HF = Horizontal Factor, proportion (see Figure 12.14)
    = $15/H$ for metric = $6/H$ for US units

H = Horizontal location (cm or in) of load center of gravity at lift origin from ankle midpoint (A convenient rule of thumb is to assume the load center of gravity is at $W/2$ where W = load width measured away from the body—i.e., perpendicular to the shoulder line. The edge of the box is assumed to be 15 cm (6 in) from the ankle midpoint. Then $H = 15$ (or 6) $+ W/2$.

VF = Vertical Factor, proportion (see Figure 12.15)
= $1 - (.004)|V - 75|$ for metric = $1 - (.01)|V - 30|$ for US units

V = Vertical location (cm or in) at lift origin
= 1 when the load is at "knuckle height"

DF = Distance Factor, proportion (see Figure 12.16)
= $.7 + 7.5/D$ for metric = $.7 + 3/D$ for US units

D = Vertical Travel Distance (cm or in) between lift origin and destination

FF = Frequency Factor, proportion (see Figure 12.17)
= $1 - F/F_{max}$

F = Frequency of lifts/min
= 0 for lifting less than once/5 min

F_{max} = Maximum lift frequency which can be sustained (see Table 12.11)

The application of the formula is clearer with an example problem. Assume a compact object 6″ × 8″ with handles (with the 8″ parallel to the shoulders) is lifted from the floor to a table 36″ above the floor. The lift would be done less than once per hour. Then the Horizontal Factor = $6/(6/2 + 6) = .67$. The Vertical Factor = $1 - (.01)|0 - 30| = .7$. The Distance Factor = $.7 + 3/36 = .78$. The Frequency Factor = $1 - 0/15 = 1$. Then:

AL = 90 (.67) (.7) (.78) (1) = 33 lb
MPL = 3(33) = 99 lb

Thus, if the load is less than 33 lb, it is acceptable for almost everyone. If it is over 99 lb, the lift is hazardous and should be modified. If it is between 33 and 99 lb, NIOSH recommends: 1) special selection of the workers and training of the workers in lifting, and 2) modifying the job to bring the load below the AL. "*The Guide does not include 'safety factors' commonly used by engineers to assure that unpredicted conditions are accommodated*" (NIOSH emphasis).

The job can be made easier by raising the initial location, by moving the initial location closer to the body, and by lowering the final location; the NIOSH factors permit quantifying the effects of these changes. Remember, of course, that most real lifts have additional variables as the load is not compact, doesn't have good handles, the lift may involve a turn, the floor may be slippery, the feet may have to move during the lift, etc. That is, there are ways to reduce lifting strain beyond the four factors in the NIOSH formula.

Table 12.12 gives a summary of the individual, technique, and task variables with appropriate principles. Selecting individuals is difficult to do and has limited effectiveness. Teaching techniques is difficult to do and has limited effectiveness. Redesigning the job, although difficult, is effective.

SUMMARY

Chapter 12 divides discussion about the human body into two systems: metabolism and the cardiovascular system, and the skeleto-muscular system.

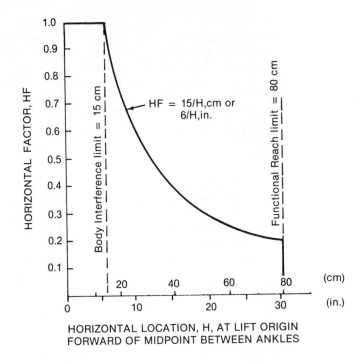

Figure 12.14 *Horizontal factor (HF)* varies between the body interference limit and the limit of functional reach.

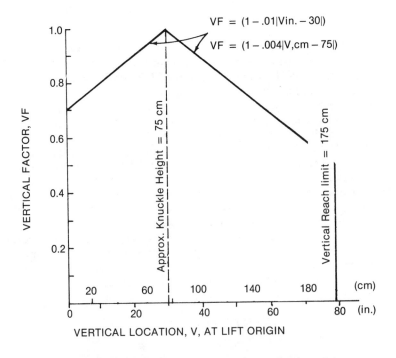

Figure 12.15 *Vertical Factor (VF)* varies both ways from knuckle height.

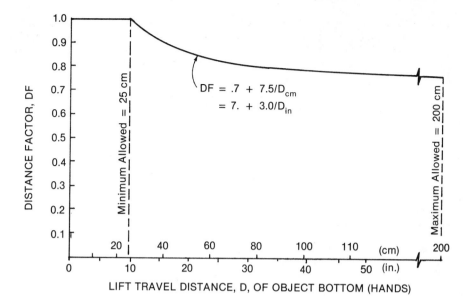

Figure 12.16 *Distance Factor (DF)* varies between a minimum vertical distance moved of 25 cm (10 in) to a maximum distance of 200 cm (80 in).

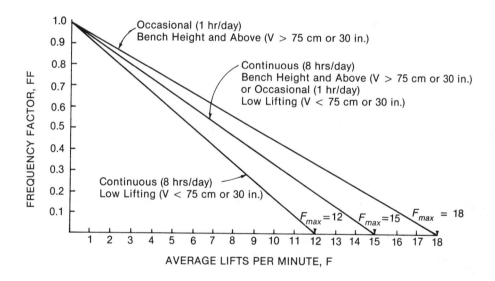

Figure 12.17 *Frequency Factor (FF)* varies with lifts/minute and the F_{max} curve. The F_{max} curve (see Table 12.12) depends upon the lifting posture and lifting time.

Metabolism, subdivided into basal and activity, is measured in laboratories but engineers usually calculate it rather than measuring it. The responses of heart, lungs, and circulation indicate the strain on the body. Cardiovascular responses are through heart rate, stroke volume, artery-vein differential, or blood distribution. Sex, age, and training affect these responses.

The skeleton has three types of levers. In addition to anthropometry differences, muscular strength varies with sex, age, and training. For engineers, the

Table 12.11 F_{max} Values Given as a Function of Work Time and Lift Origin (Posture). The Value for 8 Hours is Based on Working at 33% of Aerobic Capacity (Capacity is Assumed as 1050 W for men and 730 W (70%) for Females) While the Value for 1 Hour is Based on Psychophysical Studies.

	Initial Vertical Location *(V)* and Resulting Posture	
	$V > 75$ cm (30 in)	$V < 75$ cm (30 in)
Lifting Time, hours/day	*Standing*	*Stooped*
1	18	15
8	15	12

Table 12.12   Principles of Occasional Lifting

PRINCIPLE	COMMENT
Select Individual	
1. Don't select on stereotypes	Selection based on stereotypes is discrimination. Great individual variability even within populations (Table 12.8). Age may not affect capability as much as susceptibility to injury. Average female lifts about 60% of average male.
2. Select strong people based on tests	Strong lift more (and more safely) than weak. Select based on muscle groups used in specific job, not X-rays.
Teach Technique	
3. Use the "free-style" lift	Squat lifting needs more energy, stronger leg muscles, and is rarely used; squat is good for compact, heavy loads.
4. Don't slip	Keep feet apart; opposite foot ahead if turning; use nonslip footwear. Table 14.14.
5. Don't be a jerk	Easy does it; not too fast or too slow.
6. Don't twist during move	Bad support for disc.
Design the Job	
7. Put the compact load in a convenient container	Handles are best. Bulky containers give bad biomechanics as well as poor vision.
8. Don't put a load on the floor	Bad biomechanics both up and down plus extra metabolism of moving the body. Knuckle height best.
9. Get a good grip	Hold with hands—not fingertips. No sharp edges.
10. Keep the load close to the body	"Belly up" to objects to minimize torque. Don't wear clothes which can't get soiled by load.
11. Don't lift above shoulder	Poor biomechanics when load passes chest. Danger of falling object.

primary application of strength is manual material handling. Carrying, holding, and lifting depend upon individual, technique, and task factors but the engineer has little control over individual factors and must concentrate on technique and task factors.

SHORT-ANSWER REVIEW QUESTIONS

12.1   What is the metabolic rate (basal + activity) for an 80 kg male driving a car? What is the metabolic rate for a 60 kg female driving a car?

12.2   Sketch a schematic of the heart labeling the four chambers, the pulmonary circulation, and the systemic circulation.

12.3   Oxygen supply to a muscle from the blood is adjusted by what four techniques?

12.4   It is worse to put a load on the ground than on a knee-high platform. Give two reasons.

12.5   Using the open manhole simile, discuss the motivational, training, and engineering approaches to safety.

THOUGHT-DISCUSSION QUESTIONS

1.   One kg of body weight is equivalent to 7700 kcal. Assume you park your car so that you had to walk an extra 100 m every day. Also assume you walk at 4 km/hr instead of sitting and that you work 250 days/yr. How much weight would you lose in a year from this walking?

2.   What can an engineer do to make a job suitable for women as well as men?

REFERENCES

1.   Agan, T.; Konz, S.; and Tormey, L. Extra heart beats as a measurement of work cost. *Home Economics Research Journal,* 1, (September 1972): 28-33.

2.   American Industrial Hygiene Association. Ergonomics Guide to Manual Lifting. *American Industrial Hygiene Association Journal,* (July-August 1970): 511-516.

3.   Andrews, R. The relationship between measures of heart rate and measures of energy expenditure. *American Institute of Industrial Engineers Transactions,* Vol. 1, No. 1 (1969): 2-10.

4.   Arstilla, M. Pulse-conducted triangular exercise—ECG test. *Acta Medica Scandinavica,* Vol. 191, Supplement 529, (1972).

5.   Arstilla, M.; Wendelin, H.; Vuori, I.; and Valimaki, I. Comparison of two rating scales in the estimation of perceived exertion in a pulse-conducted exercise test. *Ergonomics,* Vol. 17 (1974): 577-584.

6.   Ashton, T., and Singh, M. The effect of training on maximal isometric back-lift strength and mean peak voltage of the erector spinae. *Biomechanics IV,* eds. Nelson and Morehouse. Baltimore: 174, University Park Press.

7.   Asmussen, E., and Heeboll-Nielsen, K. Isometric muscle strength in relation to age in men and women. *Ergonomics,* Vol. 5 (1962): 167-169.

8.   Astrand, I.; Astrand, P.; Hallback, I; and Kilbom, A. Reduction in maximal oxygen uptake with age. *Journal of Applied Physiology,* Vol. 35, No. 5 (1973): 649-654.

9.   Astrand, P., and Rodahl, K. *Textbook of Work Physiology.* New York: McGraw-Hill, 1970.

10.   Borg, G. A simple rating scale for use in physical work tests. *Kungl. Fysiografiska Sallskapets i Lund Forh.,* Vol. 32 (1962): 7-15.

11. Brown, J. Factors contributing to the development of low back pain in industrial workers. *American Industrial Hygiene Association Journal,* Vol. 36, No. 1 (1975): 26-31.

12. Brozek, J.; Grande, F.; Anderson, T.; and Keys, A. Densitometric analysis of body composition: revision of some quantitative assumptions. *Annals of N.Y. Academy of Science,* Vol. 110 (1963): 113-140.

13. Chaffin, D., and Park, K. A longitudinal study of low back pain as associated with occupational weight lifting factors. *American Industrial Hygiene Association Journal,* Vol. 34, No. 12 (1973): 513-525.

14. Chaffin, D. Human strength capability and low-back pain. *Journal of Occupational Medicine,* 16 (April 1974): 248-254.

15. Chaffin, D., and Ayoub, M. The problem of manual materials handling. *Industrial Engineering,* Vol. 7, No. 7 (1975): 24-29.

16. Chaffin, D. Functional assessment for heavy physical labor. *Occupational Health and Safety,* Vol. 50, No. 1 (Jan. 1981): 24-32, 64.

17. Coetzee, J. Modeling of forces and torques in lifting and lowering. Ph.D. dissertation, Manhattan, Kans.: Kansas State University, 1976.

18. Cooper, K. *The New Aerobics.* New York: Bantam Books, 1970.

19. Cowgill, G. A formula for estimating the specific gravity of the human body. *American Journal of Clinical Nutrition,* Vol. 5 (1957): 601-611.

20. Damon, A.; Stoudt, H.; and McFarland, R. *The Human Body in Equipment Design.* Cambridge, Mass.: Harvard University Press, 1966; pp. 226-227.

21. Datta, S., and Ramanathan, N. Ergonomic comparison of seven modes of carrying loads on the horizontal plane. *Ergonomics,* Vol. 14, No. 2 (1971): 269-278.

22. Davies, B. Better manual handling and lifting. *The Production Engineer,* Vol. 43 (1964): 394-399.

23. Davies, B. Moving loads manually. *Applied Ergonomics,* Vol. 3, No. 4 (1972): 190-194.

24. Dehn, M., and Bruce, R. Longitudinal variations in maximal oxygen intake with age and activity. *Journal of Applied Physiology,* Vol. 33, No. 6 (1972): 805-807.

25. Drury C. Predictive models for settling safe limits in manual materials handling. *Proceedings of 3rd International Congress on Production Research* (Amherst), 1977.

26. Drury, C., and Pfiel, R. A task-based model of manual lifting performance. *International Journal of Production Research,* Vol. 13, No. 2 (1975): 137-148.

27. Eie, N., and Wehn, P. Measurements of the intra-abdominal pressure in relation to weight bearing of the lumbosacral spine. *Journal Oslo City Hospitals,* Vol. 12 (1962): 205-217.

28. Emanuel, I.; Chaffee, J.; and Wing, J. A study of human weight lifting capabilities for loading ammunition into the F-86H aircraft. *WADC Technical Report 56-367,* Wright Patterson Air Force Base, Ohio, 1956.

29. Gamberale, F. Perceived exertion, heart rate, oxygen uptake and blood lactate in different work operations. *Ergonomics,* Vol. 15, No. 5 (1972): 545-554.

30. Garn, S., and Harper, R. Fat accumulation and weight gain in the adult male. *Human Biology,* Vol. 27, No. 1 (1955): 39-49.

31. Grollman, S. *The Human Body.* New York: Macmillan, 1969.

32. Guyton, A. *Textbook of Medical Physiology.* Philadelphia: W.B. Saunders, 1961.

33. Hamilton, B., and Chase, R. A work physiology study of the relative effects of pace and weight in a cartoon handling task. *American Institute of Industrial Engineers Transactions,* Vol. 1, No. 2 (1969): 106-111.

34. Henschel, A. Effects of age on work capacity. *American Industrial Hygiene Association Journal,* Vol. 31, No. 4 (1970): 430-436.

35. Hettinger, T. *Physiology of Strength.* Springfield, Ill.: Charles Thomas, 1961.

36. Hult, L. Cervical, dorsal and lumbar spinal syndromes. *Acta Orthopaedica Scandinavica* (Supplement 17), 1954.

37. Hunsicker, P. Arm strength at selected degrees of elbow flexion. *WADC Technical Report 54-548*, Wright Patterson AFB, Ohio, 1955.

38. Hunsicker, P., and Greey, G. Studies in human strength. *Research Quarterly*, Vol. 28 (1957): 109-122.

39. James W. *Memories and Studies*. New York: Longmans, Green and Co., 1911; p. 237.

40. Jorgensen, K., and Poulsen, E. Physiological problems in repetitive lifting with special reference to tolerance limits of the maximum lifting frequency. Communication 32, Danish National Association for Infantile Paralysis, Hellerup, Denmark, 1972.

41. Kalsbeek, J. Measurement of mental work load and of acceptable load: possible applications in industry. *International Journal of Production Research*, Vol. 7, No. 1 (1968): 33-45.

42. Kassab, S., and Drury, C. The effects of working height on a manual lifting task. *International Journal of Production Research*, 1976.

43. Kellerman, F., and van Wely, P. The optimum size and shape of containers for use by the flower bulb industry. *Ergonomics*, Vol. 4, No. 3 (1961): 219-228.

44. Konz, S., and Bhasin, R. Foot position during lifting. *American Industrial Hygiene Association Journal*, Vol. 35, No. 12 (1974): 785-792.

45. Konz, S., and Cahill, K. Variability of the interbeat interval of the EKG as an index of mental effort. Manhattan, Kans.: Kansas Engineering Experiment Station Report 71-4, 1971.

46. Konz, S., and Coetzee, K. Prediction of lifting difficulty from individual and task variables. *Proceedings of 4th International Conference on Production Research*, Tokyo, 1977.

47. Konz, S.; Dey, S.; and Bennett, C. Forces and torques in lifting. *Human Factors*, Vol. 15, No. 3 (1973): 237-245.

48. Kozar, A. Anticipatory heart rate in rope climbing. *Ergonomics*, Vol. 7 (1964): 311-315.

49. Lovested, G. Reducing warehousing material handling strains. *Proceedings 24th Annual Meeting of the Human Factors Society*, 1980; 653-654.

50. Meers, A., and Verhaegen, P. Arythmie sinusate, traitement de l'information et tension emotionnele. *Proceedings* (abstract), 4th International Congress on Ergonomics, Strassbourg, 1970.

51. Mitchell, D.; Strydom, N.; van Graan, C.; and van der Walt, W. Human surface area: comparison of the DuBois formula with direct photometric measurement. *Pflugers Arch.*, Vol. 325 (1971): 188-190.

52. Morris, J.; Lucas, D.; and Bresler, B. Role of the trunk in stability of the spine. *Journal Bone and Joint Surgery*, Vol. 43 A (1961): 327-351.

53. Morrissey, S.; Ayoub, M.; George, C; and Ramsey, J. Male and female responses to stoopwalking tasks. *25th Proceedings of Human Factors Society*, (1981): 445-449.

54. Muchinger, R. Manual lifting and carrying. CIS Information Sheet No. 3, Geneva: International Labour Office, 1962.

55. Mueller, E. Training muscle strength. *Ergonomics*, Vol. 1 (1956): 222-225.

56. Mueller E.; Vetter, K.; and Blumel, E. Transport by muscle power over short distances. *Ergonomics*, Vol. 1 (1958): 222-225.

57. Nachemson, A. Low back pain: its etiology and treatment. *Clinical Medicine* (January 1971): 18-23.

58. Netter, F. *Nervous System*. Summit, N.J.: Ciba Pharmaceutical Corp., 1962.

59. Neilsen, W. Some general principles of tray design. Paper at American Industrial Hygiene Association meeting, Los Angeles, 1978.

60. Parker, T. Moving weights at the same height. Master's thesis, Manhattan, Kans.: Kansas State University, 1976.

61. Peres, N. *Human Factors in Industrial Strains.* Melbourne, Australia: Tait Publishing Co., 1964.

62. Poulsen, E., and Jorgensen, K. Back muscle strength, lifting and stooped working postures. *Applied Ergonomics,* Vol. 2, No. 3 (1971): 133-137.

63. Ridd, J., and Davis, P. Industrial human factors research in a British health and safety instititute. *25th Proceedings of Human Factors Society,* (1981): 541-545.

64. Rowell, L. Human cardiovascular adjustments to exercise and thermal stress. *Physiological Reviews,* Vol. 54, No. 1 (1974): 75-159.

65. Shephard, R. *Alive Man.* Springfield, Ill.: C. T. Thomas, 1972.

66. Shock, N. The physiology of aging. *Scientific American,* Vol. 206 (January 1962): 100-110.

67. Sloan, A.; Burt, J.; and Blyth, C. Estimation of body fat in young women. *Journal of Applied Physiology,* Vol. 17, No. 6 (1962): 967-970.

68. Smith, J. A manual material handling study of bag lifting. Paper at annual meeting of the American Industrial Hygiene Association, Portland, Or., 1981.

69. Snook, S., and Ciriello, V. The effects of heat stress on manual handling tasks. *American Industrial Hygiene Association Journal,* Vol. 35, No. 11 (1974): 681-685.

70. Snook, S., and Irvine, C. The evaluation of physical tasks in industry. *American Industrial Hygiene Association Journal,* Vol. 27. No. 6 (1966): 228-233.

71. Snook, S., and Irvine C. Maximum frequency of lift acceptable to male industrial workers. *American Industrial Hygiene Association Journal,* Vol. 29, No. 6 (1968): 531-536.

72. Snook, S., and Irvine, C. Maximum acceptable weight of lift. *American Industrial Hygiene Association Journal,* Vol. 28, No. 4 (1967): 322-329.

73. Snook, S.; Irvine, C.; and Bass, S. Maximum weights and work loads acceptable to male industrial workers. *American Industrial Hygiene Association Journal,* Vol. 31, No. 5 (1970): 579-586.

74. Snook, S., and Cirello, V. Low back pain in industry. *American Society of Safety Engineers,* Vol. 17, No. 4 (1972): 17-23.

75. Sonoda, T. Studies on the compression, tension and torsion strength of the human vertebral column. *J. Koyota Prefect Medical University.* Vol 71 (1962): 659-702.

76. Stolwijk. J. Mathematical model of thermoregulation (Chapter 48) in *Physiological and Behavioral Temperature Regulation,* ed. by J. Hardy, A. Gagge, and J. Stolwijk, Springfield, Ill.: C. T. Thomas, 1970.

77. Switzer, S. Weight lifting capabilities of a selected sample of human males. *Ergonomics,* Vol. 7 (July 1964): 333.

78. Tichauer, E. A pilot study of the biomechanics of lifting in simulated industrial work situations. *Journal of Safety Research,* Vol. 3 (September 1971): 98-115.

79. Tichauer, E. Biomechanics sustains occupational safety and health. *Industrial Engineering,* Vol. 8, No. 2 (1976): 46-56.

80. Tichauer, E. Ergonomic aspects of biomechanics. Chapter 32 in *The Industrial Environment—Its Evaluation and Control.* Washington, D.C. Superintendent of Documents, 1973.

81. Tichauer, E.; Miller, M.; and Nathan, I. Lordosimetry: A new technique for the measurement of postural response to materials handling. *American Industrial Hygiene Association Journal,* Vol. 34, No. 1 (1973): 1-12.

82. Troup, J. Relation of lumbar spine disorders to heavy manual work and lifting. *Lancet,* (April 17, 1965): 857-61.

83. Troup, J., and Chapman, A. The strength of the flexor and extensor muscles of the trunk. *Journal of Biomechanics,* Vol. 12, No. 1 (1969): 49-62.

84. Ufland, J. Einfluss des lebensalters, Geschlechts, der Konstitution aud des Befufs auf die Kraft verschiedener Muskelgruppen. *Abreitsphysiol.* Vol. 6 (1933): 653-663. Cited in Fisher, M., and Mirren, J. Age and strength. *Journal of Applied Psychology,* Vol. 31 (1947): 490-497.

85. Warwick, D.; Novak, G.; Schultz, A.; and Berkson, M. Maximum voluntary strengths of male adults in some lifting, pushing, and pulling activities. *Ergonomics,* Vol. 23, No. 1 (1980): 49-54.

86. Wilmore, J., and Behnke, A. An anthropometric estimation of body density and lean body weight in young men. *Journal of Applied Physiology,* Vol. 27, No. 1 (1969): 25-31.

87. *Work Practices Guide for Manual Lifting,* NIOSH Technical Report 81-122, NIOSH, Cincinnati, Ohio 45226, 1981.

88. Van der Walt, W., and Wyndham, C. An equation for prediction of energy expenditure of walking and running. *Journal of Applied Physiology,* Vol. 34, No. 5 (1973): 559-563.

89. Yates, J.; Kamon, E.; Rodgers, S.; and Champney, P. Static lifting strength and maximal isometric voluntary contractions of back, arm, and shoulder muscles. *Ergonomics,* Vol. 23, No. 1 (1980): 37-47.

90. Zuti, W., and Golding, L. Equations for estimating percent fat and body density of active adult males. *Medicine and Science in Sports,* Vol. 5, No. 4 (1973): 262-266.

PART V
RECOMMENDED
WORK
DESIGN PRINCIPLES

ORGANIZATION OF WORKSTATIONS

PRINCIPLE 1: USE SPECIALIZATION EVEN THOUGH IT SACRIFICES VERSATILITY

Specialization is a key to progress. Use special-purpose equipment, material, labor, and organization. Seek the simplicity of specialization; thereafter distrust it, but first seek it.

Equipment

Special-purpose equipment has the advantages of greater capability and lower production cost/unit but the disadvantages of slightly higher capital cost and less flexibility.

Special-purpose equipment often can perform functions that general-purpose equipment cannot. As the designer designs specialized equipment and the user uses it, design restrictions of general-purpose equipment are eliminated and major improvements often result. For example, an ordinary grinder is used to remove very little material and give a high surface finish. A special-purpose grinder may be able to remove large amounts of material (rough cuts) as well as leave a good finish so that the job may be done on one machine in one setup instead of two machines. A general-purpose cash register will have numbers on the keys so any price product can be registered. A special-purpose register has special keys for each item (hamburger, small Pepsi) instead of prices; pressing the key actuates the proper price and perhaps may even display the complete order on a screen for the employee to use for order picking.

Lower production cost/unit results from the specialized nature of the machine's components. The components run at maximum speed, give minimum variability, require minimum labor time to operate, etc.

In theory, special-purpose equipment has fewer components since many of the components needed to make general-purpose equipment general purpose are not needed. Fewer components should mean a simpler machine and thus a lower capital cost. However, the number of copies of each special-purpose machine is small so design and build costs must be spread over a few machines instead of many. Thus, special-purpose equipment usually has a higher capital cost than general-purpose equipment.

A penalty of special-purpose equipment is its lack of flexibility—it does one job very well but only one job. What if you don't have just one job?

Material

Specialized materials have the same types of advantages and disadvantages as specialized equipment with the most common trade-off being higher material costs vs. greater capability.

For example, when you use tool steel for dies instead of the cheaper low-carbon steel you trade off greater capability and longer die life vs. a higher material cost. A titanium basket in the plating department trades longer life and less maintenance vs. higher initial material cost. A throw-away syringe in the hospital trades better sanitation and elimination of cleaning costs vs. higher initial cost/syringe. A rug in the office trades lower maintenance costs vs. higher capital cost.

Labor

Labor specialization affects both labor quality and quantity.

Quality. Quality of output of a specialist is potentially high due to the skill being in the tool and due to practice. The jack-of-all-trades is master of none.

When specialization is high, the specialist develops or purchases special-purpose machines and tools—thus the statement "the skill is in the tool."

With special-purpose tools, many hours spent at the same task, and a more restricted variety of skills required, quality should be better for the specialist. At least in theory a "Ford carburetor tune-up" operator should be more skilled at tune-ups of Ford carburetors than a "carburetor tune-up" operator (who works on all models of cars), or a "tune-up" operator (who works on all aspects of the field). Thus the specialist may be much farther out on the learning curve; that is, the specialist may have tuned 7500 Ford carburetors while the mechanic may have tuned 150. The brain surgeon may have operated on 1000 brains while the general surgeon may have operated on 1.

Quantity. Quantity of output/time usually is higher for a specialist (that is, labor time/unit is lower) for the same reasons that quality is higher. With a restricted range of skills, training time is less for the specialist.

Since the individual is trained deeply rather than broadly, the general rationale has been that a "less talented" individual is required and thus a lower rate of pay is justified. Thus, in most of industry the generalist (tool and die maker) is paid more than the specialist (turret lathe operator). Thus, specialized labor costs less both because of its greater productivity/unit and its lower wages/hr.

From the individual viewpoint, specialized work may be repetitive and monotonous. It has been difficult to recruit workers for monotonous jobs—if they have low pay. If the job has high pay, many workers can be found whether the job is monotonous or not.

Job Organization

The adjectives for specialization are rigidly structured, inflexible, disciplined, machinelike. The overwhelming characteristic is the need for high volume of a standardized product. Levitt mass-produced homes by breaking home construction into 26 steps and "reversing the assembly line" (the product stood still and the worker moved). If you are going to do nothing but brain surgery, you need many patients requiring brain surgery; if all you do is tune up Ford carburetors and it takes 30 min/carburetor, then you need about 15 Fords/day to keep busy.

If, instead of time/unit of 30 minutes, time/unit = 1 minute, then output/day is 450 (allowing for breaks), output/month is about 10,000, and output/yr is about 120,000. Can you sell 120,000 identical units/yr? For time/unit of .1 min (6 s), output is 1,200,000/yr. Can you sell that many?

Most firms don't have that amount of volume and can use specialization only as a desirable goal. One approach finding growing favor is *group technology* which attempts to get the benefits of mass production from batch production.[8] "Families" of parts are manufactured in "cells." The crux of the problem is to identify, from the large variety of components manufactured in a typical firm, the similar components (the families), so common solutions can be found for similar problems. Then "members of the family" are scheduled close together (close in time) and produced close together (close in space). Benefits include many of the advantages typically cited as advantages for specialization such as lower setup cost/unit, less paperwork cost, increased use of special-purpose fixtures, etc. However, the benefits primarily are in *avoided* costs so managers find it difficult to spend money for savings that don't show up in the cost accounting system.

PRINCIPLE 2: MINIMIZE MATERIAL HANDLING COST

Material handling does not add value, just cost. Reduce the cost by analysis of its components.

Cost of material handling can be broken down as follows:

- Material handling cost/yr = Capital cost + Operating cost
- Operating cost = (Number of trips/yr) (Operating cost/trip)
- Cost/trip = Fixed cost/trip + (Variable cost/distance) (Distance/trip)

Capital Cost of Systems

The capital costs of material handling (return on investment and depreciation) do not vary appreciably with the amount of material moved. For example, you may purchase an electric fork truck for $20,000 and a recharging station for $3000. Then $23,000 invested at 10% returns $2300; this cost occurs whether you used the truck one hr/month or 100 hr/month; whether your downtime is 10% or 90%. In addition, depreciation depends more on age of equipment than the usage. For example, resale value of the fork truck after 2 years might be $5000 if you use the truck 1 hr/day and $4000 if you used it 8 hr/day. A used conveyor probably will sell for the same amount regardless of use. Conversely, capital cost/unit moved can become very low if the equipment is kept busy.

Thus, if utilization is poor, the lowest total cost might be obtained for a system with high operating cost but low capital cost. If utilization is high, the high capital cost alternative may be best.

Eliminating peak loads by scheduling may eliminate need for some equipment. For example, schedule shipments for 5 days a week, not just Thursday and Friday; receive material from vendors 5 days/week. Move by priority—not just first-come first-served.

Number of Trips/Year

The ultimate would be to make number of trips equal zero, eliminating not only operating cost but also capital cost. Question the need for the trip; the trip may not really be required. A repair or maintenance call might be eliminated if higher-quality maintenance is used (e.g., use a component which requires service once every 120 days instead of 60 days). A sales call might be eliminated by use of a letter or electronic communication. The opposite of transportation is communication.

Reduce the number of trips by scheduling and combining trips. For example, a trip from San Francisco to New York with a stop at Chicago is less expensive than two trips, one to Chicago and one to New York. The trade-off is among reduced travel cost, increased scheduling problems, and capacity/trip. The same type of trade-off must be made in other problems. Should the clerk take each item to the duplicating machine or accumulate a batch before going? Should the money be deposited in the bank once a week, once a day, or once an hour? Should the operator send material from the machine to the next station once/min, once/hr, or once/day? Should the pallet hold 50 units and move once/day or hold 25 and move twice/day?

Fixed Cost/Trip

Fixed cost/trip has two components: (1) information transfer (mainly paper-work) and (2) start and stop.

Information transfer is a material handling cost that often is overlooked. Reduce these costs by using *line* production. In a job shop, even if the many paperwork forms are completed correctly, costs are substantial. It is not just the cost of filling out the forms; it is the cost of transporting the forms, filing the forms, transferring information from one form to another, etc. Mistakes occur. Products can get misplaced, workers can run out of supplies, trips can be made to the wrong place. One of the major advantages of line assembly is the standardization of routing and scheduling—thus reducing information transfer costs. Computer and electronic sensor technology, however, can reduce information transfer costs so that single-product assembly lines no longer are always desirable. For example, electronic sensors can "read" a box number as it moves along a conveyor, send the information to a "brain," which then consults its "memory," decides to send this box to station 14, and then moves its "arm" to put the box in station 14.

Start and stop (pick up and put down; load and unload; pack and unpack) is a substantial cost that does not vary with the distance moved. A large part of the cost of flying a commercial airplane is the takeoff and landing cost just as a large portion of a fork truck's cost and time is spent picking up and putting down the load. Thus if load time = 1.0 minute, travel time = .01 min/m, and unload time = 2.0 minutes, a 50 m trip costs 3.5 minutes and a 100 m trip costs 4.0; twice as far doesn't cost twice as much (See Figure 7.2.)

Much transportation (and communication) is "distance insensitive." Over the last 100 years technology has increased this insensitivity. Water transport especially is insensitive to distance: goods can be moved from Japan to New York for not much more than from Boston to New York. Land transport has been speeded by improved highways and air transport by improved airplanes. Improved electronics—both wired and wireless—speed communication of words and data.

These increases in insensitivity affect plant design and location as products now come from a specialized factory in Dusseldorf or Milan or Chicago rather than a number of local plants, each with a variety of products with low production volumes. Increased mobility of workers (due to the automobile) and of products (due to trucks instead of railroads) permits decentralization of places of employment and, thus, decentralization of cities.

Within a factory, fork trucks and tractor-trailer trains have increased distance insensitivity. Formerly, factories were built vertically to minimize product movement distance. Offices (paperwork factories) are still in the earlier stage of development (multistory buildings with workstations (desks) close together).

Variable cost/distance. Costs/distance are a function of energy consumed and labor cost. Energy consumption does not make much difference in many cases as it is a small portion of the total cost. Electric fork trucks do tend to be cheaper to operate than diesel, which are cheaper than gasoline, which are cheaper than liquid petroleum (propane). Low resistance to motion helps so trains use less energy than trucks; ships are cheapest of all per km-ton.

Care must be taken that low energy costs are not overcome by high labor costs. Move more product/labor hour—the ultimate is infinite volume or zero labor hours. Large oil tankers use much less labor/barrel of oil transported than small tankers as do large trucks vs. small trucks—more volume for the same labor. For distances within a plant of over 150 m, tractor-trailer trains may be more economical than fork trucks. If the route is standardized (say less than 20 destinations), the train need not have a driver as sensors and computers can replace a driver in some applications—the same volume with less labor.

Distance/trip. Reduce distances by efficient layout and arrangement. Trade off short trips for local supply vs. decreased inventories for central supply. For example, fewer micrometers are needed with a tool crib than if everyone is issued one but distance moved to get the "mike" becomes 100 m not 2 m. Micrometers have low capital costs while fork trucks have high capital cost. Thus, giving each department its own fork truck reduces distance traveled but the extra capital cost can be very high. Sharing may work; that is, department A "owns" it in the morning and B in the afternoon.

All other things being equal, a short trip costs less than a long trip. Figure 13.1 shows a "bus route" around an area. A "bus" goes around the area on a standardized route. Material going from B to A must first go to C and D before coming to A. Why use a "bus" instead of a "taxi" system (point to point service)?

First, cost of movement tends to be relatively insensitive to distance due to high pick up and put down costs in relation to movement costs (see Figure 7.2). Second, if distance moved is important, usually it is total distance moved by the *carrier* (distance/loop × number of trips around loop) which is important rather than distance moved by the objects. Third, time for the physical movement of products tends to be relatively small in relation to the time in storage at each end of the move and thus not very important. Overhead conveyors often wind for long distances around the ceiling to act as a work-in-process storage. In the international oil trade, tankers often leave port without a specified destination; their cargo is sold when they near Europe.

Emphasize minimizing total material handling cost—in itself a subset of the goal of reducing manufacturing cost—rather than minimizing distance/trip, variable cost/distance, etc.

PRINCIPLE 3: DECOUPLE TASKS

There are two reasons for decoupling:

1. The line balancing problem—that is, mean times for task A, B, C, etc., are not equal; there is a "balance delay." Assume Joe's task takes 50 s and Pete's

234

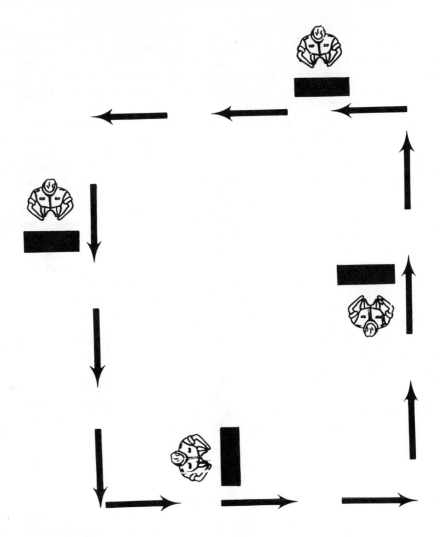

Figure 13.1 *Bus routes* may be preferred to "taxi service."

60. If Joe completes one unit and then immediately passes on just the one unit to Pete who is waiting to work on the same unit, Joe always will have 10 s of idle time ("station starvation") until Pete completes the previous unit. (Usually Joe will work slowly for 60 s rather than normally for 50 s and rest for 10.) The more tasks that are rigidly linked together the worse the balance delay becomes. Rigidly linked systems (such as gear trains) operate at the speed of the slowest component. Rigidly linked systems do not maximize the output of each component.

— People vary. Joe might take 50 s to do his task, but Bill might take 40 s and Sam 53. If Sam is assigned to the job instead of Joe, then all the stations—if rigidly linked—require 53 s. Sam may be just as motivated as Joe and just as capable physically but have 5000 cycles of experience instead of Joe's 500,000. The line's output is limited by Sam. Elimination of the slowest operator does not, in general, solve the problem; it just uncovers a new slowest operator. (There is always a slowest operator.) Davis recommends, for rigidly paced lines, putting the best operator at the start of the line rather than the end or, for equal operators and unequal station times, putting the short station time at the start of the line.[5] (See chapter 7.)

2. **The shock and disturbance problem**—that is, times vary. Although Pete's

standard time is 60 s, he may be sick today and his replacement for today takes 66 s. If the tasks are rigidly coupled then not only does Pete's task take 6 s extra but so does Joe's. This is a temporary change in the mean.

Pete's time also can have variability. His tool might become dull, his machine might break down for a minute or two, his supply of one type of component might be exhausted for several minutes, he might stop to light a cigarette, he might talk to the foreman, some parts might stick, he might watch Maria walk by on her way to the office, he might drop a part on the floor. In a rigidly linked system the speed of the system is not just the speed of the slowest operator—it is the speed of the slowest cycle of the slowest operator. Self-paced workers produce more than machine-paced workers.[15]

During the 1960s, automation came to assembly operations. "Joe" and "Pete" may be machines. Machines especially need decoupling due to their inflexibility and thus extraordinary sensitivity to even very minor disturbances. A machine will fail, for example, if the incoming part arrives upside down.

Banks (Float, Buffers)

Tasks are decoupled by *banks*. Banks permit a group to produce at the *average* cycle of the slowest station rather than at the *slowest* cycle of the slowest station.

Most jobs (secretaries, doctors, teachers, maintenance workers, police officers, many machine operations) are so decoupled that a novice might not recognize the bank before and after the task. In these tasks either the work is not standardized enough for an assembly line or material handling and work-station storage problems may not encourage an assembly line. (There are even some tasks which have very little tie to other people, time, or deadlines; examples would be sculptors, composers, novelists, etc.) Banks are most obvious where decoupling is needed most—the assembly line. Banks increase *tolerance time*.[7] (See Figure 13.2.) Working from storage to storage permits the item to be tested at each storage. Banks permit a line to produce at the *average* cycle of the slowest station rather than at the slowest cycle of the slowest station. Okamura and Yamashina discuss the amount of buffer capacity to provide.[18] The basic trade-off is between idle time and capital cost (for the buffer and the products in the buffer).

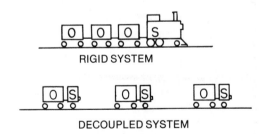

RIGID SYSTEM

DECOUPLED SYSTEM

Figure 13.2 *Two types of systems* are (1) rigid systems and (2) systems with margins. The progressive assembly line can be symbolized as a "train" with storage (S) followed by operations (O); that is, S-O-O-O. Trucks symbolize the decoupled system; that is, S-O, S-O, S-O.

Buxey, Slack, and Wild have an excellent review of line problems.[4]

If the banks are too small, stress on the operators increases. The assembly line commands; man executes. Figure 13.3 shows the distribution of cycle times for

236

(1) work with sufficient banks and (2) paced work without sufficient banks.[6] In typical circumstances there are a few long times (distribution is positively skewed) due to fumbles, delays, sneezes, and itches and scratches. As "Joe" can't keep up with the pace, he begins to "slip down the line"—interfering with the following operation—or Joe lets the unit go down the line incomplete. Joe takes fewer of his formal and informal rest breaks as he falls behind, becomes excessively tired and thus slower, takes even less of his break, becomes more tired—a positive feedback system. If "Joe" is a machine, then the line must adjust to the speed of this slowest cycle.

In the short run, Joe works during his breaks or stops the line by pushing the emergency button, or, worse, reduces quality. In the long run, Joe will transfer to other jobs giving less stress. Strikes and group slowdowns also are common.[1] Toyota, for example, has a buffer of 3 cars approximately every 10 stations. Each worker has a button which can stop the line for bad quality. In the US auto industry, in contrast, there are no buffers so only a very few people are authorized to stop the line and they rarely do so.

A machine-paced automotive assembly line usually has idle time of 8% to 15% due to the line balancing time and the shock and disturbance problem. In the U.S. auto industry, the companies also allow an extra relief of 22 min/480 min shift (about 5%) for a machine-paced line. Thus their machine-paced lines have a built-in inefficiency of 13% to 20%. Kilbridge estimated 5%-10% as a typical balance delay in the USA.[10] Figure 13.4 gives balance delay vs. cycle time in four different companies. See Figure 13.5. Consider "short" assembly lines (in the extreme complete assembly at one workstation).[20] Although the progressive assembly line is a "manager of men," it is a poor manager.[5] Although line workstations usually are connected by conveyors, they need not be. Other alternatives include manual movement, air-flotation pallets, and even automatic storage/automatic retrieval machines.

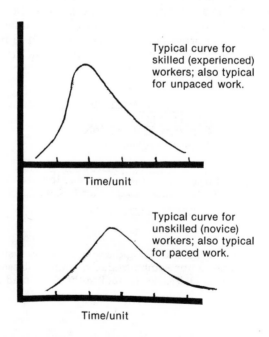

Typical curve for skilled (experienced) workers; also typical for unpaced work.

Time/unit

Typical curve for unskilled (novice) workers; also typical for paced work.

Time/unit

Figure 13.3 *Skilled workers* have a lower variance and more skew as well as a lower mean.

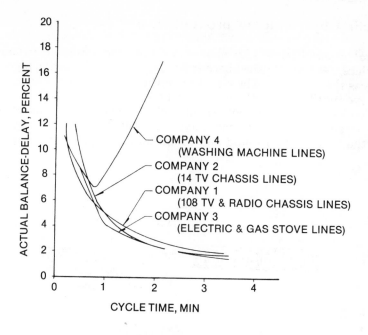

Figure 13.4   *Balance delay time* usually declines when cycle time is longer.[11]

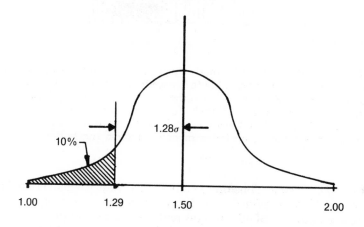

Figure 13.5   *Inefficiency of assembly lines* is increased by the typical variability of workers. The range of performance among operators is about 2 to 1. Assuming symmetry, the average worker is at 1.5. However, if a line is to run at the speed of the average worker, 50% could not keep up and 50% could work faster. (A speed too fast for an operator presents a temptation to reduce safety or quality.) Thus lines often run at a speed at which almost everyone (say 90%) can keep up. In statistical terms this is -1.28σ below the mean. If the range of performance from 1 to 2 is considered 6σ, then $\sigma = (2-1)/6 = 1/6$ and $1.28(1/6) = .21$. Then $1.5 - .21 = 1.29$ for the line speed. If decoupled individual workstations could operate at the mean speed of 1.5, then output would be $1.5/1.29 = 1.17$ of the line output.

Banking Techniques

There are two banking techniques: (1) decoupling by changing product flow, and (2) decoupling by moving operators.

Decoupling by changing product flow

Physical Bank. Figure 13.6 shows two common physical arrangements of banks. Place a simple physical barrier such as a piece of wood, a pipe, or piece of steel across the conveyor. The pieces from the previous workstation move along the conveyor until they hit the barrier and then stop. Joe lifts them up across the barrier, works on them, and puts them back on the conveyor downstream of the dam. To obtain more storage space have the input feed onto a rotary table. Parts stay on the table and go round and round until Joe picks them up to work on them.

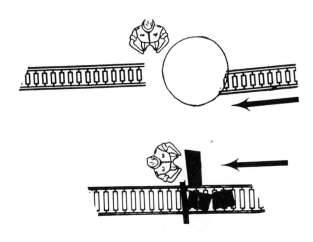

Figure 13.6 *Physical banks* can be made by putting a "dam" across the flow of parts. Another method is to use a rotary table to create the "reservoir." If there is a "flood" due to insufficient reservoir capacity, then the upstream must stop (station blocking). It may be desirable to create a "safety stock" for the incoming and outgoing buffer by putting a few of each type of unit aside at each workstation. The variability of the number of units in a buffer is an index of how effective the buffer is.

Bank by Carrier Design. Time/unit is more consistent when averaged over a container ($\sigma_{\bar{X}} < \sigma_X$). This *container* or carrier can be a pan, box, pallet, index table, or hook. There are a number of advantages and disadvantages of multiple items/carrier. It may be possible to obtain many of the advantages of multiple items/carrier even with one unit/carrier by using *batch processing* of carriers—each with one unit.

> *Labor:* • Pick up and put down of tools. In a high volume operation, say a station time of .1 minute, Joe picks up his tools at the beginning of the shift and doesn't put them down until break time. Thus his "get time" and "put away" time are prorated over many units so the cost/unit is small. If the station time is longer, say 1.0 minute, then Joe does a number of operations and uses a variety of tools. He picks up and puts down each tool. However, if the items come in a carrier with multiple items, Joe can "do" a number of units with only one "get" and one "put away." If ("get time" + "put away" time) = X and this time is prorated over N units, time/unit is X/N. For $N = 1$, time/unit = X; for $N = 2$, time/unit = $.5\,X$; for $N = 4$, time/unit = $.25\,X$; for $N = 8$, time/unit = $.125\,X$. The point of diminish-

ing returns comes fairly rapidly. In addition, as N increases, the reach and move distances in the workstation increase, so "do" times increase.

- Multiple units/carrier encourages use of both hands as two units can be worked on at the same time. When using multiple items/carrier make N an even number.

- Multiple units/carrier may restrict access to individual units on the carrier. If Joe is a person this restriction may merely increase movement time for removal and replacement from and to the carrier. If Joe is a machine, removal or replacement may be too expensive for multiple units/carrier.

Material Handling:
- Multiple units/carrier give more units/meter of the line. For a specified distance, multiple units/carrier give more float. For a specified number of units of float, the float will fit into a shorter distance for multiple units /carrier.

- Multiple units/carrier require heavier carriers. The carrier will be harder to push, pull, and lift—although there will be fewer carriers to move. Moving the heavier carrier may require motors rather than muscles. Kilbridge estimated "get time" + "put away" time as 13% of "do" time for electronics assembly when done on man-paced conveyors.[10]

- Disposal of rejected units is more difficult for multiple units/carrier. If the defective unit remains on the carrier, care must be taken to avoid additional work on it and make sure it does not get included into the "good production." The defect also fills up line space. If the defect is removed from the carrier, empty space moves up the line. In addition, a carrier must be provided for the rejects to move them to the rework station or the scrap pile.

Equipment Costs:
- Carrier cost/unit usually is less when there are multiple units/carrier. The cost/carrier is higher but a smaller number of carriers is required. For example, 1 carrier holding 8 units might cost $100 while 1 carrier holding 1 unit might cost $20.

- Automatic processing of units is more difficult with multiple units/carrier. Build duplicate heads on one machine so that all units on the carrier are done simultaneously; or do one unit and then index either the carrier or the head. The extra heads or indexing equipment may be just extra capital expense since just 1 head may have had sufficient capacity if it were not for the multiple units/carrier.

Bank by Line Arrangement. Figure 13.7 is a schematic of a "workstation." The total workstation is composed of an operator, machine, energy and information input, energy and information output, input product storage, and output product storage. Float increases the size of input storage and output storage.

Increase storage (1) by increasing the time the item is in input storage or output storage or (2) by increasing the space for input storage or output storage. Increase time by having the operator face upstream; arm motions forward are easier and, when the object can be seen, timing can be better. If the object approaches from the rear of the operator, use a rear-view mirror. Increase space

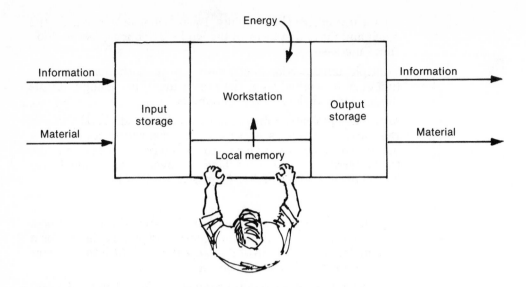

Figure 13.7 *Schematic workstation* shows that both information and material flow into input storage. Then they are transferred to processing where they are transformed using energy and local memory. Then the information and material go to output storage before being transferred to the next workstation.

for storage with a rotary table or by curving the conveyor or by using the cube of the space (see Figures 13.6 and 13.8). For a moving assembly, the components can be stored on a second conveyor (behind the operator or above the assembly conveyor) moving at the assembly conveyor speed.

Cafeterias are an example of using the assembly line concept to move product (customers) by stations (salads, desserts, drinks). Waiting lines build up at vari-

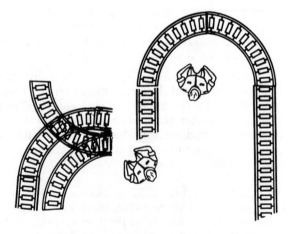

Figure 13.8 *Prevent station starvation* by increasing the storage—either by increasing the storage space or by increasing storage time. Using the cube of the space helps. Making a U in the line takes advantage of man's ability to turn his body. Increase the time the unit is available with reservoirs as in Figure 13.6. Close spacing of items on a belt and slow belt speed is preferable to wide spacing but high belt speed.

ous stations such as desserts. To decouple the stations, use the scramble system shown in Figure 13.9.

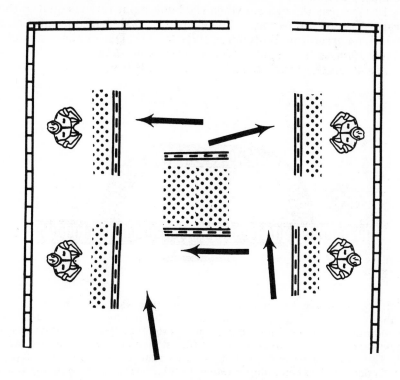

Figure 13.9 *Scramble system cafeterias* assume that customers will stop at only a fraction of the available service areas. They reduce customer throughput time by permitting customers to skip facilities at which no service is desired.

Bank Off-Line (remote in time or space). On-line banks can handle minor shocks. Major shocks (machine breakdowns, employee absences, learners, etc.) may require a larger bank. This larger bank may be off-line. Off-line banks also reduce line balancing problems. Those interested in reliability problems will recognize Figure 13.10 as a "standby" circuit. The major elements are the device,

Figure 13.10 *Off-line banks* can function as a standby component in the production system for cases of machine breakdown, employee absence, etc.

the sensor which detects circuit failure, the switch which activates the standby system, and the reliability of the standby device (in this case, how good are the units we place in storage).

Figure 13.11 shows the system when the bank input rate is equal to the bank output rate. Assume, for example, line output is 80/shift of 8 hr; station A produces 20/hr for 4 hr; station B produces 10/hr for 8 hr. During the morning, operation A sends 10/hr to B and puts 10/hr into the bank. At the end of the morning the bank has 40 units. During the afternoon, feed B from the bank at the rate of 10/hr.

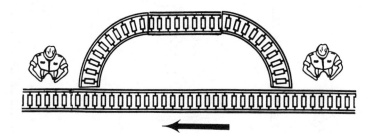

Figure 13.11 *Banks* can be used for line-balancing as well as for shock absorbers. Assume the first station produces 20 units/hr for 4 hr, putting 10/hr in a bank and sending 10/hr down the line. The second station produces 10/hr throughout the 8-hour shift, being supplied for the first 4 hr by the first station and for the second 4 hours by the bank. This example assumes bank input rate equals bank output rate. Even if bank input rate is not equal to bank output rate, banks can be used. Assume the first station produces 20 unit/hr for 6 hr; that is, 120/shift. The bank receives 5/hr and the second station 15/hr. At the end of 6 hours, station one shuts down and station two is fed from the 30 in the bank for the remaining 2 hr at the rate of 15/hr.

Figure 13.12 *Off-line processing* may be desirable as well as off-line storage. The main advantages are better utilization of equipment and people and for training of new operators. Disadvantages are increased material handling and scheduling problems.

The input and output rate of the bank need not be equal. A still produces 20/hr but for 6 hr. B, however, has a rate of 15/hr instead of 10. During the 6 hr, operator A sends 15/hr to B and puts 5/hr into the bank. At the end of 6 hr, the bank holds 30 units. During the last 2 hours of the shift, feed B 15/hr from the bank.

In Figure 13.11 it was assumed that just the bank was off-line. It is also possible to process off-line in addition to banking. Figure 13.12 shows this situation. Off-line banking and processing can be used to reduce idle time for a worker whose primary task does not require 8 hr/shift, to reduce idle time of an operator with a large machine time component, and for training purposes. Workstations often are close together to minimize distance moved. The proper criterion, however, is material handling cost and most of the cost is in load-unload, not movement. That is, while moving 10 m may cost $.10, moving 100 m may cost only $.11. Note that *remote* can be remote in time as well as space; that is, use overtime work, other shifts, and holidays to maximize output of the existing equipment. The engineer must decide if the increased material handling and more complicated scheduling required by off-line banks compensate for the benefits of the banks.

So far we have decoupled by moving product. The concept of the workstation and the assembly line have been so embedded in engineers' minds that the obvious is overlooked—that people have legs.

Decoupling by moving operators

Utility Operator. Figure 13.13 shows the first approach, the utility operator. In this concept, most of the operators work at specific work stations. One operator, operator D, is a utility or relief operator. The utility operator's assignment is to help each of the individual operators if they have trouble for a few minutes, if they want to go to the toilet, etc. The duties of this operator vary widely. In the US auto industry, for example, a common situation is one relief worker for every six stations; that is 7 people work at 6 stations with one always being off. In other industries the utility operator does not have formal times assigned to

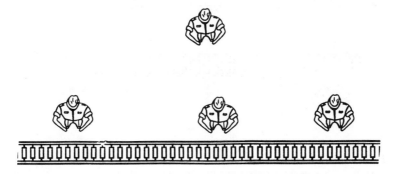

Figure 13.13 *Utility or relief operators* move from station to station while the station operators all stay at their own station. In theory all workers work at the same speed—they are interchangeable parts. In practice, put your best operators on the toughest jobs; they can cope with problems an ordinary operator couldn't handle without slowing down the line. Many supervisors do not want a line with all work times equal; they prefer one "soft" job which can be given to an older worker, a trainee, etc. The utility operator must know all the jobs and so often becomes the trainer. In many cases the utility operator is given minor management responsibility and is called a "group leader" or "working supervisor."

relieve specific operators but just helps out where needed. In these situations, the utility operator often is called a group leader (working supervisor) and has the responsibility of training new employees and making minor decisions (if the supervisor is absent from the department) in addition to working at the various stations. Minor maintenance work or product rework are other duties.

Help-Your-Neighbor. Figure 13.14 shows the second approach, "Help your neighbor." In this concept, each operator is "your brother's keeper" by management directive. That is, you help your neighbors not because you are a "good person" but as part of your job responsibility. One approach is to divide the work at each station into thirds; the middle third is the sole responsibility of the station operator, the first third is entitled to help from the "upstream" operator, and the last third is entitled to help from the "downstream" operator. "Help your buddy" is a quite useful concept since jobs are not all equally difficult, some operators are faster than others, machines break down at different times, and parts stick at different stations at different times. If management does not formally require operators to help each other out, then those helping may feel that they have been made a "sucker." Rigid pacing without buffers puts so much stress on operators that they are looking for an excuse to stop and so they may even deliberately disturb the product so that someone will push the emergency stop button and everyone can get caught up.

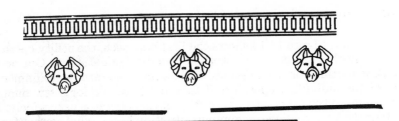

Figure 13.14 *Defined overlap* means that all workers are their "brother's keeper" as a part of the job—not because they are "nice guys." Responsibility is shared and a worker cannot ignore the problems of his fellow workers.

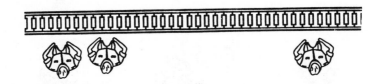

Figure 13.15 *Full float* means that there are *n* operators for the *n* stations. Specific operators are not assigned by management to specific stations. They move from station to station as the need arises. When this system is used with group incentive pay there is considerable social pressure on "slackers."

n Operators float over n stations. Figure 13.15 shows the third approach: *n* workers float among *n* stations. Stations are not given specific times/unit; just the total time required for the entire assembly is given. The operators move upstream and downstream as required. Who does what does not matter to the management as long as completed units are produced at a satisfactory rate. This type of line is used in some group incentive systems where the pay is proportional to completed units. In this situation, workers sometimes do not want substitutes when someone is absent—the substitute would interfere with their teamwork. They all work harder on that day and divide the total pay by $n - 1$ instead of n. This type of line management is very efficient since float problems are minimized (if float = 0, the operator moves to another station instead of sitting down), group pressure to produce is high (any slacker is told to "get your hand out of my pocket"), and the line runs at the average speed of the group rather than the speed of the slowest member.

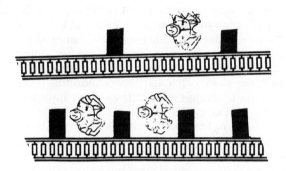

Figure 13.16 *N operators for more than n workstations* gives minimum downtime for the expensive component in the system—human labor. This system requires flexible workers but is very useful when a number of different models are produced on the same day or when production scheduling must be very flexible. Both this system and the full float design eliminate the adding of excess work to a line to reduce balance delay. (In a design where each operator is tied to only one station, it may be that a specific station has a high cycle time, thus forcing the time for every station on the line to be this high time. Then an element—say, drive 16 screws—from the high-time station is split between two workstations. This does, however, add an extra pick up and put down of a screwdriver to the required work.) Note that if the "conveyor" is eliminated from the figure, *n* operators for more than *n* stations describes many component areas such as machine shops. Assembly lines are concepts; material handling and layouts follow many patterns.

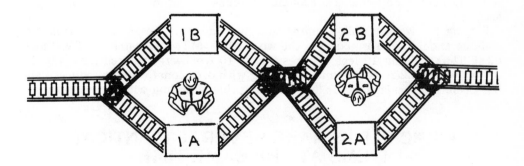

Figure 13.17 *Paired-station assembly line* is a version of double tooling. Items are removed and added at station B while the operator works at station A. Then the operator works at B while loading and unloading is done at A. That is, load and unload are done while the operator is working. Thus operator utilization is high while workstation utilization is low.

n Operators move over more than n stations. Figures 13.16 and 13.17 show the fourth approach, the *n* workers floating among more than *n* workstations. The rationale of more places than people is minimization of total cost of the production system. The human portion of the system (one shift) costs from $5000 to $20,000 per year (wages + fringes) in the USA. Many workstations can be built for $1000 as they are just a bench with some hand tools. Even with more elaborate material handling and power tools, $10,000 builds most workstations. In addition, workstation cost can be prorated over the remaining product life (say 3 years) so a $10,000 initial cost becomes $3,300/yr.

One variation of this fourth approach is to have several complete lines set up and ready to go (line A for product A, line B for product B, line C for product C). Then the workers work on, say, line A on Monday, line B on Tuesday and Wednesday, line C on Thursday morning, and then line A for the rest of the week.

An article in *Factory* describes a furnace assembly area of 9 lines for assembly of 171 models.[14] Teams of 2 rove from one line to another, doing complete assembly of a given model. Meanwhile, a setup operator converts an assembly line for the next batch of another model. Production scheduling is very flexible. Each team specializes in a range of models so that they don't have to know all 171 models. If there is a shortage of components for any line, they simply shift over to another product until the supply is sufficient. This flexibility is very useful when there is insufficient demand for just one product for a line for an entire year. Although equipment is duplicated, setup and put away costs are minimized and maintenance can be done during normal working hours instead of on weekends at premium pay.

Another version is the "one worker line." Break the total job into, say, three workstations. Joe works at station A completing those operations on a number of units, say 25. Then Joe moves the 25 units to station B, completing the required units, and then to station C. Splitting the workstation into three workstations reduces the congestion which occurs if all the components and tools are in one physical location. Since there is only one worker, he has control of all float so interruptions are not multiplied as they would be on a rigidly paced line.

More places than people helps when some operators are new.

Another version of the *n* workers and more than *n* workstations might be a single-product line with, say, 12 workstations and 4 workers. The workers work at a specific station and then send product to the next station. Then they walk to the next station and begin work again. Multiple workers do require more coordination than just one worker but float problems are minimal since they have a variety of workstations to work at. Either let one worker work at all 12 stations or let the operator specialize at a smaller number of stations (not necessarily consecutive).

Letting the worker walk requires time for walking. The amount of time for walking should be prorated over the work cycle. MTM allocates 5.3 TMU/foot (.2 s). Thus walking 10 feet requires 2 s. If 10 feet were required for every 20 units, then add $2/20 = .1$ s/unit. Standing with occasional walking reduces venous pooling in the legs and is less fatiguing than standing without walking.

PRINCIPLE 4: MAKE SEVERAL IDENTICAL ITEMS AT THE SAME TIME

Tasks can be broken into three stages: (1) get ready, (2) do, and (3) put away. Reduce cost/unit by prorating the *get ready* and *put away* over more units. Manufacturing similar parts in sequence (parts families) reduces setup time by minimizing the number of changes necessary. Items may differ only at a later stage; *B* is blue, *C* is green, and *D* is red or *B* has 1 hole, *C* has 2, and *D* has 3.

Decrease lead time and increase lot sizes by making the early operations as item *A* (unpainted part or part without holes). Then, for item *B*, pull item *A* from stock and finish. The *assembly example* analyzes physical work while the *inspection example* analyzes physical vs. mental work.

Assembly Example

In this example, the *pick up* and *put down* of handtools is minimized. Although the example could be driving nuts on a bolt, putting pickles on hamburgers, or marking cans in a store, the example is for soldering.

Trade off prorating the pick up and put down over more units vs. the increased distances moved during the *do* portion of the cycle. (See the comments about multiple items/carrier in Principle 3.) Distances moved increase because of the larger workstation. In general, it takes less time to move a greater distance than to have additional reaches and grasps during the pick up and additional move asides and releases during the put away. As the size of the workstation increases, there is a tendency to begin moving the product instead of the worker; this mechanized handling and increased worker specialization leads to the assembly line.

Inspection Example

In this application, the inspector must inspect *n* items for *m* characteristics (see Figure 13.18). The *get ready* is the mental work of fixing in the mind the quality standard for the characteristic *m*. The *do* is the sensing of the object, comparing it vs. the mental standard, making a decision, and executing the decision. The *put away* is the mental transfer of the quality standard from *working memory* to *long-term memory*. The following example could be a potato inspector looking for mold, eyes, or cuts (symbolized by circles, squares, and triangles in Figure 13.18); it could be a machine shop inspector inspecting for surface finish, concentricity, and length; it could be a teacher looking for three key words in a student's examination; it could be a typist looking for misspelled words, punctuation mistakes, and tense mistakes in a letter.

One-Item-at-a Time Method	One-Characteristic-at-a-Time Method
Recall standard for circles	Recall standard for circles
Inspect for circles on unit 1	Inspect for circles on unit 1
Dispose of standard for circles	Aside unit 1
	Get unit 2
Recall standard for triangles	Inspect for circles on unit 2
Inspect for triangles on unit 1	Aside unit 2
Dispose of standard for triangles	Get unit 3
	Inspect for circles on unit 3
Recall standard for squares	Aside unit 3
Inspect for squares on unit 1	Dispose of standard for circles
Dispose of standard for squares	
	Recall standard for squares
Etc.	Etc.

One item at a time requires less physical handling of items but considerable mental manipulation of characteristics. One characteristic at a time (less mental

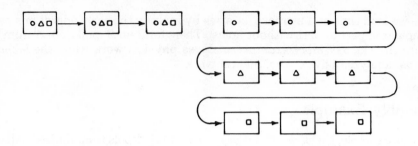

Figure 13.18 *Two inspection strategies* can be used. One-item-at-a-time is faster but one-characteristic-at-a-time probably gives better quality.

work but more physical work) ensures that a characteristic is not omitted, that the mental standard is more consistent, and that there is less halo effect of one characteristic on another characteristic. In the extreme it may cause bordom.

One-Item-at-a-Time Method	Time		Multiple Unit Method	Time	
Get soldering iron	3		Get soldering iron	3	
			Solder diode on unit 1		5
Solder diode on unit 1		5	Move iron to unit 2		1
Put iron away	2		Solder diode on unit 2		5
			Move iron to unit 3		1
Additional work		X	Solder diode on unit 3		5
			Put iron away	2	
Get soldering iron	3				
Solder diode on unit 2		5	Additional work		3X
Put iron away	2				
Additional work		X			
Get soldering iron	3				
Solder diode to unit 3		5			
Put iron away	2				
Additional work		X			

Soldering time/unit = 15/3 + 15/3   Soldering time/unit = 5/3 + 17/3
$$= 5 + 5 = 10$$
$$= 1.7 + 5.7$$
$$= 7.4$$

Konz and Osman had 24 women inspect numbers on slides.[13] For one item at a time, they made 16% errors for defect A and 9% for defect B; for one characteristic at a time, they made 5% for defect A and 7% for defect B. Effectively all the errors were type II errors. See Figure 13.19 for an explanation of type I and II errors. Total inspection time was held constant. Su and Konz found that, if inspection is "easy," there is no difference in accuracy between inspection for one or multiple characteristics at a time but a considerable time penalty for one characteristic at a time.[19] If the inspection is difficult, there is a trade-off needed as one characteristic at a time gives fewer errors but an increase in time.

PRINCIPLE 5: COMBINE OPERATIONS AND FUNCTIONS

Do several steps at the same time by using multifunction materials and equipment rather than single-function materials and equipment. Material cost/unit

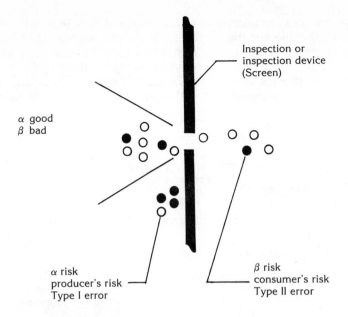

Figure 13.19 *Inspection devices* sort good and bad items into two categories. However, they are not perfect. The rejecting of good material as bad is called a type I risk, α risk, or producer's risk. The accepting of bad material as good is called a type II risk, β risk, or consumer's risk. Remember which is which by remembering that I comes before II and α comes before β; that is, I and α are on the left and II and β are on the right.

will be lower. Labor cost/unit will be reduced. Capital cost/unit generally will be higher. Total cost will be lower since material and labor cost/unit generally are more important than capital cost/unit.

Multifunction Materials

Use a compound which waxes at the same time that it cleans. Use a water-pump lubricant which also has a rust inhibitor. Although the basic chemicals will cost the same whether sold together or separately, the manufacturing and distribution costs will be reduced as one container is used instead of two, material handling will be less for one container than two, stocking expense on the store shelf will be lower for one container, and advertising expense will be lower. Labor cost will be lower and quality may be better since, if the two compounds were sold separately, one might not be used. Henry Ford had supplies delivered in special wooden boxes; the boxes became part of the Model T floor. The ultimate multifunction container is the ice cream cone.

Paperwork can fill multiple functions. Figure 13.20 shows a dividend check. By having the computer print the address as well as the name, the check can be used with a window envelope so that reproducing the address on the envelope is not required. The check also serves as a change of address form. By writing the new address on the check, the user saves the writing of a letter and a stamp; the company saves by reduced processing time (letters to the company would require several steps of internal processing before they got the proper location); and errors are reduced since the users do not need to transmit their names or account numbers since they are on the check. Use of the punched hole to signify change of address permits machine sorting of the returned checks. Another example of multiple function paper is a magnetically coded identification card which acts also as a key. Figure 13.21 shows a convention registration badge which also served as three meal tickets. If a registrant did not want to go to the meals, the appropriate numbers were crossed off with a grease pencil when he received the badge at registration.

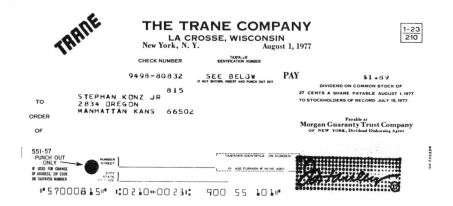

Figure 13.20 *Paperwork can fill multiple roles.* This dividend check transfers money as well as addressing the envelope and serving as a change-of-address form.

Figure 13.21 *When registering,* the convention attender indicates whether he will attend the three meals. The appropriate number is marked out for any meal not paid for. The badge then serves as an up-to-three meal ticket as well as a badge and simplifies accounting problems.

A farmer can use a fertilizer which adds nitrogen and trace compounds; he can plant two crops (each with a different growth time) at the same time. In the home, use a premixed spread of peanut butter and jelly to reduce labor cost. (Labor savings don't get much emphasis in most homes since production volume is low and labor cost is considered to be zero.) In the office, use a carbon paper to make duplicate copies at the same time as the original is made.

Multifunction Tools and Equipment

A farmer can fertilize at the same time he plows. In the home, one compressor can cool both a refrigerator and a freezer. Use a popcorn popper which melts the butter as it pops the popcorn. In the factory, a special-purpose drill can drill and countersink at one stroke; a special-purpose drill press can drill several holes at one stroke just as a multiple die can punch several holes at one stroke; a lathe tool can form as well as cut off; a fork truck can lift and move. In the office, a ruler can be used as a straightedge as well as a measuring device.

PRINCIPLE 6: VARY ENVIRONMENTAL STIMULATION INVERSELY WITH TASK STIMULATION

Low Stimulation Tasks

Many industrial tasks are quite "automatic" for the operator and require little conscious attention. Some even require little physical movement. Some examples of this minimum of mental attention and minimum of physical movement are inspecting items on a moving conveyor, monitoring an automatic pilot in an airplane, watching a chemical process indicator board, and monitoring a radar screen. If the person is seated rather than walking, if the environmental temperature, humidity, lighting, noise, and air velocity all are controlled at a constant level, the brain has minimum stimulation. Performance (measured, for example, as percent of defective units noticed) declines as the brain "goes to sleep" for periods of 1 to 20 s—even longer in some cases.[3,16] Operators strongly dislike this type of work situation—solitary confinement is the most feared of punishments. Variety is not the spice of life; it is the very essence.

Cure by adding stimulation to either the task or the environment.

Add physical movement to the task. For example, have the night security guard walk around as well as monitor a TV screen. Eliminate automatic equipment and replace it with something requiring operator movement and attention. Have inspectors dispose of rejects as well as indicate that they are rejects. Machine-paced tasks result in very poor quality if the products can pass the station while the operator is "asleep." If machine pacing must be used (and it usually need not be), then the operator should be required to make a conscious act to pass an item rather than make a conscious act to reject an item.

Add stimulation to the environment. The easiest solution is to let operators talk to each other. The managers who do not permit their employees to talk to each other are the same ones who proclaim (while sitting in their chairs) that chairs will make employees lazy. Aside from this type of feeling that only people with ties should have chairs and be permitted to talk, there is little reason not to arrange workstations so workers with low stimulation tasks be permitted to talk. Figure 13.22 shows seven schematic arrangements with my judgment of their stimulation value. More stimulation occurs when people are face to face, are close, there are no barriers, and noise is less. Reducing noise and encouraging visual and auditory contact also improves communication, especially useful for feedback of work quality.

Another common example of external stimulation is the use of background

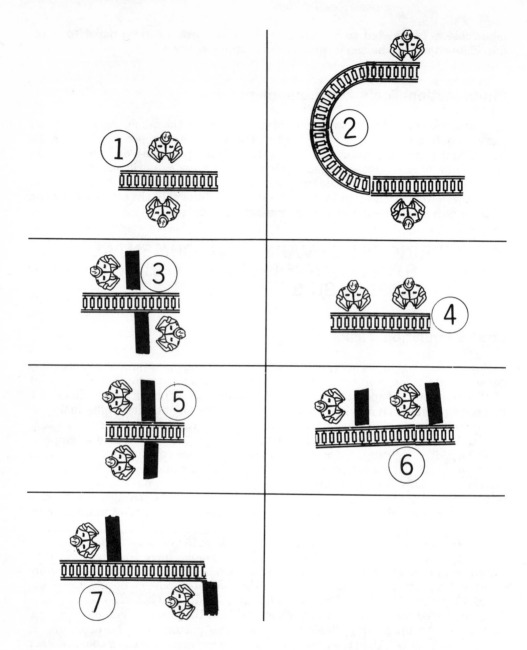

Figure 13.22 *Different arrangements* of equipment can be used to vary the amount of stimulation from fellow workers. The low stimulation situation is labeled 7.

music; it should be stimulating but not too stimulating (no vocals, played only a fraction of the time, neither "soft strings" nor "big brass")—a "soft fog".[17]

Windows also furnish stimulation. Light normally will be furnished by electric illumination and ventilation by fans so keep the size small to minimize energy losses. Many large windows have their upper portions covered with venetian blinds, curtains, or shades to reduce glare, indicating that their size is too large for their actual function. The important design feature is to have a view of the horizon with perhaps a 20° view of the ground.

Paint walls a variety of pastels, not just "industrial green." In large open office areas (say, holding over 10 people), paint each of the four walls a different color; use art work. Color-code stock racks, bins, pipes, chairs, and tables to establish "ownership" (territory) and identify function as well as add variety.

In general, the stimulation should not be expected to improve productivity over what a fresh person could do but just to prevent the decrement in performance.

High Stimulation Tasks

Many of the nonroutine tasks which require concentration are office tasks. By 1972 nearly half of the USA working population worked in an office.[2] However, even engineering design, calculations, and detailed inspection rarely require complete freedom from environmental stimulation. Konz demonstrated that background music had no effect on output or errors for a repetitive clerical task (mark sensing numbers), output or errors of hand addition of numbers, or output on a creative mental task (anagrams).[12] Workers may occasionally need freedom from high-information-content noise (conversation or vocal music) so it may be best to have two workstations—their usual one and one for concentration (a "think tank"). It could be just closing a door or an entirely separate enclosed area. Auditory privacy can be obtained by physical enclosure but also by masking the high-information-content noise with low-information-content noise such as air ventilation noise or background music. Evaluations of "office landscape" arrangements have indicated that auditory privacy, although necessary, is not sufficient. Visual privacy is needed also. The important thing about visual privacy seems to be to obscure the face—use a barrier from .6 to 1.6 m.

Many employees are perfectly willing to be distracted and will talk at length about sports, sex, politics, or the weather. If work standards are vague (as in most offices), and if conversation interferes with their work or the work of others, discourage excess conversation with desk or machine orientation (side by side rather than face to face, not facing open doors) or with head-high barriers.

SUMMARY

Chapter 13 discussed the "big picture." Chapter 14 will discuss the workstation itself; chapter 15, handtools; and chapter 16, administrative techniques.

Specialization (Principle 1) generally should be encouraged even though it presents challenges. Material handling (Principle 2) is a cost which adds no value and thus should be minimized. Avoid the common error of assuming that distance moved has a large influence on cost. Decoupling tasks (Principle 3) is done as a matter of routine for most jobs but assembly lines present special challenges. Make several identical items (Principle 4) and combine operations (Principle 5) are reminders of concepts that are "obvious" yet often forgotten. Vary stimulation (Principle 6) is based on the concept of optimum stimulation to workers—neither too much nor too little.

SHORT-ANSWER REVIEW QUESTIONS

13.1 How does labor specialization affect quality of output, quantity of output, and cost/hr?

13.2 List the two basic decoupling techniques.

13.3 How do multiple units/carrier affect labor costs?

13.4 What is the "group technology" concept?

13.5 "Make several identical items at the same time" is a principle. What is its rationale? Discuss in relation to making hamburgers, citing get ready, do, and put away.

THOUGHT-DISCUSSION QUESTIONS

1. For a specific job with which you are familiar, discuss how you could make it more specialized and how you could make it less specialized. Discuss in terms of equipment, material, labor, and organization.

2. Get ready, do, and put away are the three basic divisions of any task. For a specific job with which you are familiar, divide the total job into its get ready elements, its do elements, and its put away elements.

REFERENCES

1. Belbin, R., and Stammers, D. Pacing stress, human adaptation and training in car production. *Applied Ergonomics*, 3 (September 1972): 142-146.

2. Brookes, M. Office landscape: does it work? *Applied Ergonomics*, 3 (December 1972): 224-236.

3. Buckner, D., and McGrath, J. *Vigilance: a Symposium*. New York: McGraw-Hill, 1963.

4. Buxey, G.; Slack, N.; and Wild, R. Production flow line system design—a review. *American Institute of Industrial Engineers Transactions*, Vol. 5 (March 1973): 37-48.

5. Davis, L. Pacing effects on manned assembly lines. *International Journal of Production Research*, Vol. 4, No. 3 (1966): 171-184.

6. Dudley, N. The effect of pacing on operator performance. *International Journal of Production Research*, Vol. 1, No. 2 (1962): 60.

7. Franks, I., and Sury, R. The performance of operators in conveyor-paced work. *International Journal of Production Research*, Vol. 5, No. 2 (1966): 97-112.

8. Gallagher, C., and Knight, W. *Group Technology*. London: Butterworth and Co., 1973.

9. Johnson, S. Inter-operator variability effects on quality and productivity. *Proceedings of 25th Annual Meeting of the Human Factors Society*, Rochester, N.Y.; 1981; p. 551.

10. Kilbridge, M. Non-productive work as a factor in the economic division of labor. *Journal of Industrial Engineering*, Vol. 12, No. 3 (1966): 155-159.

11. Kilbridge, M., and Webster, L. The balance delay problem. *Management Science*, Vol. 2, No. 1 (1961): 69-84.

12. Konz, S. The effect of background music on productivity of four tasks. Ph.D. Dissertation. Urbana: University of Illinois, 1964.

13. Konz, S., and Osman, K. Team efficiencies on a paced visual inspection task. *Journal of Human Ergology* (Japan), Vol. 6 (1977): 111-119.

14. Multi-model production without set-up snags. *Factory* (April 1965): 128-130.

15. Murrell, K. Laboratory studies of repetitive work, I: paced work and its relation to unpaced work. *International Journal of Production Research*, Vol. 2, No. 3 (1963): 169-185.

16. Murrell, K. Work organization. *Applied Ergonomics*, 2 (June 1971): 79-91.

17. Musac theory and practice: music for improving work tempo. *Engineering*, Vol. 188, (Dec. 25, 1959): 689.

18. Okamura, K., and Yamashina, H. Analysis of the effect of buffer storage capacity in transfer line systems. *AIIE Transactions*, Vol. 9, No. 2 (1977): 127-136.

19. Su, J., and Konz, S. Evaluation of three methods for inspection of multiple defects/item. *Proceedings of 25th Annual Meeting of the Human Factors Society,* Rochester, N.Y., 1981.

20. Tuggle, G. Job enlargement: an assault on assembly line inefficiencies. *Industrial Engineering,* Vol. 1, No. 2 (1969): 26-31.

PHYSICAL DESIGN
OF THE WORKSTATION

In this chapter we will take a microscopic view and look at the design of individual workstations rather than the overall organization of groups of workers considered in chapter 13 and the administration of workers considered in chapter 16. The goal is to optimize the use of energy.

PRINCIPLE 1: AVOID STATIC LOADS AND FIXED WORK POSTURES

External Loads on the Body

Lifting and carrying generate torques about the body's center of gravity. These torques must be counterbalanced within the body. The greater the external torque the greater must be the internal torque. Unfortunately, most of the internal torques have short moment arms and large forces. To keep the forces down (and thus minimize strain on ligaments, tendons, and spinal discs), keep the external torque small by minimizing the moment arm. See the last section of chapter 12 for a more extensive discussion of carrying, holding, and lifting.

Table 14.1 shows the extra metabolic cost as a function of seven different carrying techniques. Best is the two-pack system (one on chest and one on back). Carrying objects on the head is very good. This is a well-known and widely used technique in Africa and Asia; it does require training. I have personally observed women in Africa with loads of over 25 kg balanced on their heads standing and conversing without putting the loads down. They feel it requires more effort to load and unload than to keep the load balanced on their heads. Third is the one-pack-on-the-back system followed by sherpa, rice bag, and yoke techniques. The pack systems and the yoke system permit the use of a staff—especially useful in uneven terrain. Worst of all is carrying weight in the hands due to the excessive torques generated, especially the "cartwheel" torque generated about the frontal axis. Soule and Goldman reported loads on the head as 1.2 times the cost of carrying a kg of your own body; loads in the hands required 1.4 to 1.9 times as much and loads on the feet were 4.2 to 6.3 times as high.[88]

Table 14.1   Comparison of Seven Modes of Carrying on a Horizontal Plane.[28] The 50 kg Subjects Carried 30 kg Weights 1 km.

	Criteria		
Carrying Method	Energy, kcal/min	Incremental Heart Beats/Min	Comments
Double pack 15 kg in two packs across shoulder—one in front and one in back	337	50	Load must be divided in two. Special harness. May not be suitable for repeated short trips without quick release harness.
Head* Basket on head	348	64	Requires training. Body movement restricted. Very suitable for repeated short distance carrying.
Rucksack High pack across the shoulder	368	62	Arms free. Not good for repeated short distance carrying.
Sherpa Load in a gunny sack supported on back by a strap across forehead	387	57	Requires practice. Arms can carry stick for hill climbing.
Rice bag Load in a gunny sack supported on back by holding two corners of sack with hands or hooks	414	60	No advantage over rucksack. Rather unsafe.
Yoke 15 kg is suspended by 3 ropes on each end of a bamboo strip across shoulder Strip held with one or two hands	434	66	Uncomfortable without practice. Contorted posture.
Hands Two canvas bags with padded handles; 15 kg in each hand	486	81	Least efficient and most fatiguing. Even worse with only one hand.

*Datta, Chatterjee, and Roy reported the energy cost for carrying weights from 0 to 50 kg on the head was well predicted for 51 kg subjects by[26]: $E, W = 6.58$ (body weight + load carried) − 152

Kellerman and vanWely investigated carrying with two hands over a 12 m distance.[48] Minimum energy expenditure/kg-meter moved was with 17 kg loads for untrained workers. Lehmann reports minimum energy per kg-meter of work was with a weight of 20 to 25 kg lifted by both hands.[62]

In addition to strain on the cardiovascular system, consider the strain on the skeletal-muscular system. *Consumer Reports* has an excellent article on backpacks on a frame.[10] The frame should fit the body size. In a good fit, when the hip belt is comfortably strapped in place, the shoulder-strap attachment on the cross member nearest the shoulders is about 25 mm above the shoulders. The hip belt (which holds the lower portion of the frame to the body and transfers most of the load to the hips) should be adjustable vertically so that the belt rests on the hips. A horizontal adjustment is desirable for comfort. The belt should go completely around the body and be padded. A well-designed backpack will cause the person to lean forward about 10°; a poor backpack will increase the angle to 20° or more. The Korean A frame has legs and the worker uses a walking stick. To rest, he squats, slips the frame off, and props it with the stick.

Even hand tools, although relatively light, can be fatiguing due to their location on the end of a long lever arm. The muscles must support the limb as well as the tool. A common solution is supporting the tool with a balancer (see Figure 14.1). Balancers can be set for weight ranges from a few grams to hundreds of kg. Another common solution is to reduce the length of the lever arm by supporting the forearm or wrist. Supporting the limb not only reduces fatigue but also reduces tremor which permits more accurate use of the tool.

Loads from the Body Posture

With normal shoes the heel of the foot is slightly higher than the ball of the foot. In this circumstance the center of gravity of the body passes in a line from the ear hole through the 5th lumbar disc; in obese people it passes forward of this line.[6] If the heel is too high (cowboy boots, women's high heels) then the center of gravity is moved forward causing an unnecessary torque. This torque can be counterbalanced but the resulting strain is unnecessary, at least in most work situations. "Earth shoes" are the opposite extreme; they have no heel at all and the ball of the foot is 50 mm above the heel. Cowboy boots have a large heel to prevent the foot from going through the stirrup; women's high heels are designed to make the woman arch her back to keep her balance. Both have a reason for the abnormal heel height. "Earth shoes" make you lean forward to keep your balance and cause completely unnecessary strain.

Need for support of the ankle depends on the person and the task. High shoe sides give better support so higher sides usually are preferred for sport activities and uneven terrain.

The sole of the shoe should give area support rather than line support. Area support is why wooden shoes are so comfortable; they mold themselves to the exact shape of the foot. The sole also should be cushioned. Crepe soles and ripple soles are favorites of people who work while standing. A nonskid sole is important in some jobs. See Table 14.4 for coefficients of friction of various types of shoes and floors. The importance of a good shoe might seem obvious except that common experience shows many people are more concerned with the shine or color of the shoe than its utility.

Hard floors cause problems. Metal gratings are the worst since they not only have little resilience but also have minimum surface area, thus acting as knives. Concrete also is bad. Plastic or cork tile are slightly better, wood is still better, and carpet is best of all. (If tile is used, make it a light color to increase the reflected light; see Table 17.8.) If management will not carpet the entire floor, workers may be able to pad their individual work surfaces with wooden boards (a favorite of lathe operators), cardboard, rubber pads, or pieces of old rugs. A seamless nonporous floor may be required for sanitation. Cold floors with high thermal conductivity (metal, concrete) cause additional problems by cooling the feet, thus causing vasoconstriction and thus restricted blood flow to the feet.

If a job has considerable standing, the nonmovement of the legs causes another problem. The veins of the body are the body's blood storage location. If the legs don't move, the blood from the heart tends to come down to the legs but not go back up; this is called *venous pooling*. Since the blood supply required by the muscles does not change, the heart tries to maintain a constant cardiac output (mL/min) by adjusting the beats/min to compensate for the lower mL/beat. The work of the heart also is increased since the heart must do the entire pumping job—normally the "milking action" of the leg muscles helps to move the blood.

Venous pooling causes swelling of the legs (edema) and varicose veins. One solution is support hose. Another is movement of the legs; thus walking is better than standing without walking. From a circulatory viewpoint, sitting is better than standing since the heart must pump against less static head. Sitting also is

Figure 14.1 *Balancers* come in a variety of sizes and shapes. A small one might suspend a handtool weighing only a few grams while large ones, mounted on a pivoted beam, can move castings weighing 500 kg. For hand tools, be sure the suspension force is aligned with the tool axis or unnecessary strain will be caused on the wrist.

better than standing or walking from a muscular viewpoint since less mass has to be supported by muscles. If continuous standing can't be eliminated, consider the advice of a Buckingham palace guard "Stood on parade four to five hours at a time. The trick is to keep the weight off your heels. That's why Guards' boots bulge in front—lots of room to wiggle your toes."

Isometric or static load is bad for the blood supply of a specific muscle as well as the body's blood supply.

$$\begin{matrix} \text{Permissible} & = & .5 \text{ permissible} \\ \text{static load} & & \text{dynamic load} \end{matrix}$$

Man is meant for movement. Diastolic blood pressure is considerably increased by isometric work (muscle does not move) but is not affected by isotonic work (muscle moves) (see Figure 14.2). Diastolic pressure is the minimum or base pressure exerted by the blood on the walls, so an increase is especially dangerous. When resting diastolic blood pressure is less than 90 torr, it is satisfactory. But from 90 to 100 is suspicious and above 100 is poor. Life insurance studies indicate that blood pressures below 110/70 are optimal for long life span. Systolic blood pressure, the peak force which occurs when the wave of blood passes down the vessel, increases with both isometric and isotonic work. Sitting without movement (e.g., microscope work) does not increase blood pressure but is fatiguing as metabolic wastes tend to concentrate in the muscles as blood flow is reduced. (See Figure 14.3.)

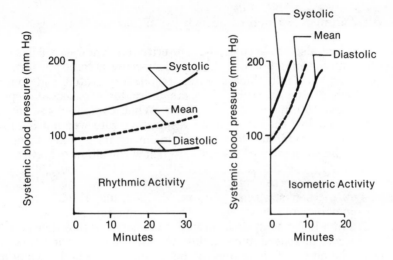

Figure 14.2 *Systolic blood pressure* increases slightly during rhythmic (isotonic) activity but more steeply during static (isometric) activity. The more dangerous diastolic blood pressure does not change when the muscle moves but increases sharply during static loads.[86]

PRINCIPLE 2: SET THE WORK HEIGHT AT 50 mm BELOW THE ELBOW

The optimum work height (-50 mm below the elbow; slightly below heart level) is based on both productivity (that is, cost to the organization paying the employee) and on physiological cost (that is, cost to the individual to achieve a specified output).

Barnes stated "with the hand allowed to work one to three inches lower than the elbow, the average height of the work surface . . . "[11] Although Barnes did not cite any evidence for his recommendation, all the following studies tend to confirm his judgment. The key points are:

- The work height is defined in terms of elbow height rather than a distance from the floor. Since individual's heights will vary, any fixed height design must be wrong.
- The optimum height is slightly *below* the elbow. Some research indicates it farther below than other research but the consensus is below.

Konz surveys the literature; the following are just a few selected studies.[51] Ellis reported optimum height for a block turning task was 75 mm below the elbow.[34] Drillis cited German studies in the 1920s which used amount of file dust produced as the criterion for optimum height. Nebel found maximum output at 22 mm below the elbow while Lysinski found the optimum at 60% of stature height.[32,64,72] In Drillis' own study of scutching of flax in Latvia, optimum length of the scutching tool was 57% of the operator's height.[31] For reference, the mean elbow height/stature height for 4062 men is 63%. In Konz' experiment one, output at 50 mm below the elbow was defined at 100%; output at 50 mm above the elbow was 99.9%, and output at 150 mm below the elbow was 97.2%. This relatively "flat" response around the optimum value also was found in Kennedy and Landesman's study.[49] That is, although the optimum seems to be about 50 mm below the elbow, output will not decrease more than a couple of percent within a range from −125 mm to 25 mm (from 0); beyond this the penalty is greater.

Physiological cost also has been used as the criterion. Knowles made a study on the proper height of ironing boards.[50] She used six criteria: force platform output, postural shifting, kcal/min, heart rate, respiratory rate, and pulmonary ventilation. Each of the standing subjects individually selected a preferred height; it averaged 150 mm below the elbow. When compared vs. "standard" height of 225 mm below the elbow, only respiratory rate was not significantly better. Konz reported minimum force platform output for moving a weight back and forth when the work surface was 25 mm below the elbow.[51] McCracken and Richardson, studying shelves at 100, 300, 500, 700, 1100, 1300, 1500, and 1700 mm above the floor, reported minimum kcal/min at 1100 mm (about 150 mm above the elbow of their subjects).[68] Agan, Tormey, and Konz, replicating the study but using incremental heart rate as the criterion, reported minimum at 700 mm (about 275 mm lower than their subject's elbows).[1] Both studies noted that the slope of the cost increased more rapidly for shelves below the elbow than for shelves above the elbow. This is due to the weight of the body. Thus a 58 kg female moving a .5 kg can to a shelf above the elbow (say, 1,300 mm) must lift a .5 kg can to a shelf, a 1 kg forearm, and a 1.5 kg upper arm. To move a .5 kg can to a shelf below the elbow requires little movement of mass until the entire body (less the feet) must be moved; then cost rises very rapidly.

An important aspect of a poor working height is the effect on the worker's posture. Floyd and Ward and vanWely point out that neglect by machine designers and supervisors in providing a work surface at the proper height can actually cause deformed bodies and is a major cause of back pain.[39,92,93] *Sense at the Bench,* a movie available from Philips, Eindhoven (and Human Factors Society in the USA) emphasizes keeping the elbow low (upper arm more than 45° from horizontal) to minimize fatigue.

The optimum height seems to be the same for both sitting and standing.

It will be emphasized that *work height is not the same as table height* since most items (hamburgers on a grill, mechanical assemblies, typewriter keyboards, etc.) have a thickness of 25 to 50 mm.

Any work height which is a fixed height above the floor, no matter who is working, is not acceptable.

Solution Techniques

There are three basic work height approaches.

- The first approach, changing the height of the machine, may not be practical if a variety of operators use the same machine over a relatively short time period. For example, an office duplicating machine might be used by five

Figure 14.3 *Source documents* for typing should be on an adjustable arm at the same distance as the display to minimize eye accommodation changes and reduce twisting of the neck and torso.

different people in a day and ten different people in a month. Many machines and operators, however, are one on one; that is, the one person is the only person to operate the machine over a long time period, say 1 year but often 10 to 20 years. In these long-term situations the machine legs or platform should be increased or decreased in height. (See Figure 14.4.) For many lightweight machines, such as a typewriter on a desk, a sewing machine on a stand, or a repair bench, this may require as much as one hour to adjust—surely a small cost for 10 years freedom from back pain. Another possibility is to have several different height workstations always available. For example, in a maintenance welding shop welding tables were built at 500, 750, and 1000 mm above the floor. Then when welders had "thin" pieces they used the 1000 mm high table; for a "thick" piece they used the 500 mm table. A welding table with several levels is also a possibility. A common example of a two-level workstation is a desk with a recessed typewriter.

- The second approach is to adjust the height of the operator's elbow rather than the height of the machine. If the operator sits, have the chair height adjustable. If the operator stands, raise the height by having the operator stand on a platform; for tall operators, remove the platform. Platforms may be merely several pieces of rug piled on top of each other; rugs reduce fatigue to the feet as well as the back.

- The third approach is to adjust the height of the work on the machine. For example, in a kitchen put items to be sliced on a cutting board of a thickness varying with the worker's height; put a spacer under a dishpan in a sink that is too low (see Figure 14.5); use shallow pans rather than deep pans to

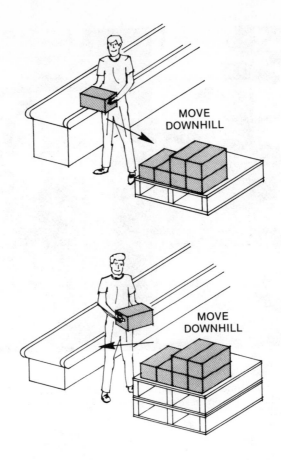

Figure 14.4 *Move downhill* when loading or unloading conveyors. When the task at a workstation is loading a conveyor from a pallet, design the conveyor to be low and have the lift truck driver place the pallet on a platform.

minimize unnecessary postural strain. Barbers raise the height of children's heads by having them sit on a board across the chair's arms.

Items also may have to be lowered. The effective height of the work may be made too high by container walls. Tip component boxes on their side or at a 45° angle or cut out the side of the box. Shallow bins filled twice a shift are better than deep bins filled once a shift since the bin filling cost is minor compared to the extra distance which must be moved over the high container wall. When "sacking" groceries, have the box or bag on its side or at an angle rather than vertical.

In agricultural labor such as planting vegetables or picking from low-growing plants such as strawberries, it often is necessary to work at or below ground level. Vos divides jobs into stationary jobs and jobs requiring movement (as along a row of beans).[95] For stationary work, sitting is best, squatting is almost as good, and bending and kneeling are poor. If bending or kneeling is necessary, then supporting the body with one hand on the knee or ground will decrease net energy required by 25% to 55% over no arm support. For movement, squatting always requires less energy than sitting on a 175 mm stool attached to a belt around the waist; picking performance was 6% to 7% higher with squatting also. Although bending required 1 kcal/min more than squatting when movement was 2 m/min, its curve "crossed-over" at 4 m/min forward speed, and at 10 m/min

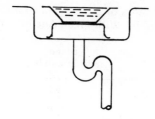

Figure 14.5 *Raising work heights* under a dish pan demonstrates Work Smart Not Hard in a home.

bending required 1 kcal/min less than squatting. Since bending causes local overload of the back muscles, bending should be used with caution.

PRINCIPLE 3: FURNISH EVERY EMPLOYEE WITH AN ADJUSTABLE CHAIR

Justification

The benefit from chairs is the reduced physiological load on the worker from sitting rather than standing. For example, dentists who work seated with the patient horizontal have fewer backaches. Table 14.2 shows the results from one of many studies in the literature. See Kroemer and Robinette for a review with 90 references on chair design.[58] Ayoub and Halcomb made an annotated bibliography (97 references) of the key anthropometric characteristics of seats, consoles, and workplaces.[9]

The cost of an adjustable industrial chair is very low. Chair cost is entirely a capital cost as operating cost is zero. Assuming a price of $100/chair, a life of 5 years, one shift operation, and 2000 working hours/yr, the cost/hr is $100/10,000 hours or 1 cent/hr. If the comparison is between a good biomechanics chair and a cheap plastic nonadjustable chair, the incremental cost will be about .2 cent/hr. Labor cost is wages plus fringes; minimum wage cost in the USA in 1982 is about $5/hr and typical wage cost is $10/hr. Thus, if output is improved .1%, .001 ($10) = 1 cent/hr. The improved productivity is more likely to occur from the workers working more minutes rather than improved productivity/min. That is, the secretary will not type more words/minute of working but will work more minutes as she will not have to get up to relieve her sore back.

For some unknown reason some managers still equate work with pain instead of productivity and refuse to provide chairs for factory workers although they give office chairs to office workers and would never consider giving up their own chairs. Although these attitudes are irrational, the engineer should recognize that they exist.

Seat Construction

When a person sits on a seat, the weight is not supported by the entire buttocks but by two small areas called the ischial tuberosities. The blood vessels in the tissue overlying the tuberosities and the heels are arranged in a special manner to reduce the effect of pressure. Figure 14.6 shows the compressive force

Table 14.2   Cardiovascular Effects of Sitting and Standing For 16 Adults Between 18 and 51 With Mean Age of 30[96]

Criterion	Stand (a little walking)	Sit	Sit/Stand %
Cardiac output, L/min	5.1	6.4	125
Stroke volume, mL/beat	54.5	78.3	144
Mean arterial pressure, torr	107.0	87.9	82
Heart rate, beats/min	97.2	84.9	87
Total peripheral resistance, dynes/cm²-s	1820.0	1207.0	66

for normal seating and for sitting cross-legged. The best design is not contoured since contouring forces the body into one standardized position—keeping the pressure on the same area. A well-designed chair permits changing the posture. Cushioning is desirable as it reduces pressure by increasing area; the upholstery should give way about 25 mm. If it gives way too much then the body is not firmly supported and must be supported by the muscles. Some lunch counters encourage rapid eating (and thus maximize seats available) by using stools with no backrest and hard seats with small diameters. The seat should slope backward at a small angle (1° to 5°). A curved front edge (waterfall front) is desirable to maximize the surface area contacting the underside of the thighs; upholstery beading in this area should be avoided. The material should be fabric since it breathes and reduces sliding of the body; plastic should be avoided since it causes perspiration problems and probably will have a shorter life due to rips.

Seat height. The height of the seat should be measured from the height of the work: a common mistake is to measure seat height from the floor. The critical factor is the location of the elbow to the work and the comfort of the upper body. Burandt and Grandjean recommend the distance between the top of the seat and the work be 275 ± 25 mm.[18] Note that this distance is occupied by three things: the work, the work surface, and the thighs. When purchasing desks and tables, insist on thin work surfaces to permit maximum dimensions for the work and the thighs. The 95% thigh thickness for both USA males and females is 175 mm.

Once the upper portion of the body is satisfied, we can be concerned with the lower portion; that is, the height of the seat from the floor. The most comfortable sitting position is one in which the thigh is approximately horizontal and the foot is supported. For any given work height, work-seat distance, and operator size the feet may not reach the floor. In these cases support the feet with a foot rest.

Since operator dimensions and work thicknesses vary, seat height should be adjustable. A range of 375 to 450 mm from the floor usually is satisfactory. A common problem is that the seat cannot be lowered far enough to permit sufficient thigh clearance.

Seat depth. A common problem is a seat that is too deep; the user must either sit forward and lose the support of the backrest or sit back and have legs supported only under the thigh. A depth of 375 to 400 mm is good; beyond this the designer has emphasized form over function. Another example of form over function is the placing of a panel on the operator's side of a machine or desk. This prevents the operator from working seated since there is no place for the legs.

Swivels. Swivels permit the chair seat to rotate. They usually are desirable as they permit small variations in posture and increase the operator's range of reach. Swivel chairs also may permit the operator to enter or leave the workspace

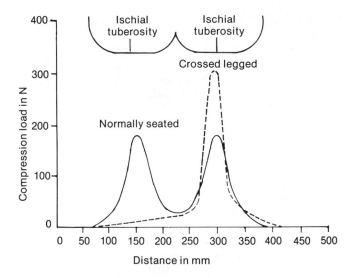

Figure 14.6 *Pressure distribution* on the buttocks is concentrated on the ischial turberosities during normal sitting. Sitting cross-legged increases the pressure on one side.[84]

without sliding the chair back and forth. Don't use swivels when operating a pedal.

Seat width. The wider the better. Wide chair seats not only accommodate a higher percent of the population but also permit more varied postures. If you wish to avoid benches, require a minimum seat width of 400 mm. Add 50 mm for clothing and pocket contents. If the chair has arm rests, the distance between the arms should be at least 475 mm.[84]

Armrests

In most industrial chairs there is little justification for armrests as they interfere with arm movements. If arm support is needed, support the arm with a pad on the table. If armrests are used they should have their tops 200 mm above the *compressed* seat. It is a common design error in lounge furniture to have armrests that are far too high.

Backrest Construction

The ideal backrest is adjustable both horizontally and vertically. In very good designs the horizontal adjustment has a spring action so that the backrest "tracks" with the back and thus moves in and out as the back moves in and out. Backrests which give some support for the shoulders as well as the lumbar region seem to be better liked by users than those which support only one of the two regions. There should be no sharp edges and the shape should be concave to give area support to the back, especially in the lumbar region. The backrest should be stiffly upholstered.

Backrest width. For work chairs, the elbows will hit a wide backrest; keep width to no more than 325 to 375 mm.

Backrest height. The height need only be about 125 mm if just the lumbar region is to be supported. If lumbar and shoulder are to be supported then length should be greater.

Angle between backrest and seat. For an industrial chair, the angle should be between 95° and 110°; 100° is a common recommendation. For a lounge chair, Yamaguchi and Umezawa give three combinations for minimum spinal disc distortion: seat inclination of 10° with seat-backrest angle of 115°, seat at 15° and seat backrest at 110°, and seat at 20° and seat backrest at 105°.[103]

Casters

Casters permit mobility but also allow the chair to move when the operator does not wish it to move and thus create safety problems. A safe alternative is to mount the chairs on rails to permit the operator to cover a wider territory yet not be concerned with chair stability or precise control of chair movements.

Figure 14.7 shows a well-designed industrial chair. In some work situations, where the worker usually stands and only occasionally sits, there may not be room for a good chair. Then use a "swing out" seat or a stool rather than nothing. Another example is a firm which does not want workers eating in their work area during breaks as it contaminates the product. In the aisle next to the wall, they have fold-down benches for use during the breaks.

Video display terminals (VDTs) should have separate housing of screen and terminal to allow each to be positioned separately for operator comfort. Ideally the screen center should be 10°—20° below the horizontal plane at the operator's eye height; the keyboard home row should be at or below elbow height. The keyboard angle should not be fixed; although a low (15°) slope is acceptable for people at normal height, a higher angle is preferred for short people or people with low seat heights.[69] For highly repetitive use of keyboards with VDTs (e.g., word processors) the operator's body is in a "straight jacket." As pointed out in Principle 1 and Figures 14.3 and 14.8, fixed postures should be avoided. A simple technique is to require the operator to leave the chair occasionally—such as to get additional material or to dispose of completed material. This movement also will permit the eyes to rest. See chapter 17 for lighting for VDTs. Cakir, Hart, and Stewart have a good handbook on Video Display Terminals (VDTs).[19]

PRINCIPLE 4: SUPPORT LIMBS

Table 14.3 shows the weight of various body segments as predicted by body weight.

Head

If you weigh 90 kg, then your head weighs about .0728 (90) = 6.6 kg.

The head is supported by the neck without apparent fatigue as long as the face is vertical or tilted forward. Figure 14.9 shows results of 1650 photographic measurements on five office workers. The mean angle is 65° below the horizontal. The angle does not differ for reading and writing. Of the two subjects analyzed in detail, one subject kept her eyes 275 mm from the desk when writing but 325 mm when reading; the other read and wrote at 325 mm.[25]

Problems occur when the line of sight is more than 15° above the horizontal. One problem is glare from artificial lighting and from windows. A second problem is fatigue and pain in the neck muscles. The solution is to redesign the task to eliminate prolonged looking upward. Sore necks occasionally occur for people with bifocals; they tilt the head backward to be able to see through the bottom of the bifocals. A solution is to use single-vision glasses at their workstation or to have bifocals with the close lens on the top. Tilt microscopes to reduce head angle tilt.

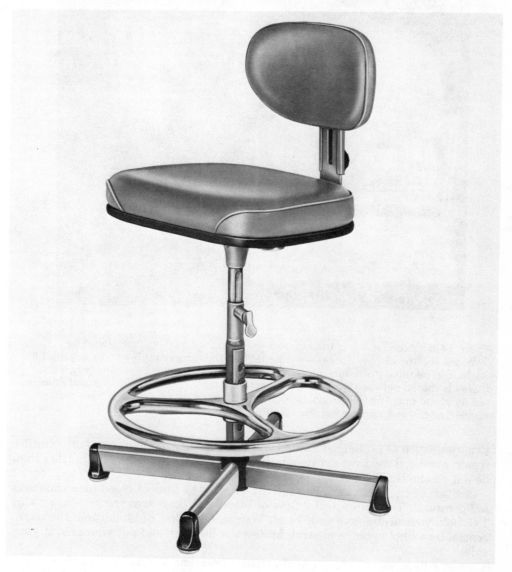

Figure 14.7 *Industrial chairs* are well designed when a person can sit for 8 hr/day. For a nonadjustable chair to be satisfactory under these severe conditions would be quite unusual. The most important adjustments are the vertical location of the seat and the horizontal location of the back rest. An adjustable footrest on the chair is not as good as an adjustable footrest on the floor, but is better than a nonadjustable footrest or none. In Germany, all chairs with casters must have 5 supports to increase stability.

Arms

If you weigh 90 kg, then one of your hands weighs about .6 kg, a hand plus forearm about 2 kg, and an entire arm about 4.4 kg.

Remember that when you hold a 25 g feather you also are holding 4500 g of bone and muscle. Even elimination of the 25 g weight doesn't help much. Avoid

Figure 14.8 *Adjustability* is the key to a well-designed VDT workstation. The keyboard and the display are on separate tables. The display table can be adjusted in height, can be tilted (to reduce glare on the screen), and can be swiveled (when the display is shared between workers). The keyboard table can be moved up and down as well as in and out. The source documents are on an adjustable holder. Note the rounded edges and corners on the furniture. Photo courtesy of IBM.

situations such as prolonged salutes and overhead use of hand tools (auto muffler repair, overhead welding, spray painting, etc.) Can the work be tilted rather than lie flat on the bench?

Position of the arm also substantially affects the flow of blood (and thus arm temperature). To minimize the flow of blood and drop arm temperature about 1°C, hold your arm above your head. To maximize flow of blood from the warm central torso, hold your arm straight down or lie down and put your arm at your side.

Tasks which require close visual attention present workers with a problem: if they hold the objects close to their eyes, they must support the weight of the arms. If they put the arms where they are comfortable, they can't see. One solution is to improve the vision—say with a 2X low power magnification lens—and let the arms be comfortable. Another possibility is to keep the work close to the eyes but support the limb weight at the wrist, forearm, or elbow with a padded support attached to the table or chair.

Legs

If you weigh 90 kg, then one foot weighs about 1.3 kg, one foot plus a calf about 5.2 kg, and an entire leg about 14.5 kg.

Support of the leg is a problem in some sitting tasks. When adjusting chair height, the first consideration is to adjust the work height to 50 mm below elbow

Table 14.3   Weight and Center of Mass for Various Body Segments
in Adult Males[21]

Body segment	Weight of Segment/ Total Body Weight	SE	Location of Center of Mass as Ratio of Segment Size
Head	7.28	.16	.46 (top of head/ht of head) .40 (back of head/head lgth)
Trunk	50.70	.57	.38 (suprastern/trunk lgth)
Hand	0.65	.02	.18 (meta 3/styl-meta 3 lgth) .56 (med aspect/hand brdth)
Forearm	1.61	.04	.39 (radiale/rad-styl lgth) .49 (ant aspect/ap at cm)
Forearm + hand	2.27	.06	.63 (radiale/rad styl lgth) .52 (ant aspect/ap at cm)
Upper arm	2.63	.06	.51 (acrom/acrom-rad lgth) .51 (ant aspect/ap at cm)
Total arm	4.90	.09	.41 (acromion/arm length)
Both arms and hands	9.80		
Foot	1.47	.03	.45 (heel/foot length) .54 (sole/sphyrion height)
Calf	4.35	.10	.37 (tibiale/calf length) .42 (ant aspect/ap at cm)
Calf + foot	5.82	.12	.47 (tibiale/tibiale ht) .33 (ant aspect/ap at cm)
Thigh	10.27	.23	.37 (trochanterion/thigh lgth) .53 (ant aspect/ap at cm)
Total leg	16.10	.26	.38 (troch/trochanteric ht) .63 (ant aspect/ap at cm)
Both legs and feet	32.20		
	99.98		

An improved estimate can be made for some segments using the following equations where X is total body weight, kg:

Segment	Equation	Standard Error
Trunk	.551 X −2.837	1.33
Head & trunk	.580 X + .009	1.36
Total arm	.047 X + .132	.23
Upper arm	.030 X − .238	.14
Thigh	.120 X −1.123	.54
Foot	.009 X + .369	.06

height. However, this may leave the legs dangling off the floor. Something is needed at the foot end of the leg to support that weight. If there is no footrest, then support comes from the pressure of the seat vs. the underside of the thigh. Unfortunately this pressure tends to cut off flow of blood to the legs. Footrests may be a separate item of furniture, be built into the workstation, or be built into the chair. A common problem is getting a footrest with a proper degree of adjustability so the leg can be in a comfortable position (thigh approximately horizontal). Figure 14.10 shows an adjustable footrest.

Figure 14.11 makes the point that sharp edges of tables and desks are poor for support. (A good knife has a sharp edge.)

PRINCIPLE 5: USE THE FEET AS WELL AS THE HANDS

The foot can react as quickly as the hand but, due to the construction of the ankle vs. the wrist and the weight of the leg vs. the arm, it is not as dexterous.

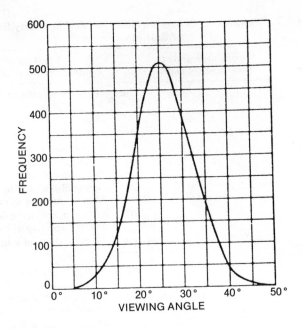

Figure 14.9 *Viewing angle* of five office workers had a mean of 25° from the vertical. Angle did not vary with reading or writing; distance from the paper to the eye had a mean of 300 mm.[25]

Figure 14.10 *Footrests* need to be adjustable as do chairs. Photo courtesy of Toledo Metal Furniture.

An example of the speed of the arms vs. the strength of the legs is the English longbow vs. the Continental crossbow. The longbow was powered by the arm and "a first rate English archer, who in a single minute, was unable to draw and discharge his bow 12 times with a range of 240 yards and who in these 12 shots once missed his man was very lightly esteemed."[46] The crossbow, which had

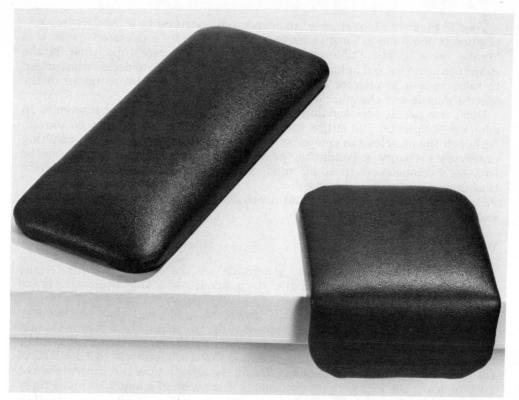

Figure 14.11 *Sharp edges* act as knives against the forearm. A rounded surface is best but pads—either commercially available ones or homemade substitutes—can be used to compensate for poor furniture selection. Photo courtesy of Toledo Metal Furniture.

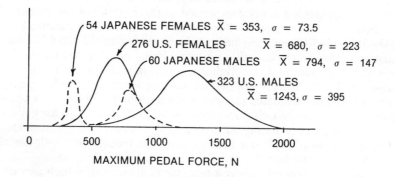

54 JAPANESE FEMALES $\bar{X}$ = 353, σ = 73.5
276 U.S. FEMALES $\bar{X}$ = 680, σ = 223
60 JAPANESE MALES $\bar{X}$ = 794, σ = 147
323 U.S. MALES
$\bar{X}$ = 1243, σ = 395

MAXIMUM PEDAL FORCE, N

Figure 14.12 *Maximum pedal force* differs for the USA and Japan.[4,70] Note that, although the absolute variability differs in all four populations, the relative variability (standard deviation/mean) is about 20% in the Japanese study and 32% in the American. There probably was more standardization in the experimental procedure in the Japanese study.

greater range, was limited in its speed of fire since it was cocked with the "belt and claw" method. The bowstring was looped over a claw attached to the belt and the bow pushed with the feet. Fortunately for the English the Chinese repeating crossbow, which was developed in the 1st century and which could fire 10 bolts in 15 seconds, was unknown in Europe. It was cocked with the arms.

Time to move the foot from a "toe depressed" position to a "heel depressed" position is about .28 s. Time to move the foot from an "accelerator" pedal to a "brake" pedal is about .55 s.[56] Both times include sensor time (time for the subject to sense the red light signal) and decision time (time to decide what action to take) as well as effector time (time for the nerve impulse to travel to the specific muscle and the muscle to act).

Don't use pedals for standing work as the body is supported unevenly. In addition, weight on the entire leg or even the entire body must be moved. Although the muscles can compensate for the unnecessary strain, the resulting unnecessary energy expenditure, pain, and fatigue are a reflection on the engineer's competence. In addition, since the operator is off-balance, reaction time in an emergency is increased.

Pedals can be used for power (continuous and discrete) and control (continuous and discrete).

Power

Examples of continuous power generation are the bicycle and the treadle sewing machine. An example of discrete power is using the automotive brake pedal for a "panic stop."

Continuous human power is usually generated by the legs since the two legs have approximately three times the power of the two arms; the arms are slightly more efficient per kg of muscle but the legs have much more muscle.[29] Both limbs are used with a rotary pedal arrangement so that each limb can rest for 50% of the cycle while output is continuous. Using the arms a man can generate about 1/25 hp; using the legs permits continuous generation of about 1/10 hp. (1 hp = 10.7 kcal/min. Since 5 kcal/min is a reasonable long-term work input rate, then work output is $1/2 \times 20\% = 1/10$ hp.) Bicycle pedaling (at 20% to 25% efficiency for an experienced cyclist) is remarkably efficient; a man on a bicycle uses about 15 kcal/g-km—this is first in efficiency among traveling animals and machines.[101] For more on the science of pedaling, see Whitt and Wilson.[100] For a man walking, see Table 12.1. For maximum efficiency, the seat to pedal distance should be adjusted so that the leg is fully extended at the bottom of the stroke. The crank length should be approximately 20% of leg length (9.5% of stature height).[93] The pedal should be in line with the axis of the lower leg so the force is exerted by the leg muscles rather than the ankle muscles. Pedal revolutions should be about 50 rev/min although people like a higher rev/min (about 60) with a lighter load.

Discrete power is usually applied by one leg since application time is usually less than 10 seconds and thus fatigue is not a problem. There does not seem to be any power advantage to using the right or left foot.[70,94] Adjusting Von Buseck's data for learning, force using both feet is 106% to 118% greater than using a single foot, but people will not always use both feet so the designer should not depend on use of both feet.[94]

Force capability depends upon a number of factors of which *percentile of the population* and *pedal location* are most important. Figure 14.12 shows that both the *work percentile* and *population* are important. First consider percentile. For most design purposes, design to exclude as few as possible. Exclusion of 10, 5, or 1% is the most common; that is, include 90%, 95%, or 99% of the population. Second, populations differ. The strength of Japanese women is not the same as American women which is not the same as American men. Leg strength declines more rapidly with age than arm strength as is shown in Figure 12.10. Women also lose strength at an earlier age than men do.

The second major factor is the *pedal location*. In general, maximal force can be exerted when there is a straight line between the pedal and back support.[79] That is, if the pedal is 250 mm below the seat, then the back support should be 250 mm above the seat. Aoki reported maximal pedal force when the calf-thigh angle was

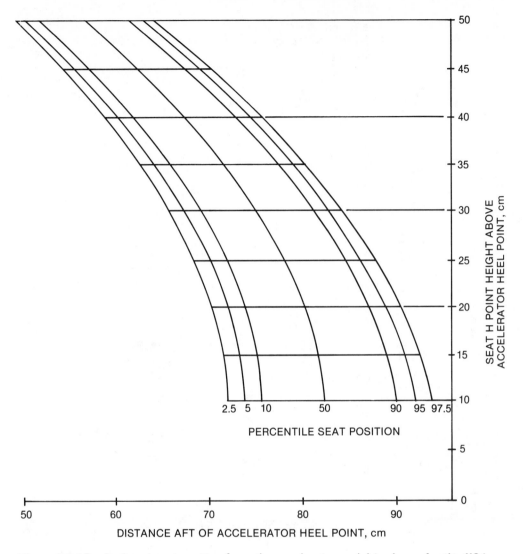

Figure 14.13 *Preferred seat position* from the accelerator pedal is shown for the USA driving population in 1973.[81]

110° and the thigh-back angle was 73°.[4] Hugh-Jones reported maximum at a knee angle of 160°.[47] Figure 14.13 shows preferred seat locations relative to the pedal for automobile drivers. Martin and Johnson showed the importance of the horizontal location of the pedal from the person.[66] The results of 155 males exerting forces at 28 different positions is depicted in Figure 14.14. The mean efficiency index is plotted at each of the 28 positions; the *efficiency index* was defined as "force at a specific position for an individual/force at the individual's best position."

If the pedal will be used repeatedly, then muscle fatigue will become a problem. A simple design solution is a wide pedal so either foot can be used at the operator's option. Another solution is to permit lateral movement of the chair or to use a wide chair (a bench) upon which the operator can change position from time to time.

Control

Example applications of a foot pedal for continuous control are the automobile

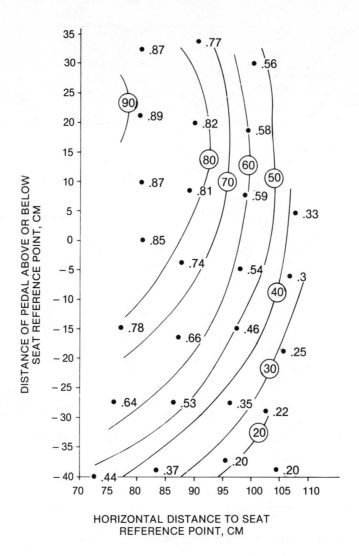

Figure 14.14 *Force* is a function of the horizontal distance from the pedal to the seat reference point (the intersection of the planes of the seat and the seat back).[66]

accelerator and, for nonpanic braking, automobile brake pedal. Examples of discrete (on-off) applications are a pedal-controlled punch press and an automobile foot switch for high-low headlight beam.

For continuous control such as the automobile accelerator, it is preferable to bend the ankle by depressing the toe rather than depressing the heel or moving the entire foot and leg. The foot of a 70 kg male weighs 1 kg vs. the 4.1 kg of the foot + calf or the 11.3 kg of an entire leg. If just the toe of the foot is moved, the heel can rest on a support so the amount of weight that must be supported by muscles is limited. The range of movement at the ankle should be between 80° and 115°.[75]

For discrete control action such as actuation of a punch press, Figure 14.15 shows the relative efficiency of various type pedals. As was pointed out in Principle 1, man is meant for motion and static positions are not desirable. Don't require an operator's leg always to be located in one standardized location—"a straight jacket." In most situations the time it takes a person to move a foot to a

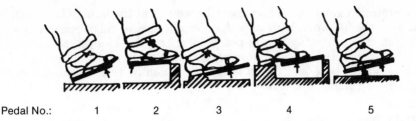

Pedal No.:   1   2   3   4   5

Figure 14.15 *Pedal strokes/minute vary.* Barnes et al. reported 187 strokes/min for pedal 1; 178 for pedal 2; 176 for pedal 3; 140 for pedal 4; and 171 for pedal 5.[12] Trumbo and Schneider reported mean response time to depress pedal through 15° of arc was 346 milliseconds for pedal 1; 395 for pedal 2; 350 for pedal 3; 370 for pedal 4; and 355 for pedal 5.[91] Using either criterion, pedal 1 is best.

control is not important as the movement can be done at the operator's leisure or can be done simultaneously with some other motion.

However, in some situations minimum reaction time is important. An example is use of an automobile brake pedal in an emergency. In this event, you can save about .27 s (about 1 car length at 100 km/hr) if you have your left foot resting on the pedal and brake with the left foot. If the left foot must move to the brake there is no advantage to using it over the right foot. Due to the present design of pedals, poising the left foot on the brake pedal is tiring, so use left foot braking only in heavy traffic.

Actuation of a control can be by a lateral motion of the knee as well as vertical motion of the foot. Many clamping fixtures are actuated by knee switches. The knee should not have to move more than 75 to 100 mm; force requirements should be light. The advantage is that the weight of the leg need not be lifted.

PRINCIPLE 6: USE GRAVITY, DON'T OPPOSE IT

Location of Work in Relation to the Elbow

The discussion of Principle 4 mentioned that holding 25 g of feathers could also require the holding of 4500 g of bone and muscle. The same situation occurs for movement as for holding.

Whenever an object is lifted by the hand, the muscles must lift the hand and arm also. In addition it requires effort to lower the limb under control. Therefore, make movements horizontal instead of vertical. (See Figure 14.4.)

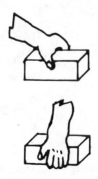

Figure 14.16 *Common method* of holding a brick; below is Gilbreth's method.

In certain circumstances, however, the weight of the human body can be used to increase the force of a lever or pedal. From a theoretical conservation of energy viewpoint, it is a substitution of potential energy for kinetic energy with no net gain; the key practical point, however, is that the potential energy can be applied over a period of time and the kinetic energy can be released all at once (e.g., a wheel with a ratchet and a release).

Orientation of the Work

Gravity can be used in some operations to move the material down to the work. Examples are paint from a paintbrush, a welding bead from a welding rod, solder on a solder joint. Gravity also can act as a "holding fixture" for components before assembly. For example, contrast driving a screw into the ceiling vs. horizontally or vertically downward.

Feeding and Disposal of Components

One technique of minimizing the up and down movement of the arm is to use drop disposal. If the item dropped is fragile, its fall can be cushioned. The most

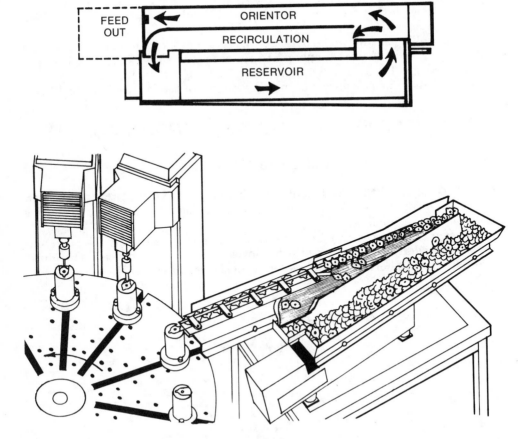

Figure 14.17 *Vibratory feeders* come in bowl and linear shapes. Covering reduces the noise level from the components. Although requiring electric power, the cost of mechanical orienting and positioning of components is small compared to human labor.

common technique is to convert potential energy to kinetic energy by using a chute. The chute also permits horizontal transportation at zero cost for labor and zero cost for mechanical energy. If greater horizontal distances must be moved, wheel or roller conveyors can be used (1 m drop for 20 m horizontal) with nominal capital costs and operating costs that are effectively zero. Gravity feeding and disposal (especially with spiral conveyors) permit very efficient use of the cube of the workspace.

If the component must have a specific orientation, drop disposal might not be considered. One common technique, however, is to use drop disposal and then use a vibratory feeder (see Figure 14.17) at the following operation to reorient the parts.

PRINCIPLE 7: CONSERVE MOMENTUM

Avoid unnecessary acceleration and deceleration. It takes time and energy to accelerate and decelerate the body, a leg, or an arm. In the Work-Factor predetermined time system, an 45 cm reach with a "change direction" is given 39% more time due to the "change direction."

First we will consider the arm and leg and then the whole body.

The Arm and Leg

Use circular rather than pumping motions. There are a number of operations in which the arm is in relatively continuous motion. An ancient example was the conversion of grain to flour with a mortar and pestle. This pounding with its acceleration and deceleration of the hand was replaced by grinding in a circular motion. Eventually water power replaced muscles for this job and now the millstone is powered by mechanical power. Figure 14.18 shows how this same principle applies to hand polishing operations, whether they are with a rag (in a factory or on your car) or whether they are on the end of a pole or hose (broom, mop, vacuum cleaner). Bicycle pedals are an application of this principle to leg motions. Writing vs. printing is another example. In a kitchen, stirring soup is the same situation. If there is insufficient mixing of the product with circular stirring in a circular container, use a rectangular container to furnish the turbulence rather than requiring a zigzag motion.

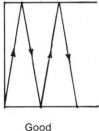
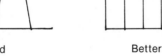

| Good | Better | Best |

Figure 14.18 *Vacuum cleaner strokes* with a looping at the end were most efficient due to minimal overlap and use of momentum.[11]

Figure 14.19 shows another example of avoiding sharp deceleration of the hands and arm. By proper alignment of the buttons on the punch press with the bins and disposal location and the die, unnecessary deceleration can be avoided.

In sports an important principle is to "follow through" to impart maximum velocity and to minimize deviations from the desired path; abrupt deceleration will give poor performance.

Still another situation in which there is unnecessary deceleration is the precise disposal of completed units rather than tossing them aside. Using MTM terminology, an 18-inch move to toss aside a part is an M18E and an RL1; a total of .63 s. A precise placement would require an M18B, a P1SE, and an RL1; a total of .89 s—an increase of 42%. In addition, the precise placement usually requires eye control so other motions cannot be done simultaneously. In MTM an R14Bm is 18% less than an R14B. If the unit might be damaged by abrupt deceleration, cushion its fall with a resilient surface or by using a chute. If the unit must be precisely oriented for the following operation, use a vibratory feeder to orient the unit automatically.

Place objects to avoid hand deceleration. The objective is to permit the hand to grasp the object "on the fly." Figure 14.20 shows good and bad front edges of parts bins. In the poor bin, the front edge of the bin is sharp sheet metal. The operators must be careful in reaching to the bin or their hands will be injured by the knifelike front edge. By making the bin edge a rolled surface the operator need not fear injury if a hand strays from its normal path by a few mm. The bin edges can be rounded by cutting a lengthwise slit in a piece of rubber or plastic tubing and placing it over the sharp edge. This is a relatively poor solution since the operator will keep hitting the tubing when sliding parts out of the bin. This slows the operator.

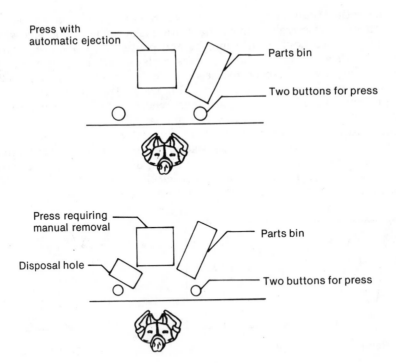

Figure 14.19 *Place bins* so the operator can grasp "on the fly"; use the same concept for disposal.

Figure 14.21 shows how the table shape can affect the speed of grasping small objects from a flat surface. In the bad design, the hand must come to a stop for a precise grasp. With a thin table top the hand can sweep the object to the edge and then grasp it while the hand continues its movement. A small front lip improves the ease of grasping the object as the hand orientation is better and less wrist twisting is required.

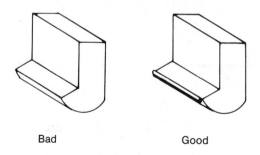

Bad   Good

Figure 14.20 *Avoid knife edges* on bins.

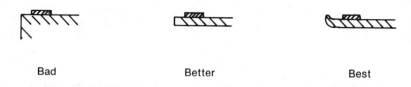

Bad   Better   Best

Figure 14.21   *Grasping small objects* (e.g., coins on a counter) is easier when the lip acts as a wedge under the object to raise it off the surface and aid grasping on both sides. The table lip also reduces the chance of objects rolling off the table as well as giving a more favorable hand orientation.

Avoid unnecessary weight. It requires extra energy per move and extra time per move to transport weight in the hand. Figure 14.22 shows the experimental points by Konz and Rode as well as the recommendations by three different predetermined time systems.[55] Moving more weight requires more time. The amount of extra time is indicated by the *slope* of the lines; the intercept indicates varying definitions of a normal pace.

Our study concerned only the *control* effects of the additional weight; any fatigue effects are additional. The increase in time/move is 3% per kg. The increase in physiological cost as measured by the integrated output of the three axes of a force platform was 6%/kg. Since the time/move increased 3%/kg, this is an increase in the acceleration-deceleration forces/kg of about 3%/move. These conclusions are compatible with Ayoub's results.[8]

Figure 14.22 gives not only our experimental times but also the times as recommended by the Methods Time Measurement (MTM) exact method, the MTM card method, the Basic Motion Time (BTM) recommendation, the Work-Factor time for men, and the Work-Factor time for women.

The MTM approach seems to give inadequate importance to weight. The slope is too flat and, in the usual method, no credit is given for any weight less than 1100 g.

The BMT approach also seems to give inadequate emphasis to weight, giving an increase of 1.9%/kg; the 900 g interval for changes in time seems better than the initial 1100 interval and then 2250 g interval for MTM.

Work-Factor seems to have an appropriate slope for males but its interval (2250 g) is too big. The slope for women of 5.5%/kg is too large.

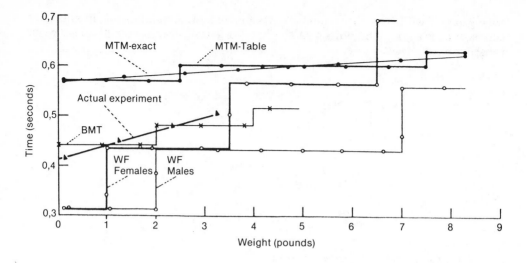

Figure 14.22 *Experimental data* by Konz and Rode are plotted vs. the recommendations of Methods Time Measurement, Basic Motion Time, and Work-Factor.[55] The slopes are not consistent. The experimental data indicate that time should be increased 3%/kg.

Although Ayoub found males took longer than females to reach maximum accelerations and maximum velocities, to my knowledge there is no evidence to support different weight allowances for men and women.[8] If there is a sex effect, it probably is due to body weight. Our experiment on males, however, indicated no difference for body weights from 47 to 73 kg. Females may be subject to more fatigue in that their muscle weight is a smaller proportion of total body weight than that of men; the additional fat is concentrated primarily on the torso, however, so it should not affect hand-arm motions. It will be repeated that our experiment concerned only control effects, not fatigue effects. The rationales and evidence for the slopes for the various predetermined time systems have never been published but they may include fatigue as well as control effects.

The Entire Body

The momentum can be horizontal or vertical. Snook, Irvine, and Bass reported that 90% of USA male industrial workers could push with an initial force of at least 26 kg and pull with a force of 24 kg; they could sustain a pushing force of 13 kg and a pulling force of 14 kg.[87] As these numbers are not very different and safety on inclines may be a problem, for cart pushing and pulling keep the load lower than the worker at all times. Hand-pull carts should have a vertical plate ahead of the front wheel so the person's heel isn't run over by the wheel when pulling. Pulling a load in a rickshaw is at least six times more efficient than carrying the same load on the head; the rickshaw studied could have been improved by making it of light metals and using better tires and bearings but that would have increased its capital cost.[27]

Table 14.4 gives the coefficient of friction between various shoe soles, both dry and soiled with an oil-water mixture, and various types of floors. For floors, a soft rubber pad has a coefficient of about 1.0, end-grain wood about .8, rough finish concrete about .7, working-decorative about .65, and steel about .4. Soiled floors have a coefficient about .2 lower than dry floors. Many slips are due to sudden changes in floor friction—from high to low and from low to high. For shoes, typical values are .5 for leather soles, .6 for neoprene, .7 for rubber-crepe, and .8 for rubber cork. Shoes with no gap ahead of the heel (a la deck shoes) put more "rubber on the road" and so reduce slipping. A tread design which permits water to escape (a la auto tires) reduces slipping on wet floors. Ramsay and Senneck report rubber soled boots give an excellent grip on clean surfaces but a poor grip on greasy or muddy surfaces. They recommend boots with tungsten carbide tipped studs.[77]

A wood block floor has excellent friction—even with oil. For decorative floors, rubber tile is best. Steel floors are slippery to stand on—especially if oil-coated.

Kroemer and Robinson studied the force that 28 males could exert "intermittently and for short periods of time" in a variety of working positions.[59] This force is not a dynamic force but a static force—a "breakaway" force to set an object in motion. Their results are shown in Table 14.5. Maximum push force = .8 (body weight).[90] Push force capability is greater when shoe/friction is high, there is enough space to position the body, and the push surface is between the hip and the shoulder.[20] *Force Limits in Manual Work* has a series of detailed diagrams and tables.[40] Man can push a surprising load if rails are used. Wyndham and Heyns reported 70 kg miners could push 900 kg cars at 3.7 km/hr at an oxygen consumption of 2.1 liters/min and mechanical efficiency of 15%.[102] They report that pushing a 700 kg car (i.e., loaded) at 2.4 km/hr and returning it (weight 400 kg) empty at 4.8 km/hr was the most efficient.

PRINCIPLE 8: USE TWO-HAND MOTIONS RATHER THAN ONE-HAND MOTIONS

Use of two hands instead of one is based on reduced physiological cost/unit of output and reduced time/unit of output.

Reduced physiological cost was studied by Nichols and Amrine; increase in heart rate was the criterion.[74] For an equal amount of pieces moved, one-hand motions had a smaller increase in heart rate than nonsimultaneous but symmetrical two-hand motions. Salvendy and Pilitsis showed kcal/min of one- and two-hand motions did not differ significantly; thus physiological cost/unit was less when using both hands simultaneously.[82] Andrews demonstrated that cranking requires about 10% less watts from the person with one arm than two for loads up to 25 watts; beyond 25 watts, cranking with two arms is about 10% more efficient than with one arm.[2] For exerting a static pull, one-arm work required 42% more watts at a 5 kg load, 18% at a 10 kg load, and 127% more at a 15 kg load.

Reduced time/unit for two hands vs. one was first reported by Barnes, Mundel, and MacKenzie.[14] Konz, Jeans, and Rathore had seven women move a stylus back and forth at 7 different angles (0°, 30°, 60°, 90°, 120°, 150°, and 180° where 0° = "3 o'clock") and 2 distances (225 and 400 mm).[53] Eighteen second trials were performed with the right hand, left hand, and all combinations of angles with both hands. The right hand was the preferred hand for all subjects. The results are summarized in Tables 14.6 and 14.7.

Fitts combined speed and accuracy into one index, bits/second.[37,38] (Shannon defined information in terms of a signal to noise ratio transmitted from a transmitter over a channel to a receiver.[85] Since the ratio was logarithmic and $\log_2$ customarily has been used, the unit of information commonly has been called bit, short for *binary digit*.) The basic concept was that any movement was limited by the amount of information to be processed. Fitts proposed that the amplitude (A) of the move was akin to the signal and the radius or width (W) of the target

Table 14.4 Coefficient of Friction* Between Various Floor and Shoe Materials (Table VIII of Kroemer and Robinson, Kroemer[57,59])

Shoe Condition	Floors 1,2	3,4	5	Group Mean	6	7,8,9	10,11	12,13	14	Group Mean	15	16	17	18	Overall Mean
Rubber cork sole, flat Dry	.65	.73	.83	.74	.70	.87	.71	.75	1.00	.81	1.03	.73	.62	.78	.83
Soiled	.70	.66	.85	.74	.30	.53	.36	.52	.95	.43	—	.39	.43	.30	.52
Rubber crepe sole, flat Dry	.65	.76	.81	.74	.75	.81	.88	.71	.80	.79	.89	.51	.49	.54	.72
Soiled	.59	.68	.75	.67	.30	.52	.49	.36	.55	.44	—	.40	.27	.34	.48
USA-USAF standard sole Dry	.81	.76	.97	.85	.78	.68	1.04	.70	.72	.64	1.04	.50	.69	.29	.76
Soiled	.83	.82	.86	.84	.57	.58	.58	.65	.62	.60	—	.57	.55	.52	.65
Rubber overshoe Dry	.72	.84	.83	.80	.48	.77	.68	.60	1.10	.73	1.05	.80	.44	.38	.72
Soiled	.47	.65	.71	.61	.44	.42	.45	.45	.39	.43	—	.34	.31	.28	.45
Neoprene heel Dry	.64	.85	.94	.81	.69	.66	.77	.59	.65	.67	.93	.72	.47	.59	.71
Soiled	.56	.88	.96	.80	.48	.51	.53	.54	.48	.51	—	.37	.44	.28	.55
Soft nylon heel Dry	.80	.73	.76	.76	.41	.79	.55	.40	.95	.62	.95	.72	.41	.27	.65
Soiled	.47	.55	.75	.59	.34	.49	.43	.30	.42	.40	—	.25	.19	.30	.41
Neoprene sole, flat Dry	.70	.85	.74	.76	.35	.73	.49	.43	.80	.56	1.13	.58	.28	.24	.61
Soiled	.62	.80	.78	.73	.35	.41	.43	.41	.44	.41	—	.27	.26	.34	.46
Leather Dry	.44	.56	.60	.53	.53	.65	.57	.47	.53	.55	.92	.33	.27	.24	.51
Soiled	.97	.90	.71	.75	.67	.60	.66	.70	.77	.68	—	.33	.38	.28	.63
Mean Dry	.68	.76	.81	.75	.61	.75	.75	.58	.82	.70	.99	.61	.46	.42	.69
Soiled	.65	.74	.80	.73	.46	.51	.50	.49	.52	.50	—	.37	.35	.33	.52

*Values over 1 indicate mechanical interlocking between the shoe and the floor
+Soiled = Oil-water mixture

Floor types

Working: appearance not important 1. Smooth concrete, finished with a trowel. 2. Painted concrete, 3 Coats of floor enamel applied to smooth concrete. 3. Rough concrete, finished with a wooden float. 4. Synthetic stone. 5. Wood block, soft end grain.

Working: decorative 6. Hardwood, oak with 2 coats of varnish. 7. Vinyl tile, smooth. 8. Vinyl tile with random decorative grain. 9. Sheet linoleum. 10. Vinyl asbestos tile, smooth. 11. Asphalt tile. 12. Vinyl asbestos tile, grain parallel to direction of motion. 13. Vinyl asbestos tile, grain perpendicular to direction of motion. 14. Rubber tile.

Special-purpose floors 15. Rubber pad (1/4 inch thick, ribbed). 16. Steel, sanded with #240 grit sandpaper. 17. Steel grid, polished with #600 grit sandpaper. 18. Steel, polished with #600 grit sandpaper.

Table 14.5 Horizontal Push and Pull Forces Exertable for Short Periods of Time by 5th Percentile Healthy Male US Adults[57,59]

Horizontal Force: At Least	Applied With	Condition (u = Coefficient of Friction)	
10 kg Push or pull	Both hands One shoulder Back	Low traction	.2 < u < .3
20 kg Push or pull	Both hands One shoulder Back	Medium traction	u about .6
25 kg	One hand	If braced against a vertical wall 500 to 1500 mm from the push panel and parallel to it.	
30 kg Push or pull	Both hands One shoulder Back	High traction	u over .9
50 kg	Both hands One shoulder Back	If braced against a vertical wall 500 to 1750 mm from the push panel and parallel to it; or feet anchored on a perfectly nonslip surface (e.g., footrest).	
75 kg	Back	If braced against a vertical wall 600 to 1100 mm from the push panel and parallel to it; or feet anchored on a perfectly nonslip surface (e.g., footrest).	

Table 14.6 Index of Performance, Bits/s, For the Left and Right Hand. Mean of Values for 225 and 400 mm movements.[53] 0 = "3 o'clock"

Hand	Angles						
	0° A	30° B	60° C	90° D	120° E	150° F	180° G
Right	13.4	13.2	13.7	13.5	12.7	12.1	11.8
Left	11.1	11.0	11.4	12.3	12.2	12.4	12.0

Table 14.7 Mean Performance, Bits/s, and Accuracy, Percent of Movements That Missed the Target, by Spread Between Targets[53]

Spread, degrees	Percent Misses	Performance, Bits/s	Performance Relative to 0° Spread (%)
0	18.0	21.75	100.0
30	28.0	21.45	98.6
60	41.3	21.04	96.7
90	44.9	20.5	94.3
120	48.7	20.7	95.2
150	54.8	20.5	94.3
180	54.8	20.4	93.8

was akin to noise. He then demonstrated experimentally that movement time for hand-arm motions can be predicted well if the information of the task is defined as:

$$I \text{ (bits)} = \log_2 \frac{A}{W/2} \tag{14.1}$$

where A = Amplitude of movement
W = Width of target in movement direction

The validity of this formula has been substantiated by other investigators in other laboratories.[3,24,98]

For the data in Tables 14.6 and 14.7, the 225 mm movement required 4.17 bits for the movement to the outer 25 mm target plus 2.37 bits for the movement for the inner 90 mm target plus 9.54 bits/movement. The 400 mm movement required 8.19 bits.

The average for the right (preferred hand) was 12.9 bits/s, the average for the left hand was 11.7, and for both hands working at the same time was 21.2. Thus, using just the preferred hand gave 12.9/21.2 = 61% of potential output and using just the nonpreferred hand gave 11.7/21.2 = 55% of potential output.

If the person is assumed to be working at maximum output in all three conditions, why isn't the rate the same for all three conditions?

The maximum output in the input data for Table 14.7 of 23.9 bits/s for 225 mm and of 22.4 for 400 mm might be assumed to be the maximum of the brain-eye-muscle system. The minimum output in the Table 14.7 data is 20.1 and 19.3. The average difference of 3.0 bits/s is the maximum difference due to different visual fields.

The reduction in performance due to use of the right hand can be estimated as the average for both hands at 225 mm of 21.4 minus the average of the right hand of 12.9; 21.4-12.9 = 8.5 bits/s. For 400 mm the estimate is 20.8-13.0 = 7.8 bits/s. Reductions for using the nonpreferred hand are 9.7 and 9.2 bits/s. Langolf, Chaffin, and Foulke reported that fingers could process 38 bits/s, wrists 23, and arms only 10.[61]

It seems that the bottleneck in the brain-eye-muscle system is neither the brain (command subsystem) nor the eyes (tracking subsystem), but the muscles and nerves (effector and feedback subsystem). In other words, the limiting factor in hand-arm movements is not the ability of the brain to command or ability of the eyes to supervise, but the ability of the nerves and muscles to carry out the orders. The spirit is willing but the flesh is weak.

PRINCIPLE 9: USE PARALLEL MOTIONS FOR EYE CONTROL OF TWO-HAND MOTIONS

Frank Gilbreth first stated[42]:

When work is done with two hands simultaneously, it can be done quickest and with least mental effort, particularly if the work is done by both hands in a similar manner, that is to say, when one hand makes the same motions to the right as the other does to the left.

Barnes was more concise[11]:

Motions of the arms should be made in opposite and symmetrical directions, and should be made simultaneously.

On the other hand, Barnes also stated:

Eye fixations should be as few and as close together as possible.

Which principle has precedence? The dilemma is posed here:

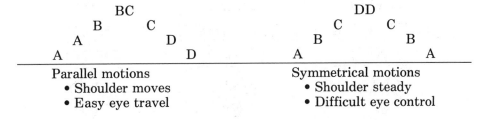

Parallel motions	Symmetrical motions
• Shoulder moves	• Shoulder steady
• Easy eye travel	• Difficult eye control

Barnes and Mundel had each of 10 students position an electrode in a hole 3600 times in 4 different conditions: 180° for the left hand and 0° for the right hand, 150° and 30°, 120° and 60°, and 90° and 90° (0° = "3 o'clock").[13] The least errors were made at 90° and 90°. For movement of a slide in a groove, the best condition was 120° and 60°.

Nichols and Amrine reported that, for an equal amount of work performed, two-hand simultaneous and symmetrical motions had a smaller increase in heart rate than two-hand simultaneous but nonsymmetrical motions.[74] Bouisset and his colleagues investigated two-hand simultaneous and symmetrical motions at (1) 90° and 150 mm, (2) 90° and 300 mm, and (3) 150° and 30° and 300 mm.[15,16] The first and third conditions have the same increase in heart rate; both were better than condition 2. They concluded that the extra 150 mm of distance in condition 2 vs. 1 was equivalent to shifting the angle from 30° to 90°. Reichard reported that simultaneous parallel motions took 8% less time and required 31% fewer eye motions than symmetrical patterns.[80]

The various predetermined time systems enable us to estimate the cost of eye control. The two eye activities are a change in viewing distance (eye focus) and a change in line of sight (eye travel). The normal area of vision is a circle with a 100 mm dia at a viewing distance of 450 mm.

Eye focus is allocated .0044 minutes by MTM. The version of MTM used by General Motors gives .0020 min for *eye focus*, .0030 for *eye reaction*, and .0030 for *eye interpretation*. Work-Factor gives eye focus from .0025 to .0100 min depending on the location of the initial and final focus location; .0050 is a typical time. In addition, Work-Factor allows extra time for *eye inspect*.

"Eye travel" is .009 × T/D minutes in the MTM system where T is the distance between the points and D is the perpendicular distance to the line of travel. General Motors gives a flat .0050. In Work-Factor, it is called *eye shift*. For a shift of 0° to 5°, time = .0004 which increases to .0015 for a 40° shift; beyond 40° is called a head or body turn. In Work-Factor you use either eye focus or eye shift but not both. White, Eason, and Bartlett give .0012 min for a 10° shift, .0015 min for a 20° shift, and .0020 min for a 40° shift.[99]

All the systems emphasize that eye control often occurs simultaneously with other activities.

Konz, Jeans, and Rathore, in the experiment described under Principle 8, calculated the number of hits and misses for various degrees of spread.[53] For motions in which the spread was held constant and symmetrical motions were possible (60° and 120°), symmetrical motions were better than nonsymmetrical. They also reported another experiment in the same paper in which physiological cost (force platform output) was 10% less when the hands moved simultaneously and nonsymmetrically than when they moved simultaneously and symmetrically. The general principle, therefore, using both time/unit and physiological cost/unit, is to minimize the degree of spread rather than worry about the symmetry of the motions.

Figure 14.23 shows an application of this principle to workstation design.

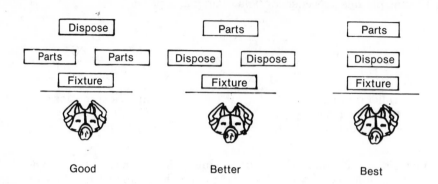

| Good | Better | Best |

Figure 14.23 *Proper location* of the disposal chute can reduce eye control as well as conserve momentum.

PRINCIPLE 10: USE ROWING MOTIONS FOR TWO-HAND MOTIONS

Simultaneous two-hand motions with 225, 1150, and 2050 g weights performed alternately increased the heart rate 1.5 beats/min over the same motions performed in a rowing manner.[73,74] Konz, Jeans, and Rathore reported that force platform output (i.e., work cost) was 10% greater when alternating motions were made in place of rowing motions.[53] In both these studies, the work output was controlled to be the same for alternating and rowing motions.

For both types of motions the hands move in a relatively flat plane with considerable accelertion and deceleration at each end of the stroke. Alternation, however, involves more movement of the shoulder and twisting of the torso. (See Principle 7.)

Note that for human generation of power (winch, bicycle) the handles (pedals) are arranged so that the arms or legs alternate strokes while the path is circular to conserve momentum. Harrison demonstrated that maximum power output came from a device in which both pedals were at the same angular position on both sides of the hub instead of the 180° out of phase used for bicycles.[44] A large flywheel was used to return the pedals for the power stroke. In addition, a "forced" motion, in which kinetic energy of the limbs was fed back into the mechanical system, had greater power output than a "free" motion, in which the kinetic energy was adsorbed by the limbs.

PRINCIPLE 11: PIVOT MOVEMENTS ABOUT THE ELBOW

The key question is: "For horizontal movements at a height, does direction of movement affect (1) the speed of movement, (2) the accuracy of the movement, and (3) the physiological cost of the movement?"

Yes. Yes. Yes.

A number of studies have investigated various combinations of this question using different criteria. In Figure 14.24, 0° is defined as "3 o'clock" and 90° as

"12 o'clock." Figure 14.24 summarizes the effect on *movement time* of studies by Briggs' experiment 3, Schmidtke and Stier, Konz's experiment 3 and 5, Konz, Jeans, and Rathore's experiment 2 and Konz and Rode.[17,51,53,55,83] The left hand data in Konz, Jeans, and Rathore have been mirrored to make them compatible to the right hand movements in all the other experiments.[53]

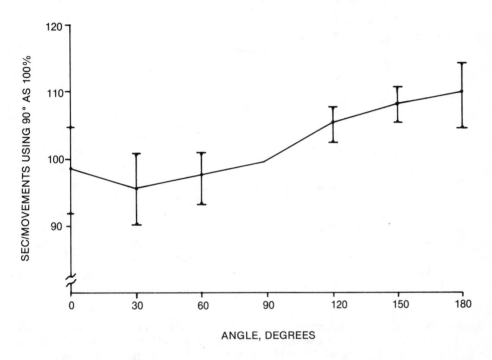

Figure 14.24 *Effect of angle movement* is shown by the mean and range of the data plotted from Konz, Jeans, and Rathore right hand and left hand (angles mirrored), Konz and Rode, Schmidtke and Stier, and Briggs (experiment 3).[17,55,83] For each study, output at 90° was defined as 100%. 0° = 3 o'clock.

Figure 14.24 shows the time benefit of pivoting the movement about the elbow rather than the shoulder.

Corrigan and Brogden measured *accuracy* as a subject moved a stylus following a target between two strips of copper; accuracy was best at 135° and worst (40% more "touches") at 45°.[23] Figure 14.25 shows the percent of styli which missed the target in two different studies.[53,55] The rather unexpected result is that cross-body movements are made more accurately than movements pivoted about the elbow.

The third criterion is *physiological cost*. Boisset et al. showed that changing the angle from 90° to 30° reduced heart rate enough to compensate for 150 mm additional move distance.[15,16] Markstrom found force platform output less when right hand movements were made to the right than when made to the left.[65] Experiment 5 had force platform cost of 2.2 for 0°, 2.5 for 45°, 2.8 for 90° and 180°, and 3.0 for 135°.[51] Konz and Rode reported a cost of 1.6 for 0°, 1.7 for 30°, 2.2 for 60°, 2.4 for 90°, 2.8 for 120° and 150°, and 2.7 for 180°.[55] Thus physiological cost is lower for movements pivoted about the elbow.

The above indicates that angle does make a difference for all three criteria.

If standard times are allocated by computer it may be desirable to assign times by the percentages given in Figure 14.24. For most operations, however, this

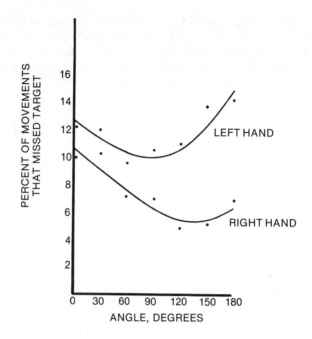

Figure 14.25 *Preferred hands* are more accurate than the nonpreferred. Maximum accuracy is for cross-body movements rather than pivoting about the elbow.[53] 0° = 3 o'clock.

degree of precision is too expensive. For a simple rule of thumb, use the standard time for movements pivoted about the elbow (say 30°-60° for the right hand; 150°-120° for the left), give 5% additional time for straight-ahead moves (say, 60°-120° for either hand), and give 15% additional time for cross-body moves (over 120° for the right hand; below 60° for the left).

The arm is pivoted about the shoulder—not the nose. Therefore put bins ahead of the shoulder instead of the nose. The primary objective is to remind engineers that cross-body movements are inefficient.

PRINCIPLE 12: USE THE PREFERRED HAND

The preferred hand is about 10% faster for reach-type motions. Konz, Jeans, and Rathore found the right hand alone processed 12.9 bits/s while the left hand alone did only 11.7.[53] The preferred hand also is more accurate; the same study reported 7% of targets were missed with the preferred hand and 12% for the nonpreferred. The preferred hand is 5% to 10% stronger.[30,57]

About 10% of the population has the left hand as the preferred hand.

Fisher studied 300 people: 88% had the right hand as the writing hand, 60% had the right eye as the dominant eye, and 54% had the right eye as the most acute eye.[36] (You can check which eye is the dominant eye by holding out your arm. Align an object with your thumb. Close one eye. If the object "moved," you closed your dominant eye.) He found a nonsignificant ($p > .25$) relationship between writing hand and dominant eye, a nonsignificant ($p > .10$) relationship between writing hand and most acute eye, and a highly significant ($p < .001$) relationship between dominant eye and most acute eye. He pointed out that although 188 of his 300 subjects had the same eye as both dominant and most acute, 112 did not. The writing hand is the best single indicator of which side of a person is dominant.

The human brain has two hemispheres; for most people the right is the analytical hemisphere and the left the intuitive hemisphere. It has been reported that musically naive people listen best with their left ear which goes first to the right hemisphere, while trained musicians listen to music best with the right ear which leads to the left hemisphere.

In general, work should come into a workstation from the operator's preferred side and leave from the nonpreferred side. The reason is that reach and grasp are more difficult motions than dispose and release. However, if the new item is obtained on the same side as the disposal, a body turn is eliminated.

PRINCIPLE 13: KEEP ARM MOTIONS IN THE NORMAL WORK AREA

Movement takes energy (cost to the individual) and time (cost to the organization). Since we are meant for movement the goal is not elimination of all movement—just elimination of unnecessary movement.

Shape and Dimensions of Normal Work Area

The first concept of the "normal" work area (as opposed to area of maximum reach) was given by Maynard.[67] It was a dimensionless sketch of an inner and outer semicircle for the right and left hand. Asa measured 30 male students to give the sketches their first dimensions.[5] Farley gave dimensions for men and for women based on average General Motors operators.[35] Farley reported his males averaged 1750 mm in height (with shoes) and females 1500 mm. Squires suggested the shape shown in Figure 14.26 as the elbow does not stay at a fixed point as assumed with the semicircles but moves in an arc as the forearm pivots.[89] The coordinates of the arc PQ are given by the equation:

$$x = A_1 \cos \Theta + A_2 \cos [65 + (73/90) \Theta]$$
$$y = A_1 \sin \Theta + A_2 \sin [65 + (73/90) \Theta]$$

where A_1 = distance EC = elbow to shoulder projection distance
A_2 = distance CP = distance from elbow to end of thumb
Θ = angle given at any instant by the radius which sweeps out the arc DC, degrees; Q = point at which $\Theta = 0$.

Konz and Goel measured 40 men and 40 women selected to be representative of the US population.[52] The subject's heights, selected to be typical of the USA population in 1960, were 1735 mm for the 50th percentile male and 1508 for the 50th percentile female. A_1 equalled 112 mm for 5th percentile males, 152 for 50th percentile males, and 198 for 95th percentile males; the corresponding values for females were 91, 145, and 188. A_2 equalled 378 for 5th percentile males, 412 for 50th percentile males, and 457 for 95th percentile males; the corresponding values for females were 356, 376, and 414. The value of AC of 211 mm for males and 194 mm for females was taken as .5 of the mean elbow to elbow distance from the National Health Survey.[97] The values for x and y are tabulated in Table 14.8 and plotted in Figures 14.27 and 14.28. As miscellaneous information, the angle that the upper arm made with the horizontal plane was 65°.

Perczel in his studies of tramway drivers in Budapest noted (1) that Hungarians had different dimensions than Americans, and (2) that there is both an *outer work limit*, corresponding to a 600 mm (24 inch) radius from the

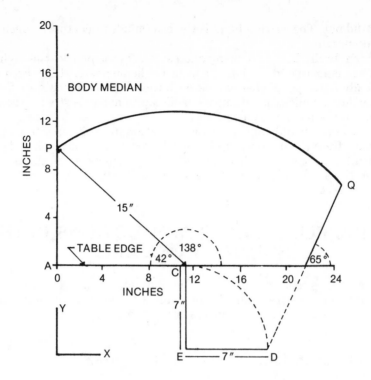

Figure 14.26 *"Windshield wiper" patterns* (where E is the projection from the elbow) show actual hand-arm motions.[89]

shoulder, for an extended arm, beyond which it was difficult to reach, and an *inner work limit*, a radius of 370 mm (15 inch), inside which controls should not be located.[76]

Dunnington reported that force platform output was 20% greater for a simulated drill press operation when one standardized workplace was used than when compared to a workplace in which dimensions were adjusted to the worker's size.[33]

Since people vary, it must be emphasized that one standard workplace regardless of the worker's dimensions is bad design. Figure 14.29 shows the importance of pallets and bins.

PRINCIPLE 14: LET THE SMALL WOMAN REACH; LET THE LARGE MAN FIT

As was pointed out in chapter 11, the designer designs for a certain range of the population rather than the mean of the population. If we place a bin at a distance which can be reached by the mean of the population, then 50% of the population cannot reach the bin. If we design a chair-workstation combination with a thigh clearance appropriate for the population mean, then 50% of the users will not be able to fit.

The design should permit "most" of the "user population" to use the design. The problem is defining *most* and *user population*.

Table 14.8 Coordinates (mm) on the Table for the 5th, 50th, and 95th Percentile Males and Females in the USA in 1960.[52] The Subject's Nose is Assumed to be at x = 0. The Subject's Elbow is Assumed to be the Distance EC From the Front Edge of the Table. The Elbow is Assumed to be 211 mm from the Body Median for males and 194 mm for Females.

	θ	5th Percentile		50th Percentile		95th Percentile	
		x	y	x	y	x	y
For Males	0.00	483	231	538	222	602	216
	11.25	440	269	488	270	544	276
	22.50	392	299	431	311	476	328
	33.75	339	327	367	344	400	369
	45.00	282	345	298	367	317	398
	56.25	222	356	225	381	229	415
	67.50	161	358	150	383	139	418
	78.75	98	352	75	375	48	407
	90.00	37	336	0	335	−41	381
For Females	0.00	436	241	498	191	557	187
	11.25	397	214	452	236	502	244
	22.50	354	292	400	245	437	292
	33.75	306	314	431	306	365	330
	45.00	255	329	292	329	286	357
	56.25	202	338	209	341	203	371
	67.50	147	340	139	346	118	372
	78.75	92	333	69	336	32	360
	90.00	38	320	66	317	29	333

User Population

Some data on various populations were given in chapter 11. The problem of selecting the specific population has become more difficult for the engineer. Some points to consider are:

- Jobs must now be designed for both sexes. Formerly a job could be considered as a male job or a female job. Changing cultural values and laws have changed that. Thus the designer must now consider the range from small woman to large man instead of the range from small woman to large woman or small man to large man.

- International populations must be considered. For example, in Switzerland over 25% of the work force is foreign, so a design to be used in Switzerland cannot use dimensions only of Swiss. Volkswagens are assembled by Turks living in Germany rather than by Germans.

   In addition, many organizations are multinational with plants in many countries. Philips, although headquartered in Holland, has factories in 59 countries. Japanese firms have plants in the USA and Brazil as well as Japan. American firms have many international plants. The result is the designer must consider a wider range of people.

- Multiperson operation of equipment is common. This means that people of different dimensions may be using the equipment—either within the same shift (e.g., ten different people use the duplicating machine) or over multiple shifts (e.g., three different police officers use the same vehicle over a 24-hour period).

- The industrial population is not the same as the general population since children and retired persons are not included, nor are those with mental or physical impairments. Military populations are especially biased as they heavily emphasize youth and physical fitness, as well as being primarily male.

Most

In general *most* of the population has been defined as 90, 95, or, very occasionally, 99% of the population. As Figure 14.30 indicates, the 90% could be the lower 90% of the population, the upper 90% of the population, or the middle 90% of the population. The concept is that by excluding midgets and basketball centers the designer's job is much easier and only a small portion of the population is hindered.

The proportion of the population to exclude depends upon the seriousness of being "designed out" and the cost of designing for everyone. For example, consider design of a tote box to be used on the shipping dock. We might design the box to carry a weight that 90% of the population could lift. In other words, 10% of the population would find it too heavy. What the designer has done in effect is to eliminate weak old ladies from jobs on the shipping dock. If the weight is reduced so that 95% of the population can lift the load (i.e., exclude only very weak old ladies), then productivity decreases as lighter loads are carried by everyone working on the dock.

Occupational safety and health standards in the USA are based on the 95% value. For example, the heat stress *threshold value* is designed to protect 95% of the population with the assumption that the 5% of the population who are "heat sensitive" (no matter how much acclimatization training they get they don't

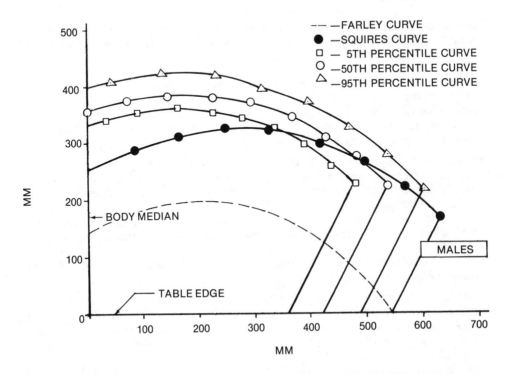

Figure 14.27 *Normal male work area* (right hand) differs from recommendations by Farley and Squires in that Farley did not consider movement of the elbow and Squires did not adjust for human variability.[52]

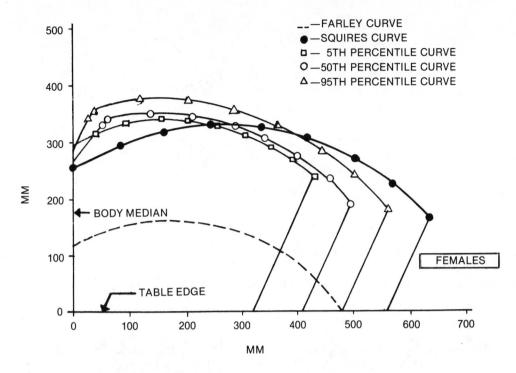

Figure 14.28 *Normal female work area* (right hand) is slightly smaller than for males.[52]

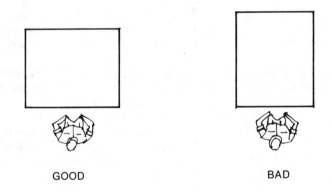

GOOD   BAD

Figure 14.29 *Pallet and bin orientation* affects reach distance. Have the lift truck driver orient the pallet or bin to minimize reaching and minimize back strain.

acclimatize) will not be working in hot jobs—either through self-selection or organization selection. To protect a higher percent would not allow people to go outside during the summer.

Automobiles are designed to the lower 90% of the American population. Tall men (upper 19% or so of men) are cramped but the cars are lower and smaller.

Design so that a small woman (e.g., Nadia Comaneci) can reach and a large man (e.g., John Wayne) can fit. The most practical design technique is to make the machine adjustable.

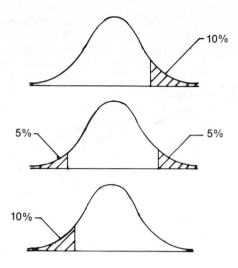

Figure 14.30 *Include most* can mean the lower, the middle, or the upper percentiles.

PRINCIPLE 15: LOCATE ALL MATERIALS, TOOLS, AND CONTROLS IN A FIXED PLACE

Performance of any task requires (1) planning and (2) doing. Locating items in a fixed place reduces the planning or information-processing portion of the task. The operator does not need to ask "Where is the screwdriver?" since the screwdriver is always in the same location.

The first level of time reduction is large savings for having a screwdriver at the workstation vs. searching several places and eventually borrowing one from a neighbor. An amazing number of people start jobs without the proper tools and materials.

The second level of reduction in time occurs as the cycle is repeated over and over. At this stage we not only have a screwdriver at the workstation but have it suspended 10 cm above the operator's right hand. The therbligs "plan," "search," "delay," and "fumble" are reduced and eventually eliminated. This reduction in time is what gives the typical learning curve discussed in chapter 24. Speed of arm movement is usually the same for novices and experienced workers; the performance difference is due to the information-processing speed of the experienced worker. For filing, use color coding.[54]

PRINCIPLE 16: VIEW LARGE OBJECTS FOR A LONG TIME

View Large Objects

The first objective is to view large objects instead of small objects. (For the effect of illumination, see chapter 17.) This can be done by keeping the objects physically close to the eyes or by bringing the object optically close to the eye.

Bringing the object physically close to the eye presents a conflict for hand-arm motions in that the work is not at elbow level. However the reason for the elbow level work-height recommendation is the need to support the weight of the arms.

The arms can be supported by pads and supports on the work surface. Older workers may not be able to focus on the work if it is brought too close to the eyes.

Another possibility is to magnify the object. One alternative is to use a low power (say 2X) lens. Be sure that the lens covers the entire field of view or the eyes become quite fatigued by shifting the accommodation back and forth. Microscopes often are used for very precise work (such as electronic assembly). However, the fixed work position acts as a "straight-jacket"; reduce muscle fatigue and eye fatigue with a "working break" every 20-30 minutes. (An example working break would be to get up and walk to a component supply area 10 m away.) TV projection of the image frees the operator's position but image resolution is a problem.

View for a Long Time

The best situation is one in which the operator works on or inspects a stationary object. To minimize misses, shoot sitting ducks.

A less desirable alternative is working on an object moving past the operator at a constant velocity. Cochran, Purswell, and Hoag found that inspection performance was significantly affected at viewing times of .25 s but not at .50 s.[22] In addition to time, consider angular velocity; values between 10 to 30 degrees/s will not affect performance too much if the worker has satisfactory dynamic visual activity.[63]

One solution technique is to have the operator face "upstream" to maximize viewing time. Remove visual obstructions. Fox[41] reported coin inspection was more accurate when presented in a standard array rather than in a random array—the good coins provided a "background" against which the defective coins stood out. If the operator is viewing randomly spaced objects on a fixed velocity conveyor, there is a very large variability in available viewing time. Some objects might be available for viewing for .2 s while others are available for 2.0 s. If the operator must occasionally make a motor movement (say to remove a defective item), the next 10 items may not be inspected at all. With this fixed pace system, the conveyor speed must be set for the worst performance level of the slowest operator.

A serious problem with machine-paced conveyors or index tables is that the operator must make a positive action to prevent the departure of a defective unit to the next station. If it is necessary to use machine pacing, have the operator make a positive action to send the units to the next station. By far the best design, however, is to use operator-paced stations so that the operator may vary viewing time in relation to the viewing requirements.

SUMMARY

Attention to detail in the physical characteristics of a workstation can reduce the problems of the workstation user and lead to improved productivity.

The first seven principles focus on various aspects of gravity, mass, and force with emphasis on reducing energy cost of the task. Principles 8 to 12 give design guides for one-hand and two-hand motions. Principles 13 to 15 cover workstation dimensions with emphasis on individual variability. Principle 16 has an inspection orientation.

Although many of the design recommendations seem to be common sense and need not be written down, if you will examine many existing workstations you will find that common sense is not very common.

SHORT-ANSWER REVIEW QUESTIONS

14.1 Why should work generally enter a workstation from the operator's preferred side and leave from the nonpreferred side?

14.2 What is the shape of the normal work area?

14.3 There are three basic techniques to achieve a desirable work surface height. Give the three and at least one example for each.

14.4 How is venous pooling avoided?

14.5 Cross-body movements should be avoided for what two reasons?

THOUGHT-DISCUSSION QUESTIONS

1. Describe the chair on which you are sitting. What are its good and bad features?

2. When should an operator work standing?

REFERENCES

1. Agan, T.; Tormey, L.; and Konz, S. Extra heart beats as a measurement of work cost. *Home Economics Research Journal*, Vol. 1 (September 1972): 28-33.

2. Andrews, R. The relative energistic efficiencies of one-armed and two-armed work. *Human Factors*, Vol. 9, No. 6 (1967): 573-580.

3. Annett, J.; Golby, G.; and Kay, H. The measurements of elements in an assembly task—the information output of the human motor system. *Quarterly Journal of Experimental Psychology*, Vol. 10 (1958).

4. Aoki, K. Human factors in braking and fade phenomena for heavy application. *Bulletin of Japan Society Mechanical Engineers*, Vol. 3, No. 12 (1960): 587-594.

5. Asa, M. A study of workplace layout. MS thesis, State University of Iowa, 1942.

6. Asmussen, E. The weight-carrying function of the human spine. *Acta Orthop. Scand.*, Vol. 29 (1960): 141-148.

7. Asmussen, E., and Heeboll-Neilson, K. Isometric muscle strength in relation to age in men and women. *Ergonomics*, Vol. 5 (1962): 167-169.

8. Ayoub, M. Effect of weight and distance travelled on body member acceleration and velocity for three dimensional moves. *International Journal of Production Research*, Vol. 5, No. 1 (1966): 3-21.

9. Ayoub, M., and Holcomb, C. Improved seat console design. Lubbock, Tex.: Institute for Biotechnology Report, Texas Tech., 1976.

10. Backpacks. *Consumer Reports* (August 1974): 572-577.

11. Barnes, R. *Motion and Time Study*, 2d ed. New York: John Wiley and Sons, 1940.

12. Barnes, R.; Hardaway, H.; and Podolsky, O. Which pedal is best? *Factory Management and Maintenance*, Vol. 100 (January 1942): 98.

13. Barnes, R., and Mundel, M. A study of simultaneous and symmetrical hand motions. Bulletin 14, *Studies in Engineering*. Iowa City: University of Iowa, April 1939.

14. Barnes, R.; Mundel, M.; and MacKenzie, J. Studies on one- and two-handed work. Bulletin 21, *Studies in Engineering*. Iowa City: University of Iowa, March 1940.

15. Bouisset, S.; Henon, D.; and Monod, H. Influence de l'amplitude de mouvement sur le cout d'un travail musculaire. *Ergonomics*, Vol. 5 (1962): 265-270.

16. Bouisset, S.; Laville, A.; and Monod, H. Recherches physiologiques sur l'economie des mouvements. *Proceedings of Second International Congress on Ergonomics, Dortmund.* London: Taylor and Francis, 1965.

17. Briggs, J. *A study in the design of work areas.* Ph.D. dissertation, Lafayette, Ind. Purdue University, 1955.

18. Burandt, U., and Grandjean, E. Sitting habits of office employees. *Ergonomics,* Vol. 6, No. 2 (1963): 217-228..

19. Cakir, A.; Hart, D.; and Stewart, T. *Video Display Terminals,* New York: Wiley & Sons, 1980.

20. Chafin, D.; Olson, M.; and Garg, A. Volitional postures during maximal push/pull exertions in the sagittal plane. *Proceedings of the 25th Annual Human Factors Society Meeting,* Rochester, N.Y., 1981: 91-95.

21. Clauser, C.; McConville, J.; and Young, J. *Weight, Volume and Center of Mass of the Human Body,* AMRL-TR-70. Dayton, Ohio: Aerospace Medical Research Laboratory, 1969.

22. Cochran, D.; Purswell, J.; and Hoag, L. Development of a prediction model for dynamic visual inspection tasks. *Proceedings of the 17th Annual Meeting of the Human Factors Society* (1973): 31-43.

23. Corrigan, R., and Brogden, W. The trigonometric relationship of precision and angle of linear pursuit-movements. *American Journal of Psychology,* Vol. 62 (1949): 90-98.

24. Crossman, E. The information capacity of the human motor system in pursuit tracking. *Quarterly Journal of Experimental Psychology,* Vol. 12 (1960): 1-16.

25. Crouch, C., and Buttolph, L. Visual relationships in office tasks. *Lighting Design and Application,* Vol. 35 (May 1973): 23-25.

26. Datta, S.; Chatterjee, B.; and Roy, B. The relationship between energy expenditure and pulse rates with body weight and load carried during load carrying on the level. *Ergonomics,* Vol. 16, No. 4 (1973): 507-513.

27. Datta, S.; Chatterjee, B.; and Roy, B. The energy cost of rickshaw pulling. *Ergonomics,* Vol. 21, No. 11 (1978): 879-886.

28. Datta, S., and Ramanathan, N. Ergonomic comparison of seven modes of carrying loads on the horizontal plane. *Ergonomics,* Vol. 14, No. 2 (1971): 269-278.

29. Davies, C., and Sargeant, A. Physiological responses to standardized arm work. *Ergonomics,* Vol. 17, No. 1 (1974): 41-49.

30. Dickson, A.; Petrie, A.; Nicholle, F.; and Calnan, J. A device for measuring the force of the digits of the hand. *Biomedical Engineering* (July 1972): 270-273.

31. Drillis, R. Flax processing (in Latvian). *Latvian Encyclopedia,* Vol. 12, Riga (1935): 23644-23661.

32. Drillis, R. Folk norms and biomechanics. *Human Factors,* Vol. 5 (October 1963): 427-441.

33. Dunnington, T. The effect of workplace dimensions upon the effort required to perform a simulated light industrial task. MS thesis, Iowa City: State University of Iowa, 1960.

34. Ellis, D. Speed of manipulative performance as a function of worksurface height. *Journal of Applied Psychology,* Vol. 35 (1951): 289-296.

35. Farley, R. Some principles of methods and motion study as used in development work. *General Motors Engineering Journal,* Vol. 2 (1955): 6.

36. Fisher, G. Handedness, eye dominance and visual acuity. *Ergonomics News Letter* (August 1974): 2.

37. Fitts, P. The information capacity of the human motor system in controlling the tolerance of the movement. *Journal of Experimental Psychology,* Vol. 47, No. 6 (1954): 381-391.

38. Fitts, P., and Peterson, J. Information capacity of discrete motor responses. *Journal of Experimental Psychology,* Vol. 67 (1964): 103.

39. Floyd, W., and Ward, J. Posture in industry. *International Journal of Production Research,* Vol. 5, No. 3 (1967): 213-224.

40. *Force Limits in Manual Work,* Guildford, Surry (England): IPC Science and Technology Press, 1980.

41. Fox, J. Quality control of coins. In Weiner, J., and Maule, H., eds. *Human Factors in Work, Design & Production.* London: Taylor and Francis, 1977.

42. Gilbreth, F. *Motion Study.* New York: D. Van Nostrand Co., 1911.

43. Gross, V., and Bennett, C. Bicycle crank length. *Proceedings 6th International Ergonomics Meeting.* College Park, Md.: 1976. See also Gross, V. Bicycle crank length and load. MS thesis, Manhattan, Kans.: Kansas State University, 1974.

44. Harrison, J. Maximizing human power output by suitable selection of motion cycle and load. *Human Factors,* Vol. 12, No. 3 (1970): 315-329.

45. Hassan, M., and Block, S. A study of simultaneous positioning. *Journal of Industrial Engineering,* Vol. 18 (December 1967): 682-688.

46. Heath, L. *The Grey Goose Wing.* Greenwich, Conn.: New York Graphic Society, 1971.

47. Hugh-Jones, P. The effect of limb position in seated subjects on their ability to utilize the maximum contractile force of the limb muscles. *Journal of Physiology,* Vol. 105 (1947): 332-344.

48. Kellerman, F., and van Wely, P. The optimum size and shape of containers for use by the flower bulb industry. *Ergonomics,* Vol. 4, No. 3 (1961): 219-228.

49. Kennedy, J., and Landesman, J. Series effects in motor performance studies. *Journal of Applied Psychology,* Vol. 47, No. 3 (1963): 202-205.

50. Knowles, E. Some effects of the height of ironing surface on the worker. Bulletin 833, Cornell University Agricultural Experiment Station, May 1946.

51. Konz, S. Design of Work Stations. *Journal of Industrial Engineering.* Vol. 18 (July 1967): 413-423.

52. Konz, S., and Goel, S. The shape of the normal work area in the horizontal plane. *AIIE Transactions,* Vol. 1 (March 1969): 70-74.

53. Konz, S.; Jeans, C.; and Rathore, R. Arm motions in the horizontal plane. *American Institute of Industrial Engineers Transactions,* Vol. 1, No. 4 (December 1969): 359-370.

54. Konz, S., and Koe, B. The effect of color coding on performance of an alphabetic filing task. *Human Factors,* Vol. 11, No. 3 (1969): 207-212.

55. Konz, S., and Rode, V. The control effect of small weights on hand-arm movements in the horizontal plane. *AIIE Transactions,* Vol. 4 (September 1972): 228-233.

56. Konz, S.; Wadhera, N.; Sathaye, S.; and Chawla, S. Human factors considerations for a combined brake-accelerator pedal. *Ergonomics,* Vol. 14, No. 2 (1971): 279-292.

57. Kroemer, K. Horizontal push and pull forces. *Applied Ergonomics,* Vol. 5, No. 2 (1974): 94-102.

58. Kroemer, K., and Robinette, J. Ergonomics in the design of office furniture: a review of European literature. AMRL-TR-68-80. Dayton, Ohio: Aerospace Medical Research Laboratories, 1968.

59. Kroemer, K., and Robinson, D. Horizontal static forces exerted by men standing in common working positions on surfaces of various tractions. AMRL-TR-70-114, Dayton, Ohio: Aerospace Medical Research Laboratory, January 1971.

60. Langolf, G., and Hancock, W. Human performance times in microscope work. *AIIE Transactions,* Vol. 7, No. 2 (1975): 110-114.

61. Langolf, G.; Chaffin, D.; and Foulke, J. An investigation of Fitts' law using a wide range of movement amplitudes. *Journal of Motor Behavior,* Vol. 8, No. 2, (1976): 118-128.

62. Lehmann, G. *Praktische Arbeitsphysiologie.* Thieme, Stuttgart, Germany, 1953.

63. Ludvigh, E., and Miller, J. Study of visual acuity during the ocular pursuit of moving test objects. *Journal of the Optical Society of America,* Vol. 48 (1958): 799-802.

64. Lysinski, E. Beitrage zu einer mathematischen theorie der korperlichen arbeit. *Psychotechnische Zeitschrift*, Vol. 1 (1925).

65. Markstrom, P. An investigation of the directional and locational effects on an unconstrained movement. MS thesis, State University of Iowa, 1962.

66. Martin, W., and Johnson, E. An optimum range of seat positions as determined by extertion of pressure upon a foot pedal. *AMRL Report 86*. Fort Knox, Ky.: Army Medical Research Laboratory, June 1952 (AD 21654).

67. Maynard, H. Workplace layouts that save time, effort and money. *Iron Age*, Vol. 134 (December 1934).

68. McCracken, E., and Richardson, M. Human energy expenditures as criteria for the design of household-storage facilities. *Journal of Home Economics*, Vol. 51 (March 1959): 198-206.

69. Miller, I., and Suther, T. Preferred height and angle settings of CRT and keyboard for a display station input task. *25th Proceedings of Human Factors Society*, Rochester, N.Y. 1981; 492-496.

70. Mortimer, R.; Segel, L.; Dugoff, H.; Campbell, J.; Jorgeson, C; and Murphy, R. *Brake Force Requirement Study*. National Highway Safety Bureau Final Report FH-11-6952, Washington, D.C. 20591, April 1970.

71. Muller, E.; Vetter, K.; and Blumel, E. Transport by muscle power over short distances. *Ergonomics*, Vol. 1 (May 1958): 222-225.

72. Nebel, W. Arbeitsstudie uber das feilen. *Psychotechnische Zeitschrift*, Vol. 4, Munich (1929): 25-40.

73. Nichols, D. Physiological evaluation of selected principles of motion economy. Ph.D. dissertation, Purdue University, 1958.

74. Nichols, D., and Amrine, H. A physiological appraisal of selected principles of motion economy. *Journal of Industrial Engineering*, Vol. 10 (September-October 1959): 373-378.

75. Nowak, E. Angular measurements of foot motion for application to the design of foot pedals. *Ergonomics*, Vol. 15, No. 4 (1972): 407-415.

76. Perczel, J. Korperstellungen bei stadtbahnfuhrern. *Proceedings 2nd International Congress on Ergonomics, Dortmund*. London: Taylor and Francis, 1965; pp. 371-376.

77. Ramsay, H., and Senneck, C. Anti-slip studs for safety footwear. *Applied Ergonomics*, Vol. 3, No. 4 (1972): 219-223.

78. Raouf, A., and Arora, S. Effect of informational load, index of difficulty, direction and plane angles of discrete moves in a combined manual and decision task. *International Journal of Production Research*, Vol. 18, No. 1 (1980): 117-128.

79. Rees, J. and Graham, N. The effect of backrest position on the push which can be exerted on an isometric foot-pedal. *Journal of Anatomy*, Vol. 86 (1952): 310-319.

80. Reichard, F. A kinesiological evaluation of parallel vs. symmetrical patterns in simultaneous hand and arm motions. MS thesis, Lubbock, Tex.: Texas Tech, 1967.

81. Roe, R. Describing the driver's workspaces eye, head, knee and seat positions. Society of Automotive Engineers paper 730 356, February 1975.

82. Salvendy, G., and Pilitsis, J. Improvement in physiological performance as a function of practice. *International Journal of Production Research*, Vol. 12, No. 4 (1974): 519-531.

83. Schmidtke, H., and Stier, F. An experimental evaluation of the validity of predetermined elemental time systems. *Journal of Industrial Engineering*, Vol. 12 (May-June 1961): 182-204.

84. Seating in industry. *Applied Ergonomics*, Vol. 1, No. 3 (1970): 159-165.

85. Shannon, C. A mathematical theory of communication. *Bell System Technical Journal*, Vol. 27 (1948): 379-423 and 623-565.

302

86. Shephard, R. *Alive Man*, Springfield, Ill. C. T. Thomas, 1972.

87. Snook, S.; Irvine, C.; and Bass, S. Maximum weights and work loads acceptable to male industrial workers. *American Industrial Hygiene Association Journal*, Vol. 31, No. 5 (1970): 579-586.

88. Soule, R., and Goldman, R. Energy cost of loads carried on the head, hands, or feet. *Journal of Applied Physiology*, Vol. 27 (November 1969): 687-690.

89. Squires, P. The shape of the normal work area. Report 275. Navy Department, Bureau of Medicine and Surgery, Medical Research Laboratories, New London, Conn.; 1956.

90. Strindberg, L., and Petersson, N. Measurement of force perception in pushing trolleys. *Ergonomics*, Vol. 15, No. 4 (1972): 435-438.

91. Trumbo, D., and Schneider, M. Operation time as a function of foot pedal design. *Journal of Engineering Psychology*, Vol. 2, No. 4 (1963): 139-143.

92. vanWely, P. *Design and Disease*. Special Report 86. Manhattan, Kans.; Kansas State University Engineering Experiment Station, 1969.

93. vanWely, P. Design and disease. *Applied Ergonomics*, Vol. 1 (December 1970): 262-269.

94. Von Buseck, C. Excerpts from maximum brake pedal forces produced by male and female drivers. Research report EM-18. Warren, Mich.: General Motors Research Department, January 1965.

95. Vos, H. Physical workload in different body postures, while working near to, or below ground level. *Ergonomics*, Vol. 16, No. 6 (1973): 817-828.

96. Ward, R.; Danziger, F.; Bonica, J.; Allen, G.; and Tolas, A. Cardiovascular effects of change of posture. *Aerospace Medicine* (now *Aviation Space and Environmental Medicine*) Vol. 37 (March 1966): 257-259.

97. *Weight, Height, and Selected Body Dimensions of Adults, United States: 1960-62.* Public Health Service Publication 1000-11-8, Washington, D.C., 1965.

98. Welford, A. Measurement of sensory motor performance. *Ergonomics*, Vol. 3 (1960): 182-230.

99. White, C.; Eason, R.; and Bartlett, N. Latency and duration of eye movements in the horizontal plane. *Journal of the Optical Society of America*, Vol. 52, No. 2 (1962): 210-213.

100. Whitt, F., and Wilson, D. *Bicycling Science*. Cambridge, Mass.: MIT Press, 1974.

101. Wilson, S. Bicycle technology. *Scientific American*, Vol. 228 (March 1973): 81-91.

102. Wyndham, C., and Heyns, A. Energy expenditures and mechanical efficiencies in pushing a mine-car at various speeds and loads. *Int. Z. angew. Physiol. einschl. Arbeitsphysiol.*, Vol. 24 (1967): 291-314.

103. Yamaguchi, Y., and Umezawa, F. The development of a chair to minimize disc distortion in the sitting posture (Abstract). *Proceedings of 4th International Congress on Ergonomics.* Strassborg, 1970.

15

DESIGN OF HAND TOOLS

Hand tools extend capability of the hand. Greater capability can be more impact (hammer), more grip strength (pliers), more torque (wrench, screwdriver), or even new functions (hand saw, soldering iron). This chapter will aid you in selecting from available tools and even, in some cases, to design a new tool. The eight principles are grouped into general principles, grip principles, and geometry principles.

GENERAL PRINCIPLES

The three general principles are (1) Use special-purpose tools, (2) Design tools to be used by either hand, and (3) Power with motors more than with muscles.

PRINCIPLE 1: USE SPECIAL-PURPOSE TOOLS

Return on investment from special-purpose tools usually is high due to a low capital cost/use. The wide variety of screwdrivers, knives, and pliers demonstrates the virtues of hand tool specialization; the right tool for the right job.

Table 15.1 gives capital cost/use for a tool costing $10 and for one costing $100. Capital cost/use almost always is less than a penny. Labor cost/use depends on the cost of labor and the time/use. A round number example of labor cost would be $7.20/hr or .2 cent/s. If a special tool saved 10 s/use, the labor savings/use would be 2 cents. Other expenses (power, repairs) might cost .1 cent/use for a net operating saving of 1.9 cent/use; compare this vs. the additional capital cost of the specialized hand tool.

Formulas may make the calculations more general and more understandable.

$$TC = CC + OC$$

where TC = Total cost/use, cents/use
   CC = Capital cost per use, cents/use
     = 100 CCT/UPL
   CCT = Capital cost total, dollars
   UPL = Uses/life of the tool
   OC = Operating cost/use, cents/use
     = $OCPS$ (SPU)
  $OCPS$ = Operating cost of tool/s, cents/s
  SPU = Seconds/use of the tool, s/use

Table 15.1 Capital Cost/Use (Cents) is Low for Most Situations; 250 Working Days/yr and a 2-year Life are Assumed

Uses/ day	Uses/ life	Initial cost	
		$10	$100
1	500	2.0	20.0
10	5,000	0.2	2.0
100	50,000	0.02	0.2
1000	500,000	0.002	0.02

Assume a nurse uses a general-purpose tool 10 times/day, 250 days/yr, and the tool lasts 2 years; the $UPL = 10 \times 250 \times 2 = 5000$ uses/tool life. If the tool cost $10, then $CC = 100\ (10)/5000 = .2$ cent/use.

If labor cost is $5/hr ($4/hr wages + 25% fringe benefits), the cost/s is .14 cent/s. If time/use is 30 seconds, then labor cost/use is 4.2 cents. If power and maintenance are zero, then $OC = 4.2$ cents/use. Total cost per use, TC, is 4.4 cents.

Consider next a special-purpose tool which can do the job in 5 s but which will cost $25. $CC = 100\ (25)/5000 = .5$ cent/use. $OC = 0.7$ cents, and total cost, TC, is 1.2 cents.

The special-purpose tool saves 3.2 cents/use or $160 over the 2-year life or $80/yr. Return on investment is $80 divided by the average investment of $12.50 or 640%.

A job is composed of *get ready, do,* and *put away.* Table 15.2 shows that a special-purpose tool saves time if it is multipurpose.

The multipurpose tool may combine two functions (such as a claw hammer combining a nail claw and a hammer or a pliers combining the grip function as well as a wire cutter). An extreme example is my camping shovel which also is a hammer, saw, ax, bottle opener, and wrench! Two-tools-in-one eliminates a reach, grasp, move, and release from the labor cost; that is, get ready and put away costs are lower.

The special-purpose tool also may save by "doing" more efficiently. Examples might be higher rpm, more precision, greater power.

The special-purpose archaeology tool shown in Figure 15.1 combined both get and put away advantages (by including a saw in one edge and by cutting off the

Table 15.2   A Combined Function Tool Saves "Get Ready" and "Put Away" Time

Combined tool	Separate tools
Reach for tool AB	Reach for tool A
Grasp	Grasp tool A
Move to point of use	Move to point of use
Use tool A	Use tool A
Regrasp for B	Aside tool A
Use tool B	Release tool A
Aside tool	Reach for tool B
Release tool	Grasp tool B
	Move to point of use
	Use tool B
	Aside tool B
	Release tool B

tip so it could be used to level the ground) and do advantages (by a better grip which permitted a better bearing surface for the hand).

When total cost is considered, a $2 tool may be more expensive than a $200 tool.

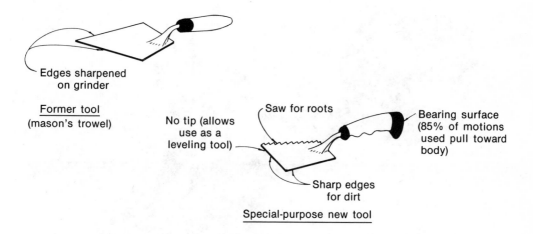

Figure 15.1 *Special-purpose tools* can give advantages in "doing." If it is multipurpose, it also may present advantages in "get ready" and "put away." The new salvage archaeology tool has both types of advantages.

PRINCIPLE 2: DESIGN TOOLS
TO BE USED BY EITHER HAND

There is considerable social stigma in "leftness." *Left* comes from *lyft*, Anglo-Saxon for weak, broken; in Latin, *sinister;* in French, *gauche* from which we also get gawk. In medieval plays, the devil entered from stage left; we give "left-hand" compliments; political radicals are "left-wing." On the other hand, valued assistants are "right-hand" men, honored guests sit at the right of the host, salutes and blessings are given with the right hand, to be correct is to be right.

Right comes from the Anglo-Saxon "rigt" which means "straight, just." In French, *a droit* means to the right; we use adroit as skillful, nimble in use of hands. Right on!

In most circumstances the tool should be in the user's preferred hand. The right hand is the preferred hand for about 90% of the population; the percent seems constant across cultures and for both sexes.

The first benefit of a tool which can be used by either hand is for the 10% of the population left out. In sports, where the emphasis is on maximum performance, both left and right hand products usually are available; the same emphasis on performance is desirable for industry.

The second benefit of a tool which can be used by either hand is that it can be used by the nonpreferred hand in special work situations where the preferred hand is not available.

On repetitive operations a tool which can be used by either hand permits switching hands to reduce local muscle fatigue. Figures 15.2 and 15.3 show food scoops.[16] The right-handed scoop has a powerful (1.2 kg) spring—the user's thumb soon fatigues and the user is either idle or working with pain since the scoop is very difficult to use with the left hand. The German design (Figure 15.3), although its sharp edges cut into the hand, permits alternating hands and so is the preferred design.

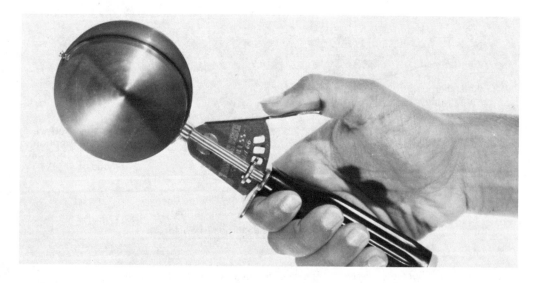

Figure 15.2 *Excessive force* by the thumb is required for the "right-hand only" tool.

Dexterity is better in the preferred hand (see Table 15.3). Without a tool, the preferred hand saves about 5% in time but with a tool the margin increases with increasing complexity of manipulation. Figure 15.4 shows a dental syringe designed to be used with either hand.

Certain tools are peculiarly right-handed in design and require a different action when used by the left hand. Capener explains the use of a right-hand scissors in the left hand[3]:

With the compressive action between the thumb and middle or ring fingers it is necessary to add extension (instead of flexor) action to the terminal phalanx of the finger, and adduction (instead of abduction) to the thumb. This is a reversal of pres-

Figure 15.3   *Local muscle fatigue* in a single finger is reduced by using the entire hand to squeeze the spring that cleans the scoop.

sure upon the rings of the instrument with a consequent tendency to twist it so that the shearing surfaces lie in the vertical rather than the horizontal plane.

Goal: Emancipate the left hand.

PRINCIPLE 3: POWER WITH MOTORS MORE THAN WITH MUSCLES

Mechanical energy is 10 to 1000 times cheaper than human energy because:

- People run 24 hr/day and 8760 hr/yr while machines are shut off when not in use. The worker's energy use (covered by wages) during the entire 8760 hours thus is contrasted vs. the 2000 or less hours for the machine.

- The worker often uses "high-cost fuel" such as meat. Cattle, sheep, and hogs require about 7 kcal of plant products for 1 kcal of meat; chickens and fish about 5; chickens and cows eat about 4.5 kcal per kcal of milk and egg output (1 kcal/hr = 1.16 W).

- The worker's "fuel cost" not only provides food for the worker but usually provides food for his family.

- The worker's wages (fuel cost) are used to purchase goods other than food (housing, entertainment, transportation).

- Man is relatively inefficient as a power source. Bicycle pedaling, the most efficient method of human-generated power, is only 20%-25% efficient.

Operating costs of most power tools are small. An air tool such as an air-powered screwdriver used on an assembly line uses .5-.7 m³/min, if run continuously. Since compressed air costs about 90 cents/100 m³, this is about 32 cents/hr. Contrast this vs. labor costs!

Table 15.3 Dexterity as a Function of Hand, Sex, and Age. In the Studies by Kellor et al. Data Were Recorded For the Right and Left Hand; the Preferred Hand was not Recorded.[15] Right is Reported Here as Preferred.

Study	Mean time for 20-yr-old		Ratio	Age Equation Coefficients*				Standard Deviation	
	Preferred	Nonpreferred	Preferred/ Nonpreferred	Preferred		Nonpreferred		Preferred	Nonpreferred
				A	B	A	B		
Kellor et al., 1971									
Nine-hole pegboard (pegs from bin to board; then back to bin. No tools.)									
Males (N=122)	15.1	16.0	.94	11.1	.200	11.7	.217	3.20	3.20
Females (N=122)	16.4	11.7	.98	13.7	.133	13.0	.183	2.71	3.34
50-hole pegboard (pegs from bin to board; no tools.)									
Males (N=122)	97.2	107	.91	68.8	1.42	77.0	1.50	23.1	20.5
Females (N=125)	103.3	108.7	.95	86.6	.883	85.7	1.15	15.9	18.6
Lowden (1977)									
12 peg-in-pliers transfers between pegboards									
Males (N=20)	15.9	19.7	.81						
Nut on bolt with open-end wrench									
Males (N=20)	94.7	108.5	.87						
Screw in and out with screwdriver									
Males (N=20)	69.6	101.2	.69						

*TIME, seconds = A + B (AGE, years) (for 50% individual) 20 < AGE < 85

SOURCE: Table 15.3 was obtained from Sister Kenny Institute's rehabilitation, *Hand Strengths and Dexterity Tests*, copyright © 1971, and is used with the permission of the Institute. Further information on rehabilitation may be obtained by writing Sister Kenny Institute, Publilcations-Audio-visuals Office of the Research & Education Department, Chicago Avenue at 27th Street, Minneapolis, Minnesota 55407.

Figure 15.4 *Duplicate controls* and an extended control shaft permit this dental syringe to be operated with either hand.[5]

GRIP PRINCIPLES

The second trio of hand tool design principles concerns grips: (4) Use a power grip for power, use a precision grip for precision; (5) Make the grips the proper thickness, shape, and length; (6) Design the grip surface to be compressible, non-conductive, and smooth.

PRINCIPLE 4: USE A POWER GRIP FOR POWER; USE A PRECISION GRIP FOR PRECISION

There are a number of different grips but the most important for hand tools are the power grip and the precision grip.[2]

Power grip. Figure 15.5 shows a power grip. The tool axis is perpendicular to the forearm axis; the hand makes a "fist" with four fingers on one side of the tool grip and the thumb reaching around the other side to "lock" on the first finger.

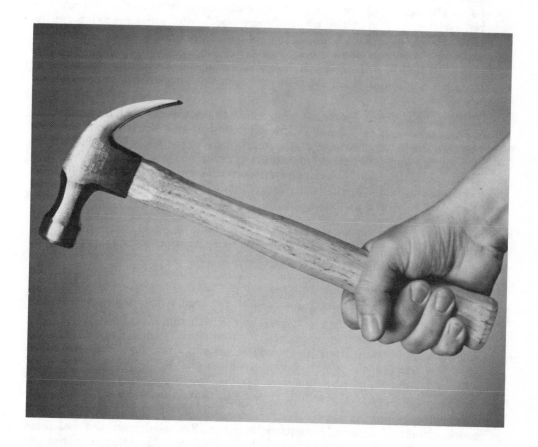

Figure 15.5   *Power grips* (shown above) and precision grips (see Figures 15.6 and 15.7) are two basic hand grips.[20]

Three subcategories of the power grip are differentiated by the direction of the line of force: (1) force parallel to the forearm (e.g., saw), (2) force at an angle to the forearm (e.g., hammer), and (3) torque about the forearm (e.g., corkscrew).

- *Force parallel to the forearm* can be typified by a hand saw, an electric iron, a hand electric drill, a "Y" shovel handle, a suitcase handle, or a pistol—the power grip is often called a *pistol grip*. The muscles can apply force along the forearm axis (electric iron, saw, shovel) or the muscles can resist force (electric drill, suitcase handle, pistol). There are two moment arms caused by (a) tool action and (b) weight of the tool. Aligning the handle under the tool center of gravity (tool balance) minimizes unnecessary torques as well as permitting sighting along the tool to improve accuracy. If the tool weight is low (a light tool, a suspended tool, or a tool sliding along a surface) and tool force is high, put the handle grip at the rear of the tool.

- *Force at an angle to the forearm* can be typified by a hammer, hand ax, ice pick, chisels, pizza cutter, and pliers (reverse grip). Although firm control is still required, the tool force angle differs from tool to tool. The tool may be above the hand (hammer) or below (ice pick, pizza cutter); the wrist may be flexed (hammer, fishing rod, tennis racquet) or locked (ice pick, pliers, power screwdriver). When the tool is above the hand, the wrist is in "high gear"—the top of the tool moves more than the bottom. When the tool is below the hand, the wrist is in "low gear"—movement at the extremity is reduced but power at the wrist is increased.[3] Additional precision can be obtained by not locking the thumb over the fingers but aligning it along the tool axis (cobbler's hammer, fly swatters). The trade-off is additional precision vs. loss in grip strength. (A variation of the power grip is the coffee-cup grip. Cup handles should have space for two fingers and should restrict heat flow.)

- *Torque about the forearm axis* while using the power grip can be typified by a corkscrew with a T-handle. The extension of the forearm axis projects through the fourth finger. A common problem for torque is insufficient bearing surface and lever arm. Doorknobs (a control related to a tool but with fixed axes rather than the unlimited axes of a tool) often are a problem as architects provide a polished sphere while expecting the user to exert the torque by relying on friction.

Precision grip. The precision grip has two subcategories. Figure 15.6 shows an internal precision grip; Figure 15.7 shows an external precision grip. The strength of a precision grip is only about 20% of a power grip.[29]

- *Internal precision grips* (table knife, blade razor, file) have three characteristics: a pinch grip by the thumb vs. first finger (or thumb vs. first + second finger); support to reduce tool tremor by the little finger and side of the hand; and, third, the shaft passes under the thumb and is thus "internal" to the hand. In many applications the hand itself is supported on a work surface or rest. The arm muscles which control the hand are very sensitive to tremor.[17] Patkin recommended surgeons should not carry suitcases for 24 hours before an operation since the seemingly minor exertion reduced hand steadiness.[22]

  If pushing or pulling is involved, the tool handle usually is parallel to the work surface. If rotation is required (e.g., small screwdriver), the tool shaft tends to be perpendicular to the work. A problem with rotation is that the end of the tool grip tends to bore a hole in the palm. Eliminate penetration by using a shaft so long that it extends beyond the palm; or reduce penetration by using a large spherical end on the grip instead of a sharp surface; or reduce the need for force along the tool axis by considering a grip with better bearing characteristics.

  As with the power grip, the thumb or little finger can be repositioned. Pointing along the top surface of a knife gives more power as well as addi-

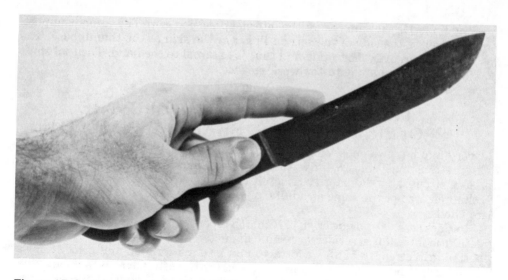

Figure 15.6   *Internal precision grips* have three characteristics: (1) a pinch grip, (2) support by the little finger or side of the hand, and (3) the shaft is "internal" to the hand.

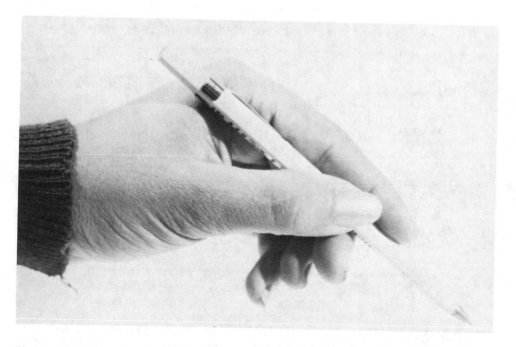

Figure 15.7   *External precision grips* have three characteristics: (1) a pinch grip, (2) support by the side of the 2nd finger or skin at the thumb base, and (3) the shaft is "external" to the hand.

tional precision. Patkin mentions pointing along a surgeon's needleholder with the thumb to gain additional precision.[22]

- *External precision grips* (pencil, spoon, chopsticks) also have three characteristics. A pinch grip by the thumb vs. first finger (or thumb vs. first + second finger) is the same as for the internal precision grip. The second support, however, is the side of the second finger or the skin at the thumb base. The shaft passes over the thumb and thus is external to the hand. The tool shaft usually is at an angle to the work surface.

PRINCIPLE 5: MAKE THE GRIP THE PROPER THICKNESS, SHAPE, AND LENGTH

Every tool has two ends: one working on the material; the other on the hand. Figure 15.8 and Table 15.4 give the key dimensions of adult hands. Note that, for size, strength, and dexterity (1) adult men are not the same as adult women, (2) children are not the same as adults, and (3) gloved hands are not the same as bare hands.

Grip thickness. For a power grip, Rubarth reported the larger the diameter the greater the force (for screwdriver handle diameters of 18 to 40 mm).[26] Hertzberg reported grip strength of 43 kg for a 40 mm grip, 65 for a 65 mm grip, 48 for a 100 mm grip, and 36 for a 125 mm grip.[11] Strength with gloves was about 20% less. Ayoub and LoPresti used electromyography and reported 40 mm was preferred to 50 and 65 mm[1]; if force at 40 is 100%, then force at 50 was 95% and 65 was 70%.

Table 15.4   Key Hand Dimensions, mm, of USA Adults[7,24]; see Figure 15.8

| Key | Dimension | Small Woman (1st Percentile) | | Large Man (99th Percentile) | |
		Garrett	Rigby	Garrett	Rigby
a	Hand length*	160	170	218	211
b	Palm length	—	97	—	119
c	Metacarpal breadth	69	79	91	99
d	Breadth with thumb	—	91	—	117
e	Tip to crotch	—	43	—	69
f	Thumb length	43	43	69	69
g	Digit 2 length	56	56	86	86
h	Digit 3 length	66	66	97	97
i	Digit 4 length	61	61	91	91
j	Digit 5 length	46	46	71	71
k	Thumb thickness	15	18	23	23
l	Digit 3 diameter (tip)	13	20	20	25

*Straight and flat. For relaxed hand, subtract 58 mm.

Greenburg and Chaffin (pp. 51 and 77) recommend that a power grip diameter should be between 50 and 85 mm with the goal toward the 50.[9] If the handles move toward each other, maximum initial span should be about 100 mm (maximum hand size) and a minimum initial span should be about 50 mm (tool force low during closure) (see Figure 15.9). If the diameter is too large, then the fingers don't overlap, there is no "locking," and strain is sharply increased. If the diameter is too small, there is insufficient friction area and the handle cuts into the hand. (You may even see workers wrap grips with tape to increase their diameter.)

Saran reported a T-handle of 25 mm was preferred to handles with either 19 or 32 mm.[27] Rigby, making recommendations for container handles, gave 6 mm diameter as the minimum for weights of less than 7 kg, 13 mm diameter for

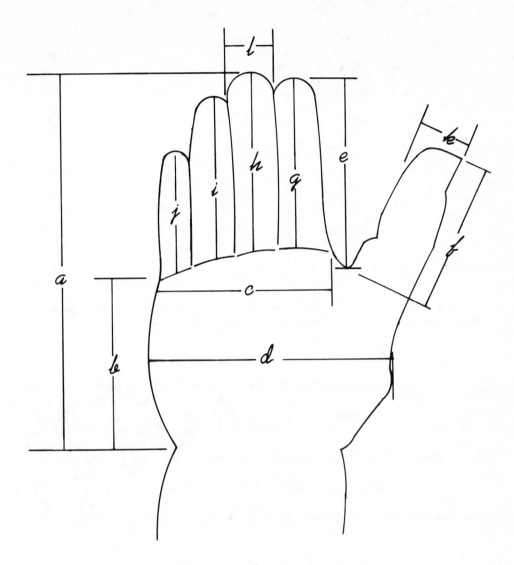

Figure 15.8 *Small woman and large male* adult dimensions are given in Table 15.4. Gloves increase the dimensions. The engineer should remember that hands differ not only in size but also in strength and dexterity.

weights of 7-9 kg, and 19 mm diameter for loads over 9 kg.[24] In summary, power grips between 25 and 50 mm diameter usually will be satisfactory. The most common mistake probably is very small (less than 13 mm diameter) handles.

For precision grips, Hunt reported time to drive one screw was 1.9 s for an 8 mm diameter screwdriver handle but was 3.6 for a 16 mm diameter handle; the larger diameter was rotated at a lower rev/min than the smaller diameter.[12] Although precision grips are not supposed to require force, diameters less than 6 mm should be avoided as they will cut into the hand if force is required. Kao reported boys had better handwriting with 13 mm diameter pens than with 10 or 6 mm diameter pens.[13]

Shape—section perpendicular to grip axis. For most situations use a captive grip (tool does not rotate in the hand). You prevent tool rotation by applying a countertorque with your hand.

A torque is a force at a moment arm. The strategy is to increase the moment

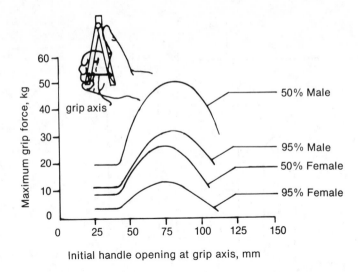

Figure 15.9 *Maximum grip strengths* for various handle openings.[9] Subjects were electronics manufacturing employees; N = 50 males and 50 females.

arm or to have the resisting surface at which the force is applied perpendicular to the force so as to minimize slippage and maximize the effective moment arm. Rubarth reported the maximum turning ratios shown in Figure 15.10.[26] His general conclusion was that, for a power grip, design for maximum surface contact so as to minimize unit pressure on the hand. A tool with a circular cross section tends to permit slippage and the effective moment arm is less; a rectangular cross section gives a good bearing surface with no decrease in moment arm due to the angle of the force application. A rectangular cross section tool does not roll when on the table. Miller, Ransohoff, and Tichauer show in Figure 15.11 how a bayonet forceps was modified to resist turning in the hand.[19]

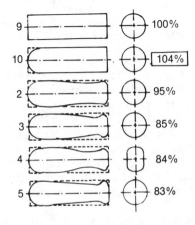

Figure 15.10 *Maximum force* which could be exerted on a screwdriver handle varies with the handle shape; the best design is one which gives maximum contact area.[26]

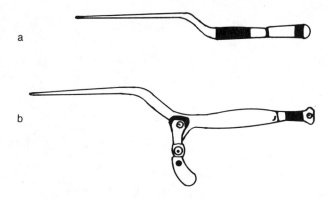

SOURCE: This illustration is from an article written by M. Miller, J. Ransohoff, and E. Tichauer, and appeared in Vol. 2, No. 4 (1971) of *Applied Ergonomics,* published by IPC Science & Technology Press Lt., Guildford, Surrey, U.K.

Figure 15.11   *Accidental twisting* was reduced by adding a bearing surface on this surgeon's forceps.

Patkin showed how the use of a thumb conoid reduced accidental twisting on a surgeon's forceps.[22] Tremor was reduced by using the thumb to rotate the tool while the forearm was supported. Pointing with the thumb aided movement precision and, since the thumb can move up and down, it could open the jaws. A thumb conoid is better than a thumb ring since (1) it fits multiple thumb diameters, (2) it has greater bearing surface, and (3) it has a wider range of up and down movement since the thumb tip, not knuckle, moves.

Workers in a cellophane plant file a flat area on the bottom of their knife handles; it permits them to orient the knife without looking at it.

Indentations parallel to the grip axis are a mixed blessing—they give a bearing surface but the sharp edges tend to dig into the hand. An indentation for the thumb is good if the thumb position is fixed; care should be taken that this does not prevent the use of either hand.

In some situations, such as a pencil, rotation is neither good nor bad. For these situations the cross section should be circular to minimize sharp edges. About 80% of wooden pencils are hexagonal rather than round, due to economics, not ergonomics. Nine hexagonal pencils can be cut from a slat that yields only eight round ones.

In some situations, such as a suture needle holder, the tool should rotate in the hand. Make the cross section within the pinch grasp circular to permit rotation by simple finger-thumb movements rather than the complex regrasps or forearm movement needed for rectangular cross sections. If the forearm is not required to move during the tool rotation, the forearm can be supported to reduce tremor.

Shape—section through the tool grip axis. Should there be any change in cross section? A change in cross section (1) reduces movement of the tool forward or back in the hand, (2) acts as a shield, (3) permits greater force to be exerted due to a better bearing surface.

Figure 15.12 shows a guard on the front of the grip. Injury from movement can come from the tool (knives, soldering irons) or simple impact of the hand on any unyielding or sharp surface. A front guard also can act as a shield vs. heat or materials (stirring soup, solder splashes). Greater force can be exerted along the tool axis since the strong muscles of the forearm and shoulder are not limited by the grip strength of the fingers vs. the grip surface.

A pommel (a shield at the rear of the grip) prevents loss of the tool when the grip is relaxed momentarily, but its most important benefit is to permit greater

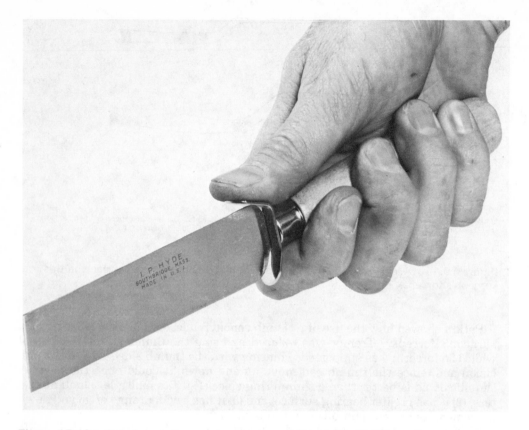

Figure 15.12 *Guards* on grips can improve capability during use as well as reduce severity of accidents.

force to be exerted when the tool is being pulled toward the body. Garrett indicated T-handles permit maximum pulling force.[7]

Finger grooves or indentations between the front and rear of the grip generally are not good since hand width (metacarpal breadth) varies about 18 mm in the population; thus the ridges between the valleys (pressure point) fit no one except the tool designer. Better is to change diameter gradually. Since digit 3 is about 25 mm longer than the thumb or digit 5, the diameter at digit 3 can be about 25/3.1 = 8 mm larger. If a tool is used by only one user, the grip could be molded to fit that specific hand. Many handles use a wedge shape to reduce hand movement.

Length. For a power grip, there must be sufficient length for all four fingers. Table 15.4 gave metacarpal breadth of 79 mm for a 1st percentile adult (small woman) and 99 for a 99th percentile. Thus 100 mm is a reasonable minimum but 125 mm may be more comfortable. If the grip is enclosed (hand saw) or a glove will be worn, use 125 mm as a minimum.

For an external precision grip, the tool shaft must be long enough to be supported at the base of the first finger or thumb. A minimum is 100 mm.

For the internal precision grip, the tool should extend past the tender palm but not so far as to hit the wrist. The "Western Electric" pliers shown in Figure 15.14 demonstrate the principle. Screwdrivers are designed to be held with the internal precision grip. If the user tries to get greater force parallel to the tool axis by pressing the butt into the palm, pain results. For screwdrivers in which considerable force must be applied, use a T-handle or a spherical handle.

PRINCIPLE 6: DESIGN THE GRIP SURFACE TO BE COMPRESSIBLE, NONCONDUCTIVE, AND SMOOTH

Compressible. Just as a compressible floor (wood or carpet) is easier on the feet and legs than noncompressible concrete, a compressible grip material is easier on the hand. Wood is the material of merit; compressible rubber or plastic are acceptable. Avoid hard plastic or bare metal. Compressible materials dampen vibration and aid the hand from slipping on the grip; compressible grips with high coefficient of friction (tape on baseball bat) may be useful. Since oil reduces the coefficient of friction, use a grip material that does not absorb oil.

Nonconductive. Grips should not conduct electricity or heat. Fortunately materials with good electrical resistance also resist the flow of heat. Note that metal rivets in a grip may conduct even if the remainder of the grip is nonconductive.

Table 15.5 gives the contact time required for pain or cell death for three different materials. As the temperature at 80 micrometers under the skin reaches 50 C, there is pain. Maximum pain occurs at 60 C; cell death (through protein denaturation) occurs at 70 C. For cell death, Table 15.5 shows that 1 s of contact vs. aluminum at 79 C (which raises the temperature at 80 μm below the skin to 70 C) is equivalent to 1 s contact of phenolic plastic at 158 C and wood at 197 C. For pain, the corresponding numbers are 53, 85, and 100 C. Steel is about the same as aluminum while concrete, marble, water, and glass are about halfway between aluminum and plastic. Wood is best for two reasons: (1) wood releases heat to the hand more slowly than plastic or metal and so can be held for a longer time before injury occurs; (2) wood gains heat more slowly than plastic or metal and so is less likely to reach a high temperature. Wu demonstrated that even a thin layer of plastic or porcelain can substantially slow down heat transfer.[33] For a 1 s contact, a .4 mm layer of plastic on steel gave a rate of rise in temperature at .1 mm below the skin of only 30% of that from uncoated steel. Faulkner recommended worksurface areas should be kept below 61 C for aluminum, 82 C for Pyrex glass, and 138 C for polystyrene to prevent first degree (reversible) burns for 1 s contacts.[6] To prevent pain, he gives values of 45 C for aluminum, 54 C for Pyrex glass, and 77 C for polystyrene. For a surface which will be held continually he recommends 42 C. Comfort is maximized at about 35 C. If a tool may become cold, give its handle a low mass so that it can be warmed by heat from the hand. A narrow "neck" between the handle and the remainder of the tool (e.g., coffee cup) reduces heat transfer.

Table 15.5   Surface Temperatures (C) of Three Materials Vs. Contact Time Required for Pain or Cell Death[32]

Contact Time, s	Pain			Cell death		
	Wood	Phenolic Plastic	Aluminum	Wood	Phenolic Plastic	Aluminum
0.1	125	104	61	260	206	96
0.3	108	92	56	218	174	92
0.5	105	89	55	204	163	81
1.0	100	85	53	197	158	79
5.0	95	82	52	183	148	76
10.0	95	82	52	180	147	76

SOURCE: Reprinted with permission from the American Institute of Aeronautics and Astronautics, Paper No. 75-713, Table 15.5.

Smooth. A knife cuts by crushing—a force over a very small area gives the high pressure. In the same way, sharp edges and corners on tools act as knives to the hand. Keep tool radii over 3 mm; 6 or 9 is better. Dip metal grips in plastic or wrap with tape to cover sharp radii as well as forging and parting lines. The plastic or tape also makes the grip compressible and nonconductive. Smoothness aids sanitation. Although grip serrations increase friction they cut into the hand. If nonslipping is important, it may be possible to increase the hand's coefficient of friction with various substances or with gloves. Be cautious of very soft grip materials as they may embed chips or splinters.

A poor grip leaves its mark—on the hand.

GEOMETRY PRINCIPLES

The third set of hand tool design principles concerns geometry: (7) Consider the angles of the forearm, grip, and tool, and (8) Use the appropriate muscle group.

PRINCIPLE 7: CONSIDER THE ANGLES OF THE FOREARM, GRIP, AND TOOL

Tools, not wrists, should bend. The underlying philosophy is that machines should adjust to man rather than requiring man to adjust to a machine. The hand is kept small and flexible by locating the muscles which power it in the bulky forearm; the fingers are moved by tendons leading from the fingers through the wrist bones (carpal tunnel) to the muscles. These "ropes" fray on the bones if moved while the wrist is bent and cause tenosynovitis. Tendon movement while the wrist is not bent is less injurious. As with any other physical dimension, carpal tunnel dimensions vary among individuals; Welch demonstrated individual predisposition to tenosynovitis can be predicted.[30]

The most comfortable hand position is the "hand-shake" position. Figure 15.13 shows a conventional hammer handle and two Bennett hammers with a 19° bend. A grapefruit knife is another example of putting the bend in the tool—in this case to reduce bending to the side. Kao, summarizing several of his studies on writing instruments, reported off-center penpoints were better than centered, straight penpoints were better than curved, and penpoints tilted at 20° down were better than straight.[14] In all three cases, improved visual feedback accompanied the best design. The "Western Electric" pliers shown in Figure 15.14 and the soldering irons shown in Figure 15.15 reduce up and down bending of the wrist; they also reduce elbow abduction (see chapter 14, Principle 4). Another alternative to changing the tool angle is to change the orientation of the work. If the tool or work cannot be modified, a wrist splint or bandage may reduce operator pain.

Sufficient clearance. Clearance is needed to minimize burns and pinch points.

- *Burns* from contact (conduction) or proximity (radiation) can be reduced by increasing the distance of the hand from the hot surface. Figure 15.16 shows various types of spatulas. On a rolling pin, for example, increase the roller diameter. The same concept of clearance applies to handles on fixed equipment such as ovens. If the handle or grip is such that the hand cannot pull away (coffee cup handle), contact time is increased (refer to Table 15.4). Viscous materials spilled on the skin and/or hot liquids or materials spilled on clothing are very dangerous due to the increased contact time.

- *Pinches* present problems when the tool is used repetitively. One pinch/100 uses in a home tool may be acceptable if that means one pinch per 10 years. Industrial tools, however, may be used 100 times/day or even 100 times/hr.

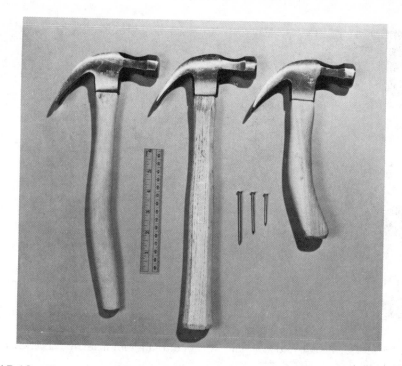

Figure 15.13 *Hammer handles* can be improved. Granada and Konz had 104 people evaluate the hammers with a semantic-differential questionnaire.[8] Defining the vote for the conventional hammer (center) as 100%, the vote for the short Bennett hammer (right) was 110% and the long Bennett hammer (left) was 113%. All three had identical heads.

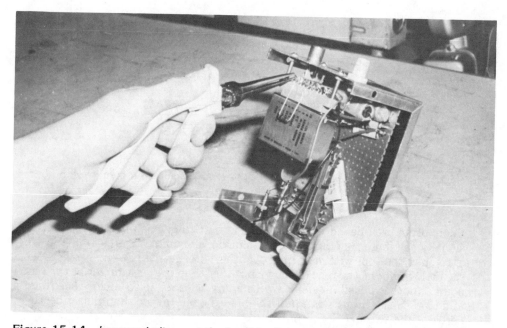

Figure 15.14 *Improved pliers* put the bend in the tool rather than the wrist, prevent slippage in the hand, fit female hands rather than male hands, and don't dig a hole in the palm. Since they were first applied at Western Electric Corp. by Tichauer, they are usually known in the USA as "Western Electric" pliers.

Figure 15.15 *Bent soldering tool* puts the bend in the tool instead of the wrist, but also has a guard.

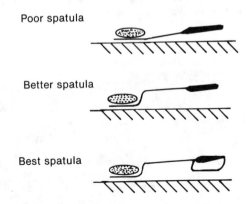

Poor spatula

Better spatula

Best spatula

Figure 15.16 *Design tools* (such as spatulas) to reduce accidental contact with hot surfaces.

Pliers often are opened by inserting the first finger behind the pivot. Occasionally the finger is pinched. One solution, shown in Figures 15.3 and 15.4, is to open the tool with a spring. Another alternative is to open pliers with the little finger at the rear of the grip rather than the first finger at the pinch point. Another possibility, used with many scissors, is to enclose the fingers in a loop so the fingers can pull the blades open rather than inserting them between the two grips and pushing.

Tool handles, especially locking handles or toggle clamps, should leave an opening of at least 25 mm in their fully closed position (Greenburg and Chaffin, p. 124).[9]

PRINCIPLE 8: USE THE APPROPRIATE MUSCLE GROUP

Hand-closing muscles are stronger than hand-opening muscles. Figure 15.17 shows the muscles used to open and close the hand. Since the muscles which open the hand are relatively weak, don't use them repetitively. Use a spring to open hand-tool blades; the strong closing muscles can easily overcome the spring resistance during the hand closing. A spring, or a spring released by a catch, also can release energy quickly even though compressed over a longer time period.

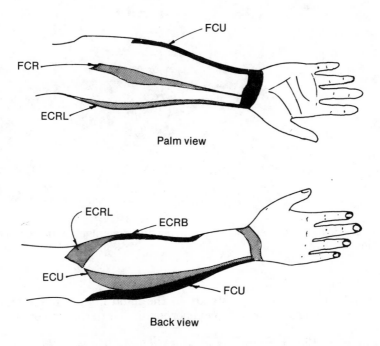

Figure 15.17 *One set* of muscles opens the hand—extensor carpi ulnaris (ECU) and flexor carpi ulnaris (FCU)—they are relatively weak. Another set closes the hand—extensor carpi radialis brevis (ECRB), extensor carpi ulnaris (ECU), and extensor carpi radialis longus (ECRL)—they are strong.[23]

Forearm muscles are stronger than finger muscles. Consider a conventional screwdriver: it is rotated by a sequence of grasp, rotate hand, release, rotate hand back, and regrasp.

In a human-powered mechanical screwdriver, the tool is rotated by movement of the grip along the tool axis. Since the hand motion is linear instead of manipulative, rotational speed is increased; since the forearm muscles replace finger muscles, power is increased. The tool usually is held in a power grip with the tool axis either parallel to the forearm or perpendicular to the forearm. Squeezing by the fingers prevents movements of the tool in the hand; eliminate this unnecessary strain with a bearing surface as in the lowest screwdriver in Figure 15.18. For pneumatic screwdrivers, cover the cold hard metal with a rubber or plastic grip with a similar bearing surface. If squeezing the tool is required, allow as much palm contact as possible.

Trigger strips are preferred to trigger buttons. Moving part of the tool inde-

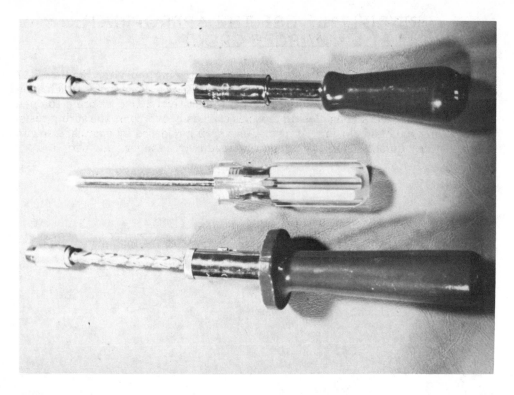

Figure 15.18 *Bearing surfaces* on a power screwdriver eliminate unnecessary finger muscle fatigue.

pendently while the tool is kept steady in the hand is known as "triggering." Triggers can be buttons or strips moved perpendicular to the grip or slides moved parallel to the grip.

Table 15.6 gives the maximum hand squeeze force; Table 15.7 gives maximum force for individual fingers. Figure 15.19 gives the effect of age.[28]

Almost all male population groups (except university men!) have a 5% grip strength value above 40 kg; females have a 5% value about 25 kg, which is compatible with the rule of thumb that females have 2/3 the strength of men.

To determine the 5% value for the individual fingers, subtract 1.64 *(SD)* from the mean. Thus, from study 1, the 5% for the second finger vs. an object would be $5.9 - 1.64 (1.3) = 3.8$. The strength of an individual finger can be estimated as 10% of the strength of the hand as a unit. This is known as a "rule of thumb"!

If only one finger is to be used, strength depends on the direction of motion and which finger will be used. The range is about 2 to 1 with the little finger the weakest and the thumb the strongest.

Trigger slides can be used for occasional on-off functions where tool steadiness is not critical and accidental actuation should be avoided.

For most situations, use a trigger strip rather than a trigger button. As shown in Figure 15.20, use a thumb button rather than a single-finger trigger since the thumb is stronger than any other single finger. However, avoid single-finger triggers if possible. If the thumb opposes other fingers (as with tweezers), have the thumb move so that it is aligned with the middle finger so that the ring and little finger can support the middle finger.

Rotate inward. The right hand and arm can rotate about 70° clockwise about the forearm axis (pronation) and about 150° counterclockwise (supination). Table

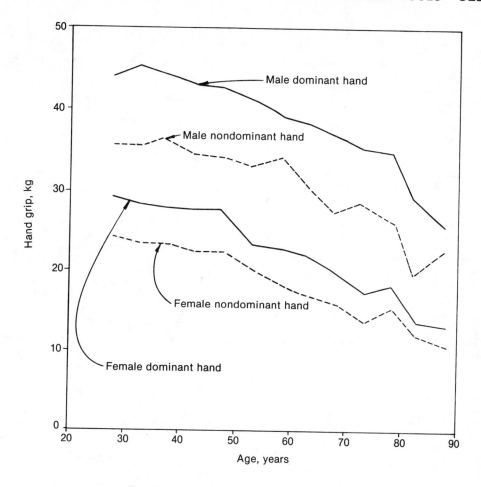

Figure 15.19 *Handgrip strength* declines with age. The top two lines are for men (solid is dominant hand; dashed subordinate hand) while the lower two lines are for women.[28]

From *The Physiology of Aging*, N. W. Shock. Copyright © January, 1962 by Scientific American, Inc. All rights reserved.

15.8 gives the time to rotate the hand about the forearm; no distinction is made for direction.

SUMMARY

Engineers can select from a wide variety of hand tools. Some tools are better than others. The eight principles are designed to aid your selection.

The three general principles emphasize the use of special-purpose tools (due to their low cost/benefit ratio), the use of a tool which can be used by either hand (to minimize local muscle fatigue), and the use of power tools (human power is expensive).

The three grip principles point out that there are two primary types of hand-tool grips (power and precision), that the grip should have the proper geometry

Table 15.6 Maximal Static Grip Strength (Hand Squeeze), kg, From Various Studies

5th Percentile	Standard Deviation	Sex	Hand	Group
48.2	9.14	M	R	Army personnel
47.8	8.18	M	R	Air force aircrewmen
44.6	7.64	M	Pref.	Air force rated officers
41.8	6.87	M	Pref.	Industrial personnel
41.4	8.23	M	R	Truck & bus drivers
40.6	9.64	M	R	Rubber industry
33.6	9.55	M	R	University men
19.1	5.46	M	R	Univ. men, force for 1 min
45.0	9.55	M	L	Army personnel
44.6	7.64	M	L	Air force rated officers
43.6	7.27	M	L	Air force aircrewmen
41.8	7.00	M	L	Industrial personnel
39.1	7.46	M	L	Truck & bus drivers
39.1	10.09	M	L	Rubber industry
29.5	8.18	M	L	University men
17.7	4.54	M	L	Univ. men, force for 1 min
26.4	4.00	F	(R+L)/2	Navy personnel
25.9	4.68	F	Pref.	Industrial workers

Figure 15.20 *Trigger strips* are better than trigger buttons since more force can be exerted. More important, local muscle fatigue can be reduced.

Table 15.7   Maximal Static Finger Force for Adult Males. Study 1 had $N = 100$.[11] Study 2 had $N = 20$.[4] Study 3 had $N = 15$.[11]

| Finger | Study 1 | | | | Study 2 | | | |
| | Right Hand | | | | Dominant Hand | | Nondominant Hand | |
	Mean, kg	%	Standard Deviation, kg	Standard Deviation/ mean, %	Mean kg	%	Mean, kg	%
Thumb vs. object	7.3	124	1.7	23				
2nd vs. object	5.9	100	1.3	22	4.6	100	4.4	96
3rd vs. object	6.4	108	2.0	31	4.3	93	3.9	85
4th vs. object	5.0	85	1.7	34	2.9	63	2.8	61
5th vs. object	3.2	51	1.1	37	2.4	52	2.4	52
Thumb vs. tip of 2nd	9.6*		2.3*	24*	7.5	163	6.9	150
vs. tip of 3rd					8.8	191	8.4	183
vs. tip of 4th					6.4	139	7.8	170
vs. tip of 5th					5.0	109	4.8	104
Thumb vs. side of 2nd	10.5*		2.2*	21*				

*Study 3

Table 15.8   Time For the Hand to Rotate About the Forearm

Angle Rotated, Degrees	Work-Factor, Minutes	Methods Time Measurement, Minutes
45	.0017	.0021
90	.0023	.0032
135	.0028	.0044
180	.0031	.0056

(thickness, shape, and length), and that grip surface is important (texture and conductivity).

The two geometry principles consider the use of the tool in relation to the user, with one principle emphasizing the effect of angles and the other the muscle group used.

SHORT-ANSWER REVIEW QUESTIONS

15.1   Describe a special-purpose tool with which you are familiar. What is its (a) Capital cost? (b) Number of applications/yr? (c) Number of applications in the tool life? (d) Capital cost/application? Does the tool save "doing" or does it save "get ready and put away"?

15.2   What are the two reasons why a tool should be usable by either hand?

15.3   Describe an internal precision grip and an external precision grip.

15.4   In addition to being compressible, wood is the material of merit for handles for what two reasons concerned with heat conductivity?

15.5   Where are the muscles which move the fingers? Why does a bent wrist cause problems?

THOUGHT-DISCUSSION QUESTIONS

1. In sports we spend a great deal of time and money getting the "tool" which will maximize our individual performance. Why don't we do the same for our jobs?

2. Analyze a tool using the principles of hand-tool design. How many times is it used in its life? What is its capital cost/use? How should the tool be modified? What does the tool user think of your ideas? What does the person who must buy the tool think of your ideas?

REFERENCES

1. Ayoub, M., and LoPresti, P. The determination of an optimum size cylindrical handle by use of electromyography. *Ergonomics,* Vol. 14, No. 4 (1971): 509-518.

2. Bendz, P. Systematization of the grip of the hand in relation to finger motor systems. *Scandinavian Journal of Rehabilitation Medicine,* Vol. 6 (1974): 158-165.

3. Capener, N. The hand in surgery. *The Journal of Bone and Joint Surgery,* 38B (February 1956): 128-151.

4. Dickson, A.; Petrie, A.; Nicolle, F.; and Calnan, J. A device for measuring the force of the digits of the hand. *Biomedical Engineering* (July 1972): 270-273.

5. Evans, T.; Lucaccini, L.; Hazell, J.; and Lucas, R. Evaluation of dental hand instruments. *Human Factors,* Vol. 15, No. 4 (1973): 401-406.

6. Faulkner, T. Comfort/safety guidelines for touching hot surfaces. Technical Report HF-74-24. Rochester, N.Y.: Eastman Kodak, 1974.

7. Garrett, J. The adult human hand: some anthropometric and biomechanical considerations. *Human Factors,* Vol. 13, No. 2 (1971): 117-131.

8. Granada, M., and Konz, S. Evaluation of bent hammer handles. *Proceedings of the 25th Annual Meeting of the Human Factors Society,* Rochester, N.Y., 1981.

9. Greenburg, L., and Chaffin, D. *Workers and Their Tools.* Midland, Mich.: Pendell Publishing Co., 1977.

10. Hertzberg, H. Some contributions of applied physical anthropology to human engineering. *Annals New York Academy of Science,* Vol. 63 (1955): 616-629.

11. Hertzberg, H. Engineering Anthropometry. Chapter 11 in *Human Engineering Guide to Equipment Design.* H. Van Cott and R. Kincaid, eds. Sup. of Documents, 1973.

12. Hunt, L. A study of screwdrivers for small assembly work. *The Human Factor,* Vol. 9, No. 2 (1934): 70-73.

13. Kao, H. Human factors design of writing instruments for children; the effect of pen size variations. *Proceedings of 18th Annual Meeting of the Human Factors Society,* 1974.

14. Kao, H. Handwriting ergonomics. *Visible Language,* Vol. 13, No. 3 (1979): 331-339.

15. Kellor, M.; Kondrasuk, R.; Iverson, I.; Frost, J.; Silberberg, N.; and Hoglund, M. *Hand Strengths and Dexterity Tests,* Manual 721. Minneapolis, Minn.: Sister Kenny Institute, 1971.

16. Konz, S. Design of foodscoops. *Applied Ergonomics,* Vol. 6, No. 1 (1975): 32.

17. Lance, B., and Chaffin, D. The effect of prior muscle exertions on simple movements. *Human Factors,* Vol. 13, No. 4 (1971): 355-361.

18. Lowden, K. Hand dexterity for three tasks. Industrial engineering report, Manhattan, Kans.: Kansas State University, 1977.

19. Miller, M.; Ransohoff, J.; and Tichauer, E. Ergonomic evaluation of a redesigned surgical instrument. *Applied Ergonomics,* Vol. 2, No. 4 (1971): 194-197.

20. Napier, J. The prehensile movements of the human hand. *Journal of Bone and Joint Surgery,* Vol. 38B (1956): 902.

21. Oh my aching back. *Factory* (April 1966): 66-70.

22. Patkin, M. Ergonomic design of a needleholder. *Medical Journal of Australia,* Vol. 2 (September 6, 1969): 490-493.

23. Radonjic, D., and Long, C. Kinesiology of the wrist. *American Journal of Physical Medicine,* Vol. 50, No. 3 (1971): 57-71.

24. Rigby, L. Why do people drop things? *Quality Progress* (September 1973): 16-19.

25. Roubal, J., and Kovar, Z. Tool handles and control levers of machines. *Annals of Occupational Hygiene,* Vol. 5 (1962): 37-40.

26. Rubarth, B. Untersuchung zur Bestgestaltung von Handheften fur Schraubenziehar und ahnlich Werkzeuge (Investigations concerning the best shape for handles for screwdrivers and similar tools.) *Industrielle Psychotechnik,* Vol. 5, No. 5 (1928): 129-142.

27. Saran, C. Biomechanical evaluation of T-handles for a pronation supination task. *Journal of Occupation Medicine,* Vol. 15, No. 9 (September 1973): 712-716.

28. Shock, N. The physiology of aging. *Scientific American,* Vol. 206 (January 1962): 100-110.

29. Swanson, A.; Matev, I.; and Groot, G. The strength of the hand. *Bulletin of Prosthetics Research* (Fall 1970): 145-153.

30. Welch, R. The measurement of physiological predisposition to tenosynovitis. *Ergonomics,* Vol. 16, No. 5 (1973): 665-668.

31. Woodruff, F., and Fox, J. Applications of biotechnology to salvage archaeology. *Proceedings of 16th Annual Meeting of the Human Factors Society* (1972): 216-219.

32. Wu, Y. Material properties criteria for thermal safety. *Journal of Materials,* JMLSA, Vol. 7 (December 1972): 575-579.

33. Wu, Y. On the control of thermal impact for thermal safety. Paper 75-713, American Institute for Aeronautics and Astronautics 10th Thermophysics Conference, 1975.

16

PRINCIPLES OF ADMINISTRATION

PRINCIPLE 1: SET GOALS

Work should be directed toward contribution rather than toward effort alone. ("We don't know where we're going but we're on our way.") What do you contribute that justifies your being on the payroll? What are your goals and objectives? What do you plan to do to achieve them?

Undirected effort is not too much of a problem with skilled trades or task-oriented workers such as machinists, bricklayers, and farmers. But knowledge workers (accountants, engineers, social workers, nurses, etc.) tend to get buried in minutiae such as weekly meetings, filling out forms, and working on projects with marginal potential for usefulness.[6]

The best technique is to (1) periodically list the major priorities of the group, then rank them; (2) periodically evaluate each assignment relative to the group's priorities; and (3) modify assignments, if necessary, to direct the assignment toward solution of the problems.

PRINCIPLE 2: REWARD RESULTS

Work smart not hard is the primary message of this book. However, reasonable effort cannot be ignored. The challenge is to get people to work hard *and* smart.

Positive motivation can be internal (self-motivation) or external. External motivation can be motivation from the job (enriched jobs), motivation from social pressure or peer groups, or rewarding results (i.e., self-interest). Figure 16.1 shows four different reward schemes.

Curve A shows pay (reward) as independent of output. Who would use such a plan? Most organizations. Pay by the month (salary) is an example; workers are paid even if they don't come to work (within limits). Salary often is used when an individual's specific contribution is difficult to count; examples are administrative and technical jobs such as deans, researchers, accountants, statisticians. Other examples are for countable but unscheduled output; examples might be patents for an engineer or articles for a professor. Salary also is used for situations where there is only an undefined relationship between effort expended and the value of the contribution; examples might be classroom teaching or radio news broadcasting.

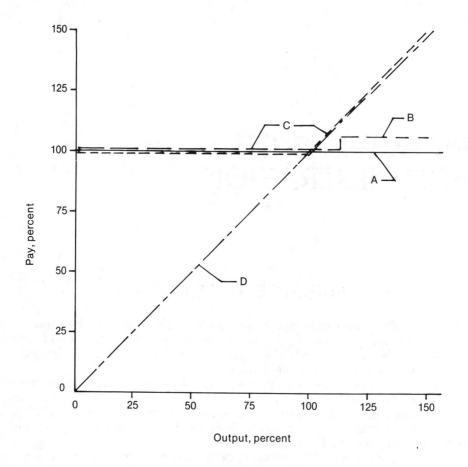

Figure 16.1 *Curve A* shows pay as being independent of output. Curve B shows pay as being independent but with a step function after a delay, such as a 5% raise at the end of the year. Curve C shows pay as proportional to output after the guaranteed base period. Curve D shows pay by the piece. Most people are paid using curve A.

A variation on curve A is pay-by-the-hour. The worker typically is not paid unless physically present at the workplace. Pay-by-the-hour plans range from pay for the entire day (once you arrive at work) to plans which pay to the nearest tenth of an hour for time worked. Although physical presence on the job does not guarantee a useful contribution, it has better odds than not being there at all.

Curve B shows pay as independent of output—except that there may be a delay in reward or punishment. The longer the delay between the stimulus and the response the less the motivation.

Delayed reward is common. Positive reward (the carrot) seems more effective than negative reward (the stick)—perhaps because negative rewards are difficult to implement fruitfully in the "real world." Some delayed "carrots" might be stars on the helmets of football players for good plays, a cash bonus for players whose teams win divisional play-offs, a bonus to an executive whose division achieved a predetermined goal (such as reduction in accidents), a raise for a teacher upon completion of an advanced degree, a raise for a newspaper photographer whose picture wins an award, a grant for a good research proposal. Some delayed "sticks" might be verbal abuse of football players who make mistakes, firing a coach whose team did not advance in the play-offs, and a salary cut for an executive whose division had an increase in accidents.

Motivation is increased and longer lasting when the size of the "step function" is greater, the relationship between the behavior and the reward is clearer, and the award is automatic (i.e., does not depend upon someone's judgment).

Curve C shows pay and performance being independent over an initial range, then related. A steelworker on piecework is an example. Just coming to work gives the base pay. But for every 1% increase in output over standard, he receives a 1% increase in pay. Output of 125% of standard? Then pay 125% of standard. Although the worker's pay increases, the cost/unit for the organization declines. First, the fringe costs (holidays, health plan, etc.) decline on a per unit basis; second, overhead and capital costs decline on a per unit basis.

Note that the arrangement is pay for output, not pay for effort. The technology of the job (machine speed, supply of parts, movement distances required) often puts a ceiling on the output. A motivated worker then tries to make methods improvements since the reward is direct—greater daily pay. Giving the bonus pay in a separate check is more effective than including the bonus in the base paycheck, as some workers consider the bonus is theirs and the base paycheck is the spouse's.

Curve D is "piecework"—pay by the piece without any guaranteed base or limitation on earnings. Do you want to be paid what you are worth? Sales workers are common applications (stockbrokers, insurance agents) as well as the self-employed (dentists, doctors). A dentist who does not come to work receives no income. Commissions are quite popular for sales work as the reward is for performance rather than effort; output for people on commissions often is twice that of people on salary.

> My object all sublime
> I shall achieve in time—
> To make the punishment fit the crime.
>
> *The Mikado, Act II*

PRINCIPLE 3: OPTIMIZE SYSTEM AVAILABILITY

Optimum availability (service to customers) is not necessarily maximum availability. Consider the cost of providing the service vs. the benefits from the service. That is, occasional delays (customer queues) probably is the lowest cost solution. If queues exist, exchange information between the customers and the servers concerning expected time in the queue and probability of being served (e.g., if the window closes in 30 min and service time/customer is 15 min, then customers after the first two will not be served). In addition, provide for the physical and mental needs of the customers by assigning numbers to queue members and providing sufficient space and chairs for longer waits.

Principle 3 discusses improving the availability of a service channel while Principles 4 and 5 discuss better matching of the service channels and service requirements.

$$\text{Availability} = \frac{\text{uptime}}{\text{total time}} = \frac{\text{uptime}}{\text{uptime} + \text{downtime}} = \frac{MTBF}{MTBF + MTR} \quad (16.1)$$

where $MTBF$ = Mean Time Between Failures, hours = reliability
MTR = Mean Time to Repair, hours = maintainability

Availability can be obtained by having reliable equipment (long interval between failures) or equipment which can be fixed quickly (short downtime). There are three strategies to improve availability:

- Increase reliability (increase uptime)
- Increase maintainability (decrease downtime)
- Make lack of availability less costly

Increase Reliability (Increase Uptime)

One possibility is to increase reliability by overdesigning the components. For example, in a specific vehicle doing a specific task, a 5-liter engine will have less strain than a 3-liter engine; reliability and capability are increased as are capital cost and operating cost. Or, use electronic circuit components which can stand temperatures from -20 to $+60$ C even though the environments usually are between 15 and 30 C.

Another related possibility is to reduce stress on the components. Thus, run equipment at less than maximum speed and load. Avoid extreme environmental conditions of temperature, pressure, and humidity. Avoid corrosive atmospheres for machines and toxic environments for humans.

A third possibility is redundant components in the design—a *parallel* circuit. In inspection, for example, each item could be inspected by two different inspectors. If the item is defective, it could be detected by either inspector. When sending a message, a telephone message could be confirmed by mail; if one message is not delivered, the other may get through.

Splitting a job between two people is an interesting approach to increasing reliability of a service. An industrial nurse's job was split between two women. Normally each worked two days one week and three days the next but they varied this schedule at their own option to cover personal desires. The organization thus got 100% coverage of the service even through illness (if one was sick the other came in for that day) and vacations.

Increase Maintainability (Decrease Downtime)

Most devices do not have redundant components, they are *series* systems. If any component of the system fails, the system fails. Examples are assembly lines, machine transfer lines, power tools, and typewriter keys. After *fault detection time* (time to find that the device doesn't work), total downtime, which usually has a log normal distribution with a positive skew, can be subdivided into *fault location time* (time to find that the air hose has a leak), *logistics time* (time to get a new air hose), and *repair time* (time to replace hose). Repair time (sometimes called inherent time) often is a minor portion of the total time.

"Standby" components (spare capacity) are a popular technique to minimize the cost of fault location, logistics, and repair time. If the system fails, a standby system is used until the main system is available. For example, if a power screwdriver fails, a spare power screwdriver or a hand screwdriver is used until the screwdriver is repaired. If a conveyor breaks down, the goods are moved by hand until the conveyor is repaired. If Joe doesn't bother to come to work today, Pete (a "spare" worker) takes over for the day (that is, if Pete is cross-trained).

Fault location time. Fault location time can be reduced by administrative procedures. For example, if a setup or maintenance technician is needed to locate the fault, what can be done to reduce the time before the technician arrives at the broken machine? One common technique is to mount a red light on a pole next to each machine. Then, an operator who detects that the machine isn't working properly just switches on the light. Factory public address systems serve the same purpose. A radio-equipped vehicle can call for help more quickly than if someone had to walk for help. Administrators can schedule maintenance personnel so that someone is on duty for more hours of the day. (Why do machine breakdowns seem to take place when all the mechanics are off duty?) One machine tool manufacturer uses relays which change color when burned out, thus reducing fault location time. Repair service is also offered by telephone—giving expert advice to the local technician—so physical travel by the service engineer rarely is necessary. Establish service priorities: first-come first-served is not the best rule for all situations.

Logistics time. Logistics time also can be reduced by administrative procedures. One technique is to cross-list spare parts between locations and organizations. For example, in a multiplant firm, one spare 10-hp motor might be sufficient rather than one spare motor for each plant. Airlines cross list parts between organizations. For example, United Airlines and Trans World Airlines might have an agreement on spare parts at airports they both serve with United stocking spares for both airlines at airports A, B, and C and TWA stocking spares for both airlines at airports D, E, and F. Reducing order time and paper shuffling time can reduce logistics time substantially.

Repair time. Repair time can be reduced by modularization. That is, technicians do not repair a failed electric motor, they replace the entire motor; they do not repair the failed automobile starter, they replace the entire starter. Modularization has become more and more common as the cost of a maintenance hour increases relative to the cost of manufactured components. In addition, with more complex manufacturing processes, maintenance becomes less feasible. For example, shafts of small motors are inserted inside their rotors by heating the bearing and freezing the shaft before inserting. Rather than give the repairman in the field the capability of replacing just a defective bearing or shaft, now the entire motor must be replaced if either fails.

Make Lack of Availability Less Costly

There are two possibilities: (a) downtime when availability is not required, and (b) partial function.

In most situations, *availability is not required* 24 hr/day, 7 days/wk. Thus, if equipment (people) are not available when they are not needed, the consequences are not as severe as if they were not available when needed. For example, if needed repairs to a fork truck can be made outside of normal working hours, then the lack of availability of the fork truck during the time is not a problem. If power must be shut off in an area for 30 min while a new machine is installed, then installation outside of the hours when people work in the area minimizes the lack of availability of power. The primary rationale for preventive maintenance is that scheduled downtime is less costly than unscheduled downtime. This is one reason most operations are not scheduled for 3 shifts per day.

Partial function means that the system can function, although not at full efficiency or output, if some components fail. A very common example is an engine with one spark plug not working: the engine works but not at full output. Another example is using a tire which can go 100 km after a puncture rather than using a tire which must be repaired within 1 km. Industrial examples are multiple plants, assembly lines, machines, and vendors: if one stops, the supply (service) from the others continues. Failures could come from strikes, fire, flood, etc. Thus an automobile company tries not to get its entire supply of tires from only one supplier; it tries to make specific models at several plants rather than just one location, it tries to have specific components made on several machines rather than just one machine, etc. In theory, one large unit is more efficient than several smaller units since equipment utilization is better and more specialized labor and equipment can be used. In practice, monopolies (one large unit) seem to be less desirable as the economies of scale are overcome by bureaucracy, lack of competition, and increased downtime. Two stations, each with an operator, may be more reliable than one station with two operators.

Machine or vendor or operator failure is now a part of facility planning. New processes or rearrangements of new facilities should have "what if" built in.

PRINCIPLE 4: MINIMIZE IDLE CAPACITY

Fixed Costs

Many machines and people have considerable idle capacity. Yet the annual cost varies little with output. For example, depreciation usually depends more on equipment age than equipment hours of use. A specific example would be the price of a used car depending more on its model year than on whether it has been driven 60,000 km or 80,000 km. Wages, and especially salaries, usually vary little with output. A secretary will receive $6,000/yr whether typing 1,000 letters/yr or 1,500 letters/yr; a cook receives the same pay whether cooking for 50 or 100; a bus driver receives the same pay whether there are 10 or 40 on the bus.

Even if wages are paid by the hour, labor costs are relatively fixed due to "fringe" benefits. A worker gets the same vacation, holidays, and medical benefits whether working 30, 40, or 50 hr/week and whether working 45 weeks/yr or 48. Labor contracts often require pay for an entire day once a worker starts work even if the work is only for 1 hour; sometimes pay is required for an entire week if one day is worked. If an employer tries to save money by "laying off" a worker, the worker receives unemployment benefits, which are obtained from a tax on each employer proportional to the amount of layoffs. In addition the workers now employed have been trained and their work quality is satisfactory. Therefore employers become reluctant to hire new employees (obtain new labor capacity) when the new capacity has such a high fixed cost. Since the incremental cost of using the existing surplus capacity is low, the goal is to get more use of the capacity.

Scheduling work for the idle capacity is a problem. Example techniques are:

- Use a resort in summer and winter.
- Have a bus carry freight as well as passengers.
- Have a restaurant open for more than one meal/day.
- Have elective surgery scheduled between 3 p.m. and 9 p.m.
- Have a bank's computer used for local small businesses as well as the bank.
- Rent the company airplane (truck) to other firms.
- Use a machine more than 1 shift/day.

- Use a machine more than 5 days/week.

- "Queue servers" often have times when no one is in the queue; give them work for this idle time.

Having secretaries receive work from more than one area (a typing pool) is a common example; with telephones, telecopiers, and dictating equipment the secretaries can be located remotely from each other.

Administrators may have some control over the peaks and valleys of customer demand. Divide the month or year into sections alphabetically, then all people whose last name starts with A, B, or C are done first, then those whose names start with D and E, etc. If demand is due to "specials" or "sales," time the specials to fill in the valleys of demand rather than amplify the peaks.[15] Retail sales often are held after Christmas or on Monday or Tuesday. Computer rentals are higher during prime time to encourage "off peak" usage. Work hours are staggered to relieve parking and traffic peaks. Building utilization was improved at Kansas State University by starting the classes on the half-hour instead of the hour. First, due to the traditional 12:00 lunch time, no one wanted a 12:00 or 12:50 class; with the classes starting at 11:30 and 12:30 the psychological adjustment was less severe and room utilization over lunch now is over 50% instead of the former 10%. Secondly, since the first class starts at 7:30 instead of 8:00 and the last day class ends at 5:20 instead of 4:50, three laboratory sections of three hours each can be scheduled in a room instead of the former two.

Some of these alternatives involve overtime for employees. There is relatively little drop in output/hr when working 6 days of 8 hr/day instead of 5 days and a little drop when working 9 or 10 hr/day. However, prolonged overtime (say, 7 days/wk for 8 weeks; 6 days/wk for 6 months) becomes "old" and workers begin not showing up for work 1 or 2 days/week. They are careful, however, to still work on premium-pay days.

Shiftwork is another way of maximizing use of capacity. Shiftwork originated with industries which could not stop for technical reasons (such as steel furnaces), but now the rationale is economic (maximizing use of expensive capacity) or social (hospitals, police, transportation).[4,12,18,19] The proportion of industrial workers on shiftwork during the period 1958-1965 was 13% in Denmark, 17% in Sweden, 20% in Norway and the United Kingdom, 21% in France and the Netherlands, and 24% in the USA.[12] Although Maurice did not have more recent figures, he felt use of shiftwork was increasing worldwide. He felt shiftwork was especially appropriate for the lesser developed countries which tend to have a surplus of labor and a shortage of capital. Shiftwork tends to be more common in larger firms and firms with more capital invested/employee.

There are very many different shiftwork plans but one choice is between two shifts and three shifts; some firms have a partial second shift—a 1.5 shift system.[20] A second choice is between semicontinuous (5 days/wk) and continuous (7 days/wk). The third choice is between fixed or rotating shifts. If the crews rotate, a fourth choice is the rotation interval and a fifth choice is the rotation direction.

From a management viewpoint, two-shift systems have an advantage of flexibility as overtime can be scheduled and repairs made during the nonworking time. Semicontinuous three-shift systems (15/week) are a little more difficult for scheduling overtime or repairs; continuous, three-shift systems (21/week) have the minimum flexibility. Villiger reported manufacturing costs in a textile factory had an index cost of 145 for 5 shifts of 8 hr/week, 109 for 10 shifts/week, and 100 for 15 shifts/week.[24] The index manufacturing costs for a machine factory were 144 for 5 shifts, 106 for 10 shifts, and 100 for 15 shifts. De Jong reported an economic model of shiftwork showed organizations should formally evaluate shiftwork as an alternative to additional capital investment as some organizations used too much shiftwork and others too little.[5]

Numerous studies of shift workers in many countries since the 1920s have shown that there is no health difference between shift workers and nonshift workers. This may be due to self-selection: about 20% to 30% of the population cannot adapt to shift work.[23] Shift workers tend to eat irregular meals and quickly prepared foods and former shift workers (now off shifts) have higher proportions of gastrointestinal problems.[19] Akerstedt feels that older workers (over 45) should not be on shift work.[1] The typical circadian rhythm (see Figure 16.2) is affected by shiftwork; Ostberg suggests "morning type" people have more difficulty adapting to shiftwork, while Horne, Brass, and Pettitt report morning and evening types inspect differently as their body temperatures change.[8,13] Reinberg uses changes in circadian rhythm as evidence for changing shifts every 2-3 days instead of weekly.[17]

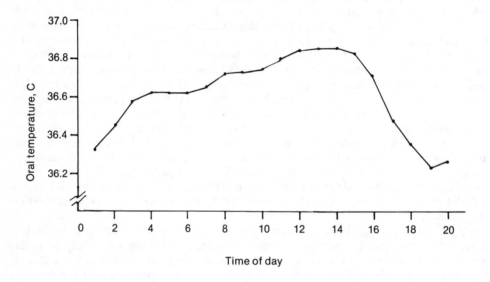

Figure 16.2 *Oral temperature* of 59 male naval ratings peaked around 4 p.m.[4] The subjects slept from 2300 to 630; performed light duties (mainly short psychological tests) from 0800 to 1600, and had meals at 0700, 1200, and 1700. McFarland says the circadian rhythm recovers from "jet lag" at the rate of one recovery day/time zone crossed.[11]

The use of shiftwork presents a conflict between the economic goals of the organization and the sociocultural goals of the workers. Other things being equal, workers prefer to work during the day and the week rather than the night and the weekend; night and weekend work tend to interfere with sleep (poor sleeping conditions at home) and make the workers live independently of their families. Leisure time of shiftworkers tends to have a lower "value" since it cannot be exchanged for many common leisure time activities. Although the German term "social death" is exaggerated, it does reflect some of the worker's social difficulties. Organizations pay premium wages to compensate for these problems. This additional income makes some workers prefer shiftwork; some want even more income and use the nonshiftwork time to "moonlight" (take a second job). As shown by a NIOSH symposium of experts, there are almost as many opinions on shiftwork as there are experts.[9]

Workers prefer the morning shift, then the afternoon shift (disturbed social and family life), with the night shift last (disturbed sleep, social life, family life). Workers prefer a permanent shift to a rotating shift (although they often will vote for whichever system they presently are using) and prefer a two-shift system to a three-shift system. If rotating is used, Walker advocates a short rotation interval (say 2 days), Pocock, Sergean, and Taylor a 7-day interval and Kripke, Cook, and Lewis and Telesky a long (30-day) interval.[10,14,22,25] The general consensus, however, is toward either the short or long and away from the 7-day cycle. If rotation is used, the schedules should be posted at least 30 days in advance so the workers can plan their personal lives. Young reports a plan, developed by shift-workers, and successfully implemented, in which workers can trade hours with workers on other shifts.[27] The most popular trade length was 4 hr; through a "bank" arrangement, trades could be ±12 hr and switches with many different people.

One technique of staffing a facility continuously is to use four permanent 12-hr shifts. Shifts A and B work 12 hr/day Monday, Tuesday, and Wednesday; shifts C and D work 12 hr/day on Thursday, Friday, and Saturday; they alternate on Sundays. Thus shifts work 36 hours one week and 48 the next; a disadvantage is possible overtime pay. Another technique is to pay 40 hours wages to a shift working 12 hr on Saturday and Sunday only; since weekends often require premium pay for other shifts, there may not be any extra cost. Because absenteeism would be critical with such a shift, severe penalties probably would need to be implemented for unexcused absenteeism.

System Cost Vs. Component Cost

Men and machines usually work in teams. A common team is one man and one machine such as a typist-typewriter or driver-truck. Also common are *multiple component* teams such as executive-dictating machine-typist-typewriter, machinist-machine-fixture, customer-clerk-cash register, customer-telephone-clerk, and driver-taxi. Although a general goal is to minimize idle time of each component, idle time in some components may reduce system cost.

In the USA, labor cost (including fringes such as holidays, vacations, etc.) varies from $8,000 to $20,000/yr. For 2,000 hr/yr this is a cost of $4 to $10/hr. On the other hand, a relatively expensive machine might have a capital cost of $10,000. For a life of 10 years and 2,000 hr/yr, this is only $.50/hr. (Human labor is expensive and machines are cheap.) It will be emphasized that some Asian and African countries have expensive machines (due to lack of local manufacturing and lack of foreign exchange) and cheap labor.

Therefore, in a multicomponent system such as a typist and typewriter, keep the expensive component (the typist) busy while not worrying as much about the utilization of the inexpensive component (the typewriter). Machining offers another example of a multicomponent system. The recommended cutting speed for a lathe tool might be given as 2 m/s. What this speed minimizes, however, is the tool cost vs. metal removed for that specific machine. If the machine is a bottleneck, then increase the cutting speed; tool cost on one machine will increase but output on the system of machines will increase and system cost/unit will decrease. "Double tooling" (two fixtures for the same machine; the operator loads one while the machine processes the other) often is profitable. Another example of low machine cost is the store with more checkout stations than are normally required; the cheap component (the extra workstations) is supplied in abundance in case they might be useful occasionally. One fast-food restaurant has four soft drink dispensers—two for brand A, one for brand B, and one for brand C. The extra cost for the second dispenser for the most popular brand A is more than overcome by the reduced labor idle time.

More examples of multicomponent systems in which one component may be idle and yet maximize system efficiency are an executive and a secretary; a twin movie theatre with one movie operator; a priest giving communion to two lines; a buffet table with two lines of customers; a central computer with multiple tape units; and a dentist, dental assistants, and customers.

Wiersma contrasted fork truck utilization vs. idle time of the production workers serviced by the trucks.[26] Minimum total cost with the original procedure was for a truck idle percentage of 40% to 60%. Reducing servicing time from 4 min to 2 min by increasing the number of in-process storage areas and reducing truck travel distances cut costs but did not change the optimum range of idleness of 40% to 60%. Combining two service areas (each with one truck) into one area with two trucks gave even greater savings and increased the optimum range of idleness to 50% to 70%.

Idle Time Distribution

A few large chunks of idle time are easier to use than many small chunks. That is, one block of 120 min of idle time is easier to use than 240 segments of 30 s each. With small segments it is difficult to assign additional work without disturbing the primary task. In addition, workers conceal the short segments as few workers tell their boss they can do 25% more work. As Parkinson said "Work expands to fill the time allocated."

Another challenge is that idle time varies with the task difficulty; some jobs are easy while others are tight. For example, Joe might be idle 15 min/hr while Pete has 10 min and Mike has 35. Fairness might indicate each could have 20 min. However, if all the work is concentrated, the 45 + 50 + 25 min = 120 min of work/clock hour; that is, just about a two-person job instead of a three-person job. On the other hand, supervisors may want one "soft" job in the department to give to "restricted" workers (a pregnant woman in the last weeks before maternity leave, a person with a broken limb, a worker with 40 years service who no longer can keep up the regular assembly line pace, brand new employees, the boss's son, etc.).

PRINCIPLE 5: USE FILLER JOBS OR FILLER PEOPLE

In repetitive factory work, cycle times generally are short; say, 2 min/unit or less. A 2-min cycle time in a typical shift of 450 min (480 − 30 min of break) then results in 450/2 = 225 units with "no" time left over.

However, many jobs have "long" cycle times—60 to 300 min. When 450 is divided by, say, 200 min, there is 50 min "left over." Using "left over" or "balance" time efficiently is the problem discussed in this principle under (1) leave partly completed work, (2) have a supply of short jobs, (3) schedule, and (4) use filler people.

Leave Partly Completed Work

Leave unfinished work to be completed the following shift. One objection might be security of the work; someone might come along and move something or take something. Security is unlikely to be a consideration if reasonable precautions are taken. In addition, the cost of an occasional disturbed job should be less than the cost of "put away" every night and "take out" every morning. The primary problem is psychological as we have been told since childhood to pick up and put away. We also have phobias against leaving "unfinished work"—particularly when the finished job is a pay point.

One solution is to redefine larger jobs into smaller jobs. Consider the task of waxing a car—it might take 6 hr. If you have just Saturday afternoon to work, you might be reluctant to start the job. Try thinking of it as four jobs, each taking a shorter time (wax top, wax hood and fenders, wax door and sides, and wax rear). Now you can do two jobs this Saturday and two next Saturday. Or consider a secretary who has an exam to type, duplicate, and assemble. If the entire job takes 40 min, the secretary may be reluctant to start it 20 min before quitting time. Think of it as three jobs: typing, duplicating, and assembling. Then just finish the typing.

Have a Supply of Short Jobs

The second approach to using left-over time is to have a supply of short jobs to be used at this time. Many routine maintenance tasks can be used as well as routine periodic clerical tasks. Make a list of short jobs. Then do the jobs only during the "balance" time, don't "waste" them by doing them early in the morning or during the normal working cycle.

Police officers in Manhattan, Kansas formerly reported back to the police station approximately 30 min before the end of the shift to dictate reports. Now each officer has a battery-operated tape recorder. Whenever a report is needed, the officer parks the car in a prominent place and dictates into the recorder—thus completing the report while the topic is still fresh and also getting more "useful" time/shift. Carefully define job descriptions to be general so that transferring workers from task to task is not a problem.

Short jobs (sometimes called "off-line" or "balance work") also should be available for service personnel while they wait for customers. For example, a grocery store checkout operator could restock the supply of paper bags or stock nearby shelves; a receptionist could type letters or answer the telephone; that is, low-priority jobs and high-priority jobs.

Schedule

The third approach considers timing, overscheduling, and short-interval scheduling.

Timing. Timing of the schedule may reduce left-over time. If a group starts work at 8:00 and a meeting is scheduled for 8:15, then 15 min of left-over time is created for everyone. Start the meeting at 8. If some people at the meeting supervise others, however, they may need a few minutes at the start of the day to get everyone organized. (Ideally the supervisor organizes the work the previous day.) For this type of group, schedule the meeting at 9:00. There also are people who enjoy long meetings since they will "have to go back to work" when the meeting is over. A meeting with 30 min work will usually take 60 min if scheduled at 9:00; it probably will take 20 if scheduled 15 min before lunch or quitting time. One executive insists that all conversations in his office be conducted while standing. He says people leave sooner and he gets to stretch his legs.

Make an analysis of customer arrival times; the most efficient staffing pattern may involve staggered starting and stopping times for the workers. For example, in a restaurant, the dishwasher can start 60 min after the waiters arrive; waiters need not stay until all the dishes are washed. A hospital emergency room staffing pattern was rescheduled to meet patient arrivals.[2] In a mail order business, the people opening the mail can start sooner than the people filling the orders.[15] Over a broader time span, schedule normal work days to coincide with maximum service requirements. For example, Joe works a 5-day week of Monday, Tuesday, Thursday, Friday, and Saturday and Pete works a 5-day week of Tuesday, Wednesday, Thursday, Friday, and Saturday so that on the slack days (Monday and Wednesday) only one is at work and both are there on the busier days. The use of a flexible schedule (Joe works 7 a.m. to 3 p.m.; Pete works 9 a.m. to 5 p.m.) can be used to keep a facility open longer without any additional labor.

An advantage of the 4/40 schedule (4 days of 10 hr each instead of 5 of 8 with half the employees off on Monday and the other half off on Friday) is that the facility is "open" 10 extra hours/week with no additional labor cost.

Overscheduling. Assign more work to your subordinates than they have time available. The reason for this is the variability of your estimate of the required time to do the job and the variability of the time the subordinates take for the job. If the job takes less than your estimated time, the worker may not report back for more work. Even if they do, you may not be able to think of a new task on short notice.

Short-interval scheduling. Short-interval scheduling is a method of controlling labor expenditure.[7] *Batches* of work, with accompanying measured time standards, are given to employees. In contrast to measured day work (which evaluates an employee's performance once/wk or once/day), short-interval scheduling evaluates performance on each batch, say, 4 times/day. The benefits come from the detailed attention the supervisor must pay to the problems and performance of each employee; the problems are labor unrest at the very close supervision, extra workload for the supervisors, and attempts to use unsound work measurement techniques. Most applications have been in clerical operations.

The following application will contrast short-interval scheduling with measured day work. Consider the aircraft service personnel at an airport. Flight 405 is scheduled to arrive at 10:45 each day; it carries from 30 to 150 passengers. Assume 1 worker is needed as a minimum plus one extra for each 40 passengers. Under measured day work, workers would be scheduled at the beginning of the month either for the average or maximum number of passengers; they would report to the gate at 10:35. Under short-interval scheduling, the supervisor would determine daily the actual arrival time and number of passengers and then assign the number of workers required and when they should report to the gate. Note the assumption that if the workers report to the gate at a different time or in different numbers than the "standard" number, the extra time and people can be used productively somewhere else.

Use Filler People

The previous sections have considered the work force as fixed and tried to adjust the work load to the fixed work force. Using filler people does the opposite in that the work force fluctuates; that is, use part-time help.

Part-time need not be part-time to the organization but just part-time to the group. For example, on Tuesday move surplus workers from the shipping dock to assembly; on Wednesday afternoon have the electrician's helper help the welder. Managements often have defined jobs much too narrowly so that there are artificial barriers to worker movement. Managements often use unions as an excuse for these barriers, but unions and workers usually are reasonable if management is.

Using part-time help is especially useful when demand is 5 to 10 times higher during peak times than base times. Part-time employees can be part-time on a day (noon meals in restaurants, late afternoon in filling stations), part-time per week (Saturday in retail stores), part-time per month (work just first two weeks of each month), or part-time per year (tax office, harvest).

Part-time workers may increase training costs and give poor quality; however, they also tend to not receive as many fringe benefits (such as paid vacations) as full-time employees. The primary benefit, however, is that they reduce the problem of surplus labor capacity during low demand times.

If neither filler jobs nor filler people can be used, then either inventory or idle time will increase.

PRINCIPLE 6: GIVE SHORT BREAKS OFTEN

Fatigue is excessive if an operator stands continuously for more than 30 minutes (walking is less fatiguing than standing), monitors or inspects continuously for more than 30 minutes, sits in one rigid position (e.g., assembly using binocular microscopes) for more than 30 minutes, or sits in one position for more than 60 minutes. (Also see chapter 23, "Allowances.")

Ramsey, Halcomb, and Mortagy reported short (0-24 min) work sessions gave better inspector performance than long (50-74 min) sessions.[16] Operators did worse when they picked their own session time length as they didn't realize when their performance began to decline. Thus predetermined short work sessions seem preferable to letting the inspectors work until they want a break.

Fatigue can be reduced by breaks. The breaks can be with or without work.

Breaks While Working (Working Rest)

Rest may be required for a specific muscle group or body organ although there is no general body fatigue. Thus a typist might shift to running the duplicating machine to give the fingers and forearm muscles a rest, to change posture, and to shift the eyes from close focusing to general vision. The movement of the legs when shifting jobs and while runing the duplicating machine is desirable since movement improves blood circulation. However, prolonged standing work is not desirable so the return to typing is also beneficial.

A task change can reduce psychological fatigue (boredom) as well as physiological fatigue.[3] Smith reviewed 50 references on boredom.[21] A change is as good as a rest.

Other examples of working rest are an inspector testing mechanical faults and then electrical faults; an operator shifting from making to maintenance; an operator shifting from making to getting or putting away the completed units; a typist also answering the telephone; a carpenter sawing and then nailing; or a painter painting walls and painting trim.

The previous examples had workers shift work within their own responsibilities. It also is possible to switch jobs with other workers—job rotation. A specific example is the control-panel operator for an order picking system. The operator scans boxes coming down the conveyor and keys a number into the computer (at the rate of 1 box/s) indicating to which shipping dock the computer should send the box. After two hours of being pushed to the limit of his capability, the operator is glad to switch with one of the truck loaders and let the muscles work instead of the brain.

Job rotation within teams is simpler for the supervisor than between teams. Job rotation also can reduce exposure to a specific environment. That is, one job may have high temperature, high noise, or high chemical concentrations. Letting two workers each be exposed to an environment for 4 hr is better than letting 1 be exposed for 8 hr.

Breaks Without Working

It may be desirable to rest the entire body as well as specific sections. The disadvantage is the short-term loss in output. At present levels of metabolic load there is very little evidence that there is much need for relief from whole body physiological fatigue. The problem seems to be relief from psychological tension and repetitive work. Breaks while working probably would be satisfactory to give this relief but breaks without working probably are more acceptable to workers.

In machine-paced repetitive work, breaks without working probably are necessary. Recovery from fatigue declines exponentially vs. time so short breaks often are the most efficient. That is, for a 4-hr period, breaks of 5 minutes after the first, second, and third hour are more efficient in relieving fatigue than one break of 15 minutes after the second hour. However, a managerial problem is that people abuse break time. If they add one minute to each end of a break, the 3 breaks of 5 minutes become 3 breaks of 7 minutes = 21 total vs. 1 break of 15 stretching to 17 minutes.

PRINCIPLE 7: GIVE PRECISE INSTRUCTIONS IN AN UNDERSTANDABLE FORMAT

A work method in the mind of the engineer must be transferred into the mind of the operator. (Chapter 26 gives more detail on communication.)

Training Requirements

New employees require training. *New,* however, means new to the job—not necessarily new to the organization. Thus someone who has worked for the organization for 10 years may require training when shifted to a new job. Workers no longer do the same job for 40 years; they are unlikely to do the same job for 5 years. Also, the addition or subtraction of one individual from an organization (through death, dismissal, quitting, transfer to another group, etc.) sets off a chain reaction of changes; a typical ratio is 3 people new to their jobs for every one person change in the organization. Thus training is required even if the number in the organization doesn't change.

Poor training shows up primarily in poor equipment utilization and poor quality rather than in reduction of output. However, cost accounting systems usually are designed to detect changes in output rather than changes in utilization or quality. In addition, poor quality is not popular; thus supervisors try to conceal decreases in quality. Thus, evidence showing the need for training is difficult to attain.

Medium of Communication

The most effective medium of communication is visual rather than auditory; pictures are better than words.

Matching physical objects (make the new one like this old one) is best. Pictures can substitute for a physical model. Color communicates better but black and white is cheaper and usually adequate. Moving pictures (videotape or movies) can show motion in either real time or slow motion; the equipment, however, is expensive for shop use. Slides or photographs permit the users to go at their own pace and even backtrack rather than have the machine pacing of motion pictures. However even slides require projectors and power so still photographs usually are best for shop use. Supplement the picture with words.

Words alone are less effective. The problem is that they must be translated. "Ein bier" would be understood by most bartenders no matter what country you were in but what do you say if the bartender replies "Ein Helles oder ein Dunkles" (a light or a dark)? The typical worker has not graduated from high school and does not understand the meaning of words such as *prior, subsequent, chartreuse,* and *incorporate;* use words such as *before, after, blue-green,* and *put in.* Especially avoid verbal orders. In addition to the translation problems, people forget what they heard (as well as not pay attention in the first place). If verbal orders must be given, have the receiver repeat them back to the giver in the receiver's own words.

Format

Paper is cheap, so don't condense instructions in order to save paper. Tabular formats communicate better than narrative formats; decision structure tables are good examples of tabular formats.

Present information in specific behavioral terms rather than generalities. For example, use "With the right hand, insert key and turn clockwise while using the left hand to hold the latch on the front" rather than "Unlock box."

SUMMARY

For an organization to be efficient it must not only have well-designed jobs, it also must manage those jobs efficiently. The seven principles of administration discuss how to get more useful output—primarily from directing the output to achieving the organization's goals.

Principle 1 (goals) points out the importance of "steering your ship." Principle 2 (reward results) concerns motivation. Principle 3 (availability) discusses reliability and maintainability while Principle 4 (idle capacity) gives some of the trade-offs and techniques to reduce capital costs. Other administrative techniques for improving utilization are given in Principle 5 (filler jobs or people). Principle 6 (breaks) discusses the problems of fatigue, and Principle 7 (instructions) gives the problems of communication.

SHORT-ANSWER REVIEW QUESTIONS

16.1 List the four components of downtime. Give an example of each.

16.2 Why is shiftwork especially appropriate for the less developed countries?

16.3 Give three multicomponent systems in which system cost is reduced by keeping one component at less than full capacity.

16.4 If a machine has a capital cost of $10,000 and a life of 10 years, what is the cost/hr? Give assumptions.

16.5 "Working rest" rests parts of the body while work continues. Give three examples.

THOUGHT-DISCUSSION QUESTIONS

1. List your goals for the next 30 days. Rank them by priority.

2. Why do you think there is so little relation between reward and results in most organizations?

REFERENCES

1. Akerstedt, T. Interindividual differences in adjustment to shift work. *Proceedings, 6th Int. Ergonomics Congress.* College Park, Md., 1976.

2. Allen, P., and Garrett, S. The emergency room as a system. *Industrial Engineering,* Vol. 9, No. 6 (1977): 46-51.

3. Bennett, C.; Marcellus, F; and Reynolds, J. Counteracting psychological fatigue effects by stimulus change. *Proceedings of 18th Human Factors Society Meeting.* Huntsville, Ala., 1974.

4. Colquhoun, W.; Blake, M.; and Edwards, R. Experimental studies of shift work, I: a comparison of rotating and stabilized 4-hour shift systems. *Ergonomics,* Vol. 11, No. 5 (1968): 437-453.

5. De Jong, J. Some outcomes of an investigation into the effects of the application of shiftwork systems. *International Journal of Production Research,* Vol. 12, No. 1 (1974): 1-19.

6. Drucker, P. Managing the knowledge worker. *Wall Street Journal,* November 7, 1975.

7. Fein, M. Short interval scheduling. *Industrial Engineering,* Vol. 4, No. 2 (1972): 14-21.

8. Horne, J.; Grass, C.; and Pettitt, A. Circadian performance differences between morning and evening types. *Ergonomics,* Vol. 23, No. 1 (1980): 29-36.

9. Johnson, L.; Tapas, D.; Colquhoun, W.; and Colligan, M. *The Twenty-Four Hour Workday* (Pub. 81-127), Cincinnati, Ohio: NIOSH, 1981.

10. Kripke, D.; Cook, B.; and Lewis, O. Sleep of night workers: EEG recordings. *Psychophysiology,* Vol. 7, No. 3 (1971):377-384.

11. McFarland, R. Air travel across time zones. *American Scientist,* Vol. 63, No. 1 (1975): 23-30.

12. Maurice, M. *Shiftwork.* Geneva, Switzerland: International Labour Office, 1975.

13. Ostberg, O. Interindividual differences in circadian fatigue patterns of shift workers. *British Journal of Industrial Medicine,* Vol. 30 (1973): 341-351.

14. Pocock, S.; Sergean, R.; and Taylor, P. A comparison of traditional and rapidly rotating systems. *Occupation Psychology,* Vol. 46, No. 1 (1972): 7-13.

15. Pomeroy, R. Adaptive methods measurement techniques to extreme fluctuations in workload. *Journal of Industrial Engineering,* Vol. 18, No. 7 (1967): 424-427.

16. Ramsey, J.; Halcomb, C.; and Mortagy, A. Self-determined work/rest cycles in hot environments. *International Journal of Production Research,* Vol. 12, No. 5 (1974): 623-631.

17. Reinberg, A.; Andlauer, P.; Guillet, P.; Nicolai, A; Vieux, N.; and Laporte, A. Oral temperature, circadian rhythm amplitude, aging and tolerance to shiftwork. *Ergonomics,* Vol. 23, No. 1 (1980): 55-64.

18. Rontos, P., and Shepard, R. (ed.). *Shift Work and Health* (HEW Publication NIOSH 76-203). Superintendent of Documents, Washington, D.C. 1976.

19. Rutenfranz, J.; Colquhoun, W.; and Knauth, P. Hours of work and shiftwork. *Proceedings, 6th International Ergonomics Congress,* College Park, Md., 1976. Also *Ergonomics,* Vol. 19, No. 3 (1976): 331-340.

20. Sergean, R. *Managing Shiftwork,* London: Gower Press, 1971.

21. Smith, R. Boredom: a review. *Human Factors,* Vol. 23, No. 3 (1981): 329-340.

22. Teleky, L. Problems of night work. *Industrial Medicine,* Vol. 12, No. 11 (1943): 758-759.

23. Thiis-Evensen, E. Shiftwork and Health. *Industrial Medicine and Surgery,* Vol. 27 (1958): 493-497.

24. Villiger, A. *Entwicklung und Sociale Probleme der Industriellen Schichtarbeit, Insbesondere in der Schweiz,* Verlag Hans Schellenberg, Winterthur, Switzerland, 1967.

25. Walker, J. Frequent alteration of shifts on continuous work. *Occupational Psychology* (London), Vol. 40 (1966): 215-225.

26. Wiersma, C. Economic utilization of fork trucks. *Industrial Engineering,* Vol. 8, No. 4 (1976): 32-35.

27. Young, W. Flexible working arrangements in continuous shift production. *Personnel Review,* Vol. 7, No. 3 (1978): 12-19.

PART VI
WORK
ENVIRONMENTS

17

THE EYE, VISION, LIGHT, AND ILLUMINATION

THE EYE AND VISION

Figure 17.1 shows the details of the eye, a 25 mm sphere.

Light enters from the air through the transparent cornea. Not only does the cornea protect the eye but the change in refraction from air to the cornea permits the distance from the lens to the fovea to be only 15 mm.

Next is the aqueous humor (watery fluid, in Latin) which nourishes both the cornea and lens. Glaucoma is a high pressure in the aqueous humor.

Next the light passes through the pupil of the biconvex lens. (Pupil is Latin for doll since you can see a small image of yourself reflected in the pupil.) The iris (rainbow, in Greek) expands and contracts to control the amount of light admitted. In northern climates where sunlight is weak a small amount of melanin is suffcient so the eyes tend to be blue; more melanin darkens the iris to a brown which is more functional in the tropics. Pupil diameter varies from about 1.5 mm to 9 mm—a factor of 6. Since light admitted is proportional to area, that is, diameter squared, the maximum amount of light admitted is about 40 times greater than minimum. The pupil expands with emotion and interest as well as for light. Belladonna causes pupils to dilate; Italian men consider any lady interested in them to be beautiful, and thus the drug got its name. Changes in pupil size reflect changes in attitude and can affect the attitude and responses of the person observing the pupil, even if the observer doesn't realize it is pupil size he is reacting to.[16]

The lens of the eye differs from a camera lens in that the lens changes shape—accommodates. Parallel rays of light from infinity are focused at a point at the rear of the eye. However, if the object is close to the eye the light rays are not parallel and the lens thickens to maintain the focus at the same point (see Figure 17.4). The closest point at which you can focus increases as the lens hardens with age; Table 17.1 shows that the change is about 12 diopters between ages 10 and 70. (Determine your own near point by bringing a printed page toward the eye until the letters blur.) Convergence, the act of aiming both eyes at a single point, is done by the eye muscles and is not needed for objects beyond 6 m. Predetermined time systems include convergence time and the time for eye focus or eye travel; see chapter 8 for some times.

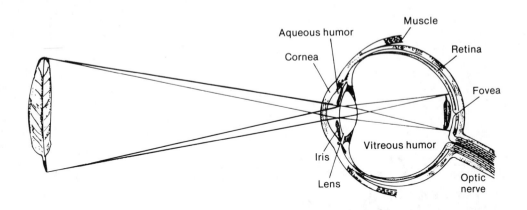

Figure 17.1 *Human eyes.* 25 mm spheres, have an iris to adjust the light admitted, a lens which adjusts focal length, and two "photographic films"—black and white and color. In addition the films change their "speed" as required.

Next the light rays pass through the vitreous humor, a jellylike substance whose primary function is to maintain the shape of the eye. If you see "moving things" when you close your eyes it may be red blood cells in the vitreous humor which escaped from the retina.

Finally the light strikes the retina (about the size and thickness of a postage stamp) which is attached to the sclera (the "white" of the eye). The retina contains two different transducer systems to convert light to an electrical signal. Below .0001 lux the rods are used, from .0001 to .001 both systems are used, and above .001 the cones are used.

The 7,000,000 cones (the daylight, color, photopic system) are concentrated about a small pit (*fovea*, if you prefer Latin) directly in line with the lens. The light hits the base of the cone and is focused on the pigment at the cone tip. The retina, an extension of the brain, processes the signal from the sensor before sending it to the brain. Each cone sends a number of impulses proportional to the light intensity. The signal is processed by a number of different types of cells all interconnected so that (1) the overall sensitivity of the retina (comparable to film speed in a camera) adjusts to ambient light; (2) adjacent cones have "recurrent inhibition" so that the eye emphasizes changes in brightness (i.e., contrast) more than the absolute intensity of the light.

Sharp vision is possible only within a visual angle of 1°; beyond this you note only movement and strong contrast. At a distance of 500 mm from the eye, this is a circle of .8 mm diameter. The larger area we actually observe is due to the saccadic eye movement as well as voluntary shifting of the area of sharp vision. The sensitivity of the entire retina varies with the amount of incoming light. If the incoming light suddenly increases (the eye sees a bright object such as an electric light, window, reflecting table top), the sensitivity of the entire retina decreases to about 20% of its previous value in about .05 s (alpha adaptation); it continues its decline in sensitivity (beta adaptation) at a slower rate for about 30 minutes.[14] Figure 17.2 shows the cumulative cone's sensitivity to light of various wavelengths. The sensitivity of each cone, the "quantum catch," determines the vectors of the Commission International de l'Eclairage (International Commission on Illumination—CIE) color triangle. Light above 620 nanometers (red light) does not affect the rods so dark vision is maintained. Figure 17.3 shows that green, red, and yellow are not seen well if they are peripheral.

Table 17.1   The Closest Point at Which the Eye can Focus, the Near Point, Increases with Age; Visual Acuity Decreases.[19] The Increase in Near Point is Called Presbyopia (Old Man's Vision, in Greek). Speed of Perception, Dark Adaptation Time, Ability to Detect Peripheral Movement, Resistance to Disability Glare, and Higher Luminance and Contrast Thresholds are Effects of Age.

Age, Years	Percent Visual Acuity (6/6 = 100)			Mean Near Point		Glare Borderline Between Comfort Discomfort, + nits
	Best Refractive Correction	Actual Correction	Without Correction	Meters	Diopters,* 1/m	
10	—	—	—	.077	13.0	6900
20	100	83	75	.091	11.0	3100
30	99	83	60	.111	9.0	1900
40	96	82	50	.167	6.0	1400
50	90	68	25	.500	2.0	1050
60	85	60	20	.833	1.2	850
70	70	40	15	1.000	1.0	700

*The power of a lens is given in diopters:

$$P \text{ (diopters)} = \frac{1}{f \text{ (meters)}} = \frac{1}{S_1} + \frac{1}{S_2}$$

where S_1 = distance from light source to lens node, meters
S_2 = distance from lens node to focal point, meters

For normal viewing, S_1 = infinity and S_2 = .015 meters, so the power of the eye is 67 diopters.

+BCD = 103,000 (age, yrs)$^{-1.17}$ from Bennett[1]

Figure 17.4 shows two common eye problems and their optical solutions. A near-sighted person (myopia) has a long eyeball—the light rays begin to diverge again before hitting the sensor. The solution is a concave external lens to bring the rays farther apart on the eye lens. A far-sighted person (hyperopia) has the opposite problem—a short eyeball; the lens becomes flatter than normal for distant objects so the muscles are in constant use. A convex lens makes the rays come together before hitting the eye lens. Some people have a line focus instead of a point focus—*astigmatism*, from the Greek for "no point." Astigmatism is corrected by an external lens with unequal curvature. Bifocals and trifocals reduce the need for accommodation. Poor vision is not caused by eyestrain, but just by the wrong shape of eyeball. College students often need glasses because their visual requirements are demanding and their eyeball shape changes—not because they studied too hard.

Spectacles were not considered manly for Prussian officers. Monocles were not very satisfactory, so the Germans developed the contact lens. Although originally developed for vanity, contact lenses now give corrections superior to those created by spectacles. In addition, since the lens supports the cornea, the typical lengthening of the eye with age is retarded so there is less visual deterioration with age. Contact lenses, of themselves, do not provide eye protection in the industrial sense. Contact lenses should not be worn where people are exposed to chemical fumes, vapors, or splashes; intense heat; molten metals; or atmospheres

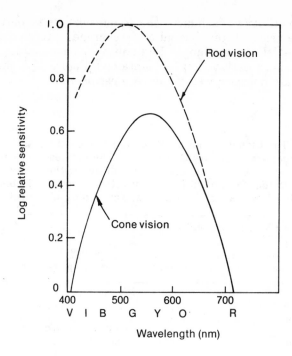

Figure 17.2 *Rods, the night vision system,* detect light about 100 times less intense than the day system (cones). The cones are much more sensitive to some colors than others with maximum sensitivity at 555 nm (yellow-green). From 400-450 light is violet and indigo; from 450 to 500, blue; from 500 to 545, yellow; from 590 to 610, orange; and from 610 to 760, red. Remember this sequence by the name Roy G. Biv; also remember to reverse the sequence since violet is at 400 nm not 760.

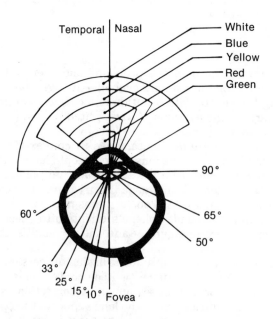

Figure 17.3 *Peripheral color detection* is difficult for red, yellow, and green. The figure shows the limits of the retinal zones in which various colors can be recognized in normal illumination.

with high dust levels. They should not be worn with a respirator in contaminated environments.

Cones specialize in detecting red, green, and blue. If your red cone pigment is defective, you cannot distinguish red or blue-green from gray; if your green cone pigment is defective, you cannot distinguish green or reddish-purple from gray. The X chromosome influences cone pigments. Since females have two X chromosomes, they have a "spare" and rarely have cone pigment deficiencies. About 8% of males and .4% of females have some color deficiency; usually color-deficient people can see all colors but need abnormal amounts of color.[33] They are "color weak" not "color blind." Only about .003% of males are "color blind."[13]

The other transducer system uses 125,000,000 rods scattered over the retina for night vision (also called black and white, or scotopic vision). At 500 nm, a quantum of light = 3.84×10^{-12} erg. For rod vision, 50 to 150 quanta must enter the eye. The pigment over the rods absorbs about 80% so 10 to 30 get to the rods.[15] Since there are no rods in the fovea, improve your night vision by looking out of the side of your eye. To prevent excitation from mechanical jarring or other nonlight events, rods are connected in a "voting circuit" with 10 to 100 rods per nerve. Each rod can be triggered by very little light (approximately 1 quantum). If light hits enough neighboring cells, the threshold of the relay cell is exceeded and the brain is signaled "I see the light." Rhodopsin, a chemical in the rods manufactured by the body from vitamin A, is bleached by light to give the photochemical effect for nerve excitation. Figure 17.5 gives recovery time for rods when exposed to bright light.

Visual acuity, the ability of the eye to distinguish detail, is defined as:

$$VA = \frac{1}{\text{visual angle of minimum object detectable, minutes of arc}}$$

The geometry of the formula is with Table 17.2. By expressing visual acuity in terms of the visual angle, any size target (dimension h) can be used at any distance from the eye (dimension d). The tangent of the visual angle, Θ, divided by 2 is equal to $d/2$ divided by h; since the angle is small, the approximation $\tan \Theta = h/d$ may be used. Table 17.2 gives minutes of arc and visual acuity for various ratios of h/d.

There are several kinds of visual acuity measurements: *minimum separable* (gap detection), *minimum perceptible* (spot detection), *vernier* (lateral displacement of two lines), and *dynamic*. Maximum gap detection visual acuity is about 2.5 (minimum visual angle = .4 min) for very good contrast, long viewing times, and high luminance. Spots (dark and light), lines, and squares are detectable even if they are very small. A star can be detected even though it subtends only .06 s of arc (a visual acuity of 960!). A dark square of 14 s of arc against a bright sky can be detected 75% of the time. Vernier acuity is 2 to 3 seconds of arc (Van Cott and Kinkade, p. 42[38]). Dynamic visual acuity, the ability to discriminate detail in a moving target, is important for inspection tasks. Unfortunately an individual's static visual acuity (gap detection) is not a good predictor of his dynamic acuity.[4,27] "Normal" vision often is expressed as 20-20 vision in the USA or as 6-6 vision in the rest of the world. That is, you see at 20 feet (6 meters) what the normal person sees at 20 feet; 20-200 vision means you see at 20 feet what the normal person sees at 200.

The effort required for seeing depends on the ratio between the visual angle of the task (Θ) and the visual angle of the smallest detectable object (Θ_0). The size ratio $R = \Theta/\Theta_0$.[10]

Visual acuity is affected by a number of factors of which contrast between the object and the background is the most important. Fortuin in Figure 17.6 plotted the threshold line of size vs. contrast for luminance of 10 cd/m².[10] Above the line

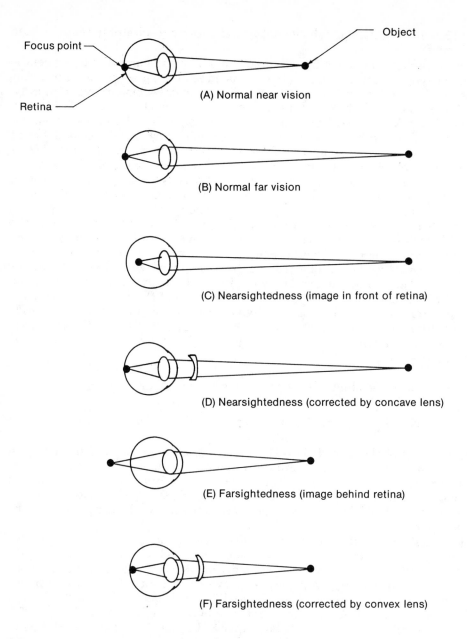

Focus point

Object

Retina

(A) Normal near vision

(B) Normal far vision

(C) Nearsightedness (image in front of retina)

(D) Nearsightedness (corrected by concave lens)

(E) Farsightedness (image behind retina)

(F) Farsightedness (corrected by convex lens)

Figure 17.4 *Eye problems and corrections* are shown. A and B show the normal eye at near and far vision. C and D show nearsightedness and its optical correction. E and F show farsightedness and its optical correction.

an object is visible; below it is invisible. Summarizing his lifetime of work at Philips, he recommends $R = 2.5$ for "easy seeing"; that is, an object should be 2.5 times threshold size. If $R > 2.5$, easy seeing; $1 < R < 2.5$, strenuous seeing resulting in eye fatigue; $R < 1$, invisible. Note that the eye is not hurt by insufficient light just as a camera isn't hurt; insufficient light just causes eye muscle fatigue, bloodshot eyes, and headaches, but not permanent damage.

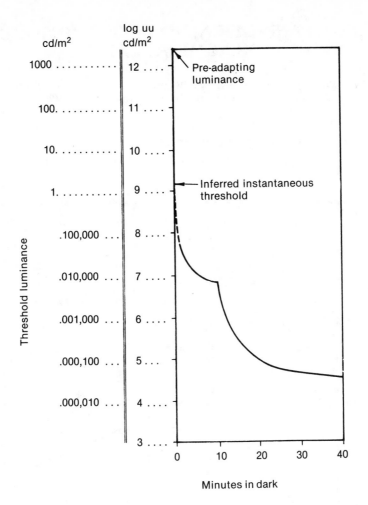

Figure 17.5 *Dark adaptation* dramatically reduces the brightness required for detection. Following initial exposure to a luminance of 3400 cd/m², the eye can detect a 3 cd/m² target. After about 2 min it can detect a .03 cd/m² object, but after 40 minutes it can detect a .000 030 cd/m² object. (Modified from Figure 13.33, *Bioastronautics Data Book,* Parker and West.[32])

Table 17.2   Visual Acuity, the Ability to Discriminate Detail, is Calculated From the Ratio of the Minimum Detectable Gap or Object *(h)* Over the Distance From the Eye to the Target *(d).* Θ = 3400 *h/d;* Visual Acuity = $1/\Theta_0$.

Tan Θ, h/d	Θ, minutes of arc	Visual acuity	
.017 455	60.00	0.02	
.001 000	3.43	0.29	
.000 500	1.72	0.58	
.000 241	1.00	1.00	
.000 200	0.68	1.45	
.000 100	0.34	2.90	
.000 050	0.17	5.81	

UNITS AND DEFINITIONS OF LIGHT AND ILLUMINATION

Table 17.3 gives the definitions of luminous intensity, luminous flux, luminance, reflectance, contrast, wavelength, and polarization. The key equation in lighting is:

$$\text{Luminance} = (\text{Illumination})(\text{Reflectance})/\pi \qquad (17.1)$$

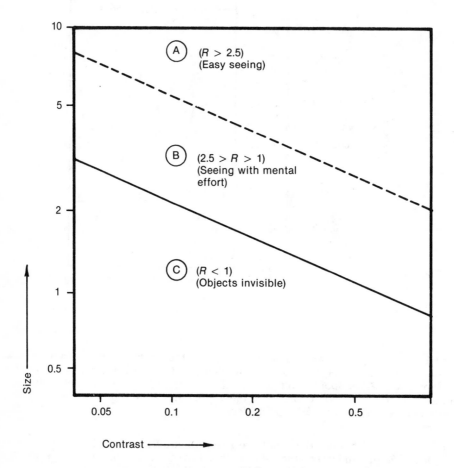

Figure 17.6 *Size* (min of arc) can compensate for lack of contrast.[10] Plot the size and contrast of a task on the figure. Then the perpendicular distance from the point to its intersection with the line represents the size ratio. The threshold line represents the line of maximum mental effort. Fortuin recommends a size at least 2.5 times threshold for easy seeing. Duncan and Konz, using light-emitting diodes and liquid crystal displays, confirmed that people do not like to use maximum mental effort.[8] Subjects could read the displays with no errors at 4 to 6 min but preferred 20 to 30 min. Note that, since legibility depends upon external factors such as size and luminance, visual acuity is not a constant factor of the individual but is subject to design.

Table 17.4 gives luminance of some common objects.

Table 17.3   Units and Definitions of Illumination

Quantity	Unit	Definition and comments
Luminous intensity	candela	Light intensity within a very small solid angle, in a specified direction (lumen/steradian). Candela = 4π lumens.
Luminous flux	lumen	Light flux, irrespective of direction. Generally used to: 1. Express total light output of source 2. Express amount incident on a surface 1 lumen/m^2 = 1 lux = 0.093 footcandle 1 lumen/ft^2 = 1 footcandle = 10.8 lux
Luminance (brightness)	nit	Luminance is independent of the distance of observation as candelas from the source and area perceived by the eye decrease at the same rate with distance. 1 candela/m^2 = 1 nit = 0.29 footLambert 1 candela/(ft^2)π = 1 footLambert = 3.43 nits
Reflectance	unitless	Percentage of light reflected from a surface.

Typical reflectances

Object	Percent
Mirrored glass	80-90
White matte paint	75-90
Porcelain enamel	60-90
Aluminum paint	60-70
Newsprint, concrete	55
Dull brass, dull copper	35
Cast and galvanized iron	25
Good quality printer's ink	15
Black paint	3-5

Recommended reflectances

	Percent
Ceilings	70-90
Walls	40-60
Furniture and equipment	25-45
Floors	20-40

Munsell Value	Reflectance %
10 =	100
9 =	78
8 =	58
7 =	40
6 =	24
5 =	19
3 =	6

Quantity	Unit	Definition and comments
Contrast	unitless	$C = \dfrac{\text{Luminance of brighter} - \text{Luminance of darker}}{\text{Luminance of brighter}}$
Wavelength	nanometers	The distance between successive waves (a "side view" of light). Wavelength determines the color. Of the 60 octaves of the electromagnetic radiation, the human eye detects radiation in the 1 octave from 380 to 760 nanometers.
Polarization	degrees	Transverse vibrations of the wave (an "end view" of the light). Most light is a mixture; horizontally polarized light reflected from a surface causes glare.

REDUCTION OF VISUAL PROBLEMS

The following section will discuss illumination from an engineering point of view. This is in contrast to illumination from an architect's viewpoint. In general, architects are not interested in the functional aspect of light but just the appearance; that is, in form not function. This remark is inserted to prepare you for the shock of dealing with architects who do not have the slightest interest in such

Table 17.4   Luminance (cd/m²) of Various Surfaces

Approximate brightness, cd/m²		
0.000 001 to 0.000 010	Absolute threshold of seeing	
0.000 010 to 0.000 100	Grass in starlight	
0.000 100 to 0.001	Snow in starlight	rod vision
0.001 to 0.010	Earth in full moon	rod + cone vision
0.010 to 0.100	Snow in full moon	
0.100 to 1.0	White paper 1 foot from a 1 candela source	rod + cone vision
1.0 to 10.0		cone vision
10.0 to 100.0	White paper in good light, indoor movie screen, TV screen	
100.0 to 1000.0	Average sky on cloudy day, full moon	
1000.0 to 10000.0	Average sky on clear day, instant-start cool white fluorescent tube (900-1150) for medium-loaded lamps, 1350 to 1800 for high-load lamps; preheat starting 500 to 1200	

things as lumens, lamp efficiencies, or lamp lives; you may be able to interest them in the color of the light or luminaire geometry but even then they will avoid calculations at all costs.

Consider the workers. Jacobson found that inspectors who performed poorly on a job inspection test tended to have poor vision; when the entire inspection department was tested, 32 of 130 either needed glasses or a change in prescription.[21] The company immediately implemented a procedure to check all inspectors' vision once a year. Lakowski reviews color vision testing.[23] A British study of industrial workers found faulty eyesight for 20% of those 15-30 years old, 25% for 31-40, 50% for 41-70, and 70% for 61-70.[37] Ferguson, Major, and Keldoulis reported that 37% of workers wearing glasses needed a new prescription and 69% of those without glasses needed glasses.[9] Maybe your inspectors *are* blind! Fairchild Camera uses visual acuity, eye muscle balance, and depth perception in job placement.[7] For microscope work, 14% don't qualify and another 13% can work with microscopes less than 4 hr/day. (There seems to be a principle that with bifocals everything of interest is on the bifocal line; its corollary is that as you get older and need trifocals, there no longer is anything of interest in the visual field!)

There are a number of techniques which the engineer can use to reduce visual problems. The contrast, the size of the object viewed, and the amount of viewing time are inherent in the task and thus are difficult to modify. Since the amount of light is relatively easy to change, the engineer tends to vary the fourth factor—amount of illumination.

Table 17.5 Cooper Tabulated the Characteristics of Various Light Sources.[5] For a 10 × 10 m Room With 1000 lux on the Task, it Cost About $.08/hr for the Lighting. Approximately 90% of the Lighting Cost was the Cost of Power.

Source	Sunlight	Incandescent			Fluorescent			Clear mercury (H33-1-CO/E)	Metal halide (MV-400/BU)	High pressure sodium (LU-400/BU)	Tungsten-halogen (2500/CL)
		Std.	Std.	Long Life	Cool-white Rapid start	Cool-white 96° output, 800 mA	Cool-white PG17 1500 mA				
Lamp color	Reference standard (excellent)	Good	Good	Good	Very good	Very good	Very good	Fair +	Very good	Fair	Good
Lamp life,	?	750-1000	800-1000	2500	25,000[c]	18,000[c]	12,500[c]	24,000	8,000	10,000	2,000
Rated life, hours					10,000	7,200	3,000	16,000	3,500	6,000	2,000
Maintained lumens					2,740	8,000	12,800	19,100	26,900	43,700	4,620
No. of lamps for 250,000 lumens					91.2	31.25	19.5	13.1	9.3	5.7	54.1
Lamp costs[b], $					.63	1.53	2.91	6.20	12.35	29.20	4.20
Luminaire cost, $					21[d]	40[d]	52[d]	60	88	160	10
Efficacy, lumens/watt — Lamp		14	16	13	69	79	75	48	67	109	20
Efficacy, lumens/watt — Lamp-ballast		14	16	13	60	64	57	42	58	93	18.5
Lamp power, watts		60	100	100	40	101	172	400	400	400	250
Lamp & ballast, watts					46	125	225	460	464	469	272

| Source | Sunlight | Incandescent | | Fluorescent | | | Clear mercury (H33-1-CO/E) | Metal halide (MV-400/BU) | High pressure sodium (LU-400/BU) | Tungsten-halogen (2500/CL) |
		Std.	Long Life	Cool-white Rapid start	Cool-white 96° output, 800 mA	Cool-white PG17 1500 mA				
Dollar cost of 1,000,000 lumen-hours of light[a]										
Lamp				.0004	.0015	.0036	.0014	.0179	.0167	.5250
Luminaire				.0763	.0500	.0406	.0640	.0655	.0729	.0472
Power				.2522	.2350	.2640	.3620	.2590	.1612	.8840
Total				.3289	.2865	.3082	.4272	.3424	.2508	1.4562

[a] 100m^2, 1000 lux on task = 100,000 lumens from lamps for typical cavity ratios, luminaires, reflectances, and dirt × 4 hrs. Power at 1.5 cents/kWh, 20 year luminaire life, no maintenance, labor or interest charges. 2500 hr/yr burning
[b] Based on $5000/yr purchase
[c] 12 hr/start. At 6 hr/start, life = 77%; at 3 hr/start life = 60%
[d] 2-lamp fixture

Table 17.6 The *Illuminating Engineering Society Lighting Handbook* Gives Recommended Lighting Levels for Interior Lighting (750 situations), Exterior Lighting (100), Sports Lighting (200), and Transportation (100).[19] A Few Selected Situations Are Given in Each Category.

Main Category	Task	Lux on Task
Interior lighting	Auto frame assy	500
	Auto chassis assy	1000
	Auto final inspection	2000
	Bakery scales	500
	Barber shop	1000
	Church pulpit	500
	Church worship area for churches with special zeal	300
	Dairy products—can washers	300
	Dance halls	50
	Foundry—pouring	500
	Foundry—shakeout	300
	Hospital autopsy table	10000
	Hospital corridor—day	200
	Hospital corridor—night	30
	Library carrels	700
	Locker rooms	200
	Office conference room	300
	Poultry egg inspection	500
	Poultry feed storage	100
	School study halls	700
	Sheet metal—punches	500
	Woodworking—fine bench	1000
Exterior lighting	Building exterior—active	50
	Building exterior—inactive	10
	Loading platforms	200
	Freight car interiors	100
	Parking—self-parking	10
Sports	Baseball—major league infield	1500
	Baseball—municipal league infield	200
	Horseshoes—tournament	100
	Table tennis—recreational	200
Transportation	Aircraft passenger compartment—general	50
	Aircraft passenger compartment—reading	400
	Automobile license plates	5
	Ships—pump room	100

Quantity of Light

Perhaps the most common technique is to add more light, especially general or area lighting. Table 17.5 gives the cost of lighting using a number of different artificial sources. The key row is the last one—cost of providing 1000 lux on the task for 100 m² for 4 hours. Assume that there is 1 employee per 10 m² and each employee has a cost (wages + fringes) of \$5/hr; that is, wage cost of \$50/hr for this area. Lighting cost for a fluorescent system is about \$.08/hr; \$.08/\$50 is less than .2% of labor cost.

The Illuminating Engineering Society (USA) has given detailed levels of illumination for many different work environments (see Table 17.6). The CIE (continental Europe) recommendations are in Table 17.7. They can be summarized as 2000 lux for precision work, 1000 lux for general office work, and 100 lux for non-work areas such as hallways. It is important to note that these standards are based on efficient seeing (95% performance where 100% is outdoor daylight) rather than safety or minimization of energy cost. The British Illuminating Engi-

Table 17.7   Unofficial CIE Recommended Illuminances for Interiors[25]

Range	Recommended Illuminance, Lux	Type of Activity
A General lighting for areas used infrequently or having simple visual demands	20 30 50	—Public areas with dark surroundings —Simple orientation (short temporary visits)
	75 100 150 200	—Rooms not used continuously for working (storage areas, entrance halls)
B General lighting for working interiors	300	—Tasks with limited visual requirements (rough machining, lecture theatres)
	500 750 1,000	—Tasks with normal visual requirements (medium machining, offices)
	1,500 2,000	—Tasks with special visual requirements (hand engraving, clothing factory inspection)
C Additional lighting for visually exacting tasks	3,000 5,000	—Very prolonged and exacting visual task (minute electronic and watch assembly)
	7,000 10,000 15,000	—Exceptionally exacting visual tasks (microelectronic assembly)
	20,000	—Very special visual tasks (surgical operations)

neering Society standards, based on 90% performance, usually have lux levels about 50% of the USA levels. Boyce reports illumination level had little effect on accuracy or vigilance; the predominant effect was on speed of working. Boyce concluded optimum illumination was between 1000 to 2000 lux (target luminances of 150 to 350 cd/m²).[3]

Figure 17.7 gives Fortuin's findings of illumination required as a function of age.[10] At age 10, a contrast of 0.1 and a luminance of 1 requires a size of about 4. Due to reduction in pupil size, yellowing of the lens, and change in the other ocular refractive media, light reaching the lens at age 60 is one-third that at age 20. For age 60, this size and contrast require a luminance of about 100. Thus, older people need much more light than do younger people.

Too much light gives glare and fatigue as well as wastes energy. Too little light will degrade visual performance. Note, however, that the eye is very adaptable and will attempt to compensate. Thus, the short-run result may not be a change in performance but an increase in fatigue.

Because light is cheap does not mean it should be wasted. In particular, there is a point of diminishing returns; beyond 500 to 1000 lux, visual performance increases very slowly. For some very severe visual tasks such as mapmaking or difficult textile work, levels should be 2000 or even 3000 lux; obtain these levels by general lighting of 500 to 1000 lux supplemented by local lighting.

Most organizations have found it most efficient to have a level of 500 to 1000 lux over a general area rather than 200 to 400 lux supplemented by local lighting. The reasoning is (1) the general lighting can be furnished by efficient sources, (2) space utilization is maximized since any workstation can be located anywhere, and (3) any workstation can be moved at any time without being concerned with lighting.

The "nonuniform" lighting technique can save energy but it requires close, con-

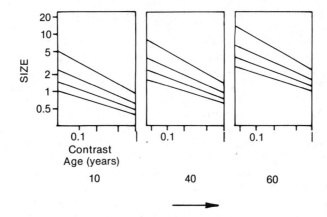

Figure 17.7 *Older people* (over 60) require as much as 100 times as much light as children.[10] The four lines are equal visiblity lines for 1, 10, 100, and 1000 cd/m^2.

tinuing coordination and cooperation between the architect, plant engineering, and maintenance.

Which source of light? The ideal illumination source—which does not exist—would be free, give the desired amount of light upon demand, show true color, and have the proper direction (glare, highlighting, and contrast considerations).

Sunlight often is considered free. It isn't. Sunlight comes through windows or skylights. These same openings pass radiation at wavelengths beyond the visible, that is, windows lose heat in the winter and admit heat in the summer. Windows light building interiors only in "small" buildings, that is, buildings with a high ratio of wall area to volume. Small buildings are inefficient to heat and cool as well as to use for manufacturing. Skylights can be used in one-floor slablike buildings if the energy penalty is acceptable. Windows, if opened, often interfere with the artificial ventilation system. Light from the sun may be too little (night, clouds, building obstructions). Sunlight also can be too great. Note the number of windows completely or partially covered with shades, blinds, curtains, or even aluminum foil. Windows admit external noise. Windows decrease privacy and security. Hopkinson has a good discussion of windows.[17]

With all these disadvantages, why do we use windows? Because people want a view. No view? Then no window.

If the purpose of a window is to furnish a view and illumination and ventilation will be furnished artificially, then windows should have minimum surface area to minimize energy losses. To minimize glare, they should not have a long vertical axis. Place windows low (i.e., a low sill) to minimize sky glare. For tall windows, use curtains which are more opaque on the top, or tint upper panes. A good design is a horizontal window, about .6 m high, placed to give a view of the horizon (1/3 sky and 2/3 world). Paint shutters and glazing bars white to reduce contrast.

If skylights are used, make them vertical rather than horizontal (minimizes dirt and breakage), a plastic instead of wired glass (better light transmission and less breakage), translucent rather than transparent (cuts glare), and able to be opened (aids ventilation).

Given the decision for artificial light, the question is *which source*. Which artificial source depends upon cost, color, and convenience.

Cost depends on cost of the lamp and luminaire (prorated over the hours of their use), the cost of the electricity, and the labor cost of replacing the burned out lamps.

As a general rule of thumb, 90% of the total illumination cost is the cost of the electricity—thus the trend to lamps which give higher lumens/watt. Since fluorescent lamps give about 60 lumens/watt vs. 15 for incandescent lamps, fluorescent lamps usually are used for new installations. High-intensity discharge lamps have several subtypes. (Gas discharge lamps require current limiting while burning, done by placing an inductance coil (a ballast) in series with the lamp.) Metal-halide lights are slightly more efficient than fluorescent and mercury-vapor slightly less efficient. High and low pressure sodium-vapor lamps, which concentrate their output where the eye is most sensitive, are the champions; the light is yellow but the lumens/watt are well over 100. Low pressure sodium has poor color. Within a lamp type, large lamps are more efficient than smaller lamps; for example, a 60-watt incandescent bulb gives 14 lumens/watt while a 100 gives 16. Long fluorescent lamps are more efficient than short lamps. A 100 mm lamp gets 25 lumens/watt, a 1 m lamp gets 65, and a 2.5 m lamp gets 72. Fluorescent lamps give optimum lumens/watt at 25 C ambient temperature. Above 25° there is a 2% drop in light for every degree increase in ambient temperature; in still air, the drop is 6% for every degree below 10 C.

The short life of incandescent bulbs means they must be replaced often—another reason for not using them. Extended-service and vibration-service bulbs have a rugged filament which gives less light. The long-life incandescent lamp shown in Table 17.4 trades off its 2400-hour life (vs. 1000 for the conventional bulb) by giving only 13 lumens/watt (vs. 16)—there is no free lunch. The life of fluorescent bulbs formerly was extremely sensitive to the number of on-off cycles. Since electricity was cheap, people were told not to turn them off when leaving the room. With improving technology, the life of present lamps is not as sensitive; in addition, electricity prices have risen. It usually is cheaper now to turn them off when leaving the room, especially with "rapid-start" circuits. Why not calculate a "switch on or off" decision table for your organization using your local lamp prices, power cost, and maintenance cost? Outside lights should be controlled with timers or photocells. In some countries even inside lights in areas such as toilets and corridors are controlled with timers. I found this out when a toilet became pitch dark five minutes after I entered it! Since fluorescent lamps tend to have similar life spans when on the same switch, it generally is cheaper to replace all the lamps in a room at the same time rather than when they individually fail. Group relamping trades off a small amount of life left in the lamps for a large maintenance cost savings.

Color is the second lamp selection criterion (see Figure 17.8). The "normal" color of light is spring outdoor daylight in northern Europe—the mind thinks of this as a "cool" light. As light intensity drops (sunset), the mind considers a light with more red as "normal."

There are about 15 different colors of fluorescent lamps available, each with a different blend of phosphors for color rendition and lumens/watt. Cool-white (75 lumens/watt) is the favorite in the USA. Cool-white deluxe, which has more red, is probably better in locations where skin color is important (hospitals, beauty parlors). Warm-white (77 lumens/watt) is a fluorescent lamp designed to match the spectrum of an incandescent lamp yet give fluorescent efficiency. Even the relatively efficient fluorescent lamps, however, convert only 15% of their energy into light; 87% of the lamp energy and 13% of the control gear energy become heat.[36] (Where there's light, there's heat.) Thus, turning off lights "to save energy" may just substitute energy generated from scarce sources such as gas or oil for energy generated from plentiful sources such as coal.

Incandescent lamps get their warm color (and low efficiency) because most of their radiation is infrared rather than in the visible spectrum. Metallic vapor and sodium lamps gain high lumens/watt by emitting light at very narrow ranges of wavelengths (spikes) in the wavelengths at which the eye is most sensitive. This narrow range gives poor color and explains why you can't identify your car in a parking lot lit with these lamps. Blends of metals can be used to improve the color at a loss in luminous efficiency.

Color is "an accidental quality." The perceived color of an object depends on the spectrum of the ambient light, the actual color of the object, and the spectral characteristics of your eyes. Metameric colors are colors which look the same under one light spectrum but different under another. When buying clothing that will be worn under incandescent lamps, purchase it under incandescent lamps; if it will be worn outdoors, inspect it under daylight. When your inspection department must match colored products from different sources, it is essential to specify the type of light under which the objects are inspected. Engineers tend to emphasize cost or color and neglect convenience. Convenience, as well as short lamp life, is why incandescent lamps were 75% of all lamps made in the world in 1968 although fluorescent lamps provided 60% of the light.[40] Since electric illumination is so cheap, even a large percent increase in a low price doesn't eliminate the use of incandescent lamps which come in a very wide variety of types and sizes—especially small sizes. On the other hand, fluorescent lamps have the convenience of rare replacement and less glare since their output comes from a rectangular area rather than a "point" source. The trend away from incandescent lamps probably will continue.

Area lighting calculations using the zonal cavity method. The basic concept is that for a lighting system with 100% efficiency:

Lux on an area = (Number of) (Number of lamps) (Lumens per)
luminaires   per luminaire   lamp

$$I(A) = N_1 \, (N_2) \, (L) \qquad\qquad (17.2)$$

where I = Illumination required in lux (lumens/m²)
A = Area illuminated, m²
N_1 = Number of luminaires used
N_2 = Number of lamps per luminaire
L = Lumens per lamp

For example, if
I = 750 lumens/m²
A = 2500 m²
N_2 = 1 lamp/luminaire
L = 22,500 lumens/lamp
then N_1 = 83 luminaires required

However, in practice, efficiency is not 100% and equation 17.2 is modified by the Coefficient of Utilization (CU), Lamp Lumen Depreciation (LLD), and Luminaire Dirt Depreciation (LDD).[12] The CU adjusts for (a) the shape and reflectance of the area lighted (the room cavity), (b) the shape and reflectance of the area above the lamps (the ceiling cavity), (c) the shape and reflectance of the area below the area lighted (the floor cavity), and (d) the light pattern of a specific lamp in a specific luminaire. The LLD adjusts for the loss of light output of the lamp with age. The LDD adjusts for the loss of light output due to dirt on the luminaire.

The equation to use is:

$$I(A) = CU \, (LLD) \, (LDD) \;\; (N_1) \, (N_2) \, (L) \qquad\qquad (17.3)$$

The calculation flow diagram is shown in Figure 17.9.

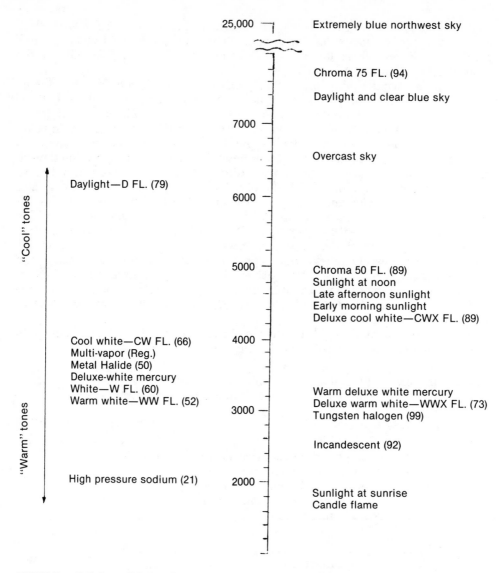

25,000 — Extremely blue northwest sky

Chroma 75 FL. (94)

Daylight and clear blue sky

7000 —

Overcast sky

Daylight—D FL. (79)

6000 —

5000 — Chroma 50 FL. (89)
Sunlight at noon
Late afternoon sunlight
Early morning sunlight
Deluxe cool white—CWX FL. (89)

Cool white—CW FL. (66) 4000 —
Multi-vapor (Reg.)
Metal Halide (50)
Deluxe-white mercury
White—W FL. (60) Warm deluxe white mercury
Warm white—WW FL. (52) Deluxe warm white—WWX FL. (73)
3000 — Tungsten halogen (99)

Incandescent (92)

High pressure sodium (21) 2000 —

Sunlight at sunrise
Candle flame

"Cool" tones

"Warm" tones

SOURCE: Terry K. McGowan, "All About Sources," reprinted from the September 1973 issue of *Progressive Architecture*, copyright 1973. Reinhold Publishing Company.

Figure 17.8 *Apparent color temperature* (chromaticity) in degrees Kelvin is shown for various light sources. The color rendering index, R_a, (1-100 scale) is given in parentheses for some sources. R_a is useful for comparing sources only if the sources have the same chromaticity. For color rendering, select a chromaticity range for its warmth or coolness, then pick a source for R_a (high R_a goes with low lumens/watt). Lamps on the left are designed for high lumens/watt and acceptable-to-good color rendering. Most incandescent lamps are in the range of 2500-3200 K.

To calculate *CU*, first divide the space into three cavities and calculate the cavity ratio:

$$\text{Cavity ratio, } CR = \frac{5H\,(LG + W)}{LG \times W}$$

where $\quad H =$ height of the cavity, m
$\qquad LG =$ length of the cavity, m
$\qquad W =$ width of the cavity, m

Figure 17.10 shows some cavity shapes. The slab is the most efficient from a lighting viewpoint while the cube, shaft, and tunnel should be avoided.

The three cavities are the room cavity (the space between the lamp and the illuminated surface), the ceiling cavity (the space above the lamps), and the floor cavity (the space below the illuminated surface). If the lamps are mounted flush with the ceiling, the ceiling cavity is zero. If the illuminated surface is the floor, the floor cavity is zero.

Assume a 15 × 15 room with the lamps 1 m below the 8 m ceiling and the work surface .75 m above the floor. Then the ceiling cavity ratio is .7, the room cavity ratio is 4.2, and the floor cavity ratio is .5.

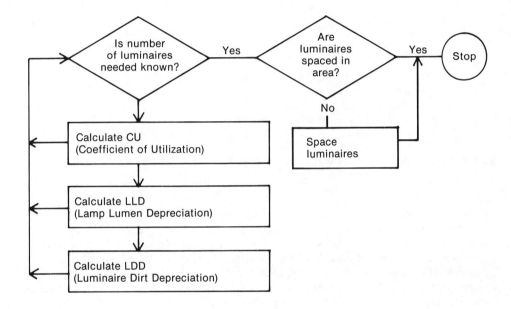

Figure 17.9 *Area lighting calculations* using the zonal cavity method use the above flow diagram.

Second, determine the reflectances of the floor cavity and the ceiling cavity. Assume, for our example, ceiling reflectance equals .8, wall reflectance equals .5, and floor reflectance equals .1. The reflectance of the cavity depends on three factors: cavity shape, wall reflectance, and ceiling (floor) reflectance. Table 17.8 gives the cavity reflectance given these three factors.

Enter the table with the closest value of ceiling (floor) reflectance and wall reflectance to select one of the 21 columns. Then move down to the appropriate cavity ratio row and read the effective reflectance for the cavity. A ceiling reflectance of .8, wall reflectance of .5, and cavity ratio of .7 give a ceiling cavity reflectance of .70. A floor reflectance of .1, wall reflectance of .5, and cavity ratio of .5 give floor cavity reflectance of .11. Use the closest value rather than interpolating since the calculations are not as precise as they seem.

Third, determine the *CU* for a specific lamp-luminaire combination in this specific room cavity, with these specific ceiling and floor cavity reflectances. Table

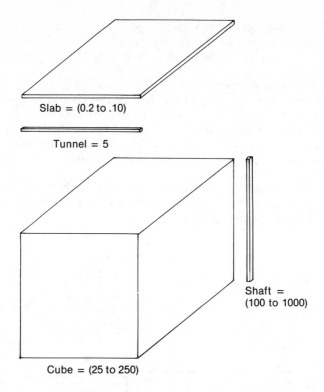

Figure 17.10 *Slabs are efficient* lighting shapes; avoid tunnels, shafts, and cubes.

17.9 gives *CUs* for some specific combinations. Assume you wished to use luminaire combination 24 (an improved color mercury lamp in a wide distribution, ventilated aluminum reflector). For a room cavity ratio of 4, ceiling reflectance of .7, and wall reflectance of 0.5, the *CU* is .61. However, the tables assume a floor cavity reflectance of .2 while our floor cavity reflectance was .11, so a correction must be made. The correction table is Table 17.10.

For a room cavity ratio of 4, ceiling cavity reflectance of .8, and wall reflectance of .5, the correction factor is 1.05. Since the floor cavity reflectance is .11, divide the .61 by 1.05 to get a corrected *CU* = .59.

Fourth, determine the lamp lumen depreciation, *LLD*. Table 17.11 gives *LLD* for various lamps. Note that the *LLD* is for *minimum* illumination between relampings based on lamp output at 70% of rated life. Rated life is 1000 hours for most incandescent lamps and 24,000 for most metallic vapor lamps. Life of fluorescent lamps vary with hr/start. Table 17.11 fluorescent values are based on 3 hr/start; for 6 hr/start add .02 to *LLD* and for 12 hr add .01. For a 400 watt *DX* lamp, the *LLD* is .71. That is, we assume the lamp is only giving 71% of its rated output. If we wish to calculate *average* illumination between relampings, use the Mean Lumen Factor, .77 in this case. That is, on the average the lamp gives 77% of its rated output. Low-pressure sodium lamps keep light output constant but increase power use.

Fifth, determine luminaire dirt depreciation *(LDD)*. Table 17.12 gives five different degrees of dirt. Assume our condition is "medium" dirt. Next refer to Figure 17.11 and select the set of curves appropriate for the specific luminaire; ours is category III (from Table 17.9). Enter at the number of months at the planned cleaning cycle. For our case assume 24 months; then *LLD* is .80.

Sixth, calculate the number of necessary luminaires, N_1, from equation 17.3. Summarizing:

$I = 750$ lumens/m^2	$LDD = .80$
$A = 50 \times 50 = 2500$ m^2	$N_2 = 1$ lamp/luminaire
$CU = .59$	$L = 22,500$ lumens/lamp
$LLD = .71$	so $N_1 = 249$ luminaires

Seventh, place the luminaires in the area (the 2500 m^2). One constraint is the maximum spacing-to-mounting-height ratio for "relatively uniform ($\pm 15\%$)" illumination. For luminaire 24, Table 17.9 indicated the maximum ratio is 1.4. Since the mounting height was 6.25 m above the work surface, the maximum spacing is $6.25 \times 1.4 = 8.75$ m between luminaires. Thus, when placing approximately 250 luminaires in a 50×50 room, the 8.75 m maximum spacing will not be a constraint.

Of the various spacing patterns, the checkerboard pattern on the "same color squares" probably is the most efficient. Light the space not as uniformly as possible but as nonuniformly as bearable. In most patterns the center of the area tends to be overlighted, so increase spacing in the center. Keep lighting near the walls satisfactory by putting extra lighting near the walls, by using a smaller spacing interval for the row next to the wall, or by using extra luminaires at the end of rows. End-to-end luminaires waste the light from the end of the luminaires and, thus, are an inefficient pattern. Lighting next to windows should be on a separate switch so it can be turned off when the sun is bright; this also saves air conditioning.[35] In a factory environment, put luminaires on adjustable tracks to permit movement due to obstructions of conveyors, catwalks, columns, etc. For our problem, a total of 250 lamps, spaced on 3 m centers with the four corner lamps 1.5 m from each wall, would be a satisfactory solution. (A common error is to control too many lamps with one switch. Fewer lamps per switch permit turning off lights in areas temporarily vacant.)

The foregoing example furnishes a *minimum* of 750 lux of area lighting. An *average* of 750 might have been more appropriate. If more than 750 lux are required on the task, consider the following before adding more general illumination: (1) local supplementary lighting, (2) lighter floors, ceilings, and walls, (3) cleaning the fixtures, walls, and ceilings more often, and (4) relamping more often.

Finally, some comments on the zonal cavity method are needed. If the engineer does not have a *CU* table for the specific lamp-luminaire combination, use a similar unit's *CU* for the same cavity ratio. Using a room cavity ratio of 2, ceiling reflectance of .5, wall reflectance of .3, the mean *CU* for 11 point source luminaires-lamps was .62 (range from .44 to .76); the mean for 11 tubes with solid top luminaires was .46 (range from .36 to .60); and the mean for 11 tubes with open top luminaires was .56 (range from .45 to .65). If lumen output/lamp is unknown, use 15 lumens/watt for incandescent, 50 for mercury, 60 for fluorescent, 85 for multivapor, and 100 for high pressure sodium.

The sketch on the left of the *CU* table shows the distribution of light from the luminaire. Note that the total is not 100%.

The tables can be used to illustrate four principles:

- A distant light is dim.
- Reuse the light.
- Low cavity ratios (slabs) are better.
- Some luminaires are more efficient.

Table 17.8   Per Cent Effective Ceiling or Floor Cavity Reflectance for Various Reflectance Combinations

Per Cent Ceiling or Floor Reflectance	90				80				70			50			30				10		
Per Cent Wall Reflectance	90	70	50	30	80	70	50	30	70	50	30	70	50	30	65	50	30	10	50	30	10
0	90	90	90	90	80	80	80	80	70	70	70	50	50	50	30	30	30	30	10	10	10
0.1	90	89	88	87	79	79	78	78	69	69	68	59	49	48	30	30	29	29	10	10	10
0.2	89	88	86	85	79	78	77	76	68	67	66	49	48	47	30	29	29	28	10	10	9
0.3	89	87	85	83	78	77	75	74	68	66	64	49	47	46	30	29	28	27	10	10	9
0.4	88	86	83	81	78	76	74	72	67	65	63	48	46	45	30	29	27	26	11	10	9
0.5	88	85	81	78	77	75	73	70	66	64	61	48	46	44	29	28	27	25	11	10	9
0.6	88	84	80	76	77	75	71	68	65	62	59	47	45	43	29	28	26	25	11	10	9
0.7	88	83	78	74	76	74	70	66	65	61	58	47	44	42	29	28	26	24	11	10	8
0.8	87	82	77	73	75	73	69	65	64	60	56	47	43	41	29	27	25	23	11	10	8
0.9	87	81	76	71	75	72	68	63	63	59	55	46	43	40	29	27	25	22	11	9	8
1.0	86	80	74	69	74	71	66	61	63	58	53	46	42	39	29	27	24	22	11	9	8
1.1	86	79	73	67	74	71	65	60	62	57	52	46	41	38	29	26	24	21	11	9	8
1.2	86	78	72	65	73	70	64	58	61	56	50	45	41	37	29	26	23	20	12	9	7
1.3	85	78	70	64	73	69	63	57	61	55	49	45	40	36	29	26	23	20	12	9	7
1.4	85	77	69	62	72	68	62	55	60	54	48	45	40	35	28	26	22	19	12	9	7
1.5	85	76	68	61	72	68	61	54	59	53	47	44	39	34	28	25	22	18	12	9	7
1.6	85	75	66	59	71	67	60	53	59	52	45	44	39	33	28	25	21	18	12	9	7
1.7	84	74	65	58	71	66	59	52	58	51	44	44	38	32	28	25	21	17	12	9	7
1.8	84	73	64	56	70	65	58	50	57	50	43	43	37	32	28	25	21	17	12	9	6
1.9	84	73	63	55	70	65	57	49	57	49	42	43	37	31	28	25	20	16	12	9	6
2.0	83	72	62	53	69	64	56	48	56	48	41	43	37	30	28	24	20	16	12	9	6
2.1	83	71	61	52	69	63	55	47	56	47	40	43	36	29	28	24	20	16	13	9	6
2.2	83	70	60	51	68	63	54	45	55	46	39	42	36	29	28	24	19	15	13	9	6
2.3	83	69	59	50	68	62	53	44	54	46	38	42	35	28	28	24	19	15	13	9	6
2.4	82	68	58	48	67	61	52	43	54	45	37	42	35	27	28	24	19	14	13	9	6
2.5	82	68	57	47	67	61	51	42	53	44	36	41	34	27	27	23	18	14	13	9	6
2.6	82	67	56	46	66	60	50	41	53	43	35	41	34	26	27	23	18	13	13	9	5
2.7	82	66	55	45	66	60	49	40	52	43	34	41	33	26	27	23	18	13	13	9	5
2.8	81	66	54	44	66	59	48	39	52	42	33	41	33	25	27	23	18	13	13	9	5
2.9	81	65	53	43	65	58	48	38	51	41	33	40	33	25	27	23	17	12	13	9	5
3.0	81	64	52	42	65	58	47	38	51	40	32	40	32	24	27	22	17	12	13	8	5
3.1	80	64	51	41	64	57	46	37	50	40	31	40	32	24	27	22	17	12	13	8	5
3.2	80	63	50	40	64	57	45	36	50	39	30	40	31	23	27	22	16	11	13	8	5
3.3	80	62	49	39	64	56	44	35	49	39	30	39	31	23	27	22	16	11	13	8	5
3.4	80	62	48	38	63	56	44	34	49	38	29	39	31	22	27	22	16	11	13	8	5
3.5	79	61	48	37	63	55	43	33	48	38	29	39	30	22	26	22	16	11	13	8	5
3.6	79	60	47	36	62	54	42	33	48	37	28	39	30	21	26	21	15	10	13	8	5
3.7	79	60	46	35	62	54	42	32	48	37	27	38	30	21	26	21	15	10	13	8	4
3.8	79	59	45	35	62	53	41	31	47	36	27	38	29	21	26	21	15	10	13	8	4
3.9	78	59	45	34	61	53	40	30	47	36	26	38	29	20	26	21	15	10	13	8	4
4.0	78	58	44	33	61	52	40	30	46	35	26	38	29	20	26	21	15	9	13	8	4
4.1	78	57	43	32	60	52	39	29	46	35	25	37	28	20	26	21	14	9	13	8	4
4.2	78	57	43	32	60	51	39	29	46	34	25	37	28	19	26	20	14	9	13	8	4
4.3	78	56	42	31	60	51	38	28	45	34	25	37	28	19	26	20	14	9	13	8	4
4.4	77	56	41	30	59	51	38	28	45	34	24	37	27	19	26	20	14	8	13	8	4
4.5	77	55	41	30	59	50	37	27	45	33	24	37	27	19	25	20	14	8	14	8	4
4.6	77	55	40	29	59	50	37	26	44	33	24	36	27	18	25	20	14	8	14	8	4
4.7	77	54	40	29	58	49	36	26	44	33	23	36	26	18	25	20	13	8	14	8	4
4.8	76	54	39	28	58	49	36	25	44	32	23	36	26	18	25	19	13	8	14	8	4
4.9	76	53	38	28	58	49	35	25	44	32	23	36	26	18	25	19	13	7	14	8	4
5.0	76	53	38	27	57	48	35	25	43	32	22	36	26	17	25	19	13	7	14	8	4

Ceiling or Floor Cavity Ratio

For a 22,500 lumen lamp in luminaire 24 and cavity ratios of 1, 4, and 10:

Room Cavity Ratio	Lamp	CU	Lumens on Task (Direct Only)
1	22,500	.60 =	13,500
4	22,500	.46 =	10,350
10	22,500	.23 =	5,175

A distant light is dim.

Table 17.9

ρ_{CC} →	80			70			50			30			10			0	
ρ_W →	50	30	10	50	30	10	50	30	10	50	30	10	50	30	10	0	Typical Luminaires and Luminaire Maintenance Category
RCR ↓	Coefficients of Utilization for 20 Per Cent Effective Floor Cavity Reflectance, ρ_{FC}																

21 — Max. S/MH$_{wp}$ = 1.5 (0% / 70%)

RCR	80/50	80/30	80/10	70/50	70/30	70/10	50/50	50/30	50/10	30/50	30/30	30/10	10/50	10/30	10/10	0
1	.78	.77	.74	.76	.75	.73	.74	.72	.71	.71	.70	.68	.68	.67	.66	.65
2	.72	.68	.66	.71	.67	.65	.68	.66	.63	.65	.64	.62	.64	.62	.61	.59
3	.66	.62	.59	.65	.61	.58	.63	.60	.57	.61	.59	.56	.60	.57	.55	.54
4	.60	.56	.52	.59	.55	.52	.58	.54	.51	.56	.53	.51	.55	.53	.50	.49
5	.55	.50	.47	.54	.50	.46	.53	.49	.46	.52	.48	.46	.51	.48	.45	.44
6	.51	.45	.42	.50	.45	.42	.49	.45	.41	.48	.44	.41	.47	.43	.41	.40
7	.46	.41	.37	.46	.41	.37	.45	.40	.37	.44	.40	.37	.43	.39	.36	.35
8	.42	.37	.33	.42	.37	.33	.41	.36	.33	.40	.36	.33	.39	.35	.33	.31
9	.39	.33	.30	.38	.33	.30	.37	.33	.30	.37	.32	.29	.36	.32	.29	.28
10	.33	.28	.25	.33	.28	.25	.32	.28	.25	.32	.28	.24	.31	.27	.24	.23

Enclosed reflector with incandescent lamp. LDD Maint. Category V

22 — Max. S/MH$_{wp}$ = 1.2 (0% / 70%)

RCR	80/50	80/30	80/10	70/50	70/30	70/10	50/50	50/30	50/10	30/50	30/30	30/10	10/50	10/30	10/10	0
1	.75	.73	.71	.74	.72	.70	.71	.69	.68	.68	.67	.66	.66	.65	.64	.62
2	.68	.64	.61	.67	.63	.61	.64	.62	.59	.62	.60	.58	.60	.58	.57	.55
3	.62	.57	.54	.60	.56	.53	.59	.55	.52	.57	.54	.51	.55	.53	.50	.49
4	.56	.52	.47	.55	.51	.47	.53	.49	.46	.52	.48	.45	.50	.47	.45	.44
5	.50	.45	.41	.49	.44	.41	.48	.44	.40	.47	.43	.40	.46	.42	.40	.38
6	.45	.40	.36	.45	.40	.36	.43	.39	.36	.42	.38	.65	.41	.38	.35	.34
7	.41	.36	.32	.40	.35	.32	.40	.35	.31	.39	.34	.31	.38	.34	.31	.30
8	.37	.32	.28	.37	.32	.28	.36	.31	.28	.35	.31	.28	.34	.30	.28	.26
9	.33	.28	.24	.33	.28	.24	.32	.28	.24	.32	.27	.24	.31	.27	.24	.23
10	.30	.25	.22	.30	.25	.22	.29	.25	.22	.29	.25	.22	.28	.24	21	.20

Medium distribution, ventilated aluminum or glass reflector with improved-color mercury lamp. LDD Maint. Category III

23 — Max. S/MH$_{wp}$ = 0.6 (0% / 80%)

RCR	80/50	80/30	80/10	70/50	70/30	70/10	50/50	50/30	50/10	30/50	30/30	30/10	10/50	10/30	10/10	0
1	.89	.87	.85	.87	.85	.84	.84	.82	.81	.81	.80	.79	.78	.77	.77	.75
2	.82	.79	.76	.81	.78	.76	.78	.76	.74	.76	.74	.72	.74	.72	.71	.69
3	.76	.72	.69	.75	.71	.69	.73	.70	.67	.71	.69	.66	.69	.67	.65	.64
4	.71	.66	.63	.70	.66	.62	.68	.65	.62	.67	.64	.61	.65	.62	.60	.59
5	.66	.61	.57	.65	.60	.57	.63	.59	.56	.62	.59	.56	.61	.58	.55	.54
6	.61	.56	.53	.61	.56	.53	.60	.55	.52	.59	.55	.52	.57	.54	.52	.50
7	.57	.52	.48	.56	.52	.48	.55	.51	.48	.54	.50	.48	.54	.50	.47	.46
8	.53	.48	.44	.52	.47	.44	.51	.47	.44	.51	.47	.44	.50	.46	.43	.42
9	.49	.43	.40	.48	.43	.40	.47	.43	.40	.47	.42	.40	.46	.42	.39	.38
10	.45	.40	.37	.45	.40	.37	.44	.39	.36	.43	.39	.36	.43	.39	.36	.35

400W   1000W. Narrow distribution, ventilated aluminum or glass reflector with clear mercury lamp. LDD Maint. Category III

24 — Max. S/MH$_{wp}$ = 1.4 (10% / 65%)

RCR	80/50	80/30	80/10	70/50	70/30	70/10	50/50	50/30	50/10	30/50	30/30	30/10	10/50	10/30	10/10	0
1	.81	.79	.77	.77	.76	.74	.73	.71	.70	.68	.67	.66	.63	.63	.62	.60
2	.74	.71	.68	.72	.69	.66	.67	.65	.63	.63	.62	.60	.60	.58	.57	.55
3	.68	.64	.61	.66	.63	.59	.63	.60	.57	.59	.57	.55	.56	.54	.52	.51
4	.63	.58	.54	.61	.57	.53	.58	.54	.51	.55	.52	.49	.52	.50	.48	.46
5	.57	.52	.49	.56	.51	.48	.53	.49	.47	.51	.47	.45	.48	.46	.43	.42
6	.53	.48	.44	.52	.47	.43	.49	.45	.42	.47	.43	.41	.45	.42	.40	.38
7	.48	.43	.40	.47	.42	.39	.45	.41	.38	.43	.40	.37	.41	.38	.36	.35
8	.44	.39	.36	.43	.38	.35	.41	.37	.34	.40	.36	.33	.38	.35	.32	.31
9	.41	.35	.32	.40	.35	.31	.38	.34	.31	.36	.33	.30	.35	.32	.29	.28
10	.35	.30	.27	.35	.30	.26	.33	.29	.26	.32	.28	.25	.30	.27	.24	.23

Wide distribution, ventilated aluminum or glass reflector with improved-color mercury lamp. LDD Maint. Category III

[a] Ratio of maximum spacing between luminaire centers to mounting (or ceiling) height above the work plane, S
[b] RCR = Room Cavity Ratio
[c] PCC = Percent effective ceiling cavity reflectance
[d] PW = Percent wall reflectance

The row for any cavity ratio shows the benefits of light colors on the walls and ceilings. For luminaire 24, a wall reflectance of .3, and a room cavity ratio of 1:

Ceiling Reflectance	Lumens/ Luminaire	CU	Lumens on Task	Percent Reuse of Light
.10	22,500	.63 =	14,175/13,500	5
.30	22,500	.67 =	15,075/13,500	12
.50	22,500	.71 =	15,975/13,500	18
.70	22,500	.76 =	17,100/13,500	27
.80	22,500	.79 =	17,775/13,500	32

Reuse the light.

Table 17.9 (Continued)

Typical Distribution and Maximum Spacing[a]	ρCC →	80			70			50			30			10			0	Typical Luminaires and Luminaire Maintenance Category[e]
	ρW →	50	30	10	50	30	10	50	30	10	50	30	10	50	30	10	0	
	RCR[b] ↓	Coefficients of Utilization for 20 Per Cent Effective Floor Cavity Reflectance, ρFC																

25 — 10% / 75% — Max. S/MH$_{wp}$ = 1.3

RCR	80/50	80/30	80/10	70/50	70/30	70/10	50/50	50/30	50/10	30/50	30/30	30/10	10/50	10/30	10/10	0
1	.87	.84	.81	.84	.81	.78	.79	.76	.74	.74	.72	.70	.69	.68	.66	.64
2	.76	.71	.66	.74	.69	.65	.69	.65	.62	.65	.62	.59	.61	.58	.56	.54
3	.67	.61	.56	.65	.59	.55	.61	.56	.52	.58	.54	.50	.54	.51	.48	.46
4	.60	.52	.47	.58	.51	.46	.55	.49	.44	.51	.47	.43	.48	.45	.41	.39
5	.52	.45	.39	.51	.44	.39	.48	.42	.38	.45	.40	.36	.43	.38	.35	.33
6	.47	.39	.34	.45	.38	.33	.43	.37	.32	.40	.35	.31	.38	.34	.30	.28
7	.42	.34	.29	.40	.33	.29	.38	.32	.28	.36	.31	.27	.34	.30	.26	.24
8	.37	.30	.25	.36	.29	.25	.34	.28	.24	.32	.27	.23	.31	.26	.22	.21
9	.33	.26	.21	.32	.26	.21	.31	.25	.20	.29	.24	.20	.28	.23	.19	.18
10	.30	.23	.19	.29	.23	.18	.28	.22	.18	.26	.21	.17	.25	.20	.17	.15

2-lamp porcelain-enameled industrial with 13° crosswise shielding
LDD Maint. Category III

26 — 20% / 65% — Max. S/MH$_{wp}$ = 1.3

RCR	80/50	80/30	80/10	70/50	70/30	70/10	50/50	50/30	50/10	30/50	30/30	30/10	10/50	10/30	10/10	0
1	.86	.83	.80	.82	.79	.77	.75	.72	.70	.68	.66	.65	.62	.61	.60	.57
2	.76	.71	.67	.73	.68	.65	.67	.63	.60	.61	.58	.56	.56	.54	.52	.49
3	.68	.62	.57	.65	.60	.56	.60	.56	.52	.55	.52	.49	.50	.48	.46	.43
4	.61	.54	.49	.58	.52	.48	.54	.49	.45	.49	.46	.42	.46	.42	.40	.38
5	.54	.47	.42	.52	.46	.41	.48	.43	.39	.44	.40	.37	.41	.37	.35	.33
6	.49	.42	.37	.47	.40	.36	.43	.38	.34	.40	.36	.32	.37	.33	.30	.29
7	.44	.37	.32	.42	.36	.31	.39	.34	.30	.36	.32	.28	.34	.30	.27	.25
8	.39	.32	.28	.38	.31	.27	.35	.30	.26	.32	.28	.25	.30	.26	.23	.22
9	.35	.29	.24	.34	.28	.23	.31	.26	.22	.29	.25	.21	.27	.23	.20	.18
10	.32	.26	.21	.31	.25	.21	.29	.23	.20	.27	.22	.19	.25	.21	.18	.16

2-lamp porcelain-enameled industrial with 35° crosswise shielding
LDD Maint. Category II

27 — 25% / 55% — Max. S/MH$_{wp}$ = 1.3

RCR	80/50	80/30	80/10	70/50	70/30	70/10	50/50	50/30	50/10	30/50	30/30	30/10	10/50	10/30	10/10	0
1	.75	.72	.71	.71	.69	.68	.65	.64	.62	.59	.58	.57	.54	.53	.52	.50
2	.67	.64	.61	.64	.61	.59	.59	.56	.54	.54	.52	.50	.49	.47	.46	.44
3	.61	.56	.53	.58	.54	.51	.53	.50	.48	.49	.46	.44	.44	.42	.41	.39
4	.55	.50	.46	.53	.48	.45	.48	.45	.42	.44	.41	.39	.40	.38	.36	.34
5	.50	.44	.40	.48	.43	.39	.44	.40	.37	.40	.37	.34	.37	.34	.32	.30
6	.45	.39	.36	.43	.38	.35	.40	.36	.32	.36	.33	.30	.33	.30	.28	.27
7	.41	.35	.31	.39	.34	.30	.36	.32	.29	.33	.29	.27	.30	.27	.25	.24
8	.37	.31	.27	.36	.30	.27	.33	.28	.25	.30	.26	.24	.27	.24	.22	.21
9	.33	.28	.24	.32	.27	.23	.30	.25	.22	.27	.23	.21	.25	.22	.19	.18
10	.30	.25	.21	.29	.24	.21	.27	.23	.20	.25	.21	.18	.23	.20	.17	.16

2-lamp porcelain-enameled industrial with 35° crosswise and lengthwise shielding
LDD Maint. Category II

28 — 15% / 65% — Max. S/MH$_{wp}$ = 1.5

RCR	80/50	80/30	80/10	70/50	70/30	70/10	50/50	50/30	50/10	30/50	30/30	30/10	10/50	10/30	10/10	0
1	.84	.83	.79	.80	.78	.76	.74	.73	.71	.69	.68	.66	.64	.63	.62	.60
2	.75	.71	.68	.72	.69	.66	.67	.65	.62	.63	.60	.59	.58	.57	.55	.53
3	.68	.62	.58	.65	.61	.57	.61	.57	.54	.57	.54	.51	.53	.51	.49	.47
4	.61	.55	.51	.59	.54	.50	.55	.51	.47	.51	.48	.45	.48	.45	.43	.41
5	.55	.48	.44	.53	.47	.43	.50	.45	.41	.46	.43	.40	.44	.40	.38	.36
6	.49	.43	.38	.47	.42	.38	.45	.40	.36	.42	.38	.35	.39	.36	.33	.32
7	.44	.38	.34	.43	.37	.33	.40	.35	.32	.38	.34	.30	.36	.32	.29	.28
8	.39	.33	.29	.38	.32	.28	.36	.31	.27	.34	.30	.26	.32	.28	.25	.24
9	.35	.30	.25	.34	.29	.25	.32	.27	.24	.30	.26	.23	.29	.25	.22	.21
10	.32	.26	.22	.31	.25	.22	.29	.24	.21	.28	.23	.20	.26	.22	.19	.18

2-lamp aluminum industrial with 35° crosswise shielding
LDD Maint. Category II

[a] Ratio of maximum spacing between luminaire centers to mounting (or ceiling) height above the work plane, S
[b] RCR = Room Cavity Ratio
[c] PCC = Percent effective ceiling cavity reflectance
[d] PW = Percent wall reflectance

The same luminaire and reflectance but a room cavity ratio of 4 instead of 1 shows the combined effect of a distant light with reuse of light.

Ceiling Reflectance	Lumens/ Luminaire	CU	Lumens on Task	Percent Reuse of Light
.10	22,500	.50 = 11,250/10,350		7
.30	22,500	.42 = 11,700/10,350		13
.50	22,500	.54 = 12,150/10,350		17
.70	22,500	.57 = 12,825/10,350		24
.80	22,500	.58 = 13,050/10,350		26

Low cavity ratios (slabs) are better.

Table 17.10   Factors For Effective Floor Cavity Reflectances Other Than 20 Percent. For 30 Percent Effective Floor Cavity Reflectance, Multiply by Appropriate Factor Below. For 10 Percent Effective Floor Cavity Reflectance, Divide by Appropriate Factor Below.

Percent Effective Ceiling Cavity Reflectance, ρCC	80			70			50			10		
Percent Wall Reflectance, ρW	50	30	10	50	30	10	50	30	10	50	30	10
Room Cavity Ratio												
1	1.08	1.08	1.07	1.07	1.06	1.06	1.05	1.04	1.04	1.01	1.01	1.01
2	1.07	1.06	1.05	1.06	1.05	1.04	1.04	1.03	1.03	1.01	1.01	1.01
3	1.05	1.04	1.03	1.05	1.04	1.03	1.03	1.03	1.02	1.01	1.01	1.01
4	1.05	1.03	1.02	1.04	1.03	1.02	1.03	1.02	1.02	1.01	1.01	1.00
5	1.04	1.03	1.02	1.03	1.02	1.02	1.02	1.02	1.01	1.01	1.01	1.00
6	1.03	1.02	1.01	1.03	1.02	1.01	1.02	1.01	1.01	1.01	1.01	1.00
7	1.03	1.02	1.01	1.03	1.02	1.01	1.02	1.01	1.01	1.01	1.01	1.00
8	1.03	1.02	1.01	1.02	1.02	1.01	1.02	1.01	1.01	1.01	1.01	1.00
9	1.02	1.01	1.01	1.02	1.01	1.01	1.02	1.01	1.01	1.01	1.01	1.00
10	1.02	1.01	1.01	1.02	1.01	1.01	1.02	1.01	1.01	1.01	1.01	1.00

As the room cavity ratio increases, direct lumens on the task get worse and the percent reuse of light generally is lower.

The effect of a different luminaire (a narrow beam luminaire [23] instead of the wide beam luminaire [24]) for the wall reflectance of .3 and room cavity ratio of 4 of the preceding table is:

Ceiling Reflectance	Lumens/ Luminaire	CU	Lumens on Task	Percent Reuse of Light
.10	22,500	.62 = 13,950/13,275		5
.30	22,500	.64 = 14,400/13,275		8
.50	22,500	.65 = 14,625/13,275		10
.70	22,500	.66 = 14,850/13,275		12
.80	22,500	.66 = 14,850/13,275		12

Some luminaires are more efficient.

Direct lighting (luminaires 23 and 25) puts more lumens on the task than direct-indirect lighting (luminaires 24 and 27). Totally indirect lighting has CU values as low as .24 for a ceiling reflectance of .5, wall reflectance of .3, and room cavity ratio of 1. Open top luminaires are kept cleaner by the flow of air; in addition the light on the ceiling gives better brightness contrast ratios.

Quality of Illumination

The distribution of brightness (color of light, its direction, diffusion, degree of glare) gives lighting quality.

Contrast. Blackwell pointed out the trade-off between quality and quantity with the "standard performance curves" used for the Illuminating Engineering Society recommendations (see Figure 17.12).[2] To move from one standard performance curve to another, change contrast, illumination, or both. Keeping illumination constant at 700 lux but improving contrast from .60 to .68 gives the same

Table 17.11 Lamp Lumen Depreciation Factors

GE LAMP TYPE		Watts	Mean Lumen Factor *(%)	LLD**(%)
Incandescent	75ER30	75	90	86
	100A	100	93	90
	150A	150	93	90
	150PAR/SP & FL	150	84	78
	150R/SP & FL	150	89	85
	300M/IF	300	91	87
	300R/SP & FL	300	94	92
	500/IF	500	91	88
	1000/IF	1000	92	89
	1500/IF	1500	84	78
Quartzline® Incandescent				
	Q250CL	250	97	96
	Q250PAR38/SP & FL	250	93	93
	Q500T3/CL	500	97	96
Fluorescent	F40CW Watt-Miser™	35	87	83
	F40LW Watt-Miser™	35	87	83
	F40CW/S Staybright™	40	90	87
	F40CW Mainlighter™	40	87	83
	F40WW Mainlighter™	40	87	83
	F40CWX Mainlighter™	40	83	73
	F40WWX (SW) Mainlighter™	40	83	73
	F96T12/CW Watt-Miser™	60	93	89
	F96T12/LW Watt-Miser™	60	93	89
	F96T12/CW Watt-Miser™	75	93	89
	F96T12/WW Watt-Miser™	75	93	89
	F96T12/CWX Watt-Miser™	75	89	85
	F48T12/CW/HO	110	87	82
	F96T12/CW/HO	110	87	82
	F96T12/CW/HO Watt-Miser™	95	87	82
	F48PG17/CW	110	72	64
	F96PG17/CW	215	76	67
	F96PG17/CW Watt-Miser™	185	76	67
	F48T12/CW/1500	110	72	64
	F96T12/CW/1500	215	76	67
High Intensity Discharge				
Mercury+	H175DX39-22 (H39KC-175/DX)	175	84	78
	H250DX37-5 (H37KC-250/DX)	250	81	75
	H400DX33-1 (H33GL-400/DX)	400	78	71
	H1000DX36-15 (H36GW-1000/DX)	1000	63	52
Multi-Vapor® Metal Halide				
	MV175	175	77	—
	MV175/C	175	73	—
	MV250	250	83	76
	MV250/C	250	78	71
	MV400	400	75	67
	MV400/C	400	72	.63
	MV1000	1000	80	.75
	MV1000/C	1000	79	.70
Lucalox® High Pressure Sodium				
	All standard lamp wattages (*LU* Types: 70, 100, 150, 200, 250, 310, 400 and 1000)	—	90	82
	LUH150	150	90	83
	LUH215	215	90	83

*Use the Mean Lumen Factor when calculations involve *average* illumination between relampings. (Based upon fraction of light output at 40%-50% of rated life depending upon lamp type.)

**Use the *LLD* (Lamp Lumen Depreciation) Factor when calculations involve *minimum* illumination between relampings. (Based upon the fraction of light output at 70% rated life.)

+Factors for mercury lamps are based upon 24,000 hours life.

Note: Factors shown for fluorescent lamps are based upon 3 hr/start. Factors for Multi-Vapor and Lucalox lamps are based upon 10 hr/start. Factors for mercury and Multi-Vapor lamps are for vertical lamp operation.

Table 17.12   Five Degrees of Dirt Conditions for Luminaire
Dirt Depreciation (Figure 17.11)

	Very Clean	Clean	Medium	Dirty	Very Dirty
Generated Dirt	None	Very little	Noticeable but not heavy	Accumulates rapidly	Constant accumulation
Ambient Dirt	None (or none enters area)	Some (almost none enters)	Some enters area	Large amount enters area	Almost none excluded
Removal or Filtration	Excellent	Better than average	Poorer than average	Only fans or blowers if any	None
Adhesion	None	Slight	Enough to be visible after some months	High—probably due to oil, humidity, or static	High
Examples	High grade offices, not near production; laboratories; clean rooms	Offices in older buildings or near production; light assembly; inspection	Mill offices; paper processing; light machining	Heat treating; high speed printing; rubber processing	Similar to Dirty but luminaires within immediate area of contamination

visual performance as keeping contrast constant at .60 but increasing illumination to 1900 lux.

To detect *shape*, maximize the contrast of the task vs. its background. If inspecting buttons to see whether they have holes or not, paint the table a color contrasting with the button color. For printed material, letters and background should have maximum contrast; white on black is slightly higher contrast than black on white. Low contrast letters (black on red, red on brown cardboard) not only have less legibility but are judged less attractive.[22] The object orientation is critical. To demonstrate this, try to recognize a print of a face turned upside down.[31]

To detect *surface characteristics* (color, texture), minimize the contrast of the task vs. background. Pearls sold at retail are displayed on black velvet; the maximum contrast makes it difficult to detect color differences between pearls. When diamond merchants buy from each other they display the diamonds on white paper to maximize color differences. Thus, if sorting green beans for color, sort on a table painted the green of a desirable bean. When inspecting the color, specify lamp color.

The light color also is important. Color differences in red material are emphasized by sources strong in blue light and in blue material by sources strong in red.[30] When inspecting for a color, specify whether the lamp used is incandescent, cool-white, high-pressure sodium, etc. as the colors are quite different.[10] In inspection tasks, the area of the visual field depends on the perceptual characteristics of the task—large area for "obvious" tasks, small for "difficult."[31]

If the eyes must view bright areas and dim areas alternately, the resulting adjustments are tiring. Tables 17.13 gives the recommended maximum brightness ratios for surrounding work surfaces, adjacent machines, and floors, walls, and windows. Office desk tops and typewriters should be relatively light in color to minimize the contrast with white paper. Paint the stationary and moving parts of machines contrasting colors; the background should be darker than the task. Use high-reflectance matte (not gloss) surfaces; avoid polished metal or glossy surfaces. Note that walls and machinery can be many colors besides the too common "battleship grey" and "industrial green." Light colors on the floors,

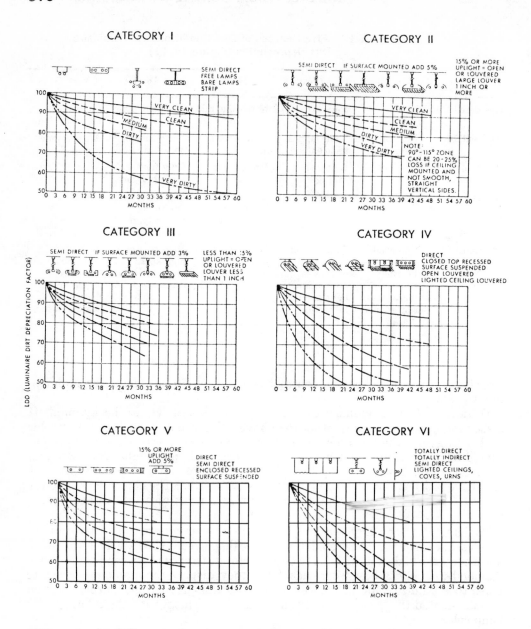

Figure 17.11 *Obtain Luminaire Dirt Depreciation (LDD)* by (1) selecting the proper category of lamp (1 to 6); (2) enter the figure at the number of months between cleaning; (3) go up until you reach the appropriate "degree of dirt" line; and (4) read *LDD* on the left hand scale.

walls, and ceilings not only reuse light but improve quality of seeing by reducing brightness contrast. Use light colors about windows (such as sashes and glass dividers) to reduce contrast.

Glare. Glare is any brightness within the field of vision which causes discomfort, annoyance, interference with vision, or eye fatigue. Sensitivity to glare increases with age; brown eyes are less sensitive than blue eyes.

Glare is subdivided into direct and indirect glare.

Direct glare is caused by a light source within the field of view. Sliney and Wolbarsht describe in 1000 pages the safety problems of lasers, bright lights,

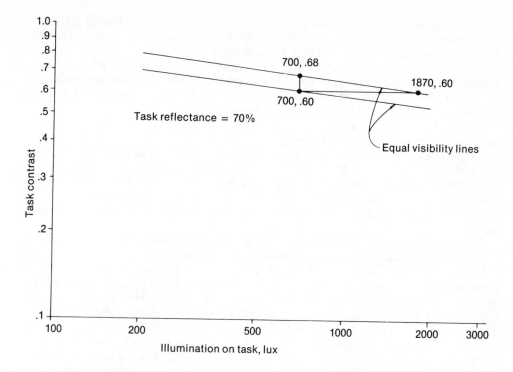

Figure 17.12 *Contrast substitution* for illumination is illustrated by showing the same visual performance either by going vertically from a contrast of .60 to .68 or by going horizontally from an illumination of 700 lux to 1900.[2]

Table 17.13 Recommended Maximum Brightness Ratios[20]

Situation	Maximum Ratio
Task and adjacent darker surroundings	3 to 1
Task and adjacent lighter surroundings	1 to 3
Task and more remote darker surfaces	10 to 1
Task and more remote lighter surfaces	1 to 10
Luminaires, windows, and skylights vs. adjacent surfaces	20 to 1
Anywhere with normal field of view	40 to 1

arcs, and hot metal.[34] Windows are a common problem. Workers should not face the window; they should have their backs and sides toward it. Change the window's transmitting material to translucent rather than transparent; use spectrally selective films. Structural overhangs can reduce upper sky glare without seriously interfering with penetration of light. Louvers, shades, curtains, and even shade trees or vines on louvered overhangs can help. Artificial sources (especially point sources) must be mounted above the normal field of view. Shield to at least 25° from the horizontal; 45° is better. *Industrial Lighting* had a calculation showing that a 21° shielded mercury luminaire had only 24% of users comfortable vs. 90% for a 33° shielded mercury luminaire.[20] In addition, the veiling glare of the 21° unit reduced contrast by 30% vs. 1.5% reduction for the 33° unit; the standard performance curve for a 30% reflectance visual task showed that the 21° unit required 28% more light for equal performance.

Filmed documents on a microfilm reader often cover only part of the screen, giving excessive contrast between the film and the screen. A mask, parabolic mirror louver reflectors for the ceiling illumination, and biofocal glasses for a 40 cm reading distance were suggested improvements.[18]

Reflected (indirect) glare ("veiling reflection") is caused by high brightness reflected from a surface. Reflected glare, horizontally polarized (vibrating left and right) light, is a special problem when it is in the line of sight for the task since the eye cannot avoid it. Lion reported more accurate manipulation and inspection under fluorescent tubes than incandescent bulbs and attributed the improvement to reduced glare from the line source rather than the point source.[25,26] Matsushita puts curtains on the inside of inspection booths to reduce glare.

Blackwell recommended filtering the light at the luminaire with a multilayer polarizer to minimize the reflected horizontal component.[2]

More common is to put the filter at the eye—sunglasses. Sunglasses reduce luminance but not contrast so they should be expected to reduce visual acuity. Mehan and Bennett reported that sunglasses do reduce visual acuity with one exception—polarized sunglasses improved visual acuity for reflected glare.[29] Their second criterion was comfort. Visual comfort is a function of (source luminance)/(background luminance)^k where k is a fraction. All the sunglasses improved comfort. In most situations visual acuity is not critical as capability far exceeds requirements of the task; in these circumstances, people trade off a reduction of surplus visual capacity for an increase in comfort.

Photochromic glasses darken when exposed to the sun or ultraviolet light (transmitting about 40% of the visible spectrum rays and 90% of the ultraviolet and infrared rays). They are satisfactory for outdoor work; don't use for welding or furnace operations as they transmit too much ultraviolet and infrared light and may give a false sense of security.

Orientation. The mind is oriented to believe light comes from above; light from below is used to make a person look "evil." Placement also causes physiological problems (such as glare). But lighting orientation (geometry) also can be beneficial—especially in inspection. Orienting lights to sharpen or blur the form and surface texture of an object is called modeling. For inspection, consider direct vs. indirect lighting. Inspect bottles of liquid with light through the bottom. Back lighting also may be useful for transparent materials; edge lighting may help for printed circuits.

Figure 17.13 gives five different luminaire locations. Table 17.14 gives five different types of luminaires. Table 17.15 gives, for different tasks, the recommended combination of luminaire and location.

Video Display Terminals (VDTs). The widespread introduction of electronic displays into the office (especially on word processors) has presented special lighting problems. Because VDTs are used relatively continuously, even small design errors cause considerable employee discomfort.

The most important step (although often forgotten) is to select a display with good readability characteristics. The character should subtend 15 or more min of arc for good viewing conditions and 20 for poor viewing conditions.[41] For small angles the radian equals the sine and the tangent and thus is a ratio of character height to viewing distance; 1 min of arc = .000 19 radian. Thus, for a 20 inch (500 mm) viewing distance, minimum character height should be 15 (.000 29) = CH/20; CH = .09 in (2.3 mm). The direction of contrast (light on dark or dark on light) is insignificant in most cases but many people strongly prefer dark on light. The display (which should be in a separate housing from the keyboard) should have a screen with an adjustable angle and should be deeply recessed (to minimize glare from ambient lights). Specular reflection can be reduced by etched glass screens or film coatings on the screen. The VDT should have brightness and contrast controls accessible to the operator. See chapter 14 for comments on workstation dimensions for VDTs.

Once the display is purchased the alternatives are to: (a) adjust the person or (b) adjust the environment.

As mentioned previously, defective vision is common—even with people wearing glasses—so visually demanding continuous work will bring out the defects. One solution is to get a proper prescription. Work glasses, ground to the focal length used on the job (say 20 in), are useful; bifocal glasses should be replaced with single-vision glasses to avoid holding the head in unnatural positions.

Adjust the environment by orienting the operator so windows are to the side rather than front or rear; see Figure 17.14. A U-shaped piece of sheet metal, see Figure 17.15, can reduce glare on a screen not properly recessed. Operators should wear dark shirts to reduce glare; walls should be dark for the same reason. A good approach is to reduce ceiling illumination (to 300-500 lux) and use adjustable (intensity, orientation) local lighting for the source documents.

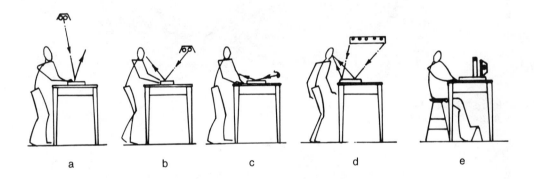

a   b   c   d   e

Figure 17.13 *Placement of supplementary luminaires* can use five techniques: (a) luminaire located to prevent reflected glare—reflected light does not coincide with angle of view; (b) reflected light coincides with angle of view; (c) low-angle lighting to emphasize surface irregularities; (d) large-area surface source and pattern are reflected toward the eye; and (e) transillumination from diffuse sources.

SUMMARY

The human eye is complex, so the engineer should be aware of its characteristics to get maximum use from it. Examples of things to remember include: older workers need more light, older workers are more sensitive to glare, and dynamic visual acuity is not necessarily related to static visual acuity.

Reduce visual problems by adding more light or by improving the quality of the light.

Although light is relatively cheap, it should not be wasted. Use efficient sources efficiently by using the zonal cavity calculations for area lighting. Supplement area lighting with local lighting if necessary.

To improve light quality, improve contrast, decrease glare, and use the proper orientation of the light.

Figure 17.14 *Windows* should be on the side to minimize glare. Note the opaque curtain.

SHORT-ANSWER REVIEW QUESTIONS

17.1 Using age as the x axis, sketch visual acuity (actual correction), near point, and glare borderline between comfort and discomfort.

17.2 How much illumination (lux) should there be for auto frame assembly, for auto chassis assembly, and for auto final inspection?

17.3 Why do most firms use 500 to 1000 lux over a general area rather than 200 to 400 lux in the area plus local sources to add light to the task?

17.4 Approximately what percent of the cost of illumination is for power?

17.5 Discuss the disadvantages of windows.

THOUGHT-DISCUSSION QUESTIONS

1. Discuss, for a specific work area, the cost/hr of illumination vs. the cost of labor.

2. Buildings should not have windows. Discuss.

REFERENCES

1. Bennett, C. Discomfort Glare: Demographic Variables, II, Special Report 116. Manhattan, Kans.: Kansas State University Engineering Exp. Station, Summer 1974.

Figure 17.15 *Reduce ambient glare* with a U-shaped piece of sheet metal painted flat black. The operator can adjust it during the day if light (from a window) changes.

Table 17.14   Luminaires for Supplementary Lighting[20]

Type	Comments
1	*Directional.* Includes all concentrating units. Examples are reflector spot lamp or units employing concentrating reflector or lenses. Also included are concentrating longitudinal units such as well-shielded fluorescent lamps in a concentrating reflector.
2	*Spread, high-brightness.* Includes high-brightness, small-area sources such as incandescent or mercury. A deep-bowl diffusing reflector with an incandescent filament lamp and without a diffusing cover is an example.
3	*Spread, moderate-brightness.* Includes all fluorescent units having a variation in brightness of more than 2 to 1.
4	*Uniform brightness.* Includes all units having less than 2 to 1 variation in brightness. Brightness usually is less than 2000 footLamberts. An arrangement of lamps behind a diffusing panel is an example.
5	*Uniform brightness with pattern.* Similar to type 4 except that a pattern of strips of lines is superimposed.

2. Blackwell, H. Visual benefits of polarized light. *American Institute of Architects Journal* (November 1963): 87-92.

3. Boyce, P. The influence of illumination level on prolonged work performance. *Lighting Research and Technology*, Vol. 2, No. 2 (1970): 74-95.

Table 17.15 Classification of Visual Tasks and Lighting Techniques

Part I—Flat Surfaces

| Classification of Visual Task | Example | | Lighting Technique | |
General Characteristics	Description	Lighting Requirements	Luminaire Type	Locate Luminaire
A. Opaque Materials				
1. Diffuse detail & background				
a. Unbroken surface	Newspaper proofreading	High visibility with comfort	S-III or S-II	To prevent direct glare & shadows (Figure 17.13a)
b. Broken surface	Scratch on unglazed tile	To emphasize surface break	S-I	To direct light obliquely to surface (Figure 17.13c)
2. Specular detail & background				
a. Unbroken surface	Dent, warps, uneven surface	Emphasize unevenness	S-V	So that image of source & pattern is reflected to eye (Figure 17.13d)
b. Broken surface	Scratch, scribe, engraving, punch marks	Create contrast to cut against specular surface	S-III or	So detail appears bright against a dark background
			S-IV or S-V when not practical to orient task	So that image of source is reflected to eye & break appears dark (Figure 17.13d)
c. Specular coating over specular background	Inspection of finish plating over underplating	To show up uncovered spots	S-IV with color of source selected to create maximum color contrast between two coatings	For reflection of source image toward the eye (Fig. 17.13d)
3. Combined specular & diffuse surfaces				
a. Specular detail on diffuse, light background	Shiny ink or pencil marks on dull paper	To produce maximum contrast without reflected glare from shiny markings	S-III or S-IV	So direction of reflected light does not coincide with the angle of view (Figure 17.13a)

b. Specular detail on diffuse, dark background	Punch or scribe marks on dull metal	To create bright reflection from detail	S-II or S-III	So direction of reflected light from detail coincides with angle of view (Figure 17.13b)
c. Diffuse detail on specular, light background	Gradations on a steel scale	To create a uniform, low-brightness reflection from specular background	S-IV or S-III	So reflected image of source coincides with angle of view (Figure 17.13b or Figure 17.13d)
d. Diffuse detail on specular, dark background	Wax marks on auto body	To produce high brightness of detail against dark background	S-III or S-II	So direction of reflected light does not coincide with angle of view (Figure 17.13a)
B. Translucent Materials				
1. With diffuse surface	Frosted or etched glass or plastic, lightweight fabrics, hosiery	Maximum visibility of surface detail	Treat as opaque, diffuse surface—See A-1	
		Maximum visibility of detail within material	Transilluminate behind material with S-II, S-III or S-IV (Figure 17.13e)	
2. With specular surface	Scratch on opal glass or plastic	Maximum visibility of surface detail	Treat as opaque, specular surface—See A-2	
		Maximum visibility of detail with material	Transilluminate behind material with S-II, S-III or S-IV (Figure 17.13e)	
C. Transparent Materials				
Clear material with specular surface	Plate glass	To produce visibility of details within material such as bubbles & details on surface such as scratches	S-V and S-I	Transparent material should move in front of Type S-V, then in front of black background with Type S-I directed obliquely. Type S-I should be directed to prevent reflected glare

D. Transparent over Opaque Materials

1. Transparent material over diffuse background	Instrument panel	Maximum visibility of scale & pointer without reflected glare	S-I	So reflection of source does not coincide with angle of view (Figure 17.13a).
	Varnished desk top	Maximum visibility of detail on or in transparent coating or on diffuse background		
		Emphasis of uneven surface	S-V	So that image of source & pattern is reflected to the eye (Figure 17.13d)
2. Transparent material over a specular background	Glass mirror	Maximum visibility of detail on or in transparent material	S-I	So reflection of source does not coincide with angle of view. Mirror should reflect a black background (Figure 17.13a)
		Maximum visibility of detail on specular background	S-V	So that image of source & pattern is reflected to the eye (Figure 17.13d)

Part II—Three-Dimensional Objects

A. Opaque Materials

1. Diffuse detail & background	Dirt on a casting or blow holes in a casting	To emphasize detail with a poor contrast	S-III or S-II or	To prevent direct glare & shadow (Figure 17.13a)
			S-I or	In relation to task to emphasize detail by means of highlight & shadow (Figure 17.13b or 17.13c)
			S-III or S-II as a black light source when object has a fluorescent coating	To direct ultraviolet radiation to all points to be checked

2. Specular detail & background				
a. Detail on the surface	Dent on silverware	To emphasize surface unevenness	S-V	To reflect image of source to eye (Figure 17.13d)
	Inspection of finish plating over underplating	To show up areas not properly plated	S-IV plus proper color	To reflect image of source to eye (Figure 17.13d)
b. Detail in the surface	Scratch on a watch case	To emphasize surface break	S-IV	To reflect image of source to eye (Figure 17.13d)
3. Combination specular & diffuse				
a. Specular detail on diffuse background	Scribe mark on casting	To make line glitter against dull background	S-III or S-II	In relation to task for best visibility. Adjustable equipment often helpful. Overhead to reflect image of source to eye (Figure 17.13b or 17.13d)
b. Diffuse detail on specular background	Micrometer scale	To create luminous background against which scale markings can be seen in high contrast	S-IV or S-III	With axis normal to axis of micrometer
	Coal picking	To make coal glitter in contrast to dull impurities	S-I, S-II	To prevent direct glare (Figure 17.13b)
B. Translucent Materials				
1. Diffuse surface	Lamp shade	To show imperfections in material	S-II	Behind or within for transillumination (Figure 17.13e)
2. Specular surface	Glass enclosing globe	To emphasize surface irregularities	S-V	Overhead to reflect image of source to eye (Figure 17.13d)

C. Transparent Materials

Clear material with specular surface	Bottles, glassware—empty or filled with clear liquid	To check homogeneity	S-II	Behind or within for transillumination
		To emphasize surface irregularities	S-I	To be direct obliquely to objects
		To emphasize cracks, chips and foreign particles	S-IV or S-V	Behind for transillumination. Motion of objects is helpful (Figure 17.13e)

4. Burg, A., and Hulbert, S. Dynamic visual acuity as related to age, sex and static acuity. *Journal of Applied Psychology,* Vol. 45 (1961): 111-116.

5. Cooper, B. Choose right to save on light. *Modern Manufacturing* (December 1970): 55-57.

6. Davis, J., and Galic, G. Welders and co-workers need eye protection. *Welding Design and Fabrication,* Vol. 54, No. 3 (March 1981): 108-109.

7. Dickerson, O. Visual criteria aid in job placement. *Occupational Safety and Health* (May/June 1976): 39-41.

8. Duncan, J., and Konz, S. Legibility of LED and liquid-crystal displays. *Proceedings of the Society for Information Display,* Vol. 17, No. 4 (1976): 180-186.

9. Ferguson, D.; Major, G.; and Keldoulis, T. Vision at work. *Applied Ergonomics,* Vol. 5, No. 2 (1974): 84-93.

10. Fortuin, G. Lighting: Physiological and psychological aspects-optimum use-specific industrial problems. *Ergonomics and Physical Factors.* Geneva: International Labour Office, 1970; pp. 237-259.

11. Freir, J., and Frier, M. *Industrial lighting systems,* New York: McGraw-Hill, 1980.

12. General Electric. *Interior Lighting Design Workbook,* TPC-42, Nela Park, Ohio, 1972.

13. General Electric. *Light and Color,* TP-119, Cleveland, Ohio, 1968.

14. Grandjean, E. *Fitting the Task to the Man.* London: Taylor and Francis, 1969; p. 94.

15. Grollman, S. *The Human Body,* 2d ed. New York: MacMillan Co., 1969.

16. Hess, E. The role of pupil size in communication. *Scientific American,* 233 (November 1975): 110-118.

17. Hopkinson, R. Glare from daylighting in buildings. *Applied Ergonomics,* Vol. 3, No. 4 (1972): 206-215.

18. Hultgren, G.; Knave, B.; and Werner, M. Eye discomfort when reading microfilm in different enlargers. *Applied Ergonomics,* Vol. 5, No. 4 (1974): 194-200.

19. Illuminating Engineering Society. *IES Lighting Handbook,* 4th and 5th ed. New York City, 1966 and 1972.

20. Illuminating Engineering Society. *Industrial Lighting.* American National Standard Practice for Industrial Lighting, RP-7. New York, 1970.

21. Jacobson, H. A study of inspector accuracy. *Industrial Quality Control,* Vol. 9, No. 2 (1952): 16-25.

22. Konz, S.; Chawla, S.; Sathaye, S.; and Shah, P. Attractiveness and legibility of various colours when printed on cardboard. *Ergonomics,* Vol. 15, No. 2 (1972): 189-194.

23. Lakowski, R. Theory and practice of colour vision testing. *British Journal of Industrial Medicine,* Vol. 26 (1969): 173-189 and 265-288.

24. *Lighting Design and Application,* 5 (November 1976): 43.

25. Lion, J. The performance of manipulative and inspection tasks under tungsten and fluorescent lighting. *Ergonomics,* Vol. 17 (January 1964): 51-61.

26. Lion, J.; Richardson, E.; and Browns, R. A study of industrial inspectors under two kinds of lighting. *Ergonomics,* Vol. 11, No. 1 (1968): 23-34.

27. Ludvigh, E., and Miller, J. Study of visual acuity during the ocular pursuit of moving test objects: 1. Introduction. *Journal of the Optical Society of America,* Vol. 48, No. 11 (1959): 799-802.

28. McGowan, T. All about sources. *Progressive Architecture,* Vol. 54 (September 1973): 108-117.

29. Mehan, R., and Bennett, C. Sunglasses—performance and comfort. *Proceedings of the 17th Annual Meeting of the Human Factors Society* (1973): 174-177.

30. Misra, S., and Bennett, C. Lighting for a visual inspection task. *Proceedings of the Human Factors Society,* Rochester, N.Y., 1981; pp. 631-633.

31. Noro, K., and Okada, Y. Constructing an industrial inspection model to predict operator's performance. *Proceedings of the Human Factors Society,* Rochester, N.Y., 1981; pp. 617-621.

32. Parker, J., and West, V. *Bioastronautics Data Book.* Washington, D.C.: Superintendent of Documents, 1973.

33. Rushton, W. Visual pigments and color blindness. *Scientific American,* (March 1975): 64-74.

34. Sliney, D., and Wolbarsht, M. *Safety with Lasers and Other Optical Sources,* New York: Plenum Press, 1980.

35. Steck, B. European practice in the integration of lighting, air conditioning and acoustics in offices. *Lighting Research and Technology,* Vol. 1, No. 1 (1969): 8-23.

36. Sherratt, A. Internal services which influence the interior environment. *Lighting Research and Technology,* Vol. 2, No. 4 (1970): 232-245.

37. Ungar, P. Sight at work. *Work Study,* (March 1971): 46-48.

38. Van Cott, H., and Kinkade, R., *Human Engineering Guide to Equipment Design.* Washington, D.C.: Superintendent of Documents, 1972.

39. Weale, R. On the eye, Chapter 16 in *Behavior, Aging and the Nervous System.* Welford, A., and Birren, J., eds. Springfield, Ill. C.T. Thomas, 1965. See also Weale, R. *The Aging Eye.* London: H.K. Lewis, 1963.

40. Willoughby, A. The evolution of electric lamps. *Lighting Research and Technology,* Vol. 1, No. 2 (1969): 69-77.

41. Winkler, R., and Konz, S. Readability of electronic displays. *Proceedings of the Society for Information Display,* Vol. 21, No. 4 (1980): 309-312.

18

THE AUDITORY ENVIRONMENT

THE EAR

Design of the Ear

Figure 18.1 gives an overview of the ear, Figure 18.2 a detail of the cochlea, and Figure 18.3 a detail of the organ of Corti. Each ear, which fits into a 25 mm cube, has an outer, middle, and inner part.

The outer part serves as a collector of sound vibrations in the air and funnels them to the ear drum. The eardrum is extraordinarily sensitive as it will move in response to changes of as little as .000 02 N/m^2 and then will move only .000 000 001 cm! (.000 000 001 cm = ½ the dia. of a hydrogen molecule.)

In the middle part, the vibration of the ear drum (tympanic membrane) is transmitted through three small bones (ossicles) known as the hammer (malleus), anvil (incus), and stirrup (stapes) to the oval window.

The hammer "handle" is connected to the eardrum and the "head" to the "top" of the anvil. The "base" of the anvil is connected to the top of the stirrup. The baseplate of the stirrup moves the oval window leading to the inner ear. The original vibration in air has now been transferred to a vibration of a second membrane. The signal is magnified by having the eardrum area 14 times larger than the oval window area (gain 23 dB) and by a 1.3 lever arm ratio for the three bones (gain 2.5 dB) for a total gain of 25.5 dB. The eardrum will be most efficient if the air pressure is the same on both sides so the Eustachian tube comes in handy. One end is in the middle ear and the other end in the top of your mouth. If this becomes plugged and pressure is changing rapidly (jet descent or elevator descent), hold your nose, close your mouth, and "blow."

The inner ear is the most interesting of all. The vibrations of the oval window set up vibrations in the fluids of the cochlear duct. The duct upper passage (scala vestibuli) starts at the oval window; the lower passage (scala tympani) ends at the round window; connecting the two is a small gap (helicotrema). The round window bulges out when the oval window bulges in—thus the vibrations are not dampened. The basilar membrane divides the two passages. Now our noise has become a wave traveling in the fluid of the inner ear.

The basilar membrane vibrates with the frequency of the fluid vibration. The basilar membrane is narrow and stiff near the oval window and wide near the gap; high pitched sounds travel only a short distance along the membrane before they die out so the end near the oval window detects high frequency sounds; low pitched sounds are detected near the gap.

On top of the basilar membrane is the organ of Corti, which has about 30,000 hair cells. As the wave goes through the tympanic chamber it deflects the vestibular membrane down into the cochlear duct. This in turn deflects the basilar membrane down into the tympanic chamber. This pulls down the hair cells in the organ of Corti. The hair cells are supported at both ends, so this pulling makes the hair cells send out an electrical pulse which goes to the brain by way of the cochlear nerve. The brain then decides that the muffler in your car either has holes in it or has a pleasant throb. The brain also is able to detect the very slight difference in time that it takes for sound to travel to each ear and so identify whether the sound came from the left or right.

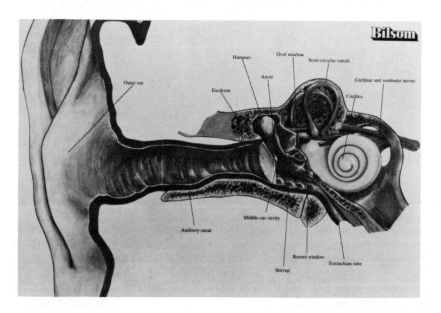

Figure 18.1 *Three parts* of the ear are the outer, middle, and inner. The outer transfers vibrations in air to vibrations of the eardrum, the middle amplifies and transmits the vibrations to the oval window, and the inner transmits vibrations in the basilar membrane into electrical pulses in the auditory nerve. Figure courtesty of Bilsom.

Conductive hearing loss occurs in the outer or middle ear from wax, punctured eardrums, "corrosion" of the bones, etc. Often it is possible to cure with medical or surgical treatment. Nerve loss in the inner ear is rarely curable. Nerve loss can be caused by old age, viruses, drugs, and noise.

Hearing Measurement

Hearing usually is measured on an automatic audiometer. Persons having their hearing tested go into a booth and put on a set of earphones. Then the audiometer automatically goes through a test cycle for the left ear and the right ear and plots the results as in Figure 18.4. During the test cycle the machine will present a tone at 250 Hz that increases in loudness until the person indicates he hears it by pushing a button. It then decreases until the person releases the button. After three or four tests for a specific frequency, it will index to the next higher frequency and repeat the tests until it has completed all seven frequencies (250, 500, 1000, 2000, 4000, 6000, and 8000 Hz); it then repeats the test for the other ear.

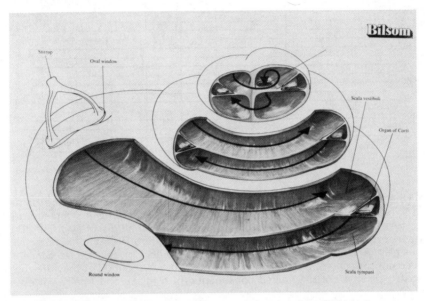

Figure 18.2 *Vibrations* on the oval window are transmitted by fluid to become vibrations in the basilar membrane which, in turn, affects the hair cells in the organ of Corti. Figure courtesy of Bilsom.

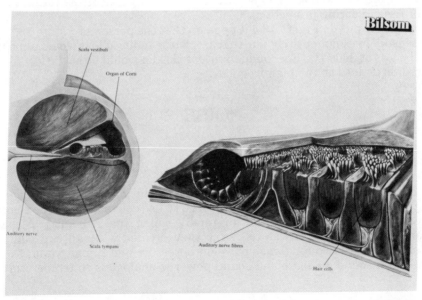

Figure 18.3 *Hair cells,* when vibrated, send electrical signals along the auditory nerve to the brain. Figure courtesy of Bilsom.

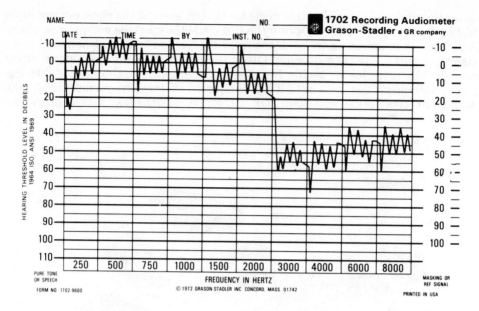

Figure 18.4 *Audiograms* show hearing ability. Plotted at each of 10 bands is the right ear of a 47-year-old male with 7700 hours of flying time. Zero dB indicates normal hearing. Hearing loss due to noise usually starts with a small dip of 20 to 30 dB at 4000 Hz before progressing to the severe loss of this pilot. Recreational noise causes problems also. A target shooter often has a 40 to 80 dB loss at 4000 Hz. Since nonvoiced consonants (p, t, k, s) have frequencies of 2000-8000 Hz, they are the first to be lost; next lost are the voiced consonants (b, d, g, z); and finally vowels.

Audiograms should be performed annually; if the audiogram shows a significant shift, the audiogram should be repeated within 30 days. If it still shows a shift, then a medical review is warranted. Each audiogram should be preceded by at least 14 hours of no noise over 80 dBA (ensure by testing before the start of the shift or use of hearing protection).[26]

From the output on Figure 18.4 it appears that the person of Figure 18.4 had a neurosensory loss, not entirely due to noise. Noise-induced loss would almost certainly be less at 8000 Hz. For conductive loss the dip would be at the low frequencies with little loss at the high frequencies.

NOISE

Noise Definitions

Absolute levels of sound pressure detectable by the human ear vary by 1,000,000,000,000 to 1. This scale is too large to use conveniently. The scale was compressed by using a log ratio (originally called the Bell after Alexander Graham Bell, but since it was too large they divided it by 10 and call it the deciBel). There are two relationships, sound pressure and sound power; the relation between sound pressure and sound power is analogous to temperature and heat.

$$SPL, \text{dB} = 10 \log_{10}\left(\frac{P}{P_0}\right)^2 = 20 \log_{10}\left(\frac{P}{P_0}\right) \qquad (18.1)$$

where	SPL	= sound pressure level, dB	0 dB was the minimum
	P	= sound pressure level of the noise, Newtons/m^2 (the 2 reflects the effect of area of the eardrum)	level of hearing when first measured. With more refined measurements the mean minimum of the population with unimpaired hearing is now given as 4 dB.

P_0 = reference sound pressure level
  = .000 020 N/m^2 = 0 dB = 20 μN/m^2
  = .000 200 dynes/cm^2 = .000 200 μbar
  (approximately the minimum level of hearing)

The power level of a noise is:

$$PWL, \text{dB} = 10 \log_{10}\left(\frac{W}{W_0}\right) \tag{18.2}$$

where PWL = power level of the noise, dB
  W = acoustic power of the noise, watts
  W = reference power level = 1×10^{-12} watt

Since 1×10^{-12} = −120 dB, equation 18.2 can be expressed as:

PWL = $10 \log W + 120$ (18.3)

That is, 120 dB of power corresponds to 1 W, 110 to .1W, 100 to .01W, etc. The decibel scale, although giving a small range, confuses the public since it is not linear; it also requires "new math." Most confusing is that 100 dB + 100 dB is not equal to 200 dB! When combining or subtracting noises with the formulas, we use the power formulas, not the pressure formula. Consider a machine generating noise with a $PWL = .01$ W (i.e., 100 dB) and with an SPL of 1 N/m^2 (94 dB) at a location 5 m from the machine. Now add another identical machine.

PWL = $10 \log .020 + 120 = 10 \log 2 \times .010 + 120$
  = $10 \log 2 \times 10^{-2} + 120 = 10 (-2 + .3) + 120$
  = $-17 + 120 = 103$ dB.

Thus adding 100 dB of power to 100 dB of power gives 103 dB; the corresponding $SPLs$ are 94 + 94 = 97.

In addition to 94 + 94 = 97, we also have 74 + 80 = 81, and 80 + 95 = 95! Ten identical sound sources have an SPL 10 dB louder than just one source! In other words, if we have two noisy machines, each generating at 80 dB, and we completely silence one machine, the noise level drops from 83 to "only" 80. Rather than use the formulas, use Figure 18.5, a quick graphical technique.

It will be emphasized that the use of Figure 18.5 is the theoretical addition in a free field. In practice most fields are not free. Outdoors, the wind can have a substantial effect, as well as reflections from reflectors such as brick walls. The noise might even be attenuated slightly by passing through intervening shrubbery. Indoors, there are many reflecting surfaces as well as solid paths to conduct the noise; occasionally some of the indirect noise is reduced by absorbent materials on the floor, wall, or ceiling.

Noise Measurement

Measuring the sound pressure level is not sufficient since the ear is more sensitive to sound at some frequencies than at others. That is, from a physiological

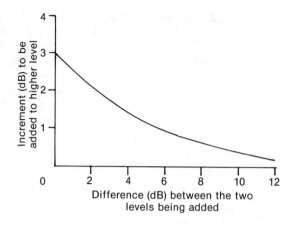

Figure 18.5 *Do addition of decibels* in a free field from the figure. Take the difference between levels, read increment from graph, and add increment to higher level. For example: 80 − 80 = 0; 80 + 3 = 83 and 86 − 80 = 6; 86 + 1 = 87.

judgment viewpoint, 80 dB at one frequency does not sound as loud as 80 dB at another frequency. The unit of loudness, the phon, equates loudness at other frequencies with the sound pressure level of a 1000 Hz tone; thus 60 dB at 1000 Hz = 60 phons and 68 dB at 100 Hz = 60 phons. See Figure 18.6. The curves show that the number of decibels to cause sounds that are equally loud varies considerably with the frequency of the sound. What makes the situation even more complicated is that the curves at various loudness levels (60, 80, 100, etc.) are not parallel.

The range of human hearing is approximately 20 to 20,000 Hz; see Figure 18.7. The ear is most sensitive from approximately 600 to 4800 Hz. The frequency range of telephones is about 200 to 3600 Hz. The range of frequencies on a piano is from 27 to 4186 Hz, "middle C" being 256 Hz. The reader can simulate this by singing "do." As you progress through the eight notes from a female deer to a drop of golden sun to a name I call myself and back to "do" you have completed an octave (doubling of frequency) to 512 Hz. Going through the notes again will demonstrate 1024 Hz (as well as a remarkable voice). When Figure 18.6 is used, the reader can see that the high pitched whine of a jet engine or saw is worse than the rumble of an engine due to the ear's greater sensitivity to high pitched noise.

As a rule of thumb, a sound will be "twice as loud" when noise increases 6 to 10 dB. Thus, elimination of one of two identical noise sources (a drop of 3 dB) will not cut "loudness" in half; subjectively, a change of 3 dB is "just noticeable"!

How then to record noise? First the instrument makers standardized on measuring noise by octave band. See Table 18.1.

With this type of instrument (an octave band analyzer), you could report to your boss that the noise level of a machine was 76 dB at 31.5 Hz, 77 dB at 63 Hz, 79 dB at 125 Hz, 83 dB at 250 Hz, 82 dB at 500 Hz, 86 dB at 1000 Hz, 85 dB at 2000 Hz, 82 dB at 4000 Hz, and 77 dB at 8000 Hz. This type of detailed analysis is quite useful when doing noise reduction work as it enables you to pinpoint exactly where the problems are.

For most work, however, you would be swamped in data. This led to the second level meter (see Figure 18.8). The sound level meter gives just one number for noise. What it does is combine—inside the meter—the various frequencies. Table 18.1 shows how the meter adjusts the actual noise for each octave band. For the A setting, 39 dB are subtracted from the octave band measurement for 31.5 Hz, 26 from the measurement for 63 Hz, etc. The adjusted band readings are then

Table 18.1   Octave Band Center Frequency (Geometric Mean), Lower Limit, Upper Limit, and Adjustment for dBA

Center Frequency, Hz	Lower Limit, Hz	Upper Limit, Hz	Phon dBA Adjustment, dB
31.5	22	44	−39
63	44	88	−26
125	88	177	−16
250	177	355	−9
500	355	710	−3
1000	710	1420	−0
2000	1420	2480	1
4000	2840	5860	1
8000	5680	11,360	−1

averaged and the result displayed on the dial. The A adjustment corresponds to the 40 phon equal loudness contour, the B to the 70 phon, and the C to 100 phon. On some meters there is a D scale which not only reduces the importance of the frequencies outside the speech range but also increases the importance within the critical 1,000 to 10,000 Hz range—it approximates the "perceived noise level" used to appraise aircraft noise. The A scale, however, has become the worldwide standard for reporting noise regardless of the intensity level; report values using the A scale as dBA.

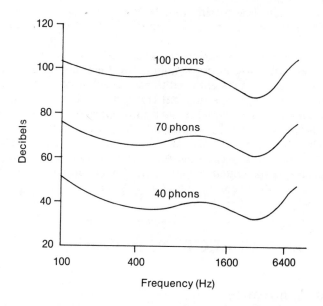

Figure 18.6 *Equal loudness curves* (free field) for pure tones are shown. The number of phons (loudness level of a sound) equates numerically with the sound pressure level of a 1000 Hz tone that sounds as loud as the sound being described. Note that not only does subjective loudness vary with frequency (curve is not horizontal) but also that the curve shapes vary with intensity (curves are not parallel). The 40 phon curve corresponds to dBA, the 70 phon to the B scale, and the 100 phon to the C scale.

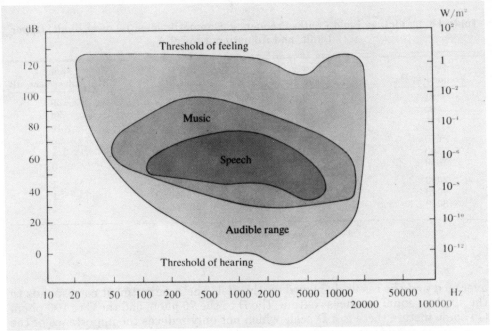

Figure 18.7 *Ear sensitivity* varies with frequency and volume. The threshold of feeling (pain) is relatively independent of frequency at about 120 dB. Very low frequencies (infrasound) may affect organs of the body other than the cochlea. The brain, for example, is especially sensitive to a frequency of 7 Hz, the frequency of the brain's alpha waves. Figure courtesy of Bilsom.

Noise is not always steady state. The speed of response of the meter can be set as "slow" or "fast." The slow setting usually used knocks the tops off spikes. It is possible to purchase special meters (impulse meters) to accurately measure peaks.

With the passage of noise legislation permitting 85 dBA for 16 hours exposure, 90 dBA for 8 hours exposure, 95 for 4 hours, 100 for 2 hours, etc., many people wonder how long an employee is exposed to each level of noise. One possibility is to make an occurrence sampling study but this tends to cost too much. Manufacturers have developed devices which record in proportion to the noise standard; that is, when noise is at 90 dB, it records at 100%, when at 95 at 200%, when at 100 at 400%, when below 85 at 0%, etc. Then, when the dosimeter (see Figure 18.9) is read at the end of any time period of exposure, it is known whether the person was exposed to too much noise. Some devices have a light that is activated if there is any exposure over the 115 dBA limit.

EFFECT OF NOISE

Comfort and Annoyance

Noise reduces comfort since the workers must increase their concentration; this tends to increase fatigue. Annoyance with noise has increased over the years as people expect technology can do anything and someone else will pay the cost. Noises which our parents experienced no longer seem tolerable. There also is some evidence that our environment is noisier than our parents' environment.

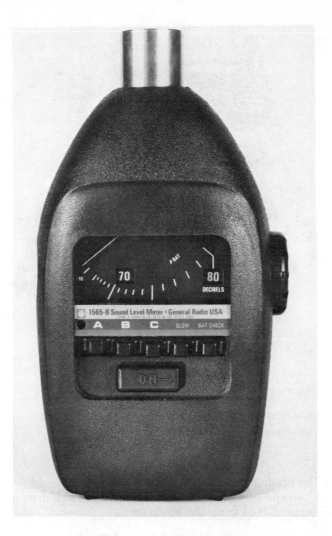

Figure 18.8 *Sound level meters* usually use the A scale and the slow setting. When recording sound, remember: (1) Don't shield the sound with your body; put the meter on an extension cable or at arm's length. (2) Microphones are directional. For a perpendicular-incidence microphone, sound waves not perpendicular to the microphone are attenuated at high frequencies; orient the microphone to get the maximum reading. For grazing-incidence microphones, sound waves not grazing the microphone can amplify or attenuate the sound, depending on the angle; orient the microphone so the dominant sound grazes the microphone.

Source: Photograph courtesy of GenRad, Inc., Concord, Massachusetts.

The point to remember is that noise reduction may be required even if the costs are high and the economic and health benefits are small or negative.

Epp and Konz made a study of annoyance and speech interference as a function of noise level for appliances[8]; Figure 18.10 shows the results. If you measure a noise level of 80 dBA for an appliance, then vote could be predicted graphically as 3.7 where 5 = extremely annoying, 4 = quite annoying, 3 = moderately annoying, 2 = slightly annoying, and 1 = not annoying. The vote could be predicted also from the equation

$$\text{Vote} = -4.798 + .1058 \text{ dBA}$$

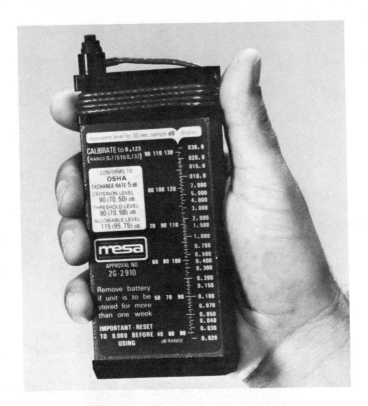

Figure 18.9 *Noise dosimeters* give exposure over a time period.

Source: Photograph courtesy of GenRad, Inc., Concord, Massachusetts.

in which case the predicted vote would be $-4.798 + .1058 (80) = 3.67$. Speech interference in percent of words missed could be predicted from the graph as 4.7% or could be predicted from the equation

$$\% \text{ missed} = 11.17 + .1989 \text{ dBA}$$

as $-11.17 + .1989 (80) = 4.74\%$. It should be noted that the subjects were college females and people with higher education levels make more complaints than those with less education. Expectations also might vary for offices, factories, stores, etc., rather than homes (See Table 18.2).

Community noise reaction to industrial noise is highly variable. The biggest factor well may be whether there is a local lawyer trying to get some free publicity! It also depends strongly on the history and background of the community—expect more complaints from those with "clout." Table 18.3 is based on the adjustments for Composite Noise Rating curves and may give the reader an idea of some of the variables. Variability of noise generally increases annoyance.[9]

The adjustments are independent so if your noise is in a heavy industrial area and only during the daytime you could (on the average!) expect the same community reaction as another factory with noise level 20 dBA lower but located in a very quiet suburb and making noise at night.

Performance

There is no firm evidence that productivity is lower when work is done in high level noise (say 100 dBA) unless the person is working at maximum mental capa-

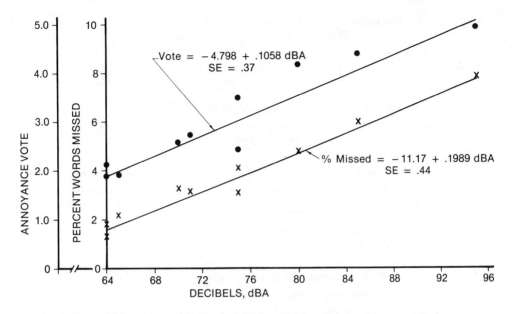

Figure 18.10 *Annoyance and speech interference* can be predicted from dBA.[8]

city. (For reference by the reader, conversation at 1 meter with a "normal" voice is 55 to 68 dBA; in a 90 dBA environment, you must shout to be understood at 1 meter.) The "no loss in productivity" has an important assumption—speech communication is not an important part of the job.

Speech interference. This is a rather complex subject. See Kryter or McCormick if you wish to dig deeper.[16,19]

The usual criterion is the percent of words missed. "Words" are quite specifically defined. Single-syllable phonetically balanced words (are, bad, bar, bask, box) or two-syllable words called spondees (airplane, armchair, backbone, bagpipe) are used.

The most refined index is noise criteria *(NC)* curves; articulation index *(AI)* is not quite as good although it uses 20 frequency bands; speech interference level *(SIL)* is the arithmetic mean of the dB readings in three octave bands centered at 500, 1000, and 2000 Hz. Figure 18.11 gives the *SIL* as a function of distance and voice level.[22] In general *SIL* is about 7 dB lower than dBA for most common noises.

Figure 18.12 gives speech interference at various levels of noise for earplugs and no earplugs.[15] Note that in high level noise (greater than 85 dB) intelligibility is improved with earplugs or earmuffs since the ear is not overloaded and it can better discriminate the signal from the noise. In low level noise, however, speech as well as noise is reduced to a level below the listener's threshold of hearing so intelligibility is reduced.

Reduction of speech interference. In addition to reducing the noise, there are four other stages at which speech transmission can be improved: the message, the speaker, the transmission system, and the listener.

If the possible vocabulary is limited, with only certain words and sequences of words being permitted, intelligibility improves. Table 18.4 gives the international aviation alphabet as interpreted in a French pilot training manual.[7]

Intelligible talkers have longer average syllable duration, speak louder, have fewer pauses, and vary pitch more often.[19]

Nontransmission of various frequencies reduces intelligibility. Filtering below

Table 18.2   Tolerable Limits (dBA) in Various Rooms for Noise Continuously Present from 7 a.m. to 10 p.m.

dBA	Type of Space
28	Broadcast studio, concert hall
33	Theaters for drama (500 seats, no amplification)
35	Music rooms, schoolrooms (no amplification). Very quiet office (telephone use satisfactory), executive offices, and conference rooms for 50 people
38	Apartments, hotels
40	Homes, motion picture theaters, hospitals, churches, courtrooms, libraries
43	"Quiet" office; satisfactory for conferences at a 5 m table; normal voice 3 to 10 m; telephone use satisfactory; private or semiprivate offices, reception rooms, and small conference rooms for 20 people
45	Drafting, meeting rooms (sound amplification)
47	Retail stores
48	Satisfactory for conferences at a 2 to 2.5 m table; telephone use satisfactory; normal voice 2 to 6 m. Medium-sized offices and industrial business offices
50	Secretarial offices (mostly typing)
55	Satisfactory for conferences at a 1 to 1.5 m table; telephone use occasionally slightly difficult; normal voice 1 to 2 m; raised voice 2 to 4 m; large engineering and drafting rooms, restaurants
63	Unsatisfactory for confrences of more than two or three people, telephone use slightly difficult; normal voice .3 to .6 m; raised voice 1 to 2 m. Secretarial areas (typing); accounting areas (business machines), blueprint rooms
65	"Very noisy," office environment unsatisfactory, telephone use difficult

Source: K. Kryter, *The Effects of Noise on Man* (New York: Academic Press), simplified from Tables 40 and 41.[16]

Table 18.3   Adjustments to Noise Levels When attempting to Predict Community Annoyance to Noise

Situation	Adjustment in dBA level, dBA
Very quiet suburban	+ 5
Suburban	0
Residential urban	− 5
Urban near some industry	−10
Heavy industrial area	−15
Daytime only	− 5
Nighttime	0
Continuous spectrum	0
Pure tone(s) present	+ 5
Smooth temporal character	0
Impulsive	− 5
Prior similar exposure	0
Some prior exposure	− 5
Signal present 20% of the time	− 5
5% of the time	−10
2% of the time	−15

600 Hz or above 4000 Hz has relatively little effect; filtering between 1000 and 3000 Hz degrades speech severely.

The listeners should have normal hearing and know the various messages they may receive. Repeating the message back to the speaker in different words is a desirable check.

Health

The most important effect of noise is on health.

Sontag reported that noise affected the fetuses of pregnant women.[25] Jansen reported that noise (say, 105 dBA) affects blood and circulation, especially for people with cardiovascular and arteriosclerosis problems[14]; this is confirmed by Canadian (Shatalov et al.), German (Jansen), and Russian (Andruikin) studies.[2,13,24] An Italian study (Carosi and Cababro) showed a much lower birth rate for couples where the husband or wife worked in a noisy environment.[5]

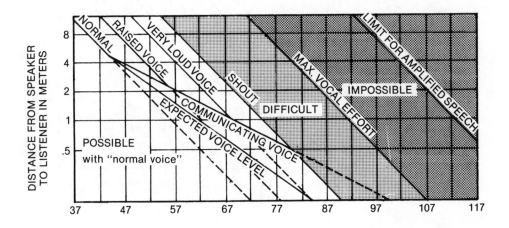

SPEECH INTERFERENCE LEVEL, dB

Figure 18.11 *Speech interference levels (SIL)* vary with the level of the speaker's voice as well as the distance from the speaker to the ear. The figure is based on males, with average voice strengths, facing the listener, no reflecting surfaces nearby, and the spoken material not being familiar to the listener. Maximum permissible speech interference level *(PSIL) = SIL + 3*. The masking effect of a sound is greatest upon sounds which are close to it in frequency. At levels above 60 dB, the masking spreads to cover a wider range, mainly for frequencies above the dominating components. For USA telephones, telephone use is satisfactory when *SIL* is less than 65 dB, difficult from 65 to 80, and impossible above 80. Subtract 5 dB for calls outside a single exchange.

Source: A.P.G. Peterson and E. Gross, *Handbook of Noise Measurement* (Concord, Massachusetts, Gen Rad, Inc.)[12]

The physiological system, however, most sensitive to noise is, as would be expected, the hearing system. Kryter says that somewhere around 67 dBA hearing loss begins to increase faster than it would from age alone; others say the value should be higher.[16] Thus if your employees are not to be deafened just because they work for your organization, noise should be kept below a certain level. What is that level? It depends.

Hearing loss depends on many factors so many adjustments have to be made to the basic data using the assumptions of independency (no interactions) and additivity. The following is a *simplified* version of the procedure given by Burns and Robinson; it applies only to nonintermittant broad-band noise (i.e., steady noise without any peaks at any frequency).[4]

Given   Noise level = 85 dBA
             Exposure   = 250 days/yr of 8 hours for 25 years

To Find Hearing loss at 1000 Hz for 25th percentile of population

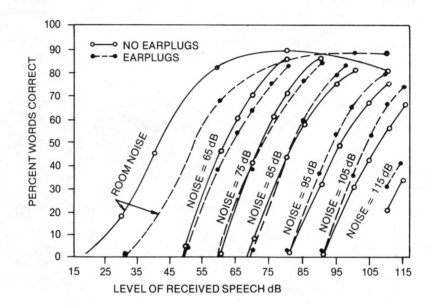

Figure 18.12 *Earplugs improve speech transmission* in loud environments. However, in environments with noise below 85 dB earplugs degrade intelligibility.

Source: K. Kryter, *The Effects of Noise on Man* (New York: Academic Press).[16]

Solution

Step	Comments and Calculations	Value
1.	Determine noise emission level	
	A. Determine dBA for environment	85 dBA

	B. Correct for *duration* of exposure, D
	     D, dB $= 10 \log_{10} T$
	        $T =$ years of exposure
		$D = 14$
		$85 + 14 = 99$

	C. Adjust for *sex*, S	

Group	S Noise Adjustment, dB
Female	−1.5
Mixed	0
Male	+1.5

                                                              $S = 0$
                                                   $99 + 0 = 99$

D. Adjust for *frequency* at which ear will lose hearing, *F*

Frequency, Hz	F Noise Adjustment, dB
500	−5
1000	0
2000	+7
4000	+15

$$F = 0$$
$$99 + 0 = 99$$

2. Predict Noise Induced Permanent Threshold Shift *(NIPTS)* measured on audiometer, dB

A. Read value *H* for population median from Figure 18.3

$$H = 2 \text{ dB}$$

B. Adjust for percentile of population, *P*

Percentile Losing Hearing	P, Audiometric Adjustment, dB Emission at 1000 Hz		
	85	95	105
10	+8	+9	+11
25	+4	+5	+ 6
50	0	0	0
75	−4	−5	− 6
90	−8	−9	−11

$$P = 5$$
$$2 + 5 = 7$$

C. Adjust for *age (A)*

$$A = cT^2$$
c = frequency constant
T = years of exposure

Audiometric Frequency, Hz	c
500	.0040
1000	.0043
2000	.0060
4000	.0120

$$A = .0043 \ (25)^2 = 3$$

$$7 + 3 = 10$$

Predicted *NIPTS* = 10 dB

As the reader can see from the example solution of the model, the really critical adjustment is *D,* duration of exposure. Burns and Robinson say the coefficient in the equation should be 10 (the equal energy principle); Kryter, summarizing many studies by many authors, says it should be 20 for noise (the equal pressure principle) but 10 for recovery.[4,16] A 10 means a doubling (or halving) of time changes *D* by 3 dB and a 20 changes *D* by 6 dB. Thus Kryter would add 28 dB instead of 14 for 25 years of exposure. On the other hand, if the noise was for 4 hr/day instead of 8 then Kryter would subtract 6 from the 85 while Burns and Robinson would subtract only 3. The USA adjustment (Occupational Safety and Health Administration) is 5; the USA Environmental Protection Agency and

Table 18.4 International Aviation Alphabet as Interpreted in a French Pilot Training Manual[7]

Lettre a Identifier	Mot de Code	Prononciation du Mot de Code*
A	Alfa	*AL* FAH
B	Bravo	*BRA* VO
C	Charlie	*TCHAH* LI
		(*CHAR* LI)
D	Delta	*DEL* TAH
E	Echo	*EK* O
F	Foxtrot	*FOX* TROTT
G	Golf	GOLF
H	Hotel	HO *TELL*
I	India	*IN* DI AH
J	Juliett	*DJOU* LI *ETT*
K	Kilo	*KI* LO
L	Lima	*LI* MAH
M	Mike	*MA*IK
N	November	NO *VEMM* BER
O	Oscar	*OSS* KAR
P	Papa	PAH *PAH*
Q	Quebec	KE *BEK*
R	Romeo	*RO* MI O
S	Sierra	SI *ER* RAH
T	Tango	*TANG* GO
U	Uniform	*YOU* NI FORM
		(*OU* NI FORM)
V	Victor	*VIK* TAR
W	Whiskey	*OUISS* KI
X	X-ray	*EKSS* RE
Y	Yankee	*YANG* KI
Z	Zulu	*ZOU* LOU

*Les syllables accentuees en caractere gras

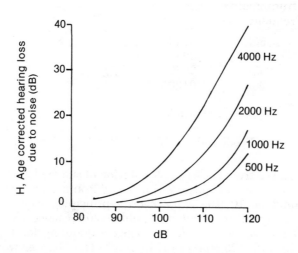

Figure 18.13 *Predicted hearing loss* caused by noise can be calculated from this simplified adaptation of Burn's and Robinson's Figure 10.17.[4] From noise emission level and audiometric frequency determine the predicted noise loss for the population median.

some other countries use 3. (There must be more pressure than energy in Washington.)

Determining P, the percentile effect, is another problem. Kryter, to protect 75% of the population, would give P a value of -10 instead of -4 to -6. Burns and Robinson, in their detailed calculations of their Figure 10.17, give $P = -6$ for noise emission level of 150 dBA and $P = -5$ at 95 dBA.[4] This sensitivity of the ears of some of the population means that some people, if exposed for many years, start to lose hearing at some frequencies at noise levels below 70 dBA; Kryter says 67 dBA.[16] There still is no technique which can identify the sensitive ears before the hearing loss. A preemployment audiogram can demonstrate whether hearing loss occurred before start of employment. Periodic audiograms also can identify sensitive ears (by noting employees who have lost some hearing) so they can be transferred before the hearing loss becomes too large. The decision of whether to ignore sensitive ears depends on whose ears are going to be ignored.

The effect of sex is controversial! Some experts agree, others don't, that noise affects females less and women have better hearing. Some feel it is reduced exposure to noise or reduced ear diseases rather than reduced susceptibility that effects the value of S.

Even the value of A is not agreed upon. For 1000 Hz, Burns and Robinson give $A = 3, 5,$ and 9 at 45, 55, and 65 years while the American Industrial Hygiene Association gives 3, 6, and 11 at the same ages.[1,4]

Another problem is to define *hearing impairment*. Everyone starts with normal hearing values from the International Standards Organization (ISO) as a reference, but from there they differ. OSHA defines material impairment of hearing "as a 25 dB hearing level at 1000, 2000, and 3000 Hz." The meaning of this obscure but vital statement is that you have "25 deductible." The hearing loss also is not in the speech frequencies! Kryter proposes "an increase of 10 percentage points on the number of people to suffer average hearing losses of greater than 25 dB at 500, 1000, and 2000 Hz because of noise rather than aging."[17] The American Academy of Ophthalmology and Otolaryngology (AAOO) and the American Medical Association specify a mean loss at 25 dB at 500, 1000, and 2000 Hz as "the beginning of slight impairment for the understanding of spoken English."[10,11] Kryter says that if the three bands were 1000, 2000, and 4000 instead of 500, 1000, and 2000 then the *NIPTS* would be 10 dB higher.[16] Kryter says the "slight impairment" of AAOO can be more precisely defined as "a person could understand but 90% of the sentences and 50% of the monosyllabic phonetically balanced (PB) words in the quiet, uttered at a normal conversational level of effort by a person one meter from him."[17] However, there is considerable redundancy in everyday speech so should we be concerned with the unexpected message?

Note that all the foregoing assumes that the noise was continuous and broad

Table 18.5   Maximum Daily Noise Level Exposures Permitted in the USA by the Occupational Health and Safety Administration. The Trade-off of Exposure Time Vs. Noise is 5 dBA for Each Doubling (Halving) of Exposure Time. Noise Beyond 115 is not Permitted; Noise Below 85 is Considered to Have Zero Effect.

Noise, dBA	Maximum Exposure/Day
85	16 hours
90	8
95	4
100	2
105	1
110	0.5
115	0.25

band; important factors such as preexposure hearing, general health, and drug effects were not included. However, one of the virtues of a math model is the explicit statement of the coefficients.

As you might surmise from the above, setting a legal standard for noise is not simple. The occupational standard in the USA is 90 dBA for 8 hours exposure/day with a 5 dBA trade-off vs. time during the day (the USA Environmental Protection Agency [EPA] uses a 3 dBA trade-off vs. time). Table 18.5 (Federal Register, 1972) gives the resulting standard. There is no adjustment for years of exposure, S, or F. All noise below 85 dBA is asumed to have 0 effect. Beyond 115 dBA, 0 exposure is permitted. Table 18.6 gives recommendations for ultrasonic noise.

Table 18.6   For Ultrasonic Noise, The American Conference of Governmental Hygienists in 1976 Recommended the Following Threshold Limit Values (8-hour Exposure) by 1/3 Octave Band

1/3 Octave Band kHz	Noise, dB SPL
50.0	115
40.0	115
31.5	115
25.0	110
20.0	105
16.0	80
12.5	80
10.0	80

NOISE REDUCTION

The following summarizes recommendations from many diverse sources: books as well as magazines and experience.[1,12,20,26]

It may be desirable to mask unwanted noise with white noise. Privacy in conversations may be the critical factor rather than the absolute noise level. Masking examples are fountains, music in restaurants, fans in telephone booths, air conditioners, and fluorescent ballast hum in offices.

If masking fails, attempt noise reduction in the following sequence: plan ahead, modify the noise source, modify the sound wave, and use personal protection.

Plan Ahead

An ounce of prevention is worth a pound of cure. The four subcategories of the *ounce* are: substitute less noisy processes, purchase less noisy equipment, use quieter materials and construction, and separate people from equipment.

Substitute less noisy processes. Impact tools are noisy. Could joining be done by compression riveting instead of pneumatic riveting? Spot, gas, or arc welding or high strength bolts may be cheaper as well as quieter. Instead of chipping castings could grinding or carbon arc air gouging be used? Substitute mechanical ejection of parts from a machine for air ejection. Electrical mowers are quieter than gasoline. Electrical fork trucks are not only quieter than internal combustion engine trucks but also have less fumes and lower operating costs.

Purchase less noisy equipment. Certain types and sizes are quieter. Squirrel cage fans are quieter than propeller; bevel gears are quieter than spur; nylon gears are quieter than metal; belt drives are quieter than gear drives; V-belt drives are quieter than toothed-type belts; reinforced rubber belts are quieter than leather or canvas belts; electric hand tools are quieter than pneumatic tools; electrically operated valves and solenoids are quieter than air-powered ones. In drilling concrete, use carbide or diamond-tipped drills since star drills and air hammers create both noise and dust. Cast aluminum vibratory feeder bowls may be 15 dBA quieter than fabricated sheet metal bowls. Bearings with less clearance and higher finish make less noise.

Properly sized fans (i.e., running at their peak efficiency) are quieter since undersize fans have high rpm and oversize fans have low rpm but separation of air flow over the blades. For the same capacity, large slow fans are quieter than small fast fans. A low-speed multibladed fan is quieter than a high-speed two-bladed fan. Thermoplastic fan blades have high inherent damping and are poor resonators and thus are quieter than metal blades.

For manually operated punch presses, a pin-type clutch contributes about 70% of the total sound energy at the operator's ear; buy presses with hydraulic drives or air-actuated clutches.[23]

Within a size and type, different manufacturer's equipment has different noise levels. Be sure to include noise levels in purchase specifications. A specification must contain three ingredients: units, levels, and conditions. For example, "*SPL* for machine with auxiliary equipment shall not exceed 85 dBA (slow response) at the operator location when installed as specified and all adjacent equipment is turned off." It is probably better, and certainly easier, to specify sound-power level. Ask venders to include silencers and sound-damping devices in their quote. Buy noise-suppression equipment (such as mufflers) from the manufacturer on the original requisition; it is much easier to get these low-cost items approved then than later as a separate requisition.

Use quieter construction and materials. Double doors and double walls with mineral wool or other insulation between the walls reduce sound transmission as well as conserve heat; double windows with a 100 mm air space do the same. Mechanical transmission of remote vibration can be reduced by expansion joints, flexible couplings, separate foundations. Fluids with low turbulance have low noise, so use surge chambers and Y joints instead of T joints. Minimizing internal welding beads decreases piping losses as well as noise.

Reduce noise by covering surfaces with resilient material or using resilient material for the container. Wood block floors in the shop are easier on the feet than concrete as well as quieter; epoxy resin terrazzo floors are easy to maintain and absorb sound; carpets in the office reduce maintenance costs as well as noise. Avoid metal-to-metal contact. Chutes and hoppers should be wood or plastic instead of metal; if metal is required for wear, back it up with wood. Cover the inside of chute tops with sound-absorbent material. Line walls of tumbling barrels. Buy plastic or fiberglass tote pans (they weigh less as well as being quieter) and wooden-top work benches. Cover metal conveyor rollers with rubber, plastic, or carpet. Replace metal wheels on vehicles with tires. For duct work carrying blasting grit and for screw machine feeder tubes, use special vibration-damping paint. In kitchens, rubber mats on sinks and drain boards reduce noise and breakage.

Separate people and noisy equipment. The key number is 6; this is the reduction in dB from a point source in a free field with each doubling of distance (provided the initial measurement is at a distance several times the longest dimension of the source). With absorbent material and good design, this can be increased to 8 dB or even more, even in an open plan office. In an open space, sound flows in all directions so the direction factor (magnification ratio), Q, equals one. On a wall, sound can flow in only 1/2 of a sphere so $Q = 2$. At a junction of two walls $Q = 4$ and in a corner $Q = 8$. Higher frequency sound waves have a high Q factor even if the source is in an open area. Therefore, don't put noisy equipment in a "megaphone" by locating it in a corner or at the end of a corridor. Supervise noisy equipment by TV rather than in person.

Modify the Noise Source

The pound of cure is more expensive than the ounce of prevention. Because of the way decibels add, always start with the loudest noise first. Three noise sources of 90, 85, and 101 dB combine to 102 dB. If the 90 dB noise is completely eliminated, the 95 and 101 still combine to 102 dB. The four subcategories of "modify the source" are: reduce the driving force, reduce response of vibration surfaces, change the direction of the noise, and minimize velocity and turbulence of air.

Reduce driving force. Step one is maintenance. Sharpen tools. Tighten screws and bolts. Lubricate bearings. Grease, oil, and replacement of worn parts reduce noise as well as wear. Rebalancing rotating equipment reduces wear and noise as well as improving quality. Replace leaky compressed air values to save air as well as reduce noise.

Use reducing valves when full shop pressure is not needed (e.g., when blowing off dirt) to improve safety as well as reduce noise. When drills, mills, and taps are under heavy load, cutting compounds reduce noise while they lubricate. A rake angle on punches lets the punch hit the work gradually; the same energy is expended over a longer time period. Stagger punch lengths when punching several holes at one stroke. Because of the reduced impact peak, noise is reduced and smaller presses can be used. The same principle works for shears. Spreading energy over a longer period also can be used to quiet diesel engines. Increase the turbulence of the fuel-air mixture (by bouncing it off the walls) so ignition occurs over a longer time span; set the ignition timing to minimize peak pressures. Turn down the paging system volume at lunch.

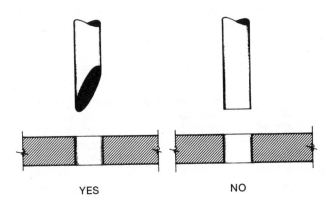

YES   NO

Figure 18.14 *Punches with rakes* (left) need less power and make less noise. The same principle applies to chipping hammers.

Reduce response of vibrating surfaces. Hitting a hammer against a concrete wall and against a brass gong illustrates the importance of resonance. Resonance can be reduced by high damping—either by material such as lead or sand or by the physical structure of the device. Damping materials change vibration into heat through internal friction. The ability of various materials to amplify vibration—given as Q, the magnification ratio—is 1000 for steel and aluminum, 500 for reinforced plastics, 200 for concrete, 100 for plywood, 50 for cast iron, and 10 for steel with polymer damping compound. The screech of subway wheels can be reduced 34 dB by adding polymer coating to the steel rim. Fill the base of equipment with sand.

A physical structure with high stiffness/weight will minimize vibration. Extra machine mass may eliminate the need for enclosure of the machine. Use ribs, corrugations, or a domed shape. The irritating whine of a circular saw can be reduced by stiffening the blade with very large washers. The Swedes have developed a multilayer saw with a layer of plastic between the outer layers of metal. Stiffen flat sections of sheet metal work (especially chutes) by backing it with wood, felt, lead, or elastomers. An outer covering of steel with the same thickness as the sheet metal helps—bolt the "sandwich" together on 150 mm centers. Use a damping material on a tool grinder rest. For items being riveted, clamp the edges to the table. Quiet machine guards by making them of perforated instead of solid material; it reduces their area of vibration (an alternative is plastic instead of metal guards).

If the area of vibration is halved, the intensity of the noise will be reduced by 3 dB; at low frequencies the reduction will be much greater. Reduce mechanical transmission of remote vibration by supporting pipes and ducts, putting bends in a pipe, and by using flexible conduit, couplings, and piping as well as vibration isolators. Vibration through concrete floors can be reduced by sawing the slab; this will probably improve accuracy on nearby machines as well as reduce noise.

Vibration damping often increases fatigue life as well as reducing noise.

Change the direction of the source. Since sound waves have a longer wavelength than light waves, they are not quite as directional as light. Yet (especially for high-pitched noise) turning a machine or exhaust 90 degrees often reduces noise as much as 5 dB. Gaseous jet noise is very directional. The best direction to turn it is up if no one is above the machine. The same 5 dB benefit can be obtained if the operators face the sound so the sound grazes their ears rather than hitting them directly. Changing the hinge location on windows and doors so they open in a different direction may help.

Minimize velocity and turbulence of air. Avoid sonic velocities. Cutting air velocity in half can reduce dB by 15. (Dispersive mufflers work by reducing air velocity.) For air drying of objects use a large number of low velocity nozzles since it is volume not velocity which is desired. Furnace noise has been cut by using additional burners to reduce gas flow per burner. For air ejection, use a multiple opening nozzle at low line pressure and placed close to the part. Since the high velocity portion of the jet is only 2 jet diameters wide, accurate aiming of the jet permits lower velocities. Halving the distance between the jet and the part permits a 30% reduction in velocity which will reduce noise by 8 to 10 dB.

Since many operations do not require full line pressure, use pressure regulators. (The OSHA requirement limiting air pressure for cleaning equipment to 30 psi is a requirement very frequently cited as a violation.) Automatic shut offs save compressed air as well as reduce unnecessary noise; since the noise is intermittent, hearing loss is less. To make a whistle, blow across a sharp edge. On jigs and fixtures and dies used with air ejection, streamline the sharp edges upstream of the part with fillets and chamfers. See Figure 18.15.

Quiet air ducts have large cross sections (to give low velocity), have several elbows, have internal insulation after elbows (before and after elbows is even better), use parallel or staggered baffles, have vibration breaks, and are not placed to produce the megaphone effect. If it is not possible to modify the duct, put a plywood baffle in front of the room air supply grille. The baffle (50% larger than the grille) should be surfaced with fiber glass on the grille side and should be placed far enough from the grille to not restrict air flow.

If you didn't plan ahead and can't reduce the noise at the source, then you can try to modify the sound wave on its path to the ear.

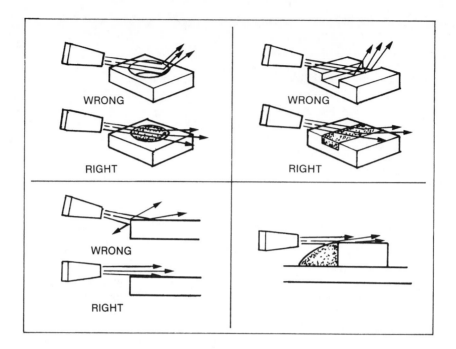

Figure 18.15 *Air-jet noise* is increased greatly whenever the jet blows against a sharp edge. Filling in cavities or redirecting the jet can reduce noise levels as much as 7 dBA.[21]

Modify the Sound Wave

Confine and absorb. Unfortunately these are expensive procedures which very often reduce dB very little.

For confinement the problem is the long wavelength; any sound will escape even through a small opening and so, for confinement to work, enclosures must be total. Holes amounting to as little as .01% of the total area of the enclosing structure will transmit more than half the sound energy, lowering the total reduction by 3 dB. For a barrier reducing noise 40 dB, openings of .01%, .10%, 1%, and 10% reduce attenuation to 37 dB, 30 dB, 20 dB, and 10 dB. Access openings are the weak points in an enclosing structure.

The first goal is to confine and absorb before the sound "gets out." Light porous materials (fiberboard, cork, foam rubber, cloth, mineral wool) absorb sound but also transmit with little attenuation of sound. Hard massive materials (brick, concrete) reflect and prevent transmission but absorb little (see Table 18.7). Transmission is less for frequencies over 1000 Hz. The ideal is a heavy wall of brick to prevent transmission with an inner lining of foam rubber to absorb. A compromise is a thin sheet of lead with a layer of foam. Note that not all foams are acoustical foams so just because a foam is cheap does not mean it is a good buy. Unglazed brick absorbs three times as much as glazed brick; unpainted contrete block absorbs 3 to 5 times as well as painted block; a wood floor absorbs 3 to 15 times as well as concrete while a heavy carpet can absorb 30 times as much as concrete at some frequencies. Carpeting office walls is effective.

Complete enclosure of single-source noise such as motors and gearboxes is sometimes worthwhile but, as the noise source gets larger, the enclosure volume (and therefore material required and therefore cost) increases rapidly. Temperature rises may be a problem. Line the enclosure inside (not outside) with material with an absorption coefficient of at least .7 (see Table 18.7). Sound flows through small cracks so use gaskets on doors and shafts. In some cases enclosure of the man is feasible but it tends to be expensive due to working space requirements, door seals, provision of air for breathing, access for maintenance, etc. Partial enclosures, such as for shop telephones, key punch machines, etc., are better than nothing but not much. In some cases complete sealing isn't necessary as a series of noise-trapping baffles can be used. A shield of auto safety glass between the noise and the employee can be used to reflect high frequency sound such as air jets while permitting vision if both the noise and employee are close to the shield; have the "acoustical shadow" extend a meter beyond the head. Higher frequency noise with its shorter, more directional waves can be absorbed if the absorbent is close to the noise source.

If the sound can't be confined and absorbed directly at the source, it can be dissipated by a muffler. Muffle intakes and outlets. Use the proper design and don't forget the mufflers wear out. Resonance mufflers are effective for specific frequencies; be sure they are designed by an expert. Check that the operator doesn't remove the muffler. (Most motorcycle violations are due to failure to use the manufacturer's muffler.) A relatively ineffective (but cheap) shield can be created by storing work in process between the noise and the man.

Absorbing the sound once it is "out" (that is, the ear is between the noise source and the absorbent) by absorbent panels on the walls or ceiling is not cost-effective. To reduce noise by 20 dB we need to drop intensity to 1 part in 100; a 30 dB drop requires reduction to 1 part in 1000; achieving these magnitudes is not economic. Reductions of 7 to 10 dB in the higher frequencies are practical if the room is reverberant but, in most, 5 dB is the best that can be accomplished. Put the absorbent on the walls if the minimum floor dimension is less than four times the room height. Carpeting a cinder block wall may be useful—especially if the wall is an outside wall and thus the carpet can act as thermal insulation also. In classrooms and offices with multiple noise sources, soft surfaces will help until about 50% of the surfaces are soft. Flush-mounted fluorescent lamp fixtures act as sound reflectors and thus negate the effect of acoustical ceilings.

If you didn't plan ahead, couldn't modify the source, and have found modifying the sound wave was too expensive, then there remains, as a last resort, protection of the receiver, the ear.

Personal Protection

Time. Higher noise levels can be tolerated if exposure time for a specific ear is short. In the USA the trade-off is 5 dBA for each doubling of time instead of the 3 dBA used in most of Europe. Since our present maximum is 90 dBA for 8 hr, an employee can be exposed to 95 dBA for 4 hr, 100 dBA for 2 hr, 105 dBA for 1 hr, 110 dBA for 1/2 hr, and 115 dBA for 1/4 hr. Beyond 115 dBA is not permitted.

Reduce exposre time by making the source noise intermittent. Then minimize the exposure of the worker. Redesign jobs so that noisy areas are inspected by TV. Provide a quiet area for work which can be done before and after work in the noisy area (blueprint reading, paperwork). Modules and quick disconnects minimize downtime as well as exposure of maintenance personnel. Consider job rotation within the day.

Equipment. Ear muffs and ear plugs attempt to totally enclose the hearing system. They are the only realistic protection against gunfire and explosive tools. See Figure 18.16 for typical attenuations of ear muffs and ear plugs.

Very rarely is the noise environment so loud that noise cannot be reduced to a safe level with personal protective equipment. Note that 15 dB attenuation is usually sufficient since most noise is below 105 dBA. Note also that since ear protectors work best in the high frequency range, it may be more desirable to reduce

Table 18.7 Sound Absorption Coefficients (α) of Typical Surfaces; α is the Ratio of Sound Energy Absorbed by the Surface to Sound Energy Incident. Note the Effect of Frequency. (Simplified From Table 37-3 of Hill)[12]

Material	Frequency (Hz)			
	125	500	1000	4000
Brick: glazed	.01	.01	.01	.02
Brick: unglazed	.03	.03	.01	.07
Concrete block: coarse	.36	.31	.29	.25
Concrete block: painted	.10	.06	.07	.08
Floor: carpet, heavy with 40-oz pad	.02	.14	.37	.65
Floor: linoleum, rubber, or cork tile on concrete	.02	.03	.03	.02
Floor: wood	.15	.10	.07	.07
Glass fiber: mounted with impervious backing, 3 lbs/cu ft, 1 in thick	.14	.67	.97	.85
Glass fiber: mounted with impervious backing, 3 lbs/cu ft, 3 in thick	.43	.99	.98	.93
Glass: window	.35	.18	.12	.04
Plaster on brick or tile	.01	.02	.03	.05
Plaster on lath	.14	.06	.04	.03
Plywood paneling, 3/8 inch.	.28	.17	.09	.11
Steel	.02	.02	.02	.02

low frequency components of the source noise if ear protectors will be worn anyway. For rockdrills, the low frequency exhaust noise is dominated by the high frequency piston and drill noise. It is worthwhile to reduce exhaust only if the high frequency noise is also eliminated (say, by earplugs).

The problem of personal protective equipment is not bad design but bad application. Workers often are fitted with the wrong size, seals no longer seal with age, and, worst of all, workers refuse to wear the device at all due to the gradual onset of the hearing disability and thus their lack of a feeling of danger. It is probably desirable to make wearing of hearing protection a part of the job description and a condition of employment. Emphasize to the employee that the employer is helping him to help himself; there is no need to lose the enjoyment of speech, music, and television. The most effective hearing protector is the one that is worn.

Earmuffs are advantageous in that one size fits almost everyone, and they are visible from a distance; a supervisor can tell who is wearing personal protective equipment. But they are hot, heavy, and muss curly locks; glasses reduce their seal, and your head is within a spring.

Earplugs should be individually fitted because ear canal sizes and shapes differ. Often the right ear is not the same size or shape as the left. Acceptance is considerably increased if employees can get earplugs molded to their ears or can try four or five sizes and models until they find plugs which are comfortable to their individual ears. You can hear better in noise with earplugs than without earplugs! (See Figure 18.12) Forbid use of useless dry cotton earplugs. Earplugs are light, compact, and do not affect appearance; therefore the supervisor can't tell if they are being worn, and they get lost. Reduce lost plugs by having the plugs connected with a cord. Be sure to have spares.

Reduction of noise is not a very difficult problem if you plan ahead. It calls for some ingenuity and expense if you modify the noise source. It requires considerable ingenuity and expense to modify the sound wave. It requires day after day selling, motivation, and supervision to get personal protective equipment used.

The goal is "Ears alive at 65!"[18]

SUMMARY

Although the ear is well designed, a noise can cause problems. The reduction in hearing ability usually is measured with an audiometer.

The noise scale confuses the general public because it is the logarithm of a ratio so that doubling the noise level only gives an increase of 3 dB. For most applications, use the dBA scale. Use a sound level meter to measure noise at a specific time and a noise dosimeter to measure noise over a period of time.

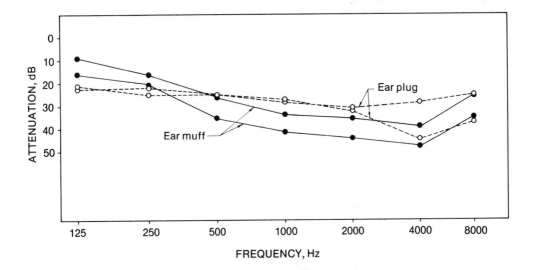

Figure 18.16 *Ear muffs* tend to give more protection than ear plugs. However, as with any other product, some models are better than others. The reported attenuations were from the ANSI Z24.22-1957 standard. The ANSI S3.19-1974 test standard uses a different test methodology and averages 5 dB less attenuation than the Z24.22-1957 test. Ear muffs can be designed to be worn with the band over the head, behind the head, or under the chin; one manufacturer reported, for its model, 2-8 dB less attenuation in the speech frequencies when behind the head and 5-8 dB less attenuation when under the chin. Wearer-molded (plastic foam) plugs probably will give attenuations in practice equal to the manufacturer's test data but pre-molded plugs may give only 50% of their attenuation potential due to poor fitting by the users.[6]

Noise control at lower levels (55 to 80 dBA) primarily is to eliminate annoyance; noise control at higher levels (90 dBA and up) is to protect the worker's hearing. Under most circumstances noise does not affect productivity.

Noise reduction is relatively inexpensive if you plan ahead; it calls for some ingenuity and expense to modify the noise source; it requires considerable ingenuity and greater expense to modify the sound wave. Use of personal protective equipment requires day-after-day selling, motivation, and supervision.

SHORT-ANSWER REVIEW QUESTIONS

18.1   What is the purpose of the Eustachian tube?

18.2   What is the total dB from 70 dB + 70 dB? 70 + 90? 70 + 75?

18.3   What is the noise level standard in the USA for 8 hr exposure? What is the trade-off for each doubling or halving of time? What is the maximum permitted level?

18.4   List in sequence the four major approaches to noise reduction.

18.5   Why would you try to eliminate the loudest noise first? Discuss using an example of a 100 dBA and a 90 dBA noise.

THOUGHT-DISCUSSION QUESTIONS

1. The noise level standard in the USA should be 85 dBA for 8 hrs/day exposure instead of 90 dBA. Discuss.

2. Workers refusing to wear ear plugs or ear muffs in a noisy environment should be fired. Discuss.

REFERENCES

1. American Industrial Hygiene Association. *Industrial Noise Manual.* Detroit, Mich., 1966.

2. Andruikin, A. Influence of sound stimulation on the development of hypertension. Clinical and experimental results. *Cor. Vassa.,* Vol. 3, (1961): 285-293.

3. Anticaglia, A. Physiology of hearing. *The Industrial Environment—Its Evaluation and Control.* Washington, D.C.: Superintendent of Documents, 1973, chapter 24.

4. Burns, W., and Robinson, D. *Hearing and Noise in Industry.* London: Her Majesty's Stationery Office, 1970.

5. Carosi, L., and Cababro, F. Prolificacy of workers in noisy industries. (La prolificita di coniugi operai di industrie reumorose) *Folia medica,* Naples, Italy, 51 (April 1968): 264-268.

6. Crawford, D., and Nozza, R. Field performance evaluation of wearer-molded ear inserts. Paper at American Industrial Hygiene Association annual meeting, Portland, Oreg., 1981.

7. David, H. French version of alphabet. *Ergonomics Research Society Newsletter* (March 1974).

8. Epp, S., and Konz, S. Appliance noise: annoyance and speech interference. *Home Economics Research Journal,* Vol. 3, No. 3 (1975): 205-209.

9. Goodfriend, L. Control of community noises from industrial sources. *The Industrial Environment—Its Evaluation and Control.* Washington, D.C.: Superintendent of Documents, 1973, chapter 46.

10. Guides to the evaluation of the permanent impairment; ear, nose, throat and related structures. *Journal of the American Medical Association,* 197 (August 1961): 489.

11. Guides for evaluation of hearing impairment. *Transactions of the American Academy of Opthalmology and Otolaryngology* (March-April 1969): 167-168.

12. Hill, V. Control of noise exposure. *The Industrial Environment—Its Evaluation and Control.* Washington, D.C.: Superintendent of Documents, 1973, chapter 37.

13. Jansen, G. Adverse effects of noise on iron and steel workers. *Stahl und Eisen,* 81 (1961): 233-239.

14. Jansen, G. Relation between temporary threshold shift and peripheral circulatory effects of sound. *Physiological Effects of Noise,* Welch, B. and Welch, A., eds. New York: Plenum Press, 1970.

15. Kryter, K. Effects of ear protective devices on the intelligibility of speech in noise. *Journal of Acoustic Society of America,* 18 (1946): 413-417.

16. Kryter, K. *The Effects of Noise on Man.* New York: Academic Press, 1970.

17. Kryter, K. Impairment to hearing from exposure to noise. *Journal of the Acoustical Society of America,* Vol. 53, No. 5 (1973): 1211-1234.

18. Maas, R. Personal hearing protection: causes of failure, reasons for success. Paper at American Industrial Hygiene Association meeting, Toronto, 1971.

19. McCormick, E. *Human Factors Engineering.* New York: McGraw-Hill, 1970.

20. Olishifski, J., and Harford, E., Ed. *Industrial Noise and Hearing Conservation,* Chicago: National Safety Council, 1975.

21. Oviatt, M. Energy conservation and noise control in pneumatic devices and systems. *Plant Engineering,* Vol. 35, No. 16 (Aug. 6, 1981): 116-118.

22. Peterson, A., and Gross, E. *Handbook of Noise Measurement.* Concord, Mass.: GenRad, Inc., 1972.

23. Roberts, T.; McFee, D.; and Hermann, E. Significance of punch press clutch noise. *American Industrial Hygiene Association Journal,* Vol. 39, No. 2 (1978): 166-169.

24. Shatalov, N.; Saitanov, A.; and Glotova, K. On the state of the cardiovascular system under conditions of exposure to continuous noise. Report T-411-R, N65-15577, Defense Research Board, Toronto, Canada, 1962.

25. Sontag, L. Effect of noise during pregnancy upon fetal and subsequent adult behavior. *Physiological Effects of Noise.* Welch, B. and Welch, A., eds. New York: Plenum Press, 1970.

26. Sound control. *Guide and Data Book-Systems.* New York: American Society of Heating, Refrigerating and Air Conditioning Engineers, 1973, chapter 35.

27. Weaver, J. Monitoring hearing loss at United Airlines. *Occupational Safety and Health,* Vol. 50, No. 8 (1981): 35-39.

19

TOXICOLOGY AND SAFETY

POISONS

Toxicology is the study of the effect "foreign" chemicals have upon the body when considering the dose/response ratio. This chapter touches very briefly upon a large and complex subject (see Patty; *The Industrial Environment;* and Peterson).[13,18,19]

Some key words in the definition might escape casual observations: effect, body, and dose/response. (French hatters in the seventeenth century used mercuric nitrate to aid fur felting; the chronic mercury poisoning led to the expression "mad as a hatter.")

Effect could be death, change in an enzyme in a test tube, or change in dark adaptation of the eye. The Threshold Limit Values (TVLs) given later in this chapter are primarily based on nonreversible functional changes in an organ (usually the liver and kidney). Maximum allowable concentrations in the USSR tend to be much stricter than those in the US; the Russians believe the standard should be set at the point where any measurable change in human physiology is effected. The problem can be demonstrated with alcohol: is a dose dangerous when it causes damage to the liver or when it causes a slowing of reaction time or change in dark adaptation? A very severe interpretative problem is that there may be a 20-year time lag between dose and response; this time lag problem also occurs in industry as the worker does not see the connection between cause and effect. A political problem is that the benefit accrues to individuals but the costs accrue to organizations. Moll and Tihansky point out that there are three critical considerations: the need to make trade-offs, the trade-off impacts are not in the same units (dollar benefits vs. loss of life), and uncertainty about the impacts of alternative decisions.[16]

Body can refer to any organism that can function by itself from a bacterium through rats to humans. What do you do if a compound causes cancer in one rat species but not in another and you must set the safe dose for humans? The USA approach has been to consider "man as responding as the most sensitive animal species" with the assumption of an 8-hour day and 40-hour week. There is a safety margin of 20% to 1000% between the dose which produces a response and the TLV.[7] In general, the magnitude of the safety factor is related to the effect to be protected against: a smaller factor for skin irritation and a larger factor for cancer.

The *dose/response* is another problem. The poison is in the dose. Water in excess will kill as will salt, sugar, alcohol, sulfanilamide, or arsenic. The problem is to define "in excess."

Think of the human as a "leaky bucket" with "components" on "shelves at various levels" (see Figure 19.1). The problem is whether the water will rise high enough to cause "corrosion." Thus poisoning depends on the rate of poison input (water input), the body size and target organ susceptibility (shelf level), and the poison removal capacity (hole size).

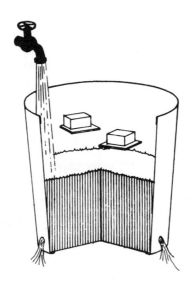

Figure 19.1 *Consider the body* as a "leaky bucket" with "components" on "shelves at various heights." Will the water rise high enough to cause "corrosion"?

Figure 19.2 gives a schematic view of the body, making the point that entrance to the mouth or lungs is not entrance to the "body"; to enter the body, a poison must enter the blood. Thus, from a toxicology viewpoint, if your little brother swallows a penny and it goes through the stomach and is eliminated with the feces, it has not been in the "body." If he swallows or breathes an object or chemical which is broken down in the intestinal tract or lungs and passes into the blood, he has the potential to be poisoned. Therefore an important characteristic of potential poisons is their ability to penetrate the body's perimeter. In general, inorganic materials are slow to penetrate the barrier, polar organic materials penetrate more quickly, and nonpolar organic materials are absorbed most quickly.

The "shelf level" depends upon the body size and the target organ susceptibility. The same amount of water will rise much higher in a small bucket than in a large bucket. Therefore, Threshold Limit Values (TLV) are based on a 70 kg adult. There is a large (5-10 fold or more) individual susceptibility to toxins (just as there is in need for vitamins, the good inputs). Once the poison enters the blood it causes its damage by attacking a target organ such as brain or heart. Toxicity is determined by the poison level in the blood being greater than the organ's sensitivity threshold. The "hole size" depends on the ability of the kidney and liver to transform the poison into a less toxic compound and to eliminate it from the body. Older people have less kidney and liver capacity; thus their "holes" are smaller.

The next sections will discuss input route, targets, and poison elimination in more detail.

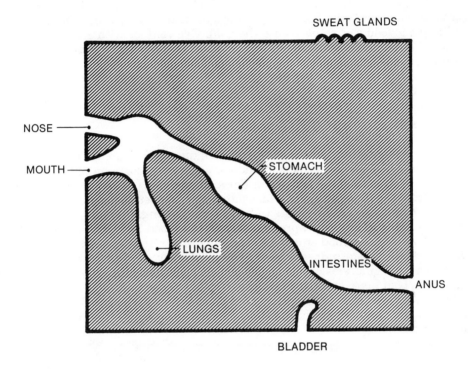

Figure 19.2 *Lungs, stomach, and intestines* are not "in the body" from a toxicology viewpoint.

INPUT ROUTE

The important potential entrance points are (1) the skin, (2) the mouth, and (3) the lungs.

The Skin

The skin is a superb barrier; Webb and Annis showed that it could even resist the vacuum of space.[27] Common experience has shown that most compounds run off the skin rather than penetrate the barrier. Thus people often insert their hands into compounds, which, if they would penetrate the skin, would kill them. Table 19.3 has the word *skin* next to compounds which penetrate the skin. (Clothing or shoes wetted with a toxic compound may prolong contact with the skin and permit the compound to enter the body.) When a compound is encountered which does penetrate the skin, such as the organic phosphates used as insecticides to replace DDT, death may occur. Agricultural workers were familiar with DDT which is not toxic to humans (not only does DDT not penetrate the skin but, even if it does, it causes no harm; you can eat a cup of DDT with no ill effects.[12]) The organic phosphates, on the other hand, kill people simply from contact. Not realizing their danger, people died from walking through cotton fields when leaves wetted with the organic phosphate brushed against their skin. The US Department of Labor estimated pesticides killed 800 migratory workers in 1972.[25]

Cuts, abrasions, or sores can break the skin barrier and permit both toxins and germs to enter the body. For example, dermatitis from a cutting oil might permit an inorganic lead salt to penetrate the skin. Eye contact with a toxin may be a problem. However, in general, poisons entering the body through the skin are not a serious toxicology problem.

The Mouth

The second entrance route is eating or drinking the poison. Although children often directly consume a poison, in industry the most common problem is toxic compounds on normal food or drink. For example, in a pesticide factory, dust in the air from the process may fall on sandwiches, in open coffee cups, or on cigarettes. The best precaution seems to be to forbid eating, drinking, or smoking in the work areas at any time; provide areas where food and drink may be stored and eaten safely. No food or drink in the work area often protects the product as well as the workers. Engineers have to be constantly alert against slack enforcement of such regulations or unexpected insertion of items into the mouth. A famous example of inserting items into the mouth was the girls who painted radium dials on watches. They occasionally pointed the tips of their brushes—by licking them! Again, however, poison entering through the mouth is not much of a problem in most industries.

The Lungs

The problem is the third route, the respiratory tract.

The lungs are a problem due to the physical nature of the poison (a gas or finely dispersed as an aerosol or mist and often invisible) and the design of the lungs (very good at moving molecules from the gas portion to the blood portion).

The most important characteristic of particles in regard to inhalation is their size. Figure 19.3 shows a typical log-normal distribution of *airborne* dust and respirable dust.[9] Wright points out that, when lungs of 50- to 60-year-olds are digested, about 50% of all particles will be .5 micrometer or less in diameter and effectively 100% will be 5 micrometers or less.[28] (1 micrometer $= 1$ μm $= 1$ micron $= .000\ 001$ meter. A human hair has a 100 micrometer diameter.) Less than .002% are over 10 micrometers. Almost all fibers (i.e., length is more than 3 times diameter) will have lengths less than 50 micrometers. Particles in the air larger than 10 micrometers are removed completely in the nose and upper airways; particles between 5 and 10 micrometers are deposited on the upper airways; only particles 2 micrometers or less are likely to penetrate the lung depths. Since fibers tend to orient their long axis to the airstream, straight fibers and short fibers penetrate deeper than curved, U-shaped, or long fibers. (Particles

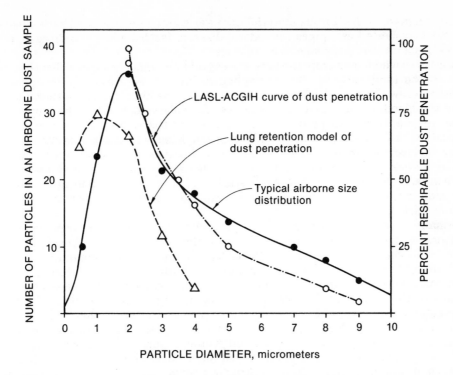

Figure 19.3 *Airborne dust* has a log normal distribution (Frazer).[9] Carson gives the Los Alamos Scientific Laboratory (LASL)—American Conference of Government Industrial Hygienist (ACGIH) curve for dust penetration as well as the lung retention curve.[6]

from .1 to 1.0 micrometers significantly reduce visibility since their diameters are comparable to the wavelength of light in the visible region.)

To understand the physical dimensions, bring the objects to "human scale" by using 1 micrometer = 25 mm. Then a .1 micrometer particle is a BB, a 1 micrometer particle is a golf ball, and a 10 micrometer particle is a basketball. The 25 mm trachea becomes 625 m in diameter with the 10 micrometer long cilia being slightly longer than a pencil. The 300 micrometer terminal bronchioles and the 75 to 300 micrometer diameter alveolar sacs are as large as the walls of a small classroom. On the enlarged scale, the three-layer membrane (alveolar epithelium, basement membrane, and capillary endothelium) separating the gas from the blood (as thin as .2 micrometer) is the thickness of a book cover.

TARGETS

An airborne particle or noxious gas might attack the respiratory system itself. One response might be an increase in air-flow resistance (asthmatic response); this response usually ends when the mucociliary escalator removes the irritant. A second response (chronic bronchitis) might occur with continuing irritation to the mucosa due to high levels of irritant gas or particles; there is an increase in coughing and sputum. A third response (acute bronchitis or pneumonia) may occur if the cilia movements are paralyzed—thus leading to inadequate lung cleaning and possible improved colonization of bacteria.

A fourth response (chronic interstitial lung disease) may occur from dusts containing microorganisms or animal proteins. In various forms, it is called bagassosis (bagasse is the fibrous material of sugar cane), mushroom picker's disease, wheat thresher's lung, pigeon breeder's disease, feather plucker's disease, malt worker's lung, and paprika slicer's disease.[5]

Particles which penetrate the lung are engulfed by macrophages when they settle on the lung surface; the macrophages, having the power of independent motion, draw the particles through the tissues of the lung wall and either directly into the blood, to surrounding bronchioles for ciliary removal to the lymph system, or simply remain permanently attached to the wall. However, free silica repeatedly kills the macrophages (silicosis) which leads to scar tissue and a loss of lung surface area (in effect, slow suffocation). Long asbestos fibers (over 5 micrometers) stick out of the macrophages and also cause scar tissue and loss of surface area. Another lung response is disturbance of the normal replication and cell division in the lungs; cancerous cells are especially fostered by particles of chromium, nickel, uranium, and asbestos.

Finally, a compound may pass into the blood and attack other organs. Different organs have different sensitivities to different compounds. Lead primarily affects nervous tissue, cadmium affects the kidneys, carbon monoxide and cyanide affect hemoglobin, and sulfur dioxide and hydrogen sulfide affect the lungs (cough). Hydrogen sulfide also causes pulmonary edema. Some anesthesia gas seems to affect the reproductive system of operating room personnel.[24] The fetus especially is endangered as the baby is a "sponge" to toxins, that is, it does not have a good ability to eliminate poisons. Thus TLV standards for normal adults often are not strict enough for pregnant women. (Surprising as it may seem, in 1975 a United Auto Worker union local sued General Motors for prohibiting women of child-bearing age from working in a certain section of a battery factory; sterile women and women past child-bearing age were permitted to work by GM. The lead concentration ranged from 1/2 to 3/4 of the TLV. Should TLVs be low enough to protect "everyone"? The judge agreed with GM.)

Although a compound may not penetrate the skin, it may cause dermatoses. Approximately 1% of the working population suffers occupational skin disease during the course of a year (Birmingham)[3]; in 1972 41% of occupational illnesses were skin ailments causing 25% of lost work days (Birmingham).[4] Causes of occupational dermatoses are:

- Mechanical and physical—abrasions or wounds, sunlight for outdoor workers, fiberglass, and asbestos

- Chemical—subdivided into strong irritants such as chromic acid and sodium hydroxide and marginal irritants such as soluble cutting fluids and acetone which require prolonged contact over time

- Plant poisons—woods such as West Indian mahogany, silver fir, and spruce may cause dermatitis when being sandpapered or polished

- Biological agents—anthrax is contracted by handlers of skins or hides from infected animals; food and grain handlers often contract grain or straw itch from handling produce infected with mites, etc.

- Some metals can sensitize the skin (i.e., cause an allergic reaction); nickel-plated earrings or stainless steel may be a problem.[26]

Young workers and males develop occupational dermatoses more readily than older workers and females due to their lack of care in handling injurious materials and lower personal cleanliness. There also may be self-selection as older workers who develop dermatitis move to other areas. Personal cleanliness is *the* most important measure in preventing occupational skin disease.[4] Personal clothing, including undergarments worn on the job, should be changed daily and washed thoroughly before reuse. It may be necessary to provide clothing and laundering to ensure daily change. Properly designed protective clothing can provide a good barrier. However, the most important factor is personal cleanliness; the engineer can help by providing adequate washing facilities near the work area.

POISON ELIMINATION

Although the lungs eliminate carbon dioxide and sweat glands eliminate some salts, the primary organs for poison elimination are the liver and the kidneys.

The liver biotransforms the toxins in the blood through oxidation, reduction, hydrolysis, and conjugation; it converts fat-soluble compounds to water-soluble compounds. Figures 19.4 and 19.5 give schematics of the liver and kidney. After the compound has been transformed in the liver it re-enters the blood; some transformed compounds are carried by the bile to the intestines for excretion. The blood with the transformed compound (as well as some of the original compound not yet transformed) goes to the various organs, including the kidney, which receives 25% of the cardiac output, takes the transformed compound and puts it into the urine, with which the transformed poison leaves the body. Not all

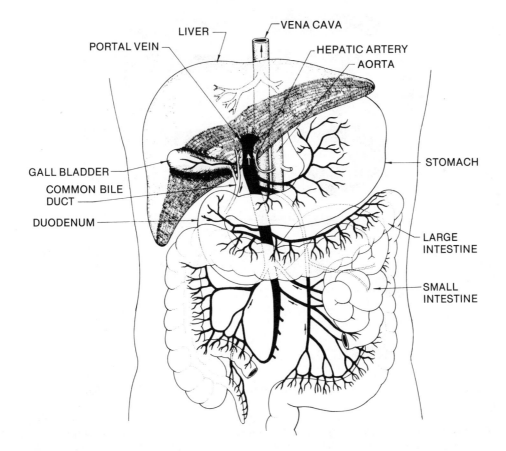

Figure 19.4 *Liver cells* are the primary source for biotransformations of poisons entering the body. About 30% of the liver's blood enters from the general circulation (through the hepatic artery) and 70% from the stomach and intestines (portal vein). Blood from the stomach and intestines must pass through the liver before entering the general circulation. Transformed materials leave the liver in bile (bile duct to small intestine) or in the blood to the general circulation (hepatic vein to vena cava).

blood goes to the liver and not all of a compound is biotransformed each time; the liver even metabolizes some compounds into compounds more toxic than the parent compound. This permits drugs to reach various target organs and cause the effects before the liver eventually detoxifies the drug. The various metabolic reactions are very complex and interactions become very important. For example, two compounds—such as alcohol and barbiturates; tobacco from cigarette smoking and cotton dust or asbestos—may both compete for the same liver enzyme (thus slowing down biotransformations) or both may act on the same or different target organs (e.g., lungs, brain). Johns-Manville Corporation, citing a study that smokers who are occupationally exposed to asbestos have a 92 times greater chance of developing lung cancer than the general population, has banned smoking in its asbestos facilities. It expects the policy to give asbestos workers the same chance of lung cancer as the general nonsmoking population. Alcohol and its products disturb liver metabolism and damage liver cells.[15] Heavy metals, such as lead and methyl mercury, inhibit enzymes.

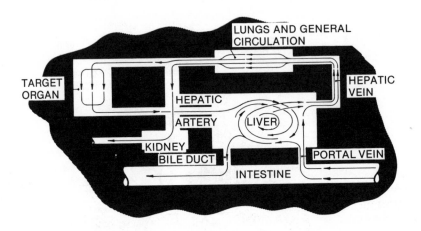

Figure 19.5 *Biotransformation* is shown schematically. The poison can directly enter the general circulation from the lungs or by penetration of the skin. If it comes from the intestines it must pass through the liver. On each passage through the liver, a proportion of the poison is transformed. The transformed compounds in the blood are water soluble and are transferred in the kidney to the urinary tract for elimination.

THRESHOLD LIMIT VALUES (TLV)

Table 19.1 gives Carson's definitions of the various types of industrial particulates.[6]

Table 19.2 gives the TLVs for *mineral* (primarily silica) dusts in the USA as of 1976 (*Federal Register*, 1974). The American Conference of Governmental Industrial Hygienists (PO Box 1937, Cincinnati, Ohio 45201) issues updated TLV recommendations for chemical substances and physical agents each year. TLVs can be expressed as a count of particles/cc of air or as the weight of the particles/m³ of air. The somewhat unusual numbers are the result of establishing the original standards in the foot-pound system and then converting later to the metric system. Mica, for example, had a TLV of 20,000,000 particles/ft³ which converts to 706,000,000 particles/m³ or 706 particles/cc. For reference, a room

Table 19.1   Definitions of Industrial Particulates[6]

Particulate	Definition	Example
Dust	Solid particles in a gaseous medium by mechanical breakup; subdivided into total and respirable	
Total (nuisance)	All dusts without respect to particle size	Portland cement, limestone, gypsum
Respirable	Inhaled into the respiratory system, and retained in the alveolar region; estimated from percent passing a filter; see Figure 19.3	Coal, silica
Fume	Formed by condensation of vaporized solid—usually metal; when from materials, called smokes	Cadmium, heavy metal oxides
Mist	Liquid droplets formed by condensation of vapor or atomization. Smaller size particles give fogs	Chromic mist sulfuric acid mist
Fiber	Particles with a length/width ratio (aspect ratio) over 3	Asbestos

with heavy cigarette smoke would have 1060 particles/cc and clean country air would have 35 particles/cc.

To convert (at 25 C):

$$\text{ppm} = \frac{24.45 \ (\text{mg/m}^3)}{\text{molecular weight}} \tag{19.1}$$

Tables 19.3 and 19.4 give the TLVs for gases, vapors, and nonmineral dusts (*Federal Register,* 1974). The primary difference between the two tables is the "excursions" above the TLVs. Table 19.3 gives the TLV by weight and by volume. However, there are certain substances, indicated by a c next to the value, for which the TLV is not for a time-weighted 8-hr day but are a ceiling which should not be exceeded even instantaneously. For example, chlorine has a TLV of 1 ppm. As long as the values average 1 ppm or less over the 8-hr day, the exposure is legal. Iodine, on the other hand, has a TLV with a c next to it so the ceiling is 0.1 ppm.

However there are certain substances for which there is an 8-hour TLV, a ceiling value, and a short-term excursion limit (STEL); see Table 19.4. Although OSHA has not given STEL values for other substances, the American Conference of Governmental Industrial Hygienists recommends STEL of 3 times the 8-hour TLV for TLVs of 0 to 1 ppm or mg/m³, STEL of 2 times the 8-hour TLV for TLVs of 1-10, STEL of 1.5 the 8-hour TLV for TLVs of 10-100, and 1.25 times the 8-hour TLV for TLVs of 100-1000.[7] Thus, using this guideline, chlorine would have a STEL of $3 \times 1 = 3$.

Table 19.4 gives materials with specific excursion recommendations. Toluene, for example, has an 8-hour TLV of 200 ppm. Excursions to 300 ppm for 10 minutes are acceptable while at no time should the worker be exposed to concentrations of 500 ppm.

The TLV is based on an 8-hour exposure. The calculations may need to be adjusted if (a) concentration varies during the day, (b) the day is not 8 hours, or (c) if exposure is to more than one compound.

First, assume a worker was exposed to isoamyl acetate for 4 hours at 50 ppm, 2 hours at 75 ppm, and 2 hours at 150 ppm. The equivalent exposure is:

$$TWA = \frac{C_a t_a + C_b t_b \dots C_n t_n}{8} \tag{19.2}$$

Table 19.2   TLVs for Mineral Dusts in the USA as of 1976. Other Dusts are in Tables 19.3 and 19.4. The S in the Formulas is the SiO_2 Determined from Air-Borne Samples, Except Where Other Methods Have Been Shown Applicable.

Substance	Count, particles/ cc or million particles/m^3	Mass, milligrams/m^3
Amorphous (including natural diatomaceous earth)	706	80/ S
Asbestos	2+ micrometers in length	
Coal dust		
*respirable fraction less than 5 SiO_2		$\dfrac{2.4 \text{ mg/m}^3}{2+S}$
*respirable fraction over 5 SiO_2		$\dfrac{10 \text{ mg/m}^3}{2+S}$
Graphite (natural)	5295	
Inert or nuisance dust		
*respirable fraction	530	5 mg/m^3
*total dust	1765	15 mg/m^3
Silica		
*crystalline		
**quartz (respirable)	$\dfrac{8825}{5+S}$	$\dfrac{10 \text{ mg/m}^3}{2+S}$
**quartz (total dust)		$\dfrac{30 \text{ mg/m}^3}{2+S}$
*cristobalite	Use 1/2 the value calculated from the quartz count or mass formulas.	
*tridymite		
Silicates (less than 1 crystalline silica)		
*mica	706	
*portland cement	1765	
*soapstone	706	
*talc		
**nonasbestos form; less than 1 quartz	706	
**nonasbestos form; over 1 quartz	Use quartz limit.	
**fibrous	Use asbestos limit.	
*tremolite	Use asbestos limit.	

Ceiling concentration = 10

where TWA = Time Weighted Average (equivalent 8-hour exposure)
$\quad\quad\quad C$ = Concentration of $a, b, c \ldots$, ppm or mg/m^3
$\quad\quad\quad t$ = time of exposure to concentration $a, b, c \ldots$, hours

For the above example, $TWA = [50\ (4) + 75\ (2) + 150\ (2)]/8 = 650/8 = 81.25$. Since 81 is less than the TLV of 100, the exposure is legal.

Next assume a worker was exposed to 125 ppm of isoamyl acetate during the entire working day but only worked 6 hr/day. Then $TWA\ [125\ (6) + 0\ (2)]/8 = 750/8 = 94$, which is below the TLV of 100; the exposure is legal.

Next assume a worker was exposed to 90 ppm of isoamyl acetate for an entire working shift of 10 hours. Then $TWA = 90\ (10)/8 = 900/8 = 112$ which is above the TLV of 100; the exposure is illegal.

The worker also might be exposed to a mixture of contaminants. Assume expo-

Table 19.3 Threshold Limit Values (8-hour Time Weighted Mean). On a Volume Basis, ppm = parts of Vapor or Gas per Million Parts of Contaminated Air by Volume at 25 C and 260 torr. On a Weight Basis, mg/m³ = Approximate Milligrams of Particulate Per Cubic Meter of Air. The Word *Skin* After a Compound Indicates That a Compound is Capable of Penetrating Intact Human Skin to Such an Extent That the Dose From Skin Absorption May Approach or Exceed the Dose Inhaled; Thus Skin Absorption Must be Prevented.

Substance	ppm	mg/m³
Acetaldehyde	200	360
Acetic acid	10	25
Acetic anhydride	5	20
Acetone	1,000	2,400
Acetonitrile	40	70
Acetylene dichloride, see 1,2-Dichloroethylene		
Acetylene tetrabromide	1	14
Acrolein	0.1	0.25
Acrylamide—Skin		0.3
Acrylonitrile—Skin	20	45
Aldrin—Skin		0.25
Allyl alcohol—Skin	2	5
Allyl chloride	1	3
Allylglycidyl ether (AGE)	10p	45p
Allyl propyl disulfide		
2-Aminoethanol, see Ethanolamine		
2-Aminopyridine	0.5	2
Ammonia	50	35
Ammonium sulfamate (Ammate)		15
n-Amyl acetate	100	525
sec-Amyl acetate	125	650
Aniline—Skin	5	19
Anisidine (o,p-isomers)—Skin		0.5
Antimony and compounds (as Sb)		0.5
ANTU (alpha naphthyl thiourea)		0.3
Arsenic and compounds (as As)		0.5
Arsine	0.5	0.2
Azinphos-methyl—Skin		0.2
Barium (soluble compounds)		0.5
p-Benzoquinone, see Quinone		
Benzoyl peroxide		5
Benzyl chloride	1	5
Biphenyl, see Diphenyl		
Bisphenol A, see Diglycidyl ether		
Boron oxide		15
Boron trifluoride	1p	3p
Bromine	0.1	0.7
Bromoform—Skin	0.5	5
Butadiene (1, 3-butadiene)	1,000	2,200
Butanethiol, see Butyl mercaptan		
2-Butanone	200	590
2-Butoxy ethanol (Butyl Cellosolve)—Skin	50	240
Butyl acetate (n-butyl acetate)	150	710
sec-Butyl acetate	200	950
tert-Butyl acetate	200	950
Butyl alcohol	100	300
sec-Butyl alcohol	150	450
tert-Butyl alcohol	100	300
Butylamine—Skin	5p	15p
tert-Butyl chromate (as CrO₃)—Skin		0.1p
n-Butyl glycidyl ether (BGE)	50	270
Butyl mercaptan	10	35
p-tert-Butyltoluene	10	60
Calcium arsenate		1
Calcium oxide		5
Camphor	2p	
Carbaryl (Sevin (R))		5
Carbon black		3.5
Carbon dioxide	5,000	9,000
Carbon monoxide	50	55
Chlordane—Skin		0.5
Chlorinated camphene—Skin		0.5
Chlorinated diphenyl oxide		0.5
Chlorine	1	3
Chlorine dioxide	0.1	0.3
Chlorine trifluoride	0.1p	0.4p
Chloroacetaldehyde	1p	3p
a-Chloroacetophenone (phenacylchloride)	0.05	0.3
Chlorobenzene (monochlorobenzene)	75	350
o-Chlorobenzylidene malononitrile (OCBM)	0.05	0.4
Chlorobromomethane	200	1,050
2-Chloro-1,3-butadiene, see Chloroprene		
Chlorodiphenyl (42 percent Chlorine)—Skin		1
Chlorodiphenyl (54 percent Chlorine)—Skin		0.5
1-Chloro, 2, 3-epoxypropane, see Epichlorhydrin		
2-Chloroethanol, see Ethylene Chlorphydrin		
Chloroethylene, see Vinyl chloride		
Chloroform (trichloromethane)	50p	240p
1-Chloro-1-nitropropane		
Chloropicrin	0.1	0.7
Chloroprene (2-chloro-1,3-butadiene)—Skin	25	90
Chromium, sol. chromic, chromous salts as Cr	0.5	
Metal and insol. salts		1
Coal tar pitch volatiles (benzene soluble fraction) anthracene, BaP, phenanthrene, acidine, chrysene, pyrene		0.2
Cobalt, metal fume and dust		0.1
Copper fume		0.1
Dusts and Mists		1
Cotton dust (raw)		1
Crag (R) herbicide		15
Creosol (all isomers)—Skin	5	22
Crotonaldehyde	2	6
Cumene—Skin	50	245
Cyanide (as CN)—Skin		5
Cyclohexane	300	1,050
Cyclohexanol	50	200
Cyclohexanone	50	200
Cyclohexene	300	1,015
Cyclopentadiene	75	200
2, 4-D		10
DDT—Skin		1
DDVP, see Dichlorvos		
Decaborane—Skin	0.05	0.3
Demeton (R)—Skin		0.1
Diacetone alcohol (4-hydroxy-4-methyl-2-pentanone)	50	240
1,2-diaminoethane, see Ethylenediamine		
Diazomethane	0.2	0.4
Diborane	0.1	0.1
Dibutylphthalate		.5
o-Dichlorobenzene	50p	300p
p-Dichlorobenzene	75	450
Dichlorodifluoromethane	1,000	4,950
1,3-Dichloro-5,5-dimethyl hydantoin		0.2
1,1-Dichloroethane	100	400
1,2-Dichloroethylene	200	790
Dichloroethyl ether—Skin	15p	90p
Dichloromethane, see Methylenechloride		
Dichloromonofluoromethane	1,000	4,200
1,1 Dichloro-1-nitroethane	10p	60p
1,2 Dichloropropane, see Propylenedichloride		
Dichlorotetrafluoroethane	1,000	7,000
Dichlorvos (DDVP)—Skin		1
Dieldrin—Skin		0.25
Dieldrin—Skin		
Diethylamine	25	75
Diethylamino ethanol—Skin	10	50
Diethylether, see Ethyl ether		
Difluorodibromomethane	100	860
Diglycidyl ether (DGE)	0.5p	2.8p
Dihydroxybenzene, see Hydroquinone		
Diisobutyl ketone	50	290
Diisopropylamine—Skin	5	20
Dimethoxymethane, see Methylal		
Dimethyl acetamide—Skin	10	35
Dimethylamine	10	18
Dimethylaminobenzene, see Xylidene		
Dimethylaniline (N-dimethylaniline)—Skin	5	25
Dimethylbenzene, see Xylene		
Dimethyl 1,2-dibromo-2,2-dichloroethyl phosphate, (Dibrom)		3
Dimethylformamide—Skin	10	30
2,6-Dimethylhelptanone, see Diisobutylketone		
1,1-Dimethylhydrazine—Skin	0.5	1
Dimethylphthalate	5	
Dimethylsulfate—Skin	1	5
Dinitrobenzene (all isomers)—Skin		1
Dinitro-o-cresol—Skin		0.2
Dinitrotoluene—Skin		1.5
Dioxane (Diethylene dioxide)—Skin	100	360
Diphenyl	0.2	1
Diphenylmethane diisocyanate (see Methylene bisphenyl isocyanate (MDI)		
Dipropylene glycol methyl ether—Skin	100	600
Di-sec, octyl phthalate (Di-2 ethylhexylphthalate)		5
Endrin—Skin		0.1
Epichlorhydrin—Skin	5	19
EPN—Skin		0.5
1,2 Epoxypropane, see Propyleneoxide		
2,3 Epoxy-1-propanol, see Glycidol		
Ethanethiol, see Ethylmercaptan		
Ethanolamine	3	6
2-Ethoxyethanol—Skin	200	740
2-Ethoxyethylacetate (Cellosolve acetate)—Skin	100	540
Ethyl acetate	400	1,400
Ethyl acrylate—Skin	25	100
Ethyl alcohol (ethanol)	1,000	1,900
Ethylamine	10	18
Ethyl sec-amyl ketone (5-methyl 3 heptanone)	25	130
Ethyl benzene	100	435
Ethyl bromide	200	890
Ethyl butyl ketone (3-Heptanone)	50	230
Ethyl chloride	1,000	2,600
Ethyl ether	400	1,200
Ethyl formate	100	300
Ethyl mercaptan	10p	25p
Ethyl silicate	100	850
Ethylene chlorohydrin—Skin	5	16
Ethylenediamine	10	25
Ethylene dibromide, see 1,2-Dibromoethane		
Ethylene dichloride, see 1,2-Dichloroethane		
Ethylene glycol dinitrate and/or Nitroglycerin—Skin	0.2p	1p
Ethylene glycol monomethyl ether acetate, see Methyl cellosolve acetate		

Substance	ppm	mg/m³	Substance	ppm	mg/m³
Ethylene imine—Skin	0.5	1	Pentachloronaphthalene—Skin		0.5
Ethylene oxide	50	90	Pentachlorophenol—Skin		0.5
Ethylidine chloride, see 1,1-Dichloroethane			Pentane	1,000	2,950
N-Ethylmorpholine—Skin	20	94	2-Pentanone	200	700
Ferbam		15	Perchloromethyl mercaptan	0.1	0.8
Ferrovanadium dust		1	Perchloryl fluoride	3	13.5
Fluoride (as F)		2.5	Petroleum distillates (naphtha)	500	2,000
Fluorine	0.1	0.2	Phenol—Skin	5	19
Fluorotrichloromethane	1,000	5,600	p-Phenylene diamine—Skin		0.1
Formic acid	5	9	Phenyl ether (vapor)	1	7
Furfural—Skin	5	20	Phenyl ether-biphenyl mixture (vapor)	1	7
Furfuryl alcohol	50	200	Phenylethylene, see Styrene		
Glycidol (2,3 Epoxy 1 propanol)	50	150	Phenylglycidyl ether (PGE)	10	60
Glycol monoethyl ether, see 2-Ethoxyethanol			Phenylhydrazine—Skin	5	22
Guthion (R), see Azinphosmethyl			Phosdrin (Mevinphos) (R)—Skin		0.1
Hafnium		0.5	Phosgene (carbonyl chloride)	0.1	0.4
Heptachlor—Skin		0.5	Phosphine	0.3	0.4
Heptane (n-heptane)	500	2,000	Phosphoric acid		1
Hexachloroethane—Skin		0.2	Phosphorus (yellow)		0.1
Hexane (n-hexane)	500	1,800	Phosphorus pentachloride		1
2-Hexanone	100	410	Phosphorus pentasulfide		1
Hexone (Methyl isobutyl ketone)	100	410	Phosphorus trichloride	0.5	3
sec-Hexyl acetate	50	300	Phthalic anhydride	2	12
Hydrazine—Skin	1	1.3	Picric acid—Skin		0.1
Hydrogen bromide	3	10	Pival (R) (2-Pivalyl-1,3-indandione)		0.1
Hydrogen chloride	5p	7p	Platinum (soluble salts) as Pt		0.002
Hydrogen cyanide—Skin	10	11	Propargyl alcohol—Skin	1	
Hydrogen peroxide (90%)	1	1.4	Propane	1,000	1,800
Hydrogen selenide	0.05	0.2	n-Propyl acetate	200	840
Hydroquinone		2	Propyl alcohol	200	500
Iodine	0.1p	1p	n-Propyl nitrate	25	110
Iron oxide fume		10	Propylene dichloride	75	350
Isoamyl acetate	100	525	Propylene imine—Skin	2	5
Isoamyl alcohol	400	980	Propylene oxide	100	240
Isopropylamine	5	12	Propyne, see Methylacetylene Pyrethrum		5
Isopropylether	500	2,100	Pyridine	5	15
Isopropyl glycidyl ether (IGE)	50	240	Quinone	0.1	0.4
Ketene	0.5	0.9	RDX—Skin		1.5
Lead arsenate		0.15	Rhodium, Metal fume and dusts, as Rh		0.1
Lindane—Skin		0.5	Soluble salts		0.001
Lithium hydride		00.025	Ronnel		10
L.P.G. (Liquified petroleum gas)	1,000	1,800	Rotenone (commercial)		5
Magnesium oxide fume		15	Selenium compounds (as Se)		0.2
Malathion—Skin		15	Selenium hexafluoride	0.05	0.4
Maleic anhydride	0.25	1	Silver, metal and soluble compounds		0.01
Manganese		5p	Sodium fluoroacetate (1080)—Skin		0.05
Mesityl oxide	25	100	Sodium hydroxide		2
Methanethiol, see Methyl mercaptan			Stibine	0.1	0.5
Methoxychlor		15	Stoddard solvent	500p	2,950p
2-Methoxyethanol, see Methyl cellosolve			Strychnine		0.15
Methyl acetate	200	610	Sulfur dioxide	5	13
Methyl acetylene (propyne)	1,000	1,650	Sulfur hexafluoride	1,000	6,000
Methyl acetylene-propadiene mixture (MAPP)	1,000	1,800	Sulfuric acid		1
Methyl acrylate—Skin	10	35	Sulfur monochloride	1	6
Methylal (dimethoxymethane)	1,000	3,100	Sulfur pentafluoride	0.025	0.25
Methyl alcohol (methanol)	200	260	Sulfuryl fluoride	5	20
Methylamine	10	12	Systox, see Demeton (R) 2,4,5T		10
Methyl amyl alcohol, see Methylisobutyl carbinol			Tantalum		5
Methyl (n-amyl) ketone (2-Heptanone)	100	465	TEDP—Skin		0.2
Methyl bromide—Skin	20p	80p	Tellurium		0.1
Methyl butyl ketone, see 2-Hexanone			Tellurium hexafluoride	0.02	0.2
Methyl cellosolve—Skin	25	80	TEPP—Skin		0.05
Methyl cellosolve acetate—Skin	25	120	Terphenyls	1p	9p
Methyl chloroform	350	1,900	1,1,1,2-Tetrachloro-2,2-difluoro ethane	500	4,170
Methylcyclohexane	500	2,000	1,1,2,2-Tetrachloro-1,2,difluoro ethane	500	4,170
Methylcyclohexanol	100	470	1,1,2,2-Tetrachloroethane—Skin	5	35
o-Methylcyclohexanone—Skin	100	460	Tetrachloroethylene, see Perchloroethylene		
Methyl ethyl ketone (MEK), see 2-Butanone			Tetrachloromethane, see Carbon tetrachloride		
Methyl formate	100	250	Tetrachloronaphthalene—Skin		0.075
Methyl iodide—Skin	5	28	Tetrahydrofuran	200	590
Methyl isobutyl carbinol—Skin	25	100	Tetramethyl lead (as Pb)—Skin		0.07
Methyl isobutyl ketone, see Hexone			Tetramethyl succinonitrile—Skin	0.5	3
Methyl isocyanate—Skin	0.02	0.05	Tetranitromethane	1	8
Methyl mercaptan	10p	20p	Tetryl (2,4,6-trinitrophenyl-methyinitramine)—Skin		1.5
Methyl methacrylate	100	410	Thallium (soluble compounds)—Skin as Tl		0.1
Methyl propyl ketone, see 2-Pentanone			Thiram		5
a Methyl styrene	100p	480p	Tin (inorganic compounds, except oxides)		2
Methylene bisphenyl isocyanate (MDI)	0.02p	0.2p	Tin (organic compounds)		0.1
Molybdenum:			Toluene-2,4-diisocyanate	0.02p	0.14p
Soluble compounds		5	o-Toluidine—Skin	5	22
Insoluble compounds		15	Toxaphene, see Chlorinated camphene		
Monomethyl aniline—Skin	2	9	Tributyl phosphate		5
Monomethyl hydrazine—Skin	0.2p	0.35p	1,1,1-Trichloroethane, see Methyl chloroform		
Morpholine—Skin	20	70	1,1,2-Trichloroethane—Skin	10	45
Naphtha (coal tar)	100	400	Titaniumdioxide		15
Naphthalene	10	50	Trichloromethane, see Chloroform		
Nickel carbonyl	0.001	0.007	Trichloronaphthalene—Skin		5
Nickel, metal and soluble compounds, as Ni		1	1,2,3-Trichloropropane	50	300
Nicotine—Skin		0.5	1,1,2-Trichloro 1,2,2-trifluoroethane	1,000	7,600
Nitric acid	2	5	Triethylamine	25	100
Nitric oxide	25	30	Trifluoromonobromomethane	1,000	6,100
p-Nitroaniline—Skin	1	6	2,4,6-Trinitrophenol, see Picric acid		
Nitrobenzene—Skin	1	5	2,4,6-Trinitrophenylmethylnitramine, see Tetryl		
p-Nitrochlorobenzene—Skin		1	Trinitrotoluene—Skin		1.5
Nitroethane	100	310	Triorthocresyl phosphate		0.1
Nitrogen dioxide	5	9	Triphenyl phosphate		3
Nitrogen trifluoride	10	29	Turpentine	100	560
Nitroglycerin—Skin	0.2	2	Uranium (soluble compounds)		0.05
Nitromethane	100	250	Uranium (insoluble compounds)		0.25
1-Nitropropane	25	90	Vanadium:		
2-Nitropropane	25	90	V205		0.5p
Nitrotoluene—Skin	5	30	V205 fume		0.1p
Nitrotrichloromethane, see Chloropicrin			Vinyl benzene, see Styrene		
Octachloronaphthalene—Skin		0.1	Vinylcyanide, see Acrylonitrile		
Octane	500	2,350	Vinyl toluene	100	480
Oil mist, mineral		5	Warfarin		0.1
Osmium tetroxide		0.002	Xylene (xylol)	100	435
Oxalic acid		1	Xylidine—Skin	5	25
Oxygen difluoride	0.05	0.1	Yttrium		1
Ozone	0.1	0.2	Zinc chloride fume		1
Paraquat—Skin		0.5	Zinc oxide fume		5
Parathion—Skin		0.11	Zirconium compounds (as Zr)		5
Pentaborane	0.005	0.01			

Table 19.4   Substances Which OSHA Assigns 8-hour TLVs, Short-Term Exposure Limit TLVs, and a Ceiling TLV. Benzene, for Example, Has an 8-hour TLV of 10 ppm, a TLV-STEL of 25 ppm, and a Ceiling of 50 ppm. A Worker May Not Be Exposed to Over 50 ppm For Any Length of Time, But Can Be Exposed to 25 ppm For a Maximum Time of 10 min as Long as the Average Exposure for the 8 Hours is Less Than 10 ppm.

	TLV		STEL			Ceiling	
	ppm	mg/m³ ppm	mg/m³		Maximum duration, min	ppm	mg/m³
Benzene	10	25		10		50	
Beryllium & beryllium compounds		.002	.005	30			.025
Cadmium dust		.2	.6				
Cadmium fumes		.1	.3				
Carbon disulfide	20	30		30		100	
Carbon tetrachloride	10	25			5 (in any 4 hrs)	200	
Chromic acid and chromates			.1				
Ethylene dibromide	20	30		5		50	
Ethylene dichloride	50	100			5 (in any 3 hrs)	200	
Fluoride as dust		2.5					
Formaldehyde	3	5		30		10	
Hydrogen fluoride	3						
Hydrogen sulfide		20		10*		50	
Lead & its inorganic compounds		.2					
Mercury			.1				
Mercury, organo (alkyl)		.01	.04				
Methyl chloride	100	200			5 (in any 3 hrs)	300	
Methylene chloride	500	1000			5 (in any 2 hrs)	2000	
Styrene	100	200			5 (in any 3 hrs)	600	
Tetrachlorethylene	100	200			5 (in any 3 hrs)	300	
Toluene	200	300		10		500	
Trichlorethylene	100	200			5 (in any 2 hrs)	300	

*Once only if no other measurable exposure occurs.

sure (for 8 hours) to acetone at 500 ppm, 2-butanone of 45 ppm, and toluene of 40 ppm. The three TLVs are 1000, 200, and 200 ppm. Then

$$TWA_{mixture} = \frac{C_1}{TLV_1} + \frac{C_2}{TLV_2} + \cdots \frac{C_n}{TLV_n} \qquad (19.3)$$

where $TWA_{mixture}$ = Equivalent TWA mixture exposure (maximum of 1 permitted)

$C_{1,2,3}$ = Concentration (8 hours) for a particular contaminant

$TLV_{1,2,3}$ = TLV for a particular contaminant

For the numbers above, $TWA_{mixture}$ = 500/1000 + 45/200 + 40/200 = .500 + .225 + .200 = .925. Since .925 is less than 1, the exposure is legal.

The TLVs by themselves are not always indicators of relative hazard.[11] Other factors such as volatility may be important. At a specific temperature a compound with greater volatility may produce more gas. For example, the TLV for methyl alcohol is 4 times higher than furfuryl alcohol but its vapor pressure at room temperature is 18 times higher. Anhydrous ammonia, which has the same TLV as furfuryl alcohol, is 1100 times more hazardous in case of a spill due to its vapor pressure.

The above calculations have been performed legalistically. The engineer should remember, however, the accuracy of the measuring equipment and the unknowns the scientists faced when trying to establish the standards. That is, the TLV numbers are guides rather than precise numbers. In general, TLV standards are getting lower as more evidence is obtained. Also, exercise makes a difference; Astrand found that the air in the lungs contained twice as much toluene during light work as during rest.[23] People with sedentary occupations inhale about 5 m^3 of air/8 hr while those working very hard may inhale 20/8 hr (see equation 12.15).

In general, repeated exposures to values close to the TLVs are not desirable and extra precautions should be taken even though TLVs are supposed to protect workers exposed to the chemical 40 hours/week for their working life. (Providing pure air for breathing may also benefit your product quality.) The TLVs also assume good health; be especially careful for people with liver or kidney problems. Women who might become pregnant present a critical problem due to poisoning of the baby; the baby may become poisoned during the first month or two before the worker even knows she is pregnant.

CONTROL OF RESPIRATORY HAZARDS

Measurement of the contaminant concentration should be representative of the worker's breathing zone. The standard technique is to use a battery-powered pump with the sampling device in the breathing zone. Use of a direct readout unit (where sampling and analysis occur in one step) has the advantage that the second-by-second results of a worker's motions can be related to exposure. This is much more useful for changing the workstation or the method than just knowing the total exposure over an 8-hour period.

Reduce exposure (say, to toluene which has a TLV = 200 ppm) with two approaches: (1) engineering controls to reduce concentration in the worker's breathing area, and (2) administrative controls to reduce worker sensitivity. Eight engineering controls and four administrative controls are given in Table 19.5 As a last resort, use personal protective devices. Unfortunately, managements often tolerate working conditions for workers that they would not accept for themselves. A key question is "Would you let your daughter work there?"

Table 19.5   Controls for Respiratory Hazards can be Subdivided into Engineering Controls and Administrative Controls.[22] Engineering Controls are More Desirable Than Administrative Controls.

Engineering Controls	Administrative Controls
1. Substitute a less harmful material	1. Screen potential employees.
2. Change the machine or process.	2. Periodically examine existing employees (biological monitoring).
3. Enclose (isolate) the process.	3. Train engineers, managers, and workers.
4. Use wet methods.	4. Reduce exposure time.
5. Provide local ventilation.	
6. Provide general ventilation.	
7. Use good housekeeping.	
8. Control waste disposal.	

Engineering Controls

Substitute a less harmful material. Use methyl chloroform (TLV = 350 ppm) as a solvent instead of carbon tetrachloride (TLV = 10 ppm); use toluene (TLV = 200 ppm) to strip paint instead of benzene (TLV = 10 ppm); use water-base cleaning compounds instead of organic-base.

Change the machine or process. Reduce carbon monoxide by using electric powered fork trucks instead of gasoline. Use safety cans instead of glass bottles. Reduce dust by using low-speed oscillating sanders instead of high-speed rotary sanders. Paint with a brush or roller instead of spray painting. Remove grinding particles with a vacuum cleaner rather than blowing them away with compressed air.

Enclose (isolate) the process. Usually it is less expensive to capture materials and vapors before they "get out" rather than once they have dispersed. For example, ventilate laboratory storage cabinets. The plastic strips used for strip doors also can be used to enclose machines or processes, such as solvent tanks, which require passage of product. Enclosure also reduces the number of workers exposed. Isolate pumps which could leak toxic compounds. During cleaning, maintenance, and filling of enclosed machines, take special care to prevent escape of the materials and care for the protection of the maintenance workers. (An automotive plant found its maintenance workers were exposed to beryllium when repairing copper welding tools.) A sealer coat on concrete floors reduces dust; less dust means less air changes/hr and thus lower heating costs. Factories are not the only location of toxic compounds. Asbestos fiber concentrations are high in many offices. Asbestos has been used as a fire retardant on steel beams. Over time, vibrations shake fibers loose to be circulated by the ventilation system. The problem is greatest when the area between the dropped ceiling and the floor above is used as a ventilating plenum. Plunkett and Barbela suggest embalmers may be exposed to too much formaldehyde.[21]

Use wet methods. Water alone may not be enough; use a wetting agent and dispose of the wetted particulate before it dries. Wet floors before sweeping. In rock drilling, use hollow drills through which water is passed. Steam cotton. Use moistened flint in potteries. Use a high-pressure water jet instead of abrasive blasting to clean castings.

Provide local ventilation. Capturing vapors before they disperse is more efficient than after dispersal, since, as a first approximation, cost of air pollution control is proportional to the volume of air handled. Two principles are: (1) physically enclose the process or equipment as much as possible, and (2) remove air from the enclosure (hood) fast enough so that air movement at all openings is *into* the enclosure.[19] A mechanical supply of air usually is superior to depending on infiltration of air. Remove solder fumes with "vacuum cleaner" hoses mounted on the work surface. Dumping the exhaust "out the window" is not satisfactory; clean the air with filters, cyclones, vapor traps, precipitators, etc. In some cases, this trapped "waste product" can be sold for a profit. One example is spraying alfalfa dust in a cyclone with liquid lard from a rendering company; the resulting mixture is sold at a profit as cattle feed. Paper mills formerly dumped sulfite liquor into the river. When forbidden to do this, they found they could sell it at a profit as a dust suppressor on roads.

Provide general (dilution) ventilation. Use general ventilation when the contaminant is released from nonpoint sources; for point sources, local exhaust is more efficient. Dilution ventilation tends to exhaust large volumes of heated air. Local ventilation (fans, blowers) is preferrable to natural ventilation (open doors, windows) since air direction, volume, and velocity can be controlled. Inadvertent recirculation of exhausted air is a problem; discharge exhaust air so that it escapes from the "cavity" which forms as a result of wind movement around buildings.

Use good housekeeping. Remove dust from floors and ledges to prevent dust generation by traffic, vibration, and air currents. Eliminate piles of open containers of chemicals. Immediately clean up spills of volatile materials. Fix leaking containers. Dermatoses often can be prevented by good personal hygiene but that requires conveniently located wash basins and disposable hand towels.

Control waste disposal. Follow specific procedures under strict supervision. In one General Motors plant, management had the sewer covers tack-welded shut to prevent employees from dumping wastes into the sewer. Each disposal problem should be considered separately and specific procedures established for safe disposal of unused dangerous substances, toxic residues, contaminated wastes, material containers which are no longer needed, and containers with missing labels. Some plants have a special remote area for collection of materials before disposal.

Administrative Controls

Screen potential employees. Murphy, for example, found cigarette smokers were especially susceptible to byssinosis from the bracts and trashy materials of the cotton plant.[17] For jobs with toxicology hazards, screen out people with impaired livers or kidneys.

Periodically examine existing employees. BLVs (Biologic Limit Values) are measurements of a specific individual's exposure through measurement of the amount of a substance in the person (from blood, urine, hair, nails, or exhaled breath as may be appropriate). These periodic examinations also are known as "biological monitoring" to contrast with the physical monitoring of the environment. These backup examinations may catch problems that somehow have been overlooked or that may be caused by off-the-job exposure. (One electronics firm noted an employee's hearing was deteriorating rapidly. His job was in a quiet area but, on weekends, he played in a band. The firm noted the information on the employee's record in case the employee later tried to sue for occupational hearing loss.)

Train the engineers, managers, and workers. Give managers overviews with emphasis on costs and legal aspects; give details to the engineers. Informing the worker of the dangers often is neglected. Be sure the information is communicated in pictures and symbols—words often are misunderstood. Give specific behavior rather than glittering generalities ("Change the filter once/4 hr" rather than "Change filter if dirty").

Reduce exposure time. Have two workers each work 4 hr/day instead of one for 8; in addition to cutting exposure in half, each worker's recovery time increases from 16 hr/day to 20. Mines schedule blasting at the end of the shift so dust can settle down before the next day's shift arrives. Maintenance can be scheduled during off hours so the minimum number of people are exposed.

Personal Protection Equipment

Figures 19.6 and 19.7 show two pieces of personal protective equipment. Personal protective equipment is a last line of defense. It often fits poorly, workers abuse the equipment, equipment is not maintained, workers are not informed how to use the equipment, workers get a false sense of security from the equipment, etc.

Protective equipment generally should be furnished by the organization as employees tend to buy the cheapest product with no attention to quality or protection. Work clothing used in toxic environments should be washed by the organization to prevent contamination of family wearing apparel. The families of workers can die prematurely due to contaminants brought home by the worker in clothing and on the skin; beryllium, asbestos, lead, PCBs, and chlorinated hydrocarbons are especial problems.[2] Require changing and washing of clothes at work as well as showering. Take precautions that the commercial laundry personnel are not the new recipients of the toxin.

SAFETY

While toxicology deals with long-term effects (health), safety deals with the short-term (injuries). The goal is to: (1) reduce accident frequency, (2) reduce the proportion of accidents that become injuries, and (3) reduce lost days/injury. See Table 10.2. Because injuries are relatively rare events, it is easiest to work on the accidents.

Accidents are caused by unsafe conditions ("equipment" failure) and by unsafe acts ("human" failure). Managers tend to blame an accident on an unsafe act rather than an unsafe condition. Thus no blame can be attached to them. However, it has been demonstrated many times that the best way to improve safety is to reduce unsafe conditions. Commercial aircraft safety is a good example of what can be done in an inherently unsafe situation. The open manhole problem is the easiest way to remember which approach to use. The slogan approach to the problem is to put up signs "don't step into open manholes." The training-motivation approach is to have people practice walking around open manholes. The engineering approach is to put a cover on the open hole. The engineering approach is the best.

REDUCTION OF ACCIDENTS: UNSAFE CONDITIONS

Unsafe conditions will be subdivided into: reduce equipment failure rate and hazard, use protective equipment, and design the correct displays, controls, and environment. See Table 12.12 for techniques to improve safety of manual material handling.

Reduce Equipment Failure Rate and Hazard

Failure rate. Reduce the failure rate (increase the Mean Time Between Failures or *MTBF*) through safety factors, redundant equipment, and maintenance policies.

Safety factors involve designing so that the unit can take greater stress than the anticipated load. Examples are extra-thick electrical insulation, a larger than required motor, a stronger than required brace. Derating is using a component with a design life greater than the equipment life (such as an auto transmission which would last 50,000 miles beyond the expected vehicle life). Redundant equipment can be in parallel (e.g., two batteries in a car, either of which will start it) or in standby (diesel-powered electrical generator in a hospital). The parallel system has the advantage that the redundant unit is instantly available when needed, but the disadvantage that the redundant unit is in service and is wearing out (although stress on the unit may be very low due to the two units, thus giving a long *MTBF* for each component). The standby design unit isn't wearing out from use (although some items deteriorate even without use); the disadvantage is that the standby unit must be switched in and out of the system (giving time delays plus possibility of switching failure). Humans using the "buddy system" can use either a parallel or standby mode. Maintenance can replace a component before failure. This preventive maintenance can be based on hours of use (engine overhaul), built-in monitors (noise emitted by a smoke alarm when the battery is failing; noise from metal in brake pad when pads are worn), or just operator observation ("It's running hot today" or "It seems to shake more than usual").

Hazard. Hazard can be eliminated or reduced even if equipment fails. For example, a hand drill can replace an electric drill—thus eliminating a potential shock hazard; however, there probably would be an unacceptable loss in performance. Fail-safe is another way of eliminating the hazard (e.g., fuse in electrical circuit, deadman throttle on lift truck). Failure consequences can be reduced. For example, a battery-powered drill has a lower shock hazard than 110 V drill; a compressed-air-powered drill has less danger than an electrical drill; chemical A might be less caustic than chemical B and thus less dangerous to skin and eyes.

Use Protective Equipment

Distance. Distance is a powerful protective technique. Separating people from accidents reduces the chance of injury. Examples are: putting a machine which emits sparks or chips away from the aisle (or at least changing its orientation so that thrown objects can't hit people), putting a dike around a chemical tank so leaking compounds are contained, sufficient vertical distance under beams, pipes, and conveyors, clearance between vehicles and obstacles (trucks, cranes, RR cars), monitoring equipment by a TV instead of personally, not permitting people to walk under cranes which are carrying a load, automating a process (e.g., loading a press; automated warehouse), and loading-unloading a process with a push-pull jig (i.e., use mechanical feeding, not hand feeding). A fence (barrier, wall) can increase the effective distance without changing the actual distance.

If it is impossible to protect the worker with distance, then use fixed or mobile guards. Guards are not acceptable if (a) the guard can be defeated easily or (b) the guard defeat or failure is not easily detectable by the user *and* the using organization. For example, if a press is designed to be operated with two separated buttons to keep both hands out of the die, but one button is tied down, then this defeat should be obvious to any observer. Another example is an electrical guard (such as insulation) which might fail and so pass a lethal current. A poor guard may be a hazard in that it gives an impression of protection which doesn't exist.

Fixed guards (machine guards). Prevent contact of a person close to a potential accident with a guard on the machine or tool. Examples are: a barrier to prevent people from reaching into run-in or nip points (rotating machinery, shears), guards on sharp objects (saws), guards to prevent electrical shock, guards to catch falling objects (under overhead conveyors or on lift trucks), guards to intercept flying objects (chips, sparks), and guards (enclosures) to prevent vapor or liquid escape. Guards and handrails also prevent falls to a lower level (stairs, ramps, docks). Fixed guards should be purchased from the machine manufacturer when the machine is purchased as homemade guards generally are less effective and may not even be installed. Someone from the purchasing function should be on the safety committee. Purchase guards which permit oiling and maintenance without removing the guard. Guards should be sufficiently rugged to withstand predictable events (impact from lift trucks, bursting of abrasive wheels, stock kickback).

Mobile guards (people guards). Supplement machine guards with guards on the individual. Examples are: gloves and gauntlets (nonpermeable for chemicals; tough for abrasion resistance), aprons and leggings, helmets, safety shoes (see Table 14.4), goggles and face masks to protect the eyes and face, specialized clothing to protect against heat, cold, chemicals, and radiation (welding), earplugs and earmuffs to protect against noise (see Figure 18.12), hairnets and form-fitting clothing to protect against rotating machinery, and respirators to protect the respiratory system (see Figure 19.6). There are two problems with personal protective equipment—the last line of defense. First, it *is* the last line; distance between it and the person is minimal so, if it fails, an injury results. Second, personal protective equipment often decreases the comfort and performance of the user—thus there is the temptation not to use it. Thus safety-conscious managements impose severe penalties on employees not using safety equipment (e.g., 1 day off without pay on the first offense, 1 week off on the second offense, dismissal on the third offense). Unions strongly support such policies (as long as they are impartially administered), as unions are very safety-conscious. Since workers rarely have the technical knowledge to properly select safety equipment and might be tempted to purchase inferior protection to save money, organizations should purchase personal protective equipment and give it to the employees at no cost.

Design the Correct Control, Display, and Environment

Controls. Operators make machines do their bidding through controls. Most controls are operated by the hands and fingers although a few are operated by the feet or by voice. A safety problem is making the control operation clear to unpracticed users such as trainees and maintenance operators. Controls should conform to population stereotypes (common expectations). For example, in the USA we expect that moving a control up will cause the device to turn on or increase output; down is for off or decrease. Other motions such as push or pull, turn clockwise or counter-clockwise are not as firmly stereotyped. In addition, up doesn't mean on in all countries. Thus the designer can not *depend* upon users following the stereotyped direction. However, if a control operates in a nonstereotyped manner, then the chances of a mistake by someone are much greater. Thus controls should operate in predictable directions.

Another control problem is someone turning on the machine while someone else is maintaining it. A lockout system in which the maintainer has the only key is the best solution. Remember that electrical lockout is not sufficient as there may be energy stored in hydraulic lines, compressed air, springs, or potential energy of suspended parts; the machine should be in a "zero mechanical state." Emergency stop controls should be properly located. Start buttons should be guarded so the device is not started accidentally.

Displays. Displays—either audio or visual—present machine information to the operator or maintainer. Danger generally is presented through an auditory alarm. The auditory alarm can be of several different types (e.g., one sound for fire and a different sound for a lift truck). However, although audio alarms are attention-getting, the detailed danger information should be presented on visual displays. Visual displays alone are not sufficient because they are too easy to overlook (as when the operator's back is turned). As a convention, red (a light or the display background) means danger, yellow means caution, green means normal or safe, and white and blue have a neutral meaning. Flashing lights mean danger or problems. Table 19.6 gives some of the practical tips on effective emergency warning messages reported by Johnson.[14]

Environment. The environment problems seem to be housekeeping, distractions, and stress. Housekeeping can be grease and oil on the floor, projecting nails, sharp edges. Poor housekeeping causes fires. The key to good housekeeping is prevention, not cleanup. Solutions include "a place for everything and everything in its place" and organized scrap and waste disposal. Shelving and drawers use the cube of the space so there is less congestion. Drains reduce standing liquids and thus falls. Be careful not to make the floor too slick with wax. The distraction can be a girl with a new dress, trying to keep track of several things at the same time, or even an alarm occurring elsewhere. Stress can be caused by climate (too hot or cold), excessive noise, glare, or too rapid a work pace for the operator.

REDUCTION OF ACCIDENTS: UNSAFE ACTS

Lack of operator knowledge. An operator may cause an accident through lack of knowledge. The famous Three Mile Island nuclear accident occurred when an operator turned off a safety device which was bringing the system back to normal. In that case, the display did not present the proper information to the operator. A more common problem is when the displays are presenting the situation correctly but the individual (who may be the operator but also may be a maintainer or even a passer-by) does not know what to do. Or when there is a failure with no display (spilled battery acid or chemical or flammable liquid), do people know what to do? First, does anyone know *exactly* what actions should be taken to return the system to normal for *each* type of failure? A decision structure chart (see chapters 7 and 26) is a useful tool. Another useful technique is to present the instructions in three columns with the left column giving step-by-step instructions (what to do), the second column the details (how), and the third the reasons (why). Second, don't assume that, because management made a decision what to do five years ago, the present operator knows what to do today. Are the operating and maintenance instructions permanently attached to the machine or are they in some long-lost booklet? Thus knowledge and practice (a la fire drills and aircraft simulators) make perfect.

Table 19.6   Tips for Emergency Warning Messages[14]

- Use pictures not just words
- Use three-dimensional pictures, not two-dimensional
- Use pictures rather than symbols (which require learning)
- Number sequential drawings
- Show both initial and final locations of controls
- Present what not to do as well as what to do
- Use drawings, not photographs (less irrelevant information, can emphasize important items, easier to change)
- Show time pressure with clockface; time changes in each picture
- Evaluate message on typical users (*not* equipment designer)

Deliberate risk. Workers may take a deliberate risk because the risk is low and the rewards are large and immediate. Low probability events (accidents) are ineffective in controlling present behavior. Management must reinforce safe behavior with "carrots" and punish unsafe behavior with "sticks." Some carrots are praise, public recognition of a group's behavior, free T-shirts or hats with safety slogans, etc. Sticks are verbal abuse and disciplinary action.

SUMMARY

Accidents are caused by unsafe conditions and unsafe acts. Reduction of unsafe conditions (putting the cover on the hole) is the most important approach.

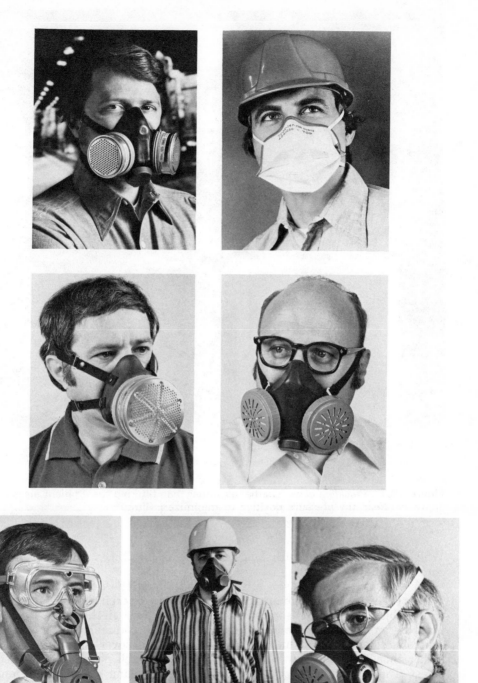

Figure 19.6 *Respirators* put a filter at the last line of defense—the nose and mouth. The OSHA respirator standard is: atmospheric oxygen content, maximum of 5 mg of condensed hydrocarbons per cubic m of gas, maximum of 5000 ppm of CO_2, and 20 ppm of CO.

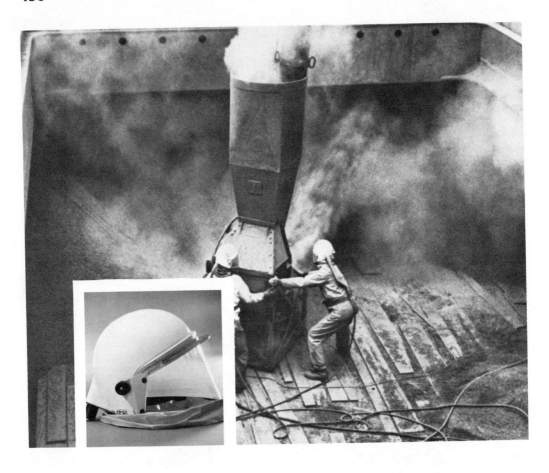

Source: Photograph courtesy of 3M Company, St. Paul, Minnesota.

Figure 19.7 *Providing clean external air* rather than filtering the ambient air is another approach; keep the pressure positive to minimize infiltration.

While toxicology deals with long-term effects (health), safety deals with the short-term (injuries).

For toxicology, the poison effect depends upon the dose. Threshold Limit Values (TLVs) give the maximum permitted dose for each chemical. Most industrial problems concern poisons entering the body through the lungs. Respiratory hazards can be reduced by engineering controls and administrative controls. To control accidents, reduce unsafe conditions (put the cover on the open manhole) rather than depend upon motivation or training (tell people not to step into holes). There are a number of techniques for reducing unsafe conditions and unsafe acts.

SHORT-ANSWER REVIEW QUESTIONS

19.1 Threshold Limit Values are based on 70 kg adults and an 8-hour day. How are the values determined in the USA?

19.2 Unsafe conditions can be reduced through reducing equipment failure rate and hazard, by using protective equipment, and by designing the correct displays, controls, and environment. Give two examples for each category.

19.3 The TLV is based on an 8-hour exposure. What is the TLV for acetone for a 4-hour exposure?

19.4 For toluene, what are the TLV, ceiling with its time, and STEL?

19.5 Why should organizations furnish personal protective equipment rather than having the employees buy it?

THOUGHT-DISCUSSION QUESTIONS

1. Toxicology is the study of the effect "foreign" chemicals have upon the body when considering the dose/response ratio.
   a. Discuss the word *effect*.
   b. Discuss the word *body*.
   c. Discuss the words *dose/response*.
2. Should TLVs be low enough to protect "everyone" or only "almost everyone"? Why?

REFERENCES

1. American Society of Heating, Refrigeration and Air Conditioning Engineers. Air contaminants. *Handbook of Fundamentals*. New York, 1972, chapter 10.

2. Bellin, J. Don't take your work home with you. *Occupational Health and Safety*, Vol. 5, No. 6 (1981): 39-42.

3. Birmingham, D. Occupational dermatoses: their recognition, control and prevention. *The Industrial Environment*. Washington, D.C.: Superintendent of Documents, 1973, chapter 34.

4. Birmingham, D. *The Prevention of Occupational Skin Disease*. New York City: Soap and Detergent Association, 1975.

5. Buechner, H. Organic dust: critical emerging health hazard. *Health and Safety* (January 1975): 22-24.

6. Carson, G. Sampling particulates in the industrial environment. *ASHRAE Journal*, Vol. 16, No. 5 (May 1974): 45-49.

7. Dinman, B. Medical aspects of the occupational environment. *The Industrial Environment*. Washington, D.C.: Superintendent of Documents, 1973, chapter 17.

8. _____. Principles and use of standards. *The Industrial Environment*. Washington, D.C.: Superintendent of Documents, 1973, chapter 8.

9. Frazer, D. Sizing methodology. *The Industrial Environment*. Washington, D.C.: Superintendent of Documents, 1973, chapter 14.

10. General Industry Safety and Health Regulations, Part 1910. *Federal Register*, June 27, 1974.

11. Hammer, W. *Occupational Safety Management and Engineering*, No. 377. Englewood Cliffs, N.J.: Prentice-Hall, 1976.

12. Hays, W.; Dale, W.; and Pirkle, C. Evidence of safety of long-term, high, oral doses of DDT for man. *Archives of Environmental Health*, Vol. 22 (1971): 119-135.

13. *The Industrial Environment—Its Evaluation and Control*. Washington, D.C.: Superintendent of Documents, 1973.

14. Johnson, D. The design of effective safety information displays. *Proceedings of Human Factors and Industrial Design in Consumer Products*, Tufts University, 1980.

15. Lieber, C. The metabolism of alcohol. *Scientific American*, Vol. 234, No. 3 (1976): 25-32.

16. Moll, K., and Tihansky, D. Risk-benefit analysis for industrial and social needs. *American Industrial Hygiene Association Journal*, Vol. 38, No. 4 (1977): 153-161.

17. Murphy, A. A respiratory screening and surveillance program in a textile industry. *Occupational Health Nursing* (March 1972): 12-15.

18. Patty, F. *Industrial Hygiene and Toxicology.* New York: Interscience Publishers, 1963.

19. Peterson, J. Principles for controlling the occupational environment. *The Industrial Environment.* Washington, D.C.: Superintendent of Documents, 1973, chapter 35.

20. _____. *Industrial Health.* Englewood Cliffs, N.J.: Prentice Hall, 1977.

21. Plunkett, E., and Barbela, T. Are embalmers at risk? *American Industrial Hygiene Association Journal,* Vol. 38, No. 1 (1977): 61-62.

22. Revoir, W. Control of respiratory hazards. *Safety Sentinal,* Southbridge, Mass.: American Optical Company, 1973.

23. Soderlund, S. Exertion adds to solvent inhalation danger. *Health and Safety* (January 1975): 42-43.

24. Study indicates anesthesia gas imperils health. *Wall Street Journal,* October 15, 1974.

25. Temporary rules to shield farm workers. *Wall Street Journal,* May 1, 1973; p. 4.

26. Watt, T., and Vaumann, R. Nickel earlobe dermatitis. *Archives of Dermatology,* Vol. 98 (1968): 155-158.

27. Webb, P., and Annis, J. The principle of the space activity suit. NASA Contractor Report, NASA CR-973, December 1967.

28. Wright, G. The influence of industrial contaminants on the respiratory system. *The Industrial Environment.* Washington, D.C.: Superintendent of Documents, 1973, chapter 33.

20

CLIMATE

AIR VOLUME

Ventilation requirements formerly were set on the basis of odor removal. With increasing personal cleanliness, knowledge of contaminants, and attention to energy costs, required ventilation has decreased and more attention is paid to contaminants.[3]

For a sedentary person (met = 1.0), oxygen requirements are about .0062 L/s. Assuming oxygen is 21% of the air and 25% of the air breathed is consumed, air requirements are only .12 L/s.

Carbon dioxide is the usual limiting contaminant. Outside air is about .03% carbon dioxide. The carbon dioxide percent inspired air equals $.03 + .52/V$ where V = ventilation rate, L/s-person. ASHRAE uses .25% as the maximum allowable carbon dioxide. For a 1.0 met task, $V = 2.5$ L/s for each person. For 1.5 met, $V = 3.5$ L/s; for 2.0 met, $V = 5$ L/s; for 2.5 met, $V = 10$ L/s. Recognizing the health dangers of "passive smoking" (breathing tobacco smoke fumes), if there is smoking permitted in the room, then, for 1.0 met, $V = 10$ L/s and, for 1.5-2.5 met, $V = 17.5$ L/s. This is the minimum value of outside air for *occupied* spaces. If a space is not occupied, it does not need ventilation (unless it generates contaminants). Thus a typical office needs to be ventilated only 40 of the week's 168 hours. Many older systems have considerable potential for volume reduction and thus energy savings. The three common methods of controlling the mass quality of indoor air are source control, removal control, and dilution control.[58] Source control minimizes the generation of contaminants (such as prohibiting smoking and local exhaust with hoods). Source control generally is the most cost-effective method. Removal control includes particle removal devices (mechanical filters and electronic air cleaners) and gas and vapor removal (activated charcoal). Most expensive is dilution control (infiltration, open windows, and forced ventilation). Since contaminant generation (people, processes) generally varies throughout a 24 hour day and 7 day week, ventilation should not be the same at all times.

Higher humidities reduce intensity of odors; for minimum odor perception and irritation, keep air water vapor pressure between 10 and 15 torr. (1 torr = 1 mm Hg.)

Air volume sufficient to remove explosive substances such as acetone, ethanol, or xylene usually is not sufficient to prevent health problems as the Threshold Limit Values (TLVs) are 1% to 3% of the Lower Explosive Limit (LEL). For example, the LEL for acetone is 25,500 ppm while the TLV is 1000 ppm. Thus the TLV usually is limiting. See chapter 19 for TLVs.

Air exhausted from a space must be replaced. Recycling or reuse of the air is desirable if contamination by toxic substances or odors is not a problem. Ventilation design is beyond the scope of this book but it generally is desirable to remove contaminants locally (such as with exhaust hoods) rather than after they have spread. Reuse of previously conditioned air generally saves money since the recycled air is closer to the desired values of temperature and vapor pressure than outside air. Dumping contaminated air "out the window" is not considerate of those downwind so, since you are going to remove the contaminants, you may as well reuse the air yourself. For complete coverage of ventilation, see McDermott, ASHRAE Systems Handbook (especially chapters 21 and 33) and ASHRAE Handbook of Fundamentals (especially chapters 11 and 12).[2,31]

COMFORT

ASHRAE defines comfort as "that state of mind which expresses satisfaction with the thermal environment"; specifying a comfortable environment is difficult since comfort is influenced by six major factors. Four are environmental (dry bulb temperature, water vapor pressure, air velocity, and radiant temperature) and two are individual (metabolic rate and clothing). In the following, minor factors—barometric pressure, percent body fat, sweat glands/mm^2, etc.—are neglected.

The psychometric chart, Figure 20.1, gives the relation between dry bulb temperature (horizontal axis) and water vapor pressure (right-hand vertical axis). Vapor pressure is given in absolute units (torr) on the right-hand vertical scale and in relative units (relative humidity) on the curves coming from the left-hand vertical scale. The top of the figure (the curve, not the truncated horizontal portion) gives maximum amount of water vapor pressure which can be obtained for any dry bulb temperature. This curve's temperatures are called "psychometric wet bulb temperatures" or "wet bulb temperatures" if they are determined by following the wet bulb temperature lines (solid slanting lines). However, they also can be "dew point temperatures" if they are determined by reading across horizontally. They are "effective temperatures" if determined by the dashed line.

Comfort for Standard Conditions

This simplified psychometric chart (mechanical engineers use a more complicated one) has a cross-hatched area giving the American Society for Heating, Refrigeration and Air Conditioning Engineers (ASHRAE) standard comfort zone.

The standards are based on studies on several thousand subjects.[10,26,35,45,47] Subjects voted their comfort: $TS = 1$ = cold; 2 = cool; 3 = slightly cool; 4 = comfortable; 5 = slightly warm; 6 = warm; 7 = hot. This vote can be predicted from[46]:

$$TS = -1.047 + .158\, ET^* \qquad ET^* < 20.7 \qquad (20.1)$$

$$TS = -4.444 + .326\, ET^* \quad 20.7 < ET^* < 31.7 \qquad (20.2)$$

$$TS = 2.547 + .106\, ET^* \qquad ET^* > 31.7 \qquad (20.3)$$

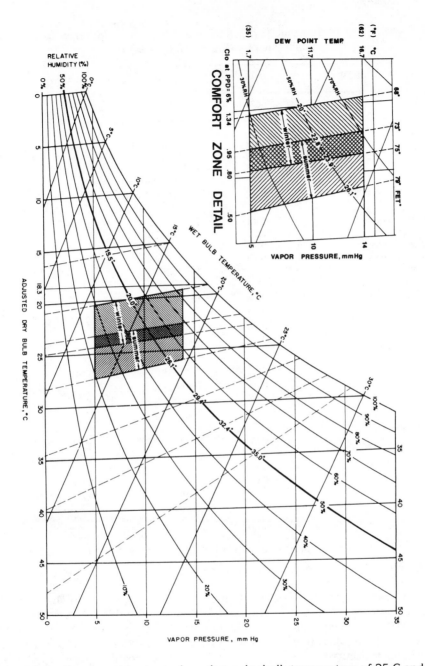

Figure 20.1 *Psychometric charts* show that a dry bulb temperature of 25 C and vapor pressure of 15 mm of Hg (i.e., 15 torr) identify a specific point.[49] Use the curve coming from the left vertical axis to read 64% relative humidity (23.5 torr is the maximum water vapor pressure in 25 C air; 15/23.5 = 64%). Move horizontally to the left to read dew point temperature of 17.5 C. Move to the left using the slanting lines to read psychometric wet bulb of 19.8 C. Move up on the dashed line at a steep angle to read effective temperature *(ET)* of 24 C when the dashed line reaches 100% humidity. Any point along a dashed line has approximately the same skin wettedness (see eq. 20.16) and gives approximately the same comfort.[18] In 1971 the number attached to each dashed line was changed to be the intersection with the 50% rh line (rather than the 100% line); the line was named "new effective temperature" or SET or *ET** (rather than effective temperature which is *ET*). Move down on the dashed line to read *ET** = 26 C.

Table 20.1 *Clo* Values *(ICLI)* for Individual Clothing Items[48];
Ensemble *clo, ICL* = .82 (Σ *ICLI*).

MEN		WOMEN	
Clothing	*Clo*	**Clothing**	*Clo*
Underwear			
Sleeveless	0.06	Bra and panties	0.05
T-shirt	0.09	Half-slip	0.13
Briefs	0.05	Full-slip	0.19
Long underwear upper	0.10	Long underwear upper	0.10
Long underwear lower	0.10	Long underwear lower	0.10
Shirt			
Light, short sleeve	0.14	Light blouse	0.20
long sleeve	0.22	Heavy blouse	0.29
Heavy, short sleeve	0.25		
long sleeve	0.29	Light dress	0.22
(Plus 5% for tie or		Heavy dress	0.70
turtleneck)			
Light vest	0.15	Light skirt	0.10
Heavy vest	0.29	Heavy skirt	0.22
Light trousers	0.26	Light slacks	0.10
Heavy trousers	0.32	Heavy slacks	0.44
Light sweater	0.20	Light sweater	0.17
Heavy sweater	0.37	Heavy sweater	0.37
Light jacket	0.22	Light jacket	0.17
Heavy jacket	0.49	Heavy jacket	0.37
Socks		Stockings	
Angle length	0.04	Any length	0.01
Knee high	0.10	Panty hose	0.01
Shoes		Shoes	
Sandals	0.02	Sandals	0.02
Oxfords	0.04	Pumps	0.04
Boots	0.08	Boots	0.08

where TS = thermal sensation vote for sedentary activity and .60 *clo*

            ET^* = new effective temperature, C

It will be emphasized that you can't satisfy all the people any of the time. For the conditions within the cross-hatch, none of the 1600 subjects voted hot (7) or cold (1) but 3% were warm (6) or cool (2); 94% were slightly warm (5), slightly cool (3), or comfortable (4); the mean vote was 4.0 and standard deviation was .7.[47] This variability occurred in spite of three hours of exposure, standard clothing (0.6 *clo*), and standard metabolic rate (sedentary sitting).

$$\left(1\ clo = \frac{.155\ \text{m}^2\text{-C}}{\text{W}} = \frac{.18\ \text{m}^2\text{-h-C}}{\text{kcal}}\right)$$

The percent people dissatisfied (PPD) can be calculated.[48]

    PPD = percentage of people dissatisfied (voting other than 3, 4, 5) corresponding to cumulative area from negative infinity for $CSIG$ or $HSIG$

    $CSIG$ = number of standard deviations from 50% for cold conditions ($< 25.3\ ET^*$) for sedentary activity and .5-.6 *clo*   (20.4)

           = $10.26 - .477\ (ET^*)$

$HSIG$ = number of standard deviations from 50% for hot conditions
$\qquad$ ($> 25.3\,ET^*$) for sedentary activity and .5-.6 *clo* $\qquad$ (20.5)
$\qquad$ = $-10.53 + .344\,(ET^*)$
ET^* = new effective temperature, C

For example, for ET^* of 18 C, $CSIG = +1.67$; from a normal table, 95% are dissatisfied. For ET^* of 30 C, $HSIG = -.21$, and 42% are dissatisfied. At 25.3 ET^*, the minimum (6%) are dissatisfied.[48]

The comfort temperature with a PPD of 6% and amount of clothing are related as follows:

$$ET^*6 = 29.75 - 7.27\,(ICL) \qquad (20.6)$$

where $\quad ET^*6 = ET^*$ temperature (C) at which 6% will be dissatisfied

$\qquad ICL$ = Insulation value of clothing ensemble, *clo* $\qquad ICL < 1.1$
$\qquad$ = .82 $(\Sigma\,ICLI)$

$\qquad ICLI$ = Insulation values of individual clothing items, *clo* (see Table 20.1)

For example, 6% will be dissatisfied with .8 *clo* at 23.9 and with .95 *clo* at 22.8 C.

The former standard was valid for persons dressed with 0.7 to 1.0 *clo* with a metabolic rate of "office work." The new standard (parallelogram) is valid for persons dressed with 0.5 to 0.7 *clo* and a metabolic rate of "sedentary sitting." Tables 20.1 and 20.2 give the *clo* values for various clothing ensembles; Figure 20.2 shows *clo* values for three typical ensembles for men and women.

Engineers should remember that sweating to remove heat is not considered "comfortable"—that is, a "wet skin" is not acceptable for "comfort." (Ingredients in old ink recipes include salt as mold-retardant and a few drops of brandy as antifreeze—reminder of the days when climate control was crude.)

Fanger reported that during comfort[9]:

$$TSKIN \quad = 35.7 - .027\,6\,M \qquad (20.7)$$
and $\quad ESWEAT = .42\,(M\text{-}58) \qquad (20.8)$

where $\quad TSKIN \quad$ = mean skin temperature, C
$\qquad M \qquad\quad$ = metabolic rate, W/m^2
$\qquad ESWEAT$ = evaporative sweating rate, W/m^2

Adjustments for Nonstandard Conditions

Nevins and Gorton give the following recommendations for adjustments to achieve comfort (also see Fanger).[8,36]

Clothing. Nevins says Dry Bulb Temperature *(DBT)* decreases .6 *clo* for every .1 *clo* increase[37]; Nevins and Gorton say *DBT* decreases .6 C for every .1 *clo* increase from .6 *clo* when total metabolism is less than 225 *W* but *DBT* decreases 1.2 C/.1 *clo* for over 225 *W*.[36]

Activity. For each 30 *W* increase in total metabolism above 115 *W*, decrease *DBT* by 1.7 C. To permit sweat evaporation, keep relative humidity below 60% (15 torr).

Air velocity. For each .1 m/s increase in velocity up to .6 m/s, increase *DBT* by .3C; for each .1 m/s between .6 and 1.0, increase *DBT* by .15 (Nevins).[37] Rosen, using a box fan (i.e., turbulent flow) at .8 and 1.3 m/s, found .1 m/s offsets a 0.3 C increase in temperature.[50] Keep maximum velocity less than .7 m/s for sedentary occupations. The air can come from the front, rear, side, or above or below.[8] Air

Table 20.2   *Clo* Values for Various Clothing Ensembles.[51] All Ensembles Include Low Quarter Shoes and Underwear. Decrease *clo* Value by .13 for Every 1 m/s Increase in Air Velocity Above .25 m/s. Men's Underwear can Change *clo* by 10% but Variations in Women's Underwear have Little Effect on *clo*.

Male Ensembles	*Clo* Range
Woven shirt, trousers	.51- .65
Knit shirt, trousers	.48- .76
Sweater, trousers	.60- .75
Woven shirt, sweater, trousers	.81- .90
Woven shirt, jacket, trousers	.89-1.00
Female Ensembles	
Blouse, skirt	.33- .51
Dress	.21- .71
Dress, blouse	.32- .73
Sweater, skirt	.40- .68
Blouse, sweater, skirt	.42- .80
Jacket, blouse, skirt	.45- .80
Blouse, slacks	.51- .82
Sweater, slacks	.58- .89

Reprinted from *ASHRAE Transactions* 1972 Vol. 78, Part I, by permission of the American Society of Heating, Refrigerating and Air-Conditioning Engineers, Inc.

flow from above interferes with the heat rising from the body, causing turbulent flow; air from below gives a more laminar flow. Bring warm air for heating in at the bottom of the people as they will be comfortable at a lower air temperature; bring air for cooling from above as they will be comfortable at a higher air temperature. Discomfort may be more affected by velocity fluctuations than the velocity mean so, for comfort, decrease fluctuations. For fixed work positions with light activity, velocity should be about .2 to .3 m/s. Maximum velocity for continuous exposure is 1.0 m/s. For intermittent exposure and high work levels, use 5 to 20 m/s; 5 to 10 are common (ASHRAE, p. 21.6[2]); 0.5 to 1.5 m/s are most common for spot cooling of work places.

Mean radiant temperature (MRT). For each 1 C deviation of *MRT* from dry bulb temperature, change the *DBT* 1 C in the opposite direction. For example, 25 C is a comfortable *DBT* when *MRT* also equals 25. But if *MRT* equals 27, then *DBT* should be 23.

Time of exposure. The standard values are based on the vote after three hours of exposure. Males voted .5 vote warmer and females .2 warmer at the end of the first hour than they did after 3 hours. At present, it is recommended that you use the 3-hour values and not compensate for actual occupancy time.

Time of day. Even though core temperature is on a 24-hour cycle, thermal comfort conditions do not differ with time of day.[12]

Season of year. To most people's surprise, comfort temperature does not vary with season of the year even for different ethnic groups—if clothing and metabolic rate are standardized. Fanger found that Nigerians who had just arrived by jet in Copenhagen had the same comfort temperatures as Danes, Danes who swam in the sea in winter, Danes in cold meat packing jobs, and Americans.[8]

Humidity control becomes a problem in winter. Vapor pressures above 12 torr may cause condensation problems and affect the glue from boxes in the warehouse. A vapor pressure of 7 torr can still fit within the comfort zone. Cold outside air could have an 80% relative humidity but still only have 7 torr. When this warmed air with the low vapor pressure passes a nose or respiratory tract (with

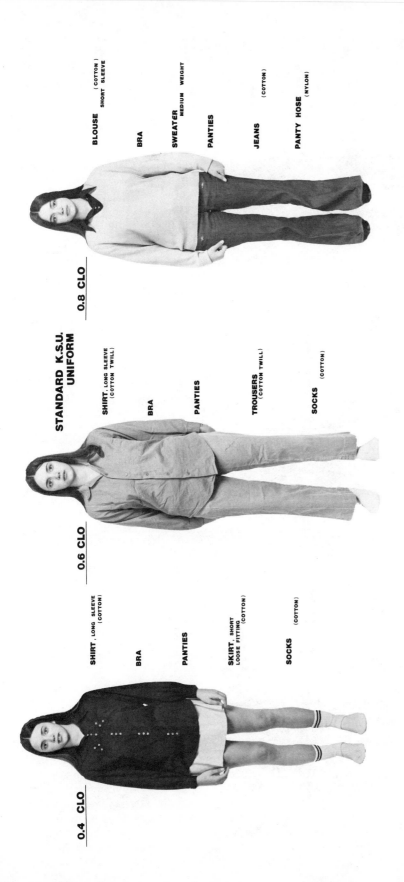

Figure 20.2 *Clothing insulation* can be given in *clo* units (1 *clo* = (.155 m² − *C*)*W*. As heating and air conditioning technology have improved, people have tended to wear less clothing. Zero clothing equals 0 *clo*; arctic clothing equals about 4 *clo*.

448

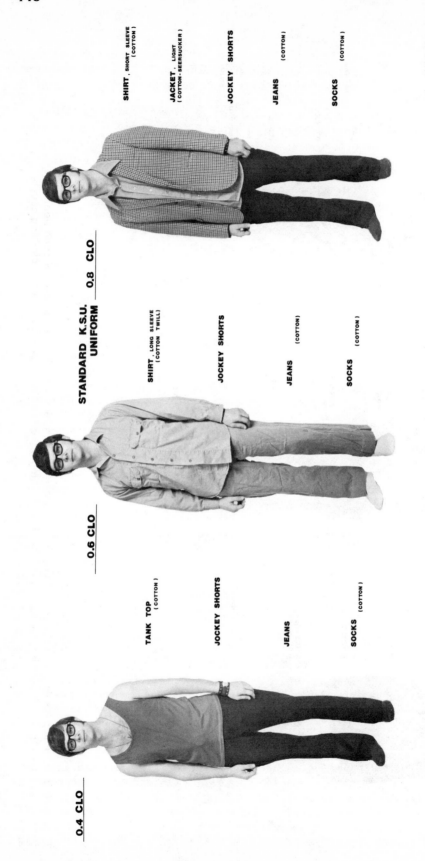

STANDARD K.S.U. UNIFORM

0.8 CLO

SHIRT, SHORT SLEEVE (COTTON)

JACKET, LIGHT (COTTON-SEERSUCKER)

JOCKEY SHORTS

JEANS (COTTON)

SOCKS (COTTON)

0.6 CLO

SHIRT, LONG SLEEVE (COTTON TWILL)

JOCKEY SHORTS

JEANS (COTTON)

SOCKS (COTTON)

0.4 CLO

TANK TOP (COTTON)

JOCKEY SHORTS

JEANS (COTTON)

SOCKS (COTTON)

moisture at about 45 torr), the 38 torr driving force transfers water from the person to the air. The person gets a sore throat although the room temperature is comfortable. Green found, for indoor climates in the winter, that colds and upper respiratory infections increase as humidity drops below 50% (about 12 torr).[20] In general, for comfort, .1 torr is equivalent to .10 CDB.

Sex of occupant. If men and women have the same clothing insulation values, they have the same comfort temperature. Women's skin temperature at comfort is about 0.2 C lower and their evaporative loss during comfort is about 4 g/m²-hr less; these balance women's lower basal metabolic rate.[10] Table 20.1 shows that women tend to have clothing with lower *clo* values.

Age of occupant. If the occupants have the same clothing insulation values and the same metabolic rates, they have the same comfort temperature.

The decline in basal metabolic rate with age seems to be compensated for by a lower evaporative loss during comfort (19 g/m²-hr at age 23; 15 at 68; and 12 at 84).[9]

HEAT STRESS

Criteria of Stress

In the previous section, comfort was the criterion. For more extreme environments the problem becomes the effect on performance (both physical and mental) and on health.

Wyndham and Strydom gave Figure 20.3 as the temperature-time trade-off for physical work (8.7 W/kg).[60] Wing and Touchstone made a 162-reference bibliography on the effects of temperature on human performance.[56] Wing, summarizing 15 different studies of sedentary work in heat, gave Figure 20.4 as the temperature-time trade-off for mental performance; he noted that human performance deteriorates well before physiological limits have been reached.[57] Hancock shows the effect on different tasks in Figure 20.5.[23] Ramsey, Dayal, and Ghahramani report their own data plus others support Wing's curve.[44] A difficulty of using performance criteria from short-run situations (such as experiments) is that motivation seems to increase during experimental stress situations; performance is *higher* in a heat stress environment than in comfortable conditions. I have confirmed this in my own experiments.[28] Thus the short-run studies probably underestimate the effect of heat. It can't be entirely coincidence that all the developed countries are in the temperate zone—not the tropics. Thus, for productivity, design using comfort as the criterion, realizing that motivated people can work in hot environments but the ordinary worker in a hot environment will work at a low efficiency.

For health, increase in heart rate or systolic blood pressure could be the criterion. During physical exercise in the heat, the effect of heat is reflected in increased skin blood flow up to about seven times basal skin blood flow; Pirnay, Petit and Deroanne report the heart rate increases 32 beats/min for every C increase in body temperature.[41] Most experts, however, use maximum body core temperature as the best single criterion. The National Institute for Occupational Safety and Health (NIOSH) set 38 C as the maximum "core" temperature. Core temperature usually is measured as rectal temperature although ear canal or esophageal temperature also are used. Rectal temperature = brain temperature = liver temperature = right heart atrium + .6 C = esophageal + .6 C = mouth + .4 C. Rectal temperature represents the temperature of 20% of body mass while esophageal represents the temperature of 80% of the body.[32]

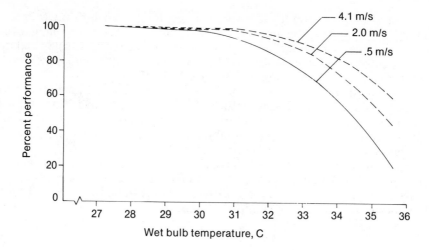

Figure 20.3 *Air velocity* affected shoveling performance; 27.2 C *WB* was 100% (Wyndham and Strydom).[60]

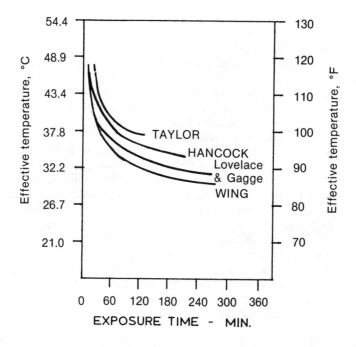

Figure 20.4 *Sedentary ("mental") performance* was summarized in 15 studies.[56,57] The curves give the upper limit—above the curve some performance decrement should be expected. Grether gives 35 C *ET** (29.5 *ET*) as the point at which performance begins to deteriorate.[21]

Environmental Limits

Measuring rectal temperatures of thousands of workers on the job is not practical. Environmental limits were needed that could predict the stress beforehand.

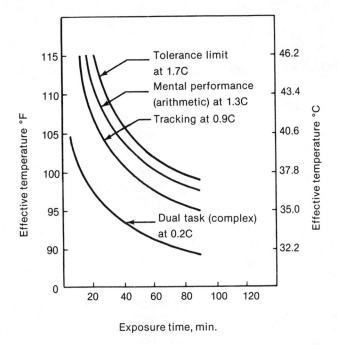

Figure 20.5 *Performance in heat* declines as body temperature rises.[23] The curves give the length of time at a temperature to achieve the indicated body temperature rise. Thus 90 min in a 35 C effective temperature environment will give a rise of body temperature of about 0.9 C; at this rise, tracking performance will begin to decline.

Lind presented the critical concept of the prescriptive zone (see Figure 20.6).[29] At lower environmental temperatures the body remains in thermal equilibrium although the set point temperature is higher for higher metabolic rates. The point at which rectal temperature begins to rise is a function of metabolic rate as well as the environment. In addition, at higher metabolic rates there is little margin before rectal temperature reaches 38 C.

Wyndham and Strydom remark that under compulsion (e.g., military training), heat stroke may occur at 28.5 C *ET* while industrial workmen in Australian mines have worked without heat stroke at 32 C *ET* since they slow down in the heat.[60]

NIOSH took Lind's concept and the idea that a small amount of heat storage is not harmful (i.e., as long as core temperature is below 38 C) to get the heat threshold values given in Table 20.3.[43] The values protect 95% of the workforce; it was assumed that the 5% of heat-intolerant individuals would not be working on hot jobs. It is essential to note that these are not heat *limits;* these are values at which precautions (provision of adequate drinking water, annual physical examinations, training in emergency aid for heat stroke) should *begin* to be taken.

The recommendations are expressed in a specific index called Wet Bulb Globe Temperature *(WBGT)* which attempts to combine into one number the effect of Dry Bulb Temperature *(DBT)*, water vapor pressure, air velocity, and radiant temperature. *WBGT* "probably exaggerates the danger of heat stress in medium and low humidity environments."[17]

For an environment in which radiant temperature is close to air temperature:

$$WBGT = .7\,NWB + .3\,GT \tag{20.9}$$

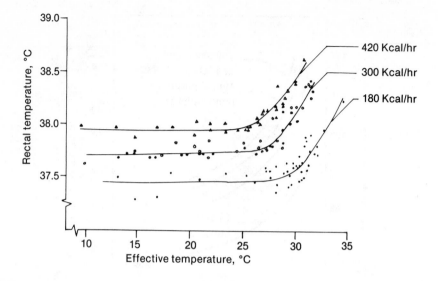

Figure 20.6 *Rectal temperature* remains constant at 37.4 C (at lower metabolic rates such as 180 kcal/hr) until environmental temperature reaches 29 C *ET* (Lind).[29] Then rectal temperature begins to rise. At a higher metabolic rate of 300 kcal/hr, rectal temperature remains constant at 37.7 C until environmental temperature rises to 27 C *ET*. At 420 kcal/hr, rectal temperature remains constant at 37.9 C until environmental temperature rises to 26 C *ET*. The environmental temperatures of 26 to 29 C *ET* are called the "prescriptive zone." Rectal temperature in the nonstressed zone = 37.0 + .0038 (metabolic rate, *W*) (Berenson and Robertson).[4] Nielsen made the original discovery that rectal temperature varied with metabolic rate.[39]

Table 20.3 Threshold *WBGT* Temperatures (C) as a Function of Air Velocity and Metabolic (Basal + Activity) Rate.[43] A Velocity of 1.5 m/s is a "Noticeable Breeze"; Paper Blows at 0.8 m/s.

Metabolic Rate, *W*	Low air velocity (up to 1.5 m/s)	High air velocity (1.5 m/s or above)
Light (up to 230)	30	32
Moderate (230 to 350)	27.8	30.5
Heavy (over 350)	26	28.9

For an environment in which radiant temperature is not close to air temperature:

$$WBGT = .7\,NWB + .2\,GT + .1\,DBT \qquad (20.10)$$

where   *WBGT* = Wet Bulb Globe Temperature

　　　　NWB　= Natural Wet Bulb Temperature (temperature of a sensor with a wet wick exposed to natural air currents). Don't confuse it with *WB* which is wet bulb, also called psychometric wet bulb, which is the temperature of a sensor with a wet wick exposed to "high" (3 m/s) air velocities. $WB = NWB$ for air velocity > 2.5 m/s; for $.15 < V < 2.5$, $NWB = .1$ $DBT + .9 \, WB$ (Gagge and Nishi).[17]

GT = Globe Temperature (temperature at the center of a 15 cm diameter black sphere)

DBT = Dry Bulb Temperature (sensor shielded from radiation)

Rather than measuring $WBGT$ directly with the expensive and cumbersome $WBGT$ apparatus, many people use the inexpensive and simple Botsball device. Then either the Botsball temperature can be used directly or the $WBGT$ temperature calculated. A good conversion equation ($r = .95$) is[5]:

$$WBGT, C = 0.80 + 1.07 \,(\text{Botsball Temp.,C}) \qquad (20.11)$$

See Principle 6 of chapter 15 for a discussion of burn temperatures.

Reduction of Heat Stress

Figure 20.7 depicts the human thermoregulatory system as a closed loop system. Body temperature is not maintained constant about a fixed set point (homeostasis) but rather is maintained about a fluctuating set point (homeokinesis) as set point changes with exercise and 24-hour rhythm. (In Greek, homoios = similar and stasis = position, standing.) The schematic has two basic systems: the controlling system and the controlled system. The negative feedback signal of temperature (skin + internal) is fed back to the comparator which weights the signal from the skin and core. The difference (error) between the desired temperature (set point) and actual temperature actuates one or more of the three controllers (sweat, shivering, or vasomotor activity). The controllers act through the control elements (sweat glands, muscles, skin blood vessels) to modify heat transfer. The resulting temperature then is fed back to the comparator where the cycle repeats. External events (metabolic work or change in environmental conditions) act as disturbances.

Solutions for various environments are complex due to the many feedback loops but solutions can be done by computer simulation.[4,27] For the person concerned with reducing heat stress, examination of the heat storage equation (with the realization that the equations treat the body as an open loop system in a static environment) will solve most problems. For the student interested in digging deeper, consult chapter 8, "Physiological Principles," in ASHRAE *Handbook of Fundamentals* and Slonim.[3,53]

The key equation is the heat balance equation:

$$S = M - (\pm W) + (\pm R) + (\pm C) + (\pm E) + (\pm K) \qquad (20.12)$$

where S = heat storage rate, watts

M = metabolic rate, watts

W = mechanical work accomplished rate, watts (walk up steps = +; down = −)

R = radiation rate, watts (gain = +; loss = −)

C = convection rate, watts (gain = +; loss = −)

E = evaporation rate, watts (gain = condensation = +; loss = −)

K = conduction rate, watts (gain = +; loss = −)

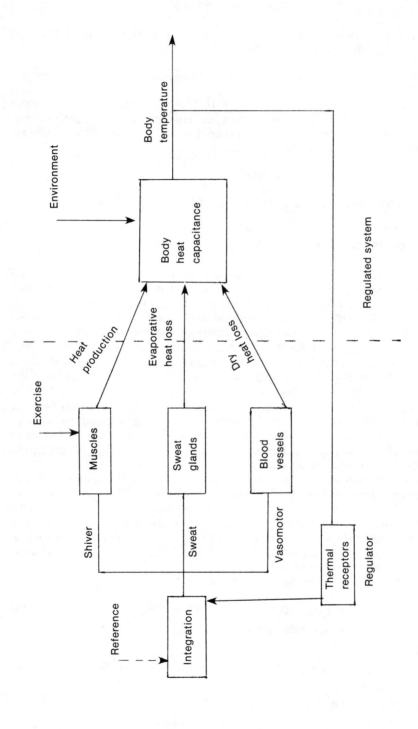

Figure 20.7 *Human thermoregulation*, a closed-loop system, can be divided into the controlling system and the controlled system (body temperature). Many of the problems associated with setting work limits in hot environments are due to using open-loop equations rather than closed-loop equations with individual variation in equation coefficients.

Storage. Human body heat storage is:

$$S = 1.15 \text{ m } C_p \text{ } (MBT_f - MBT_i)/t \tag{20.13}$$

where S = storage gain (+) or loss (−), watts

m = weight of body, kg

C_p = specific heat of body = .83 kcal-kg/C

MBT_i = initial mean body temperature, C

MBT_f = final mean body temperature, C

t = time, hours

Mean body temperature is calculated by weighing skin temperature and core temperature—usually with coefficients of .33 and .67 although in the heat some experts recommend .3 and .7 and others .2 and .8 as the skin "shell" becomes thinner.

$$MBT = .33 \text{ (skin temperature)} + .67 \text{ (core temperature)} \tag{20.14}$$

Metabolism. Table 12.1 gave metabolic rates for various activities. (Although you have more heat than a light bulb, you're not so bright!) (100 kg-m/min = 723 ft-lb/min = 16.35 watts = 14.04 kcal/hr.) Reducing metabolic rate can be done by mechanization or by working slower. Since working slower usually reduces productivity, look for other alternatives first.

Work. When the body accomplishes mechanical work such as walking up stairs or pedaling a bicycle, this energy must be subtracted from M to determine the net heat within the body core. The ratio of W/M is the mechanical efficiency of the body when performing work; it is about .2 for pedaling, .06 for repetitive lifting, and is considered 0 for most activities.[25]

Radiation. Radiant heat transfer is:

$$R = \sigma A \text{ } f_{eff} f_{clr} F_{clr} \text{ } e \text{ } (T^4_{mrt} - T^4_{skin}) \tag{20.15}$$

R = radiant gain (+) or loss (−), watts

σ = Stefan-Boltzmann constant = 5.67×10^{-8} watts/(m^2-K^4)

A = skin surface area, m^2 (see equation 12.4)

f_{eff} = effective skin radiation area factor (.725 for standing, .696 for sitting)[11]

f_{clr} = increase in radiant area due to clothing = $1 + .155 \text{ } I_{clo}$

I_{clo} = insulation value of clothing, clo (see Table 20.1)

F_{clr} = multiplier to radiant heat transfer coefficient to adjust for clothing barrier = $1/(1 + .155 \text{ } (5.2) \text{ } I_{clo})$ where 5.2 $W/$(m^2-C) is the reference radiant heat transfer coefficient.

e = emissitivity (skin = .99; clothing in nonvisible radiation = .7)

T_{skin} = temperature in K of the skin, K = C + 273)

T_{mrt} = mean radiant temperature in K of environment, K = C + 273

The "driving force" is the difference between the two temperatures—each raised to the 4th power. The important number is 35 C—skin temperature in the heat. For T_{mrt} above 35 C, the body gains heat; below 35 C it loses. For a typical task with T_{mrt} = 43 C, R = 60 watts; with T_{mrt} = 37.8 C, R = 5.

Reduce radiant load by "working in the shade" since heat radiation behaves much as light radiation. Working in the sun can add 170 watts to a clothed person and 220 to an unclothed person.

Clothing (a mobile shield) is the first line of defense; use hats and long sleeved shirts. As a first approximation, clothing temperature is halfway between skin temperature and environmental temperature. For visible radiation (as the sun), use light-color clothing; clothing color does not matter for nonvisible radiation. Clothing material, density, and thickness matter.

A fixed shield between the person and the source is a second line of defense. Use with ovens, welding torches and arcs, and molten glass. Reflecting heat shields tend to be more effective and economical than absorbing shields or water-cooled shields. The air on the source side of the shield becomes heated and rises—giving a welcome current of "makeup" air to the worker. See Figure 20.8. Aluminum is a very good shield since it has high reflectivity and doesn't corrode. If the operator must see the source, use a screen of chains or coated glass (glass, although more effective, tends to get broken or dirty). Cover an oven conveyor entrance and exit with a screen of hanging chains.

Convection. Convection heat transfer is:

$$C = h_c A f_{clc} (t_{air} - t_{skin})$$ (20.16)

where C = convection gain (+) or loss (−), watts

h_c = convective heat transfer coefficient, watts/(m² − C)
= 4.5 W/(m² − C) for standing adults with velocity .05 to 2 m/s
= 8.3 V^6 for seated adults

A = skin surface area, m² (See equation 12.4)

f_{clc} = multiplier to h_c for clothing
= 1/(1 + .155 (2.9) I_{clo})
where 2.9 W/(m² − C) is h_c in still air (.15 m/s)

I_{clo} = insulation value of clothing, clo (see Tables 20.1 and 20.2)

V = air velocity, m/s

t_{air} = air temperature, C

t_{skin} = skin temperature, C

The "driving force" is the difference between the two temperatures. As with radiation, keep temperature of the environment below 35 C. Second, increase air velocity *on the skin*. Figure 20.9 shows that air velocity above 2 m/s has little additional benefit. It is important to note that air velocity drops very, very rapidly with distance. The primary effect of clothing is not obvious in the equation—it drops air velocity next to the skin to effectively zero. Therefore, to maximize convective loss, wear little clothing—if the temperature is below 35 C and insects, radiation, and social mores are not a problem. If clothing is worn, don't restrict air circulation at the neck, waist, wrist, or ankles.

Convective cooling is effective if air temperature is sufficiently below 35 C, say at 15 C. The problem is to get the cool air. One solution is to pass air through water which will normally be about 10-15 C. The air then is 15 C but humidity is 100%, say 12 torr. The worker then is cooled by convection. The technique works best with relatively high air velocities (.5 m/s).

Evaporation. Evaporation heat transfer is:

$$E = h_e A W F_{pcl} (VP_a - VP_s)$$ (20.17)

where E = evaporative gain (+) or loss (−), watts

h_e = evaporative heat transfer coefficient = 2.2 h_c

A = skin surface area, m²

W = E/E_{max}. Mislabeled the porportion of the skin that is wet, it

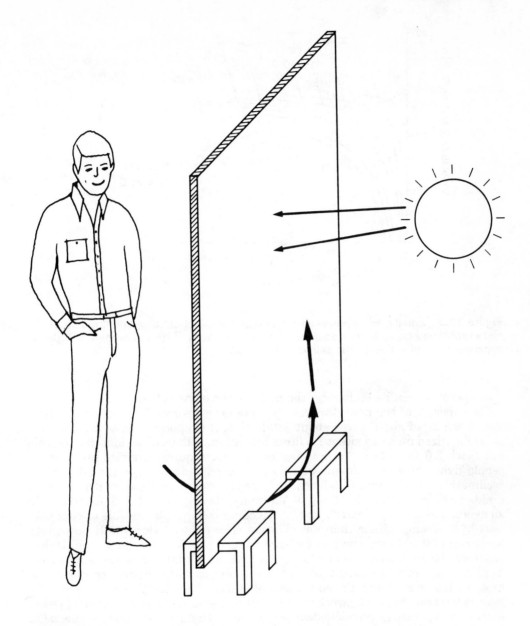

Figure 20.8 *Radiant heat shields* should have a gap at the bottom to aid convective air flow.

actually is the proportion of actual sweat to maximum possible sweat. Values of .7 are a reasonable maximum and .5 are much more common.

F_{pcl} = decrease in evaporative efficiency for permeable clothing
= $1/(1 + .143 (2.9) I_{clo})$ where 2.9 $W/(m^2\text{-}C)$ is h_c in still air (.15 m/s) for a sedentary person.

I_{clo} = insulation value of clothing, *clo* (see Tables 20.1 and 20.2)

VP_s = vapor pressure of water on skin (45 torr if t_{skin} = 35 C)

VP_a = vapor pressure of water in air, torr

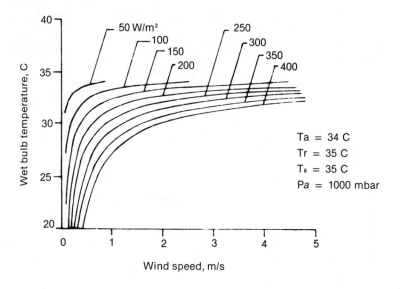

Figure 20.9 *Equal cooling power lines* combine the effect of convection, radiation, and evaporation on nude men as a function of air velocity.[33] Note that there is little benefit for air velocities on the skin over 2 m/s.

Evaporation can be limited by the body or the environment.

The capacity of the body to sweat is quite large, especially after acclimatization. Each kg of sweat gives about 580 kcal/kg if evaporated *from the skin.* An unacclimatized man can sweat 1.5 liters/hr; within 10 days of acclimatization this can reach 3.0 liters/hr.[22] Acclimatization to heat requires *exercise* in the heat; people living in a hot climate who do not exercise do not get the benefits of acclimatization. Vitamin C deficiency slows the rate of acclimatization; Vitamin C aids sweating.[54] Use of 250 mg of Vitamin C/day permitted acclimatization in an average of 5.2 days instead of 8.7.[55] Acclimatized people are able to eliminate heat by sweating rather than vasodilation. The reduced skin blood flow gives heart rates 30-40 beats/min less for acclimatized people when exercising. Unfortunately acclimatization can be lost in as little as 10 days so workers returning from vacation must be careful of heat stress. For each day of nonexercise in the heat (such as vacations), the equivalent of .5 day of acclimatization is lost.[19] As long as water is replaced, sweating can continue without health problems. Dehydration of 3% causes physiological performance changes, 5% gives evidence of heat exhaustion, and at 7% hallucinations occur. Losses totaling 10% are extremely hazardous and lead to heat stroke; if not treated immediately, death will result. Dehydration potential is greater when under forced ventilation (fans, vortex tube cooling, air jets). However, the thirst drive is not sufficient to replace water loss during heavy sweating. Supervisors must insist that workers drink water; frequent, small amounts are better than occasional large amounts. Drink from a container as volume drunk from a water fountain tends to be small (30-60 mL/drink) if in a comfortable temperature environment. "Tank up" before heat exposure.

Sweat is .2% to .4% salt (.4% at high sweat rates) whether you are acclimatized or not, as long as you are in positive salt balance. If salt balance becomes negative due to lack of salt intake or heavy sweating, the kidneys start to decrease urine salt content within 30 minutes; sweat salt content declines after several days.[6] Salt tablets rarely are desirable; people take them excessively and get stomach problems and high blood pressure. Salt on food usually is sufficient.

Whenever more than 4 liters of water/day are required to replace sweat loss, provide extra salt. Give 2 g of salt for each liter over 4; a person doing heavy work under hot conditions may need an extra 7 g/day.[34] The normal adult diet in the USA has 4 g of sodium and 10 g of salt/day; requirements are 2 of sodium and 5 of salt. A typical well-salted meal has 3-4 g of sodium; 1 cup of beef or chicken broth, consomme, or bouillon has about 1 g of sodium.[1] If additional salt is absolutely necessary during work, add salt to a lime drink to reduce its concentration and increase palatability.

Evaporation also can be limited by the environment. It isn't the heat, it's the humidity. Wyndham, Williams, and Bredell pointed out that when water vapor pressure is over 32 torr a 1 C increase in wet bulb is equal to a 10 C increase in dry bulb.[59] Reduce environmentally limited evaporation by (1) increasing air velocity or (2) decreasing water vapor pressure.

Figure 20.9 showed that beyond 2 m/s air velocity had little benefit. However, achieving 2 m/s at the skin is difficult. Figure 20.10 shows a plan view of air velocity for a typical fan. Figure 20.11 makes the same point: velocity drops off very rapidly with departure from the fan's axis and rapidly even along the fan's axis. To be useful a fan must be pointed directly at a worker and be close. Clothing, since it reduces air velocity next to the skin, hinders evaporation as well as convection. In addition, evaporation from clothing instead of the skin gives a cooling to the person of about 250 kcal/kg instead of 580.

The second approach to improve evaporation is to decrease water vapor pressure in the air. Dehumidification, although expensive, gives a pleasant environment since sweat evaporates rapidly (discomfort is related to skin wettedness).

Conduction. Astronauts had metabolic heat to be removed. The early spacesuits used evaporative cooling but blowers took considerable power and there was no place to dump the sweat-laden air. The solution was a network of small tubes on the torso through which small (1 liter/min) volumes of cool (10 C) water were passed. Heat removal was excellent and there were no physiological problems from removing heat from only a localized area of the body. For earthbound applications the problems with this personal cooling technique are the high capital cost/unit (about $500) and the weight of the associated pump, compressor, and battery. If the weight is not kept on the back, then the worker must be connected with a lifeline to the source of water and power.

Another technique is cooling with dry ice. The dry ice is placed in pockets which are held next to the skin. Figure 20.12 shows a dry ice jacket. One "charge" of ice lasts about 4 hours. Cooling with water ice also has been used. Water ice has about 80 kcal/kg of phase change vs. the 137 for dry ice, but water can be refrozen. Dry ice gets 23 kcal/kg additional benefit as the CO_2 warms from −79 C to skin temperature.

COLD STRESS

Criteria of Stress

At low levels of cold stress, we experience discomfort. Additional cold gives loss of mental performance and manual dexterity. Then comes pain and potential loss of extremities such as fingers, ears, feet. Finally death. (At very low body temperatures (20 C), vital signs disappear, but the brain oxygen supply may still be sufficient; death from accidental hypothermia should be defined as "failure to revive upon rewarming."[24]

Use the formulas in the section on comfort and Figure 20.1 to evaluate whether people will consider themselves slightly cool, cool, or cold.

If the core is warm, dexterity does not drop until hand skin temperature

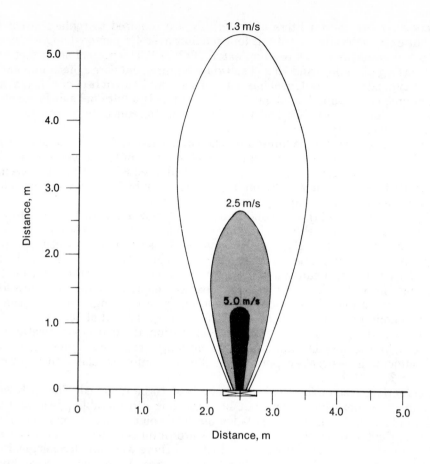

Figure 20.10 *Fan air velocity contours* show a very rapid drop in velocity with distance, both along the axis and with departure from the axis.

reaches about 15 to 20 C.[42] Fox, reviewing 68 references, says hand skin temperature *(HST)* is critical; tactile sensitivity declines when *HST* is below 8 C; manual dexterity when *HST* is below 12-16 C.[14] Poulton (p. 150)[42] reported less vigilance on ships when oral temperature dropped to 36.5 C. At body temperature of 35 C, dexterity is reduced to the point where you cannot open a jackknife or light a match. When body temperature drops below 35 C, the mind becomes confused; at about 32 C there is loss of consciousness. At deep body temperature of 26 C, death occurs from heart failure. Local tissue such as fingers and ears freezes at −1 C instead of 0 due to osmotic pressure.

Environmental Limits

Just as people want one number to combine all factors for heat and use *WBGT*, they want one number for cold and use the "wind chill index," *I*.

$$I, \text{kcal/(m}^2 - \text{hr)} = h\,(t_{skin} - t_{air}) \tag{20.18}$$

where h, kcal/(m^2 − hr − C) = $10.45 - V + (100\,V)^{.5}$

V = air velocity, m/s

t_{skin} = 33 C

t_{air} = dry bulb air temperature, C

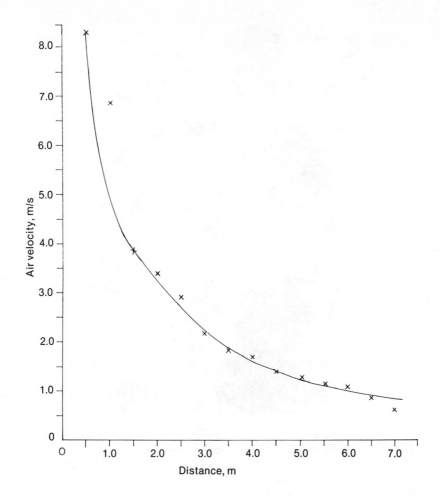

Figure 20.11 *Fan placement* is extremely critical because air velocity drops so rapidly with distance.

Table 20.4 gives the sensation for various values of *I*. The concept of wind chill also can be presented in the format of Table 20.5. Values give the 1.8 m/s air temperature that has the same values of *I* as the moving air.

Be cautious about rigorous use of the wind chill index as it is based on cooling a liter cylinder of 33 C water, not a clothed human with a metabolism.[4] It is useful as a popular index for precautions against frostbitten hands and ears.

Protection vs. Cold Stress

Clothing is the primary defense in a heat-hungry environment. Fourt and Hollies have a good book on clothing.[14] As clothing thickness increases, the insulation value increases to a point at which the increase in insulation is overcome by the increase in surface area—thus thin gloves may increase heat loss from a child's fingers! (If dexterity is not a problem, use mittens with liners rather than gloves.) The best insulator is air so the Eskimos used two layers of fur. The Russian army since the Napoleonic War has issued winter boots one size too

Figure 20.12 *Dry ice cooling* combines unrestricted mobility with low cost cooling.

Table 20.4   Sensations for Various Values of *I* From Equation 20.18
(Newburgh, p. 423)[38]

I, kcal/(m²-hr)	Sensation
50	Hot
100	Warm
200	Pleasant
400	Cool
600	Very Cool
800	Cold
1000	Very cold
1200	Bitterly cold
1400	Exposed flesh freezes

large—the soldiers were to stuff the extra space with straw or newspapers. The German army in World War II issued boots which fitted exactly and thus suffered many frostbitten feet.

Floors conduct heat (and cold) to the feet. For normal shoes, floors should be above 18 C and below 29 C.[11] Because of vasoconstriction, foot skin temperature usually is the lowest body skin temperature. (Normal skin foot temperature =

33.3 C for males but 31.2 for females; the feet are the only skin locations in which temperatures differ between males and females.[40]) Thus cold feet may be due to a cold environment rather than a cold floor.

Provide windbreaks near the work. For clothing, a number of wind-proof layers tends to be more efficient than a single thick layer. It is especially important to protect the head since it does not vasodilate or constrict. Froese and Burton estimated[16]:

$$H = 285 - 7.55\,T_a \tag{20.19}$$

where H = Head heat loss, kcal/(hr−m² of head)
(the head was approximately .12 m²)

T_a = air temperature, C

Therefore at −4 C, heat loss from the head may be 50% of resting metabolism. Adding 2.4 *clo* units over the head gives the same result as increasing the insulation over the 1.7 m² of body by 4 *clo* units.

Another technique is to warm the hands by putting on a jacket. The jacket reduces heat loss on the torso and thus lets warm blood rather than cold blood flow to the hands.

Table 20.5   Wind-chill "Equivalent" Temperatures Predict the Effect of Air Velocity at Various Temperatures. The Number in the Table gives the Temperature at 1.8 m/s Which has the Same Wind-Chill as the Dry Bulb Temperature at an Air Velocity. The Wind-Chill Index is Based on Cooling of a Water Container Rather than a Person With Clothing.

Dry bulb temperature, C	Air Velocity, m/s							
	2	4	6	8	10	12	14	16
+ 4	+ 3.3	+ 1.8	− 5.0	− 7.4	− 9.1	−10.5	−11.5	−12.3
+ 2	+ 1.3	− 4.2	− 7.6	−10.2	−12.0	−13.5	−14.6	−15.4
0	+ 0.8	− 6.5	−10.3	−12.9	−15.0	−16.5	−17.6	−18.5
− 2	− 2.8	− 8.9	−12.9	−15.7	−17.9	−19.5	−20.7	−21.6
− 4	− 4.9	−11.3	−15.5	−18.5	−20.8	−22.5	−23.8	−24.8
− 6	− 6.9	−13.7	−18.1	−21.3	−23.7	−25.5	−26.9	−27.9
− 8	− 9.0	−16.1	−20.0	−24.1	−26.6	−28.5	−29.9	−31.0
−10	−11.0	−18.5	−23.4	−26.9	−29.5	−31.5	−33.0	−34.1
−12	−13.1	−20.9	−26.0	−29.7	−32.4	−34.5	−36.1	−37.2
−14	−15.1	−23.3	−28.6	−32.4	−35.3	−37.5	−39.1	−40.4
−16	−17.2	−25.7	−31.3	−35.2	−38.2	−40.5	−42.2	−43.5
−18	−19.2	−28.1	−33.9	−38.0	−41.1	−43.5	−45.3	−46.6
−20	−21.2	−30.5	−36.5	−40.8	−44.0	−46.5	−48.3	−49.7

One serious problem with clothing is exercise—the wearer becomes a tropical man in Arctic clothing. Exercise causes sweating; the sweat may accumulate in the clothing and freeze. The Eskimo avoids this in two ways; he constantly pulls off his outer fur parka so that much of the time his skin is cool. More importantly, his clothing is designed with many ventilation areas, gaps, and drawstrings. Eskimos bending over while wearing a parka show the bare skin of their backs. Wool is superior to cotton: (1) it doesn't absorb water (wet clothes lose 90% of their *clo* value), and (2) the weave permits sweat to move from the skin to the outside of the garment (evaporation on the skin removes more heat from the body than evaporation from clothes).

Relative humidity makes little difference since a 10% rh may give an absolute vapor pressure of 1 torr vs. a 100% rh giving 7 torr. Skin vapor pressure is about 44 torr so 44 − 1 = 43 is not much more driving force than 44 − 7 = 37. A cold rain, however, puts cold water on the skin. Then raising each kg takes 1 kcal/C. Thus raising 1 kg of water from 5 C to the 33 C of skin temperature would require 28 kcal; if the water then is evaporated it takes 80 more kcal/kg from the body.

Respiratory heat loss is one reason you can eat more calories in the cold weather; for 0 C air, dry loss is 3 kcal/hr and wet loss is 10 kcal/hr so at the end of the day you can have an extra 8 spoonfuls of sugar, or 1 beer. In the cold, urine production increases to about three times normal—leading to dehydration. Counter with warm drinks such as coffee, tea, or bouillon.[52]

There is acclimatization to cold just as there is acclimatization to heat. The exact physiology is relatively unknown but probably involves improved non-shivering thermogenesis (increased heat generation without shivering). The location of the nonshivering thermogenesis in man is debated but is probably the liver, the muscles, and fat. Metabolism in the muscles increases up to 50% before the physical movement of shivering begins. The cold-acclimatized man retains his ability to use shivering also.

SUMMARY

The volume of air required at a work station depends primarily on the purity of the air in the workspace—more pollution requires more air to dilute the contaminants.

A comfortable climate is not especially difficult to calculate since the human variables of age, sex, geographic location, and time of the year make little difference. Remember, however, that a single climate will not satisfy all the people all the time.

Performance effects of heat and cold stress interact with motivation and are very difficult to isolate—especially for time periods of over one day. Physiological effects, slightly easier to predict, depend on the amount of heat stored or lost. Storage usually is measured by mean body temperature. Prediction of the effect of a specific disturbance (such as an increase in environmental temperature) is complicated by the body's homeostasis—that is, the body makes compensating adjustments (such as shivering or sweating) to reduce the effect of any external disturbance.

In most industrial environments, the primary defense against heat stress is adequate air velocity to improve evaporation of sweat. The primary defense against cold stress is properly designed clothing.

SHORT-ANSWER REVIEW QUESTIONS

20.1 Do air volume requirements depend on oxygen requirements, carbon dioxide removal requirements, or odor requirements?

20.2 For a dry bulb temperature of 30 C and relative humidity of 50%, what is absolute humidity, dew point, wet bulb temperature, effective temperature, and new effective temperature? Give to nearest degree or mm.

20.3 What is the *clo* value for a man wearing a woven shirt and trousers? Wearing a woven shirt, jacket, and trousers? For a woman wearing a blouse and skirt? A sweater and slacks? Assume they are wearing shoes and underwear.

20.4 What is the adjustment of the comfort zone for time of day, for season of the year, for sex of the occupant, and for age of the occupant?

20.5 Why are salt tablets a poor idea?

THOUGHT-DISCUSSION QUESTIONS

1. Should an employer furnish a "shirt sleeves" environment for everyone (heated in winter and cooled in summer)? Who should be required to sweat in summer? Who should be required to wear a jacket in winter?

2. Discuss water and salt replacement in the heat.

REFERENCES

1. American Association for Health, Physical Education, and Recreation. *Nutrition for Athletes*. Washington, D.C.: American Association for Health, Physical Education, and Recreation, 1971, p. 36.

2. *ASHRAE Systems Handbook*. New York: American Society of Heating, Refrigerating, and Air Conditioning Engineers, 1976.

3. *ASHRAE Standard Ventilation for Acceptable Indoor Air Quality (62-1981)* Atlanta: American Society of Heating, Refrigeration and Air Conditioning Engineers, 1981.

4. Berenson, P., and Robertson, W. Temperature in *Bioastronautics Data Book*. Washington, D.C.: National Aeronautics and Space Administration, 1973.

5. Beshir, M. A comprehensive comparison between WBGT and Botsball. *American Industrial Hygiene Association Journal*, Vol. 42, No. 2 (1981): 81-87.

6. Collins, K. Endocrine control of salt and water in hot conditions. *Federation Proceedings*, Vol. 22 (1963): 716-720.

7. Committee on Nutritional Misinformation. *Water Deprivation and Performance of Athletes*. Washington, D.C.: National Academy of Sciences, 1974, p. 2.

8. Fanger, P. Assessment of man's thermal comfort in practice. *British Journal of Industrial Medicine*, Vol. 30 (1973): 323-324.

9. Fanger, P. Conditions for thermal comfort—a review. *Thermal Comfort and Moderate Heat Stress*. London: Her Majesty's Stationery Office, 1973.

10. Fanger, P. *Thermal Comfort*. New York: McGraw-Hill, 1972.

11. Fanger, P.; Angelius, O.; and Kjerulf-Jensen, P. Radiation data for the human body. *ASHRAE Transactions*, paper 2168, part I, 1970.

12. Fanger, P.; Hojbjerre, J.; and Thomsen, J. Thermal comfort conditions in the morning and in the evening. *International Journal of Biometeorology*, Vol. 18, No. 1 (1974): 16-22.

13. Folk, G. *Textbook of Environmental Physiology*, 2d ed. Philadelphia: Lea and Febiger, 1974.

14. Fourt, L., and Hollies, N. *Clothing Comfort and Function*. New York: Marcel Dekker, 1970.

15. Fox, W. Human performance in the cold. *Human Factors*, Vol. 9, No. 3 (1967): 203-220.

16. Froese, G., and Burton, A. Heat losses from the human head. *Journal of Applied Physiology*, Vol. 10, No. 2 (1957): 235-241.

17. Gagge, A., and Nishi, Y. Physical indices of the thermal environment. *ASHRAE Journal*, Vol. 18, No. 1 (1976): 47-51.

18. Gagge, A.; Stolwijk, J.; and Nishi, Y. An effective temperature scale based on a simple model of human physiological regulatory response. *ASHRAE Transactions*, Vol. 77 (1971): 247-262.

19. Givoni, B., and Goldman, R. Predicting effects of heat acclimatization on heart rate and rectal temperature. *Journal of Applied Physiology,* Vol. 35, No. 6 (1973): 875-879.

20. Green, G. The effect of indoor relative humidity on absenteeism and colds in schools. *ASHRAE Transactions,* Vol. 80, Part 2 (1974).

21. Grether, W. Human performance at elevated environmental temperatures. *Aerospace Medicine,* Vol. 44, No. 7 (1973): 747-755.

22. Guyton, A. *Textbook of Medical Physiology,* 4th ed. Philadelphia: W.B. Saunders, 1971.

23. Hancock, P. The limitation of human performance in extreme heat conditions. *Proceedings of The Human Factors Society,* Rochester, N.Y., 1981; pp. 74-78.

24. Jessen, K., and Hagelsten, J. Search and rescue service in Denmark with special reference to accidental hypothermia. *Aerospace Medicine,* Vol. 43, No. 7 (1972): 787-791.

25. Jorgensen, K., and Poulsen, E. Physiological problems in repetitive lifting. *Ergonomics,* Vol. 17, No. 1 (1974): 31-39.

26. Koch, W.; Jennings, B.; and Humphreys, C. Environmental study II, sensation responses to temperature and humidity under still air conditions in the comfort range. *ASHRAE Transactions,* Vol. 66 (1960): 264-287.

27. Konz, S.; Hwang, C.; Dhiman, B.; Duncan, J.; and Masud, A. An experimental validation of mathematical simulation of human thermoregulation. *Computers in Biology and Medicine,* Vol. 7 (1977): 71-82.

28. Konz, S., and Nentwich, H. A cooling hood in hot-humid environments. Kansas State University Eng. Experiment Station Report 81, 1969.

29. Lind, A. A physiological criterion for setting thermal environmental limits for everyday work. *Journal of Applied Physiology,* Vol. 18, No. 1 (1963): 51-56.

30. Lovelace, W., and Gagge, A. Aero medical aspects of cabin pressurization for military and commercial aircraft. *Journal of Aeronautical Science,* Vol. 13 (1946): 143-150.

31. McDermott, H. *Handbook of Ventilation for Contaminant Control.* Ann Arbor, Mich.: Ann Arbor Science Publishers, 1976.

32. Minard, D., and Copman, L. Elevation of body temperature in health. *Temperature—Its Measurement and Control.* C. Hertzfield, ed. New York: Reinhold, 1963, chapter 46.

33. Mitchell, D., and Whillier, A. Cooling power of underground environments. *Journal of South African Institute of Mining and Metallurgy* (October 1971): 93-99.

34. National Academy of Sciences. *Recommended Dietary Allowances.* Washington, D.C.: National Academy of Sciences, 1974, p. 90.

35. Nevins, R.; Rohles, F.; Springer, W.; and Feyerherm, A. Temperature humidity chart for thermal comfort of seated persons. *ASHRAE Transactions,* Vol. 72 (1966): 283-291.

36. Nevins, R., and Gorton, R. Thermal comfort conditions. *ASHRAE Journal,* Vol. 16 (January 1974): 90-93.

37. Nevins, R. Energy conservation strategies and human comfort. *ASHRAE Journal,* Vol. 17, No. 4 (1975): 33-37.

38. Newburgh, L., ed. *The Physiology of Heat Regulation and the Science of Clothing.* New York: Hafner Publishing Co., 1968.

39. Nielsen, M. Die Regulation der Korpertemperatur bei Muskelarbeit. *Skandinavian Archives Physioloque,* Vol. 79 (1938): 193-230.

40. Olesen, B., and Fanger, P. The skin temperature distribution for resting man in comfort. *Archives des Sciences Physiologigues,* Vol. 27, No. 4 (1973): A385-393.

41. Pirnay, F.; Petit, J.; and Deroanne, R. A comparative study of the evolution of heart rate and body temperature during physical effort at high temperature (in French). *Internationale Zeitschrift fur Angewandte Physiologie Einschliesslich Arbeitsphysiologie,* Vol. 28 (December 1969): 23-30.

42. Poulton, E. *Environment and Human Efficiency.* Springfield, Ill.: C. T. Thomas, 1970.

43. Ramsey, J. Heat stress standard. *National Safety News* (June 1975): 89-95.

44. Ramsey, J.; Dayal, D.; and Ghahramani, G. Heat stress limits for the sedentary worker. *American Industrial Hygiene Association Journal,* Vol. 36, No. 4 (1975): 259-265.

45. Rohles, F. The revised modal comfort envelope. *ASHRAE Transactions,* Vol. 79, part V (1973): 52-59.

46. Rohles, F.; Hayter, R.; and Milliken, G. Effective temperature (ET*) as a predictor of thermal comfort. *ASHRAE Transactions,* Vol. 81, part 2 (1975): 148-156.

47. Rohles, F., and Nevins, R. Thermal comfort: new directions and standards. *Aerospace Medicine,* Vol. 44 (July 1973): 730-738.

48. Rohles, F.; Konz, S.; and Munson, D. Estimating occupant satisfaction from effective temperature (ET*), *Proceedings of 24th Annual Meeting of the Human Factors Society,* 1980; pp. 223-227.

49. Rohles, F.; Konz, S.; and Munson, D. New psychometric chart for thermal comfort. *ASHRAE Journal,* Vol. 24, No. 1 (1982): 85-87.

50. Rosen, E. Comfort and cooling with box fans. MS Thesis, Kansas State University, 1982.

51. Seppanen, O.; McNall, P.; Munson, D.; and Sprague, C. Thermal insulating values for typical clothing ensembles. *ASHRAE Transactions,* Report 2219 RP-43, Vol. 78, part I, (1972).

52. Sjostrom, S. Unexplained accidents may be due to cold weather. *Working Environment 1981,* (1981): 20-21.

53. Slonim, N. ed. *Environmental Physiology.* St. Louis: C. V. Mosby, 1974.

54. Strydom, N., et al. Effect of ascorbic acid on rate of heat acclimatization. *Journal of Applied Physiology,* Vol. 141, No. 2 (1976): 202-205.

55. Strydom, N.; van der Walt, W.; Jooste, P.; and Kotze, H. Note: a revised method of heat acclimatization. *Journal of South African Institute of Mining and Metallurgy,* Vol. 76 (1976): 448-452.

56. Wing, J., and Touchstone, R. A bibliography of the effects of temperature on human performance. Technical Report AMRL-TDR-63-13, Wright Patterson AFB, Ohio (February 1963).

57. Wing, J. Upper thermal tolerance limits for unimpaired mental performance. *Aerospace Medicine,* Vol. 36 (October 1965): 960-964.

58. Woods, J.; Maldonadio, E.; and Reynolds, G. How ventilation influences energy consumption and indoor air quality. *ASHRAE Journal,* Vol. 24, No. 9 (1981): 40-43.

59. Wyndham, C.; Williams, C.; and Bredell, G. The physiological effects of different gaps between wet and dry-bulb temperatures at high wet-bulb temperatures. *Journal of the South African Institute of Mining and Metallurgy* (September 1965): 52-57.

60. Wyndham, C., and Strydom, N. The effect of environmental heat on comfort, productivity and health of workmen. *The South African Mechanical Engineer,* (May 1965): 209-221.

PART VII
DETERMINING THE
TIME FOR THE TASK

21

MEASURING THE TIME REQUIRED

WHY DETERMINE TIME/JOB?

The emphasis in this book so far has been on designing the job. In the following chapters the emphasis will shift to determining the amount of time that a job takes. The first question is why is the time determined at all? In most organizations there are four answers: cost allocation, scheduling, evaluation of alternatives, and acceptable day's work. In some organizations there is a fifth: incentive pay.

Cost Allocation

Cost of a good or service has three major components: materials, labor, and burden.

$$\text{Labor cost/unit} = (\text{Time/unit})(\text{Cost/time})$$

For example a barber might calculate the cost of a haircut as:

Materials (hair oil, etc.)	$ .06
Labor (.25 hr × $8/hr)	$2.00
Burden (light, heat, rent, etc.)	$2.00
Total Cost	$4.06

Without the time per haircut, the barber could not calculate the labor cost. In addition, many cost accounting systems use the labor cost as a means of apportioning burden. If you don't know the cost, how can you determine the price?

In the barber's case, the price might be $5/haircut, as the $4.06 assumes the barber is busy 100% of the working time. In other firms (hospitals determining the cost of a meal, hotels determining the cost of the room, or a car dealer determining the cost of your car) the procedure is the same: (1) determine the cost, (2) then determine the price. Without the cost of labor you can't determine the cost of the product or service; without the time/unit, you can't determine the cost of labor.

Scheduling

Without the time/unit, you can't schedule. If a customer wants 10,000 widgits, when will they be ready? If each widgit takes 1 hr, then you need 10,000 work hours. Then, if there are, say, 25 workers assigned to making widgits and output is at the rate of 25/hr, the 10,000 will be completed in 400 hr or approximately 50 days. Thus time/unit is necessary for deciding how many workers should be assigned to a job, how many machines are needed, when the units can be delivered, etc.

Evaluation of Alternatives

Closely related to scheduling is the selection of alternative methods. Should a bushing be made on a turret lathe or an engine lathe? Should a building be built of brick or wood or steel? Should the defective starter on your car be repaired or replaced with a new one? Whether the decision is on a simple level (compare a price of $40 for repair of your starter vs. $100 for a rebuilt starter) or on a complicated level (where operation research analysts are feeding a_{ij} values into the jaws of their computers), decision making is based on knowing the labor cost/unit. Incorrect times give incorrect decisions.

Acceptable Day's Work

Managers want to judge the performance of individuals working for them. George made 29 widgits today; is that good or bad? Betty Jo sewed 500 sleeves today; is that good or bad? Juan installed carpet in six houses last week; is that good or bad? To evaluate performance a standard is necessary. Thus managers want to know the typical time/unit so they can determine the relative performance of George, Betty Jo, and Juan. The manager's boss in turn evaluates the manager's performance by using the group's performance vs. the group's standard time.

Incentive Pay

A minority of organizations pay by results; that is, more output means more pay. For example, a textile firm might conclude that a typical worker could sew 450 sleeves/hr. If a typical wage in that locality was $4.50/hr, the firm might pay $.01/sleeve instead of $4.50/hr. Thus if Betty Jo sewed 500 sleeves, her pay would be $5.00; if she sewed 475 it would be $4.75. Pay for less than standard performance depends on the specific policy used. A common plan is a guaranteed base at 100%. That is, for output up to 100%, pay equals 100%. For output over 100%, there is a 1% increase in pay for a 1% increase in output. But, to implement pay by results, the firm must know the required time/unit.

In the USA between 1945 and 1968, about 26% of the US production work force was on incentive. Coverage by industry varied widely with 80% coverage in men's garments, 66% in basic steel, 31% in automotive parts, meat packing, and textiles, and practically none in the chemical and process industries. In England in 1961 the overall percent for all industries was 33%; it is reported to be about 60% in the USSR.[1] Rice, using 1500 US responses, reported 95% of manufacturing firms and 69% of nonmanufacturing firms used work measurement.[6] Incentive wages were used by 59% of the manufacturing firms but only 1 nonmanufacturing firm used incentive wages.

The question now becomes "How do you determine time/unit?"

The following will discuss time *measurement* as opposed to mere opinions on the amount of time needed. Although guesses may occasionally be correct, build your time requirements upon a solid foundation of measurements rather than upon unsupported conjecture.

TIMING TECHNIQUES

Record the amount of time to do the task while counting the number of items produced during that time.

Four time measuring instruments can be used:

- The calendar—accurate to about 1 day
- The stopwatch—accurate to about .5 s
- The TV or movie film—accurate to about .01 s
- Electronic circuit breaking—accurate to about .001 s

Inaccuracy of the measuring instrument and measuring techniques are random errors added to the variability of the data.

Another technique is to not measure time at all but to reuse times from similar tasks—standard data.

The Calendar

First take an easily identifiable block of time (usually one shift although it could be one week or one day with several shifts). Assume that only one task is being done over and over. Count the units reportedly produced during that period; then calculate the standard. Although a calendar only measures in days, if the time worked/day is relatively constant, the time may be counted in minutes. For example, assume that Harry, a mechanic, did nothing but truck tune-ups, that he worked 450 minutes per day, 5 days/wk, and that during a specific week he did 18 tune-ups. Then time/tune-up was (450 × 5)/18 or 125 min/tune-up. Another example might be the number of calls a salesperson could make. Mary worked 21 days for an average of 3 hr/day and made 92 calls. Then time/call would be (21 × 3)/92 = .7 hr/call.

A second possibility is that different items are produced. Harry may tune up 5 trucks and 13 cars. Occurrence sampling (see chapter 9) may indicate he spent 40% of his time on truck tune-ups and 60% on car tune-ups. Therefore truck tune-ups take (450 × 5) (.4)/5 = 180 min while car tune-ups take (450 × 5) (.6)/13 = 104 min.

Inaccuracy increases for less standardized situations such as a part-time salesperson such as Mary. Using the calendar as a measuring instrument has an advantage of low cost; its primary disadvantage is low accuracy.

The Stopwatch

With a stopwatch, an observer starts the watch at the start of the work cycle and stops it at the end. The work cycle, therefore, is recorded within a second or so. As an example, assume that Jones cuts risers off castings. During the .67 hr of the study, he cut risers off 125 castings. Time per casting then would be .67/125 = .0054 hr/casting. Instrument error of less than a second/cycle usually is more than sufficiently accurate. The primary disadvantage is the cost of using the instrument; that is, the wages of the person using the watch.

For many situations, a stopwatch is the standard measuring instrument; therefore it is discussed in more detail in the next section of this chapter, Timing Procedure: Sequential Observations.

Film

Make a film (usually TV but sometimes movie) of the work. For example, a TV tape might be made of Betty Jo while she stitched pockets on shirts. During the 24 minutes of the tape, Betty put on 197 pockets for a time of (24/60)/197 = .002 hr/pocket. Or, rather than take the overall time/overall output, the film might be analyzed pocket by pocket. Time for each pocket then can be determined by putting a large clock in the picture (see Figure 21.1 for a chronocyclograph picture of a person lifting). An alternative is to use a movie camera which exposes the film at a very constant rate so that the bulky and potentially disturbing clock is not needed. Then the analyst counts the number of frames per pocket and multiplies by the time/frame. For example, stitching on the first pocket might be from frame 75 to frame 197, the next from frame 196 to 325, the next from 326 to frame 440, etc. The average might be 120 frames/pocket; then multiply 120 by the time/frame for that camera (960 and 1000 frames/min are common speeds). If it was 1000 frames/min, .001 min/frame $\times$ 120 frames = .12 min/pocket or .002 hr/pocket.

Movie film, the traditional method, has approximately the same timing accuracy as TV tape and low initial cost for equipment. However, besides being immediately available, TV tape has the lowest cost after approximately 20 hours of recording, due to its reusable tape and low tape cost/hr.[4] TV tapes play for an hour; this same length of movies either requires constant observer changes of film or use of time-lapse photography. TV also is amenable to multiple cameras feeding into a central monitor which records only the most relevant view, especially useful for crew operations.[4] Schantz gives the following shooting tips[9]: (1) Make a sketch of the location of the camera, subject, and lights ahead of time. (2) Test your equipment before going to the shooting scene. (3) Keep the camera style slow and deliberate. Begin the scene with a slow zoom-in; end with a slow zoom-out. Avoid short shots; stay on a shot until the action is complete. Use a tripod. (4) Don't overlook the audio; it can be done while filming or dubbed in later. TV also has an advantage of being accepted by the workers due to their being able to see what is being recorded. "Low light" TV cameras operate in 50 lux vs. 300 lux for a standard TV camera.

Electronics

You may want to measure times extremely accurately. The most common method is to put a flashlight battery (1.5 V DC) in a circuit with the person and a target and a strip chart recorder. When the person touches the target a 1.5 V signal is sent to the recorder which registers a signal until the contact is broken, at which time the pen moves. The paper speed (mm/s) times the distance between spikes (mm) gives the time. In a study on positioning accuracy by one of my students, he made the target 1.5 V and the surrounding error surface 3 V and so was able to record not only times but which area was hit.

The most common measuring instruments are the calendar and the stop-

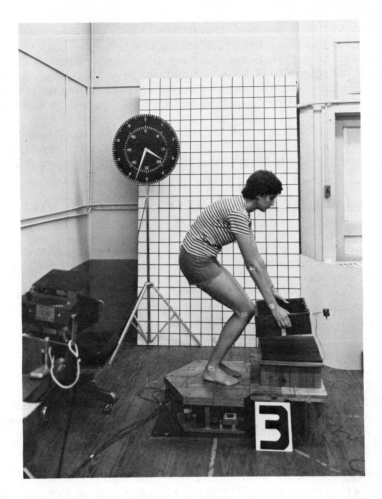

Figure 21.1   *Time* between events on a film can be determined by putting a large clock with a fast moving hand (chronocyclograph) in the picture. Another technique is to mount a light bulb on the person's limbs and interrupt the circuit periodically; the result is a "dot to dot" film.

watch—primarily the stopwatch. The very high precision of film and electronics rarely are economical since there are very few operations which have not been mechanized which have the millions of cycles necessary to cover the cost of precise *time* analysis. For example, consider an operation done 1,000,000 times/yr which takes .01 hr/unit (36 s). Total hours/yr would be 10,000 so intensive *methods* analysis would be worthwhile if it were to save even 10% or 1,000 hr/yr. The principles of workstation design (i.e., intensive *methods* analysis) can be applied to the job to get the best method. Once the proper method has been developed, then time/unit can be determined inexpensively with stopwatches or predetermined time systems.

Standard Data Systems

A very popular technique is not to measure the time to do the task at all. The task is described in a specified amount of detail; then the times for similar elements are used to predict the time for the new job.

The task can be broken down into very fine detail—reach 40 cm to object jumbled in a pile, grasp $10 \times 10 \times 5$ mm object from jumbled pile, move object 15 cm to other hand, etc.—just as you use letters of the alphabet to make words for a message. For more information on this level of analysis and timing, see chapter 8.

The task also can be broken into quite large units: load truck, drive truck 3000 km, unload truck. Here we have selected "words" to make the message. For more information on this macro level of analysis and timing, see chapter 25.

TIMING PROCEDURE: SEQUENTIAL OBSERVATIONS

The most common technique is to record the time of a series of sequential observations. Occurrence sampling, the use of nonsequential observations, is discussed later in this chapter and in chapter 9.

The timing procedure will be discussed in eight sections: methods analysis, operator selection, preparation for timing, stopwatch timing techniques, number of observations, foreign elements, rating, and allowances.

Methods Analysis

The first step is to develop a good method to be timed. Simply recording the time of an existing method without analysis is more than a mere waste of money—it will lead to endless future problems for a poor method will be protected by a "grandfather clause," inequities will appear between poor standard times and good standard times and thus cause the operators to doubt the integrity and honesty of the standard setters, and grief will follow grief. (A "grandfather clause" is a clause, in an agreement, that exempts present participants from meeting the requirements. For example, all engineers in a department might be required to have a BS in engineering; a "grandfather clause" would exempt present employees.)

If an existing method must have a time attached to it without the method being analyzed, then the resulting standard must be labeled temporary. The temporary standard must be replaced with a permanent standard as soon as possible, say within one week. The reader might wonder why such emphasis is made on this simple concept but the road to hell is paved with good intentions. Perhaps the best way to cause labor problems is to have inequitable work standards and the best way to get inequitable work standards is to make a time study without a methods study first.

Even if a detailed method analysis is made before a detailed time analysis, the resulting standard should not be considered 100% accurate forever. Salvendy and McCabe discuss reasons for inaccurate standards. Some of these standards are[7]:

1. Variability from day to day in organizational variables—quality and quantity of materials, mechanization and machinery speed, supervisory effectiveness, working conditions

2. Dependability of fatigue and delay allowances

3. Individual effectiveness on specific jobs at specific times

4. Variations in workplace layout, tools, and equipment

5. Product design modifications

Thus standards should be audited periodically to see how much has changed since the standard was set. Use the Pareto principle in auditing; that is, audit the "mighty few" and ignore the "insignificant many." In addition, to reduce the decrease in operator productivity and satisfaction from inequitable standards, fix an expiration date for each standard, say 5 years from the time first installed. The actual time and the standard time will gradually diverge over the years. Eventually it will be recognized that the standard no longer is valid but it will be difficult to change suddenly as the workers will say "Why me?" "Why now?" and there won't be a good answer. It is better to have the standard automatically expire after 5 years so that the revalidation is not pointing a finger at any individual. After all, when you point a finger at someone else, three fingers point at yourself.

Design of the job has been covered in the previous chapters of this book. Note the relative proportion of this book allocated to design of the job and the proportion to timing the job. The critical task is job design rather than job timing since a poor job design can easily cause a 50% inefficiency while it is difficult to mistime a well-designed job with an error of much over 10% and even a 10% error in the time standard may not be reflected in the operator's actual performance.

Thus, in general, methods engineers and technicians should spend 75% to 95% of their time in job design and only the remainder in job timing. This obvious allocation of resources is mentioned because many organizations use the reverse proportions and spend all their resources in time determination rather than job design. As Santayana said: "Those who ignore history must relive it."

Operator Selection

Once the proper method has been developed, the next question is who should be studied. Under no circumstances should the timing be done without the knowledge of the worker and the supervisor. In most cases, there is no choice as there is only one worker for the job. But in some cases more than one person does the task. Who should be timed?

If there is a choice, first select someone who has experience on the task studied rather than an inexperienced person. The standard will apply to experienced workers; therefore a time sample from an experienced worker is more likely to be representative than one from an inexperienced worker. Work methods of novices have an unusually high percentage of delays, fumbles, hesitations, and slow decisions; it is quite difficult for the time study technician to determine precisely the exact correction for these items.

If there are several experienced workers, select an average worker rather than someone who is unusually fast or slow. (It is a good idea to study a fast operator when making a *methods* analysis since the speed is likely to be due to a good method rather than a fast work pace. After developing the good method though, make the *time* analysis on an average worker who has been taught the good method.) There are two reasons for selecting an average operator: rating accuracy and worker acceptance.

Rating (discussed in more detail in chapter 22) requires the time study technician, say John, to "normalize" the times he records. For example, assume the problem is to find the standard time for college males to run 1500 m. If timing a world-class distance runner, John might observe a time of 3.95 min; if timing "Joe Student" he might get 6.5 min; if timing Professor Konz he might get 13.0 min. If John had timed the expert, he would say "better than normal" and might assign a value of 60% better; the resulting normal time would then be $3.95 \times 1.60 = 6.32$ min. If he had studied "Joe Student," he might estimate 100% as the pace and get a normal time of $6.50 \times 1.00 = 6.50$ min. If he had studied Professor

Konz, he might estimate the pace as 50% so the normalized time would be 13.0 ×
.5 = 6.75 min.

Ability to normalize is easier if the performance is close to 100%; as performance gets farther and farther from 100% absolute error of rating increases.
Thus, to improve rating accuracy, study, if possible, an average operator.

The second reason to study an average operator is acceptance of the standard.
Workers really don't trust rating. The more a time is adjusted the less they trust
it. As pointed out in the previous paragraph, their mistrust of extreme adjustments is well founded.

Preparation for Timing

Assume the task is the polishing of a pair of shoes. The task consists of getting
the shoes, polishing both shoes, and putting the shoes away. Rather than time
the entire process as a unit, break the task into elements. In this case the
elements might be: (1) get the shoes, (2) polish the shoes, (3) put the shoes away.

There are four reasons to break a job into elements:

1. Elements make it possible to reuse the data. Assume the next job varied
   only in that he got 4 shoes at a time instead of 2, or that someone else
   brought the shoes. Then a new standard would be required only for element
   1 instead of the entire task. In some tasks, elements may be done in different sequences. Consider elements as "bricks" which are used to build a
   "structure"—the bricks of time are used to build a structure of times. We
   will reuse bricks instead of making new bricks.

2. Elements give good internal consistency checks during the time study. For
   example, if John, the time study technician, sees that the first pair of shoes
   takes 1.2 minutes to get, the second 1.1 minutes, but the third takes 2.1 minutes, the difference becomes obvious. If only the task total was timed, then
   times of 4.1, 4.2, and 5.2 are not so obviously different. It might even be
   that, by chance, the polish and put away time for the third pair of shoes
   might be shorter than normal so the three total times might be 4.1, 4.2, and
   4.2 and the unusually long "get time" of 2.1 for the third pair may not be
   detected.

   Elements also permit external consistency checks. If, in another study,
   getting shoes took .3 minutes, then methods in this study which require 1.1
   minutes should be checked very closely.

3. Elements permit different ratings for different elements. John can rate "get
   shoes" as 90%, "polish shoes" as 100%, and "put shoes away" as 105%—if
   they are elements. For an unbroken task, he must use one overall rating.

4. Elements improve methods descriptions. A constant problem is to get the
   method described in enough detail. Breaking the job into elements improves
   the odds that a good methods description will be recorded. See Table 21.1.

After the task is broken down into elements, write the element description on
the time study form with a description of its end point (also called termination
point or *TP*). For example:

No.	Element Description	*TP*
1	Get 1 pair of shoes from closet in bedroom	Drop shoes
2	Polish shoes with dauber, brush, and rag	RL rag
3	Put shoes away in closet	RL shoes

Table 21.1   The Ideal Methods Description Can be Checked Years Later by Another Person at Another Site with no Confusion Concerning the Exact Method. This Skill Primarily Requires Giving *Detail* (Example from Quick, Duncan, and Malcomb, p. 276).[5]

Work To Be Done			Methods Description
Pick up bolt from random pile in lip tray. Move to hole in assembly fixture, turning bolt end for end.	Poor:	R	to bolt
		Gr	bolt
		PP	bolt
		M	to fixture
	Better:	R	to bolts
		Gr	bolt
		PP	bolt
		M	to hole
	Best:	R	to bolt in L/tray
		Gr	bolt
		PP	bolt EFE
		M	bolt to fixture hole
Pick up single, isolated bolt from table top. Move to hole in assembly fixture.	Poor:	R	to bolt
		Gr	bolt
		M	to fixture
	Better:	R	to bolt on table
		Gr	bolt
		M	bolt to fixture hole

Select a very definite, easily defined point as the *TP;* a *TP* with a sound usually is best.

Stopwatch Timing Techniques

There are four alternatives: one watch with continuous hand movement, one watch with "snapback," three watches with "snapback," and an electronic watch with a "hold" circuit.

First consider using one stopwatch, starting the watch at the start of the study (say, 1:02) and letting it run until the end of the study (say, 1:32). Thus the elapsed time was 30 minutes or .5 hours. As a check, the sum of all the elements should total .5 hours. The observer aligns his or her eye, the watch, and the task and, when an element is completed, writes down the time where the moving hand was at the completion of the element. (Element terminal points also can be detected by sound in some cases.) The resulting record is shown in Figure 21.2. Rather than write down the entire number, simplified codes often are used (instead of 95 102 121 187 206, omit the "redundant" first numbers giving 95 102 21 87 206). Then, later in the office, the numbers are subtracted (using ink or a contrasting color) and the analyst learns what the times for each element are. See Figure 21.3. The advantage of continuous recording is that the clock never stops so that no time is omitted; the disadvantages are that the observer does not know, at the job, how individual elements vary and the observer is trying to read a moving target (the watch hand).

An alternative is to use one watch with "snapback." The observer aligns the eye, the watch, and the task. When the element ends, the time is read while simultaneously snapping the hand back to zero where it starts moving forward again. Then the element time is written on the form. See Figure 21.4. Advantages are

NO.	ELEMENTS	SPEED FEED	UPPER LINE : SUBTRACTED TIME 1 2 3 4 5 6 7 8 9 10 11 12 13 14 15 LOWER LINE : READING	MIN. TIME	AV. TIME	Std. Dev.	OCC. PER CYCLE	EFFORT RATING	NORMAL TIME
1	Pick up part TP piece at vise		14 22 48 74 501 29 55 81 1008 34				1	90	
2	Put part in vise TP RL vise handle		42 51 76 405 29 57 84 909 35 62				1	90	
3	Mill slot TP power feed disengages		98 706 332 59 85 713 889 66 91 218				1	100	
4	Aside part to pan TP RL part		108 34 60 87 613 41 67 94 119 46				1	95	

FOREIGN ELEMENTS :

TOOLS, JIGS, GAUGES, PATTERNS, ETC. : Vise 407

OVERALL EFFORT RATING	BEGIN	END	ELAPSED	UNITS FINISHED	ACTUAL TIME PER PIECE
	1:14	1:22		10	

Figure 21.2 *Watch hands never stop* in the continuous one-watch method. The form after the study would give the sketch of the workstation, the element breakdown, the recorded times, and the rating by element. Times usually are coded; that is, 14 instead of .0014.

NO.	ELEMENTS	SPEED FEED	UPPER LINE : SUBTRACTED TIME 1 2 3 4 5 6 7 8 9 10 11 12 13 14 15 LOWER LINE : READING	MIN. TIME	AV. TIME	Std. Dev.	OCC. PER CYCLE	EFFORT RATING	NORMAL TIME
1	Pick up part TP piece at vise		14 14 14 14 14 14 14 14 14 15	14	14.1		1	90	12.69
2	Put part in vise TP RL vise handle		28 29 28 29 28 28 29 28 27 28	27	28.2		1	90	25.38
3	Mill slot TP power feed disengages		56 55 56 56 56 56 55 57 56 56	55	55.9		1	100	55.90
4	Aside part to pan TP RL part		10 28 28 28 30 28 28 28 28 28	10	26.4		1	95	25.88
									1.1905
									=.0119 hrs/piece

FOREIGN ELEMENTS :

TOOLS, JIGS, GAUGES, PATTERNS, ETC. : Vise 407

OVERALL EFFORT RATING	BEGIN	END	ELAPSED	UNITS FINISHED	ACTUAL TIME PER PIECE
	1:14	1:22	8 min	10	.0133 hr

= .133 hr.

Figure 21.3 *Subtraction* of the times later in the office is one of the disadvantages of the continuous one-watch system. The times are totaled, averaged, and multiplied by the rating to get normal time/element. Use a contrasting color such as red or ink for the subtracted times. The sum of the element times (14.1 + 28.2 + 55.9 + 26.4) per unit × 10 units studied = .1246 hr or 7.5 min. This agrees with the 8-min difference (1:22 − 1:14) from the start to the end of the study recorded on a wristwatch. The normal time is .001 190 hr/unit.

that the observer can note the variability within an element and eliminate the subtraction. Disadvantages are the need to read a moving target, the difficulty of

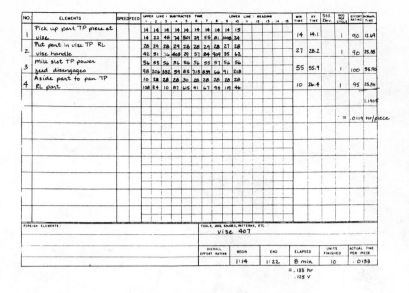

NO.	ELEMENTS	SPEED	FEED	UPPER LINE : SUBTRACTED TIME ／ LOWER LINE : READING																MIN TIME	AV. TIME	Std. Dev.	OCC PER CYCLE	EFFORT RATING	NORMAL TIME
1	Pick up part TP piece at vise			14 14 14 14 14 / 14 14 14 14 15 ／ 14 22 48 74 501 / 29 55 81 1008 34																14	14.1		1	90	12.69
2	Put part in vise TP RL vise handle			28 29 28 29 28 / 28 29 28 27 28 ／ 42 51 76 405 29 / 57 84 909 35 62																27	28.2		1	90	25.38
3	mill slot TP power feed disengages			56 55 56 56 56 / 56 55 57 56 56 ／ 98 206 332 59 85 / 713 839 66 91 218																55	55.9		1	100	55.90
4	Aside part to pan TP RL part			10 28 28 30 28 / 28 28 28 28 ／ 108 34 10 87 415 / 41 67 94 119 46																10	26.4		1	95	25.86

1.190 S

= .0119 hr/piece

FOREIGN ELEMENTS :	TOOLS, JIGS, GAUGES, PATTERNS, ETC.
	vise 407

	OVERALL EFFORT RATING	BEGIN	END	ELAPSED	UNITS FINISHED	ACTUAL TIME PER PIECE
		1:14	1:22	8 min	10	.0133

= .133 hr
.125 V

Figure 21.4 *No subtraction* is necessary in the snapback one-watch system as element times are recorded directly on the form.

reading two successive, short (say, 3 s) elements in a row, the fact that the moving hand takes a finite amount of time to come to zero and start again (about .2 s) which shortens every observed time, and the possibility that the observer may not record each and every cycle since observers tend to mentally judge foreign elements and irregular elements (discussed later in this chapter) as delays and stop the watch when it should be recording. There also is a tendency to stop the watch when the observer becomes confused (as when an operator changes the sequence of elements). The lack of a complete accounting of all the times makes it quite difficult to "sell" the resulting time standard to the worker. If a time study is to be made with only one conventional watch (an obsolete, expensive technique), my recommendation is to use the continuous system rather than the snapback system.

A better alternative is the three-watch system. See Figure 21.5. All three watches are controlled by the same lever. Initially the hands of the first watch are moving, the second are stopped at some time value, and the third are stopped at zero. At the end of element A, the observer depresses the lever. Watch one moves to zero, watch two starts recording, and watch three stops at a time. The observer now writes the element A time down from the stationary hand of watch three. When element B ends, depress the lever again. Now watch one starts, watch two stops at a time, and watch three goes to zero. The observer writes the element B time down from the stationary hand of watch two; then one, three, two, one, etc. The data form will be similar to Figure 21.6. The only disadvantage of this system is that it requires a watch actuated by the crown (about 10% more expensive) and requires three watches instead of one. However, the cost of the watches amortized over their life is trivial compared to the extra clerical cost of the one conventional watch snapback system as well as being less susceptible to observer errors.

In 1976, electronic stop watches with a display hold feature became available. They can be used in either the snapback or continuous mode. The key feature is that, when the user depresses the button, the display time is "frozen" while the clock continues timing. Thus they eliminate the "moving target" and "watch hand movement time to zero position" problems.

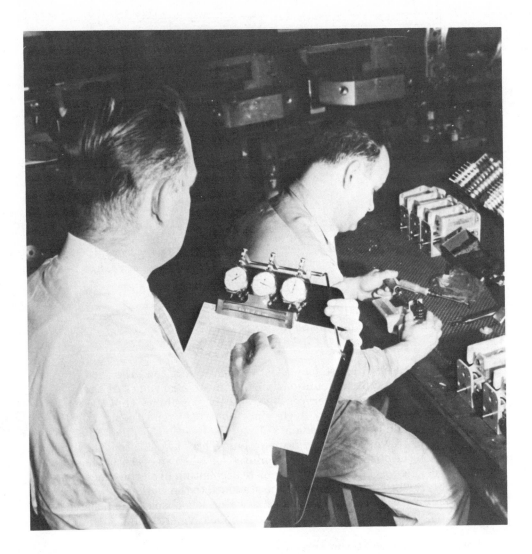

Figure 21.5 *Three-watch systems* give improved recording accuracy and eliminate the clerical labor of subtraction required in the continuous system.

Number of Observations

The number of observations to record depends on (1) accuracy desired, (2) confidence desired, and (3) variability of the data. The times vary due not to *moving* faster or slower but to fine differences in the micromethod (grasp, position, fumbles). A time study of sequential observations is a *sample* from a *population* of times. The goal is to estimate the population mean from the sample mean. Sample means are not precise (unbiased) estimators of the population mean. That is, if the mean time from a time study (a sample) is 2.0 min/unit, we cannot say that the population of times (time/unit over the weeks, months, and years) will have a mean of exactly 2.0 min/unit. To minimize the difference between the two means, increase the sample size; that is, make more observations. Unfortunately, more observations increase the cost of taking the sample. (Sample cost = fixed cost of making the study + cost/observation, so sample cost/observation declines with more observations even though total cost increases.)

Accuracy desired. Accuracy can be stated in relative or absolute terms. For a 2.0 min element, $\pm 5\%$ relative accuracy is the same as absolute accuracy of ± 0.1 min. For a .2 min element, $\pm 5\%$ relative accuracy is the same as ± 0.01 min of absolute accuracy. The limits of precision for these two cases would be 1.9 to 2.1 min and .19 to .21 min—both having $\pm 5\%$ accuracy (relative accuracy, that is). Consider the limits of precision as the "size of the target" that we wish to hit. Although it seems too obvious to state, it is more difficult to "hit" a smaller target than a bigger target—we need more "shots." Thus to hit the small target (.19 to .21 min) we will need more observations than to hit the big target (1.9 to 2.1 min) since the small target has a width of .02 while the big target has a width of .20.

Confidence desired. Continuing with the target analogy, our series of shots may not always hit the target. We may miss occasionally. For "90% confidence," we want our shots to hit the target 90% of the time; for 95%, we want to hit 95% of the time. In terms of the time study, 90% confidence with $\pm 5\%$ accuracy for a 2.0 minute element means that if we made 100 different time studies, then 90 times the sample mean would be between 1.9 min and 2.1 min, if the population mean is 2.0 min. Actually we are only going to make one time study, so the confidence refers to the "long run" confidence. We could be wrong on any one specific study. For the .2 min element, 95% confidence and $\pm 5\%$ accuracy says 95 of 100 sample means would be between .19 and .21 min, if the population mean is .2 min.

Variability of the data. The more variability there is in the data the more observations are required to hit the target with confidence (that is, fewer shots are required with a good rifle, a good shooter, and no wind gusts than with a poor instrument, unskilled technician, and random variations in the task). To the time study technician, the variability of the data is unknown when the study begins. Therefore it must be estimated.

Estimate population variability by taking a subsample (say, 10 to 20 cycles), calculate the subsample's variability, and, from the subsample's variability, estimate the population variability.

The standard deviation of the subsample is:

$$\sigma_X = \sqrt{\frac{\Sigma X^2}{N} - \overline{X}^2}$$

where X = an individual time

N = number of times (observations) in subsample

Since $\overline{X} = \dfrac{\Sigma X}{N}$

it can be shown that:

$$\sigma_X = \frac{1}{N}\sqrt{N\Sigma X^2 - (\Sigma X)^2}$$

Accuracy desired, confidence desired, and population variability are related by the following equation:

$$A = z\sigma'_{\overline{X}} \tag{21.1}$$

where A = Amplitude of the target (i.e., precision)

z = number of standard deviations corresponding to the confidence desired

$\sigma'_{\overline{X}}$ = standard deviation of the population times

The standard deviations of population times and subsample times are related by:

$$\sigma'_{\bar{X}} = \frac{\sigma'_X}{\sqrt{N'}} \tag{21.2}$$

where N' = Number of observations required to meet the criteria of precision and accuracy

Substituting and solving for N' gives:

$$N' = \left(\frac{z\,\sigma'_X}{A}\right)^2 = \left(\frac{z\frac{1}{N}\sqrt{N\Sigma X^2 - (\Sigma X)^2}}{A}\right)^2 \tag{21.3}$$

Equation 21.3 can be modified a number of ways. For example, precision level, A, can be expressed in terms of X. Then 5% relative accuracy becomes .05 $(\Sigma X/N)$ and 10% becomes .10 $(\Sigma X/N)$. The normal distribution can be assumed and 95% confidence rounded to 2 standard deviations instead of 1.96.
Then for 5% precision and 2σ confidence:

$$N' = \left[\frac{40\sqrt{N\Sigma X^2 - (\Sigma X)^2}}{\Sigma X}\right]^2 \tag{21.4}$$

For 10% precision and 2σ confidence:

$$N' = \left[\frac{20\sqrt{N\Sigma X^2 - (\Sigma X)^2}}{\Sigma X}\right]^2 \tag{21.5}$$

Using the subsample size $= N$ as an estimate of population variability gives a "biased estimate." For an "unbiased estimate," use N-1 instead of N.

For unbiased estimates, equation 21.3 becomes:

$$N' = \left[\frac{z\sqrt{\dfrac{\Sigma X^2 - (\Sigma X)^2/N}{N\text{-}1}}}{A}\right]^2 \tag{21.6}$$

For 5% precision and 2σ confidence and unbiased estimates:

$$N' = \left[\frac{40N\sqrt{\dfrac{\Sigma X^2 - (\Sigma X)^2/N}{N\text{-}1}}}{\Sigma X}\right]^2 \tag{21.7}$$

For 10% precision and 2σ confidence and unbiased estimates:

$$N' = \left[\frac{20N\sqrt{\dfrac{\Sigma X^2 - (\Sigma X)^2/N}{N\text{-}1}}}{\Sigma X}\right]^2 \tag{21.8}$$

Engineers might think of the unbiased estimate as a factor of safety, since the formulas using N-1 make the variance look larger and thus the user will be more cautious in using the data.
To show the difference between the biased and unbiased estimates, equations 21.4 and 21.7 will be calculated using the following example data. Assume 10

times were obtained: 20, 22, 20, 22, 20, 18, 18, 20, 19, and 21 s. Then $N = 10$, $\Sigma X = 200$, $\Sigma X^2 = 4018$, and $\overline{X} = \Sigma X/N = 200/10 = 20$.

$$\text{(biased) } \sigma_X = \sqrt{\frac{\Sigma X^2}{N} - \overline{X}^2} = \sqrt{\frac{4018}{10} - (20)^2} = \sqrt{401.8 - 400}$$

$$= 1.34 \text{ s}$$

$$\text{(unbiased) } \sigma_X = \sqrt{\frac{\Sigma X^2 - (\Sigma X)^2/N}{N\text{-}1}} = \sqrt{\frac{4018 - (200)^2/10}{9}} = \sqrt{\frac{18}{9}}$$

$$= 1.41 \text{ s}$$

$$\text{(biased) } N' = \left[40\sqrt{\frac{10(4018) - (200)^2}{200}}\right]^2 = \left[\frac{40\sqrt{180}}{200}\right]^2 = \left[\frac{40(13.4)}{200}\right]^2$$

$$= (2.68)^2 = 7.2$$

$$\text{(unbiased) } N' = \left[\frac{40(10)\sqrt{\frac{4018 - (200)^2/10}{9}}}{200}\right]^2 = \left[\frac{400\sqrt{\frac{18}{9}}}{200}\right]^2$$

$$= \left[\frac{400(1.41)}{200}\right]^2 = (2.82)^2 = 8$$

Thus, equation 21.4 says a standard can be set from a sample of 7 while equation 21.7 says a sample of 8 is required. Since 10 have already been taken, no more observations are needed. If N' had been 20, then 10 more observations would be required.

Although the above is perfectly clear to any engineer, it is difficult to implement on the shop floor due to the complex calculations.

We have already shown how equations 21.3 and 21.6 can be simplified by specifying the accuracy and precision in advance. Another popular simplification is to estimate the population variability, not from the standard deviation of the sample but from the range of the sample. Although the range is not as efficient an estimator of variability it is much simpler to calculate.

$$\overline{R} = d_2 \sigma_X \text{ or } \sigma_X = \frac{\overline{R}}{d_2} \tag{21.9}$$

where $\overline{R}$ = mean range of a subgroup of a specified size

d_2 = number of standard deviations that the mean range includes for a specified subgroup sample size

σ_X = standard deviation of individual times

For example, for a subgroup sample of 10 times, $d_2 = 3.078$; that is, the range of the times from the sample of 10 will, on the average, include 3.078 standard deviations. Table 21.2 gives the d_2 factor for various subgroup sizes.

Repeating equation 21.1:

$$A = z\sigma_{\overline{X}}'$$

Table 21.2   The Range of a Sample Includes d_2 Standard Deviations, on the Average

Subgroup Number, N	Number of Standard Deviations in Range
5	2.326
6	2.534
7	2.704
8	2.847
9	2.970
10	3.078
15	3.472
20	3.735
25	3.931

But from equation 21.2 and 21.9:

$$A = \frac{z\overline{R}}{\overline{d}_2 \sqrt{N'}} \qquad \text{so} \qquad N' = \left(\frac{z\overline{R}}{\overline{d}_2 A}\right)^2 \tag{21.10}$$

For 5% precision and 2σ confidence:

$$N' = \left(\frac{2\overline{R}}{.05\overline{X}d_2}\right)^2 = \left(\frac{40\overline{R}}{\overline{X}\overline{d}_2}\right)^2 \tag{21.11}$$

For 10% precision and 2σ confidence:

$$N' = \left(\frac{2\overline{R}}{.10\overline{X}d_2}\right)^2 = \left(\frac{20\overline{R}}{\overline{X}d_2}\right)^2 \tag{21.12}$$

If the technician is told to take a subsample size of 10, then equations 21.11 and 21.12 become:

For 5% precision, 2σ confidence and $N = 10$:

$$N' = \left[\frac{40\overline{R}}{3.078\overline{X}}\right]^2 = \left[\frac{13\overline{R}}{\overline{X}}\right]^2 \tag{21.13}$$

For 10% precision, 2σ confidence and $N = 10$:

$$N' = \left[\frac{20\overline{R}}{3.078\overline{X}}\right]^2 = \left[\frac{6.5\overline{R}}{\overline{X}}\right]^2 \tag{21.14}$$

In practice R is used instead of the $\overline{R}$ of the equations. Using equation 21.13 instead of equation 21.4 or 21.7 for the data following equation 21.8 gives N' of $6.7 = 7$ instead of the $7.2 = 7$ from equation 21.4 or the 8 from equation 21.7. In general, the standard deviation equations give a smaller N' than the range methods.[2]

Since different elements may have different variability, solve the equation for each element. Then select your N. For example, solving equation 21.13 for the data on Figure 21.4 gives R of 1, 2, 2, and 20 for the four elements; N' becomes less than 1 for three elements and 97 for element 4. The problem with element 4 is the very low value of 10 in cycle 1. If the 10 is eliminated as an outlier (see the next section), then element 4 also needs only 1 observation. Thus the 10 observations already made are sufficient and the study is complete as is.

The above discussion assumes that the number of cycles should be determined solely from probability. However, in real life, you can drown in a river that is only a meter deep on the average. Telling your boss that you will be correct "most of the time" doesn't help when you have made a mistake on an important decision. The same problem was faced in statistical sampling for quality control. The Military Standard 105 tables recommend larger sample sizes for larger lot sizes, even though they give the same operating characteristic curve regardless of sample size; that is, the plans give more protection for larger (i.e., assumed more important) lots. Table 21.3 gives two approaches to a larger sample for more important decisions. Note that two companies in the same industry don't agree so the values should not be expected to apply exactly to other industries. Generally there is a certain "fixed cost" of making a time study (analyzing the job, contacting the supervisor, making the calculations, and writing it up). Thus it is probably worthwhile to spend at least 15 minutes recording times, regardless of the statistical calculations of the number of cycles required. This seems to be the general approach in Table 21.3 (although the minimum time seems to be 20 min for GE and 6 min for Westinghouse). In general, increase sample size when (1) cycle time is shorter, (2) activity/yr is larger, (3) cost of an inaccurate time is high. The goal is to make the benefit from the time study exceed the cost of the time study.

Foreign Elements, Irregular Elements, and Outliers

When an observer misses a reading, mark an M on the form where the time

Table 21.3   Minimum Number of Cycles to Study[10,11]

IF	AND Westinghouse Electric value			OR General Electric values
Time/piece or cycle is over	Activity/yr is under 1,000	Activity/yr is from 1,000 to 10,000	Activity/yr is over 10,000	Activity/yr is any value
.002 hrs (under)	60	80	140	200
.002	50	60	120	175
.003	40	50	100	125
.004	35	45	90	100
.005	30	40	80	85
.008	25	30	60	60
.012	20	25	50	40
.020	15	20	40	30
.035	12	15	30	18
.050	10	12	25	15
.080	8	10	20	13
.120	6	8	15	10
.200	5	6	12	9
.300	4	5	10	8
.500	3	4	8	5
.800	2	3	6	3
1.000	2	3	5	3
2.000	1	2	4	3

should have been. Do not guess at the time as this will bias the time of both the reading and the following reading. See column 2 of Figure 21.6.

If the worker omits an element, put a dash where the time would have been placed. See column 4 of Figure 21.6. An omitted element might indicate an inexperienced operator. It might indicate an operator who is trying to confuse the observer. It also might indicate that the element might not be needed for 100% of the cycles. For example, if "brush off chips" is omitted on some cycles, the observer might consider investigating the proper frequency. Before a time study, workers usually describe many different activities that are necessary to complete a unit; then during the study they may become engrossed in doing the job and do the job the way the usually do it and forget to add the embellishments for the observer.

Operators also may do an element out of the order the observer has on the form (say, 5-4 instead of 4-5) because the sequence isn't critical or because they are trying to confuse the observer. See column 6-8 of Figure 21.6. The simplest technique is to put a dash down for the omitted element (4), put the time down for the next element (5) one column over and then, when the skipped element (4) occurs, move over a column on the form and record it. Then element 5 is marked with a dash and the normal sequence resumes.

NO.	ELEMENTS	SPEEDFEED	UPPER LINE - SUBTRACTED TIME 1 2 3 4 5 6 7 8	LOWER LINE - READING 9 10 11 12 13 14 15	MIN TIME	AV. TIME	Std. Dev.	OCC PER CYCLE	EFFORT RATING	NORMAL TIME
1	Pick up part TP piece at vise		14 34 52 89 16 56	95 35 1109 51				1	90	
2	Put part in vise TP RL / vise handle		42 63 80 417 45 84	823 999 38 80 Ⓐ				1	90	
3	Mill slot TP power / feed disengages		98 210 335 73 601 739	80 1005 95 335				1	100	
4	Aside part to pan TP / RL part		108 38 63 502 29 —	80 908 83 223 63				1	95	
5	Brush chips TP RL / brush		20 M 75 - 42	52 - 21 96 37 76				?	80	

FOREIGN ELEMENTS: Ⓐ watched Maria
TOOLS, JIGS, GAUGES, PATTERNS, ETC: vise 407

OVERALL EFFORT RATING	BEGIN	END	ELAPSED	UNITS FINISHED	ACTUAL TIME PER PIECE
1:	1:	1:45	8	10	.0133 hr

.133
.137 ✓

Figure 21.6 *Omitted times* are shown by an *M* as in column 2. When the worker omits an element, put down a dash as in column 4. When an element is done out of order, as in column 6-8, put a dash down for the omitted element, record the next element over a column, and, when the omitted element occurs, move over a column. If an unexpected element occurs, as in column 10, record the time and next to it a letter. Then elsewhere on the form write a short note describing the event.

The operator may have delays, both avoidable and unavoidable, and unexpected elements. Avoidable delay would be stopping to light a cigarette, nose-blowing, or watching Maria walk by. Unavoidable delay would be talking to the supervisor or breaking a tool. An unexpected element would be going for a supply of parts. The observer should record the time for these events; do not stop the watch. As these events occur, move over a column and code them A, B, C, etc., next to the time and write a short note (1-3 words) defining the meaning of the A, B, C, etc., elsewhere on the form. See column 10, Figure 21.6. These times must be recorded because later analysis may show them to be valid events for which the operator should receive credit. The observer should never determine what should be allowed and what not allowed during the course of the study by starting and stopping the watch. Times can always be omitted later but, if a time was not recorded when it should have been, there is no way to recover the missing data.

Later, in the office, the events can be classified as foreign elements or irregular elements.

Foreign elements are not allowed directly as part of the time standard. Indirectly they may be considered as part of personal, fatigue, and delay allowances (see chapter 23).

Some of these unexpected events, however, may turn out to be elements which are part of the work cycle but which occur relatively rarely—irregular elements. The observer should have put them down on the data form but forgot to do so. The element time then is determined from the time available and the element is allowed at whatever frequency is appropriate (1/10, 1/250, 1/500, etc.). If the observer had stopped the watch during the study, the worker would never get all the time deserved.

Upon examination of the times for an element, there may be a suspected *outlier*. For example, times for element A might be 10, 10, 12, 8, 12, 13, 10, 8, 11, 9, 10, 10, 21, 11, and 9. Is the 21 representative?

There are two valid ways of eliminating an outlier: for cause and statistically. A time value should never be eliminated (red-circled) due to a whim of the observer. The key issue here is the use of standardized procedures which do not depend on personal judgment.[3] The workers do not trust management in the first place so any arbitrary, whimsical treatment of data will be long remembered; management may win the "battle" (a lower time standard) but will lose the "war" (actual performance on the job).

Cause was discussed before. If the 21 had a letter next to it and the identification "watched Maria," then it can be omitted.

The second technique is statistically. The argument for exclusion of a potential outlier is that the nonuse of the outlier decreases the sample size by some small amount but inclusion biases both the mean and variance of the sample by unknown amounts. Therefore 5% and 10% risks of throwing out a valid time (because you thought it was an outlier when it really was a valid time) are commonly used.

Different outlier tests can be used depending on the statistical distribution assumed. My recommendation is to use the Dixon test, a distribution-free test: it works for any data distribution. A ratio is calculated from two "distances"; the numerator is the distance of the outlier from a "near part" of the data and the denominator is the distance from a "far part" of the data.

The Dixon test requires three steps: (1) read the maximum ratio (r) which could occur by chance (Table 21.4), (2) compute the ratio which occurred in the sample, and (3) compare the two ratios.

- Table 21.4 gives the maximum ratio which could occur by chance for various sample sizes and risks. In our example, $n = 15$, so an r of .525 could occur 5% of the time by chance and an r of .472 could occur 10%.

- Second, compute the ratio which occurred in the sample. In our case, we had an $n = 15$ so we use the formula (from Table 21.4) $r = (X_n - X_{n-2})/(X_n X_3)$. To use the formula, put all the values in sequence starting with the smallest (8, 8, 9, 9, 10, 10, 10, 10, 10, 11, 11, 12, 12, 13, 21). The smallest is X_1 and the largest X_n. Therefore the observed ratio is $(21 - 12)/(21 - 9) = 9/12 = .75$.

- Third, compare the two ratios. The maximum chance ratio (assuming a 10% risk) was .472. The observed ratio was .75. Therefore eliminate the 21 as an outlier.

Examine unusual times—especially low times—for information for possible methods improvements.

Table 21.4 Maximum Ratio of the Dixon Test Which Could Occur by Chance by Number of Observations and Alpha Risk Levels

Number of Observations	Ratio to Calculate If: Largest (X_n) Is Suspect	Ratio to Calculate If: Smallest (X_1) Is Suspect	Alpha Risk .10	Alpha Risk .05
3	$\dfrac{X_n-X_{n-1}}{X_n-X_1}$	$\dfrac{X_2-X_1}{X_n-X_1}$	.886	.941
4			.679	.765
5			.557	.642
6			.482	.560
7			.434	.507
8	$\dfrac{X_n-X_{n-1}}{X_n-X_2}$	$\dfrac{X_2-X_1}{X_{n-1}-X_1}$	.479	.554
9			.441	.512
10			.409	.477
11	$\dfrac{X_n-X_{n-2}}{X_n-X_2}$	$\dfrac{X_3-X_1}{X_{n-1}-X_1}$	.517	.576
12			.490	.546
13			.476	.521
14	$\dfrac{X_n-X_{n-2}}{X_n-X_3}$	$\dfrac{X_3-X_1}{X_{n-2}-X_1}$	.492	.546
15			.472	.525
16			.454	.507
18			.424	.475
20			.401	.450
25			.360	.406

Rating

The previous steps give the time a specific worker spent working on the task. The next step is to adjust the recorded time by a rating factor to obtain the time that a "normal" worker would take. In actual practice, rating is done on the shop floor *before* the data are refined by considering foreign elements and outliers. Rating has been briefly discussed in the section on Operator Selection and is discussed more fully in chapter 22.

$$(\text{Recorded time})(\text{Rating}) = \text{Normal time}$$

To summarize the procedure, assume that Jones took .0054 hr/casting during the time study. But he is very good at his job and worked rapidly during the time study, 25% faster than an ordinary worker. A normal worker needs more time. Normal time therefore would be .0054 × 1.25 = .00675 hr/casting. If Betty Jo was rated at 95% when she stitched pockets at a rate of .002 hr/pocket, then a normal worker would need less time; normal time would be .002 × .95 = .0019 hr/pocket.

Allowances

Normal time is not sufficient time for the job; it is the time necessary under ideal conditions and without any breaks.

$$\frac{\text{Normal time}}{(1 - \text{Allowance percent})} = \text{Standard time}$$

Allowances are discussed more fully in chapter 23. They generally include personal allowances, fatigue allowances, and delay allowances. Personal allowances include such interruptions of the work cycle as blowing your nose, coughing, itching, and getting a drink of water. Fatigue allowances are usually given at a low percent (say, 5%) for all jobs with a larger percent for physically or mentally demanding work. Delay allowances include both avoidable delay (stopping work while discussing who will win the game this weekend) and unavoidable delay (wait for the material handler to bring a supply of parts).

Therefore, if allowances were 25% for cutting off casting risers, standard time would be .00675/.75 = .009 hr/casting. If allowances were 15% for stitching pockets, the standard time would be .0019/.85 = .0022 hr/pocket. Allowances probably will differ by element and so standard time should be calculated by element and then added rather than adding normal times and then considering allowances.

As was discussed in the section on Methods Analysis, no standard time should be considered permanent. Creeping changes in methods (called manufacturing progress) or improved skill due to many cycles of practice (called learning) make any standard time subject to change. See the section, Acceptable Day's Work, in chapter 24 for a more complete discussion of the effect of learning and manufacturing progress curves on time standards. Table 21.5 summarizes the calculations.

Table 21.5   Summary of Time Calculations Using Data of Figure 21.4

Step	Example
1. Calculate mean measured time/element	.001 410 (for element 1)
2. Calculate normal time/element	.001 410 × .90 = .001 269
3. Consider element frequency/unit	.001 269 × 1 = .001 269
4. Calculate standard time/unit by elements using elements' allowance	.001 269/.85 + .002 538/.90 + .005 590/.85 + .002 588/.80 ———— .014 124 hr/unit
5. Sum for standard time/unit	.014 100 hr/unit

TIMING PROCEDURE: NONSEQUENTIAL OBSERVATIONS

The previous section discussed the case where the sample of times was sequential from the universe of times. That is, an observer goes out on, say, Tuesday morning and times a worker for, say, 50 cycles between 9:15 and 9:50. From these 50 times the observer predicts the time the worker will need. The occurrence sampling technique, however, is to sample the universe a number of times (say, 100) but to spread these observations over a period of days or weeks. The number of observations will be larger for occurrence sampling than sequential sampling since the sampling theory is based on a discrete distribution (the binomial) instead of a continuous distribution (the normal). Chapter 9 discusses occurrence sampling in detail.

The primary advantage of using occurrence sampling to set time standards is that neither production nor data recording need be continuous. For example, for a maintenance standard, changing a light bulb may occur only a few times per week so it would be difficult to obtain an adequate sample size using stopwatch time study. The primary disadvantage is that it is difficult to get observers to make a careful definition of the method and sketch of the workstation and to question the method. Thus there tends to be unquestioning acceptance of whatever method the operator uses. The job also is not broken down into small elements when using occurrence sampling so the data are less useful for standard data. The observed time probably is more accurate using occurrence sampling due to the many observations over a long time period but, since rating usually is not used, normal time may not be estimated as accurately as with stopwatch studies. Fatigue allowances are "built into" the observed time. Thus, don't use occurrence sampling standards for applications requiring considerable accuracy. (Each organization will need to make its own definition of considerable.)

See chapter 9 for a detailed discussion of occurrence sampling. Two examples will be given below of occurrence sampling for time standards; see also Table 9.5. The key difference (vs. not setting time standards) is that, in addition to making the occurrence sampling study, you need to record units produced during the time of the study.

In Figures 21.2, 21.3, 21.4, and 21.6, a stopwatch time study was made. In the stopwatch technique, 10 cycles were observed in sequence. Observed time was .012 46 hr/unit, normal time was .011 90 hr/unit and time (using 15% allowances) was .014 hr/unit. Assume that an occurrence sample of 500 observations was taken over a 20-day period. The observer needs to separate time for personal allowances and delay allowances from work time; fatigue allowances are built into the work times; see chapter 23 for more on allowances. Assume the study showed the worker worked 88% of the shift time—that is, the worker took 12% for personal and delay time. Assume further that 10% is the specified allowance for personal and delay time. Then normal time = .88(8 hr/day)(20 days) = 140.8 hr. If the person produced 11,300 units during the 20 days, then observed time = normal time = 140.8/11,300 = .012 460 hr/unit. Standard time would be .012 460/.90 = .013 845 hr/unit.

As a second example, consider Harry, a mechanic who does tune-ups. The organization wants to know how much to charge per tune-up. Observe him 100 times over a 10-day period. Record the output during the 10 days (say 5 trucks and 13 cars) as well as the scheduled work time (say 8 hr/day). Assume the study showed idle was 9%, truck tune-up was 36%, and car tune-up was 54%. Then normal time was $(.91)(80 \text{ hr}) \times .36/5 = 5.24$ hr/truck and $(.91)(80 \text{ hr}) \times .54/13 = 3.02$ hr/car. If the allowances are assumed to be 10%, then standard time is $5.24/.9 = 5.8$ hr/truck and 3.4 hr/car. The organization then might charge 5.8 hours for every truck tune-up and 3.4 hours for every car tune-up. Total administrative cost and paperwork cost would be decreased by charging everyone the same although some would be overcharged and some undercharged. It might be desirable in a real application to use finer subdivisions of work (say dividing tune-ups into simple, normal, and major and vehicles into cars, pickups, and trucks so there are nine time charges to select from instead of the two of this simplified example.)

SUMMARY

The most popular timing technique is to time sequential observations with a stopwatch. Use of a calendar for timing gives less accurate standards but at a savings in cost of determining the time. Standard data, when available, may give both more accuracy and less cost.

Sequential timing with a stopwatch should follow the following steps:

1. Determine the best method
2. Break the task into elements
3. Pick an average, experienced operator
4. Use the three-watch or electronic watch system
5. Record enough observations
6. Adjust data if necessary
7. Rate operator pace
8. Give allowances

Nonsequential timing (occurrence sampling) is becoming more popular because it gives a representative sample at a reasonable cost.

SHORT-ANSWER REVIEW QUESTIONS

21.1 In organizations without incentive pay, what are the four reasons for determining time/task?

21.2 For methods engineers and technicians, what proportion of their time should be spent on methods analysis and what proportion on job timing? Why?

21.3 The number of observations to take in a time study depends on what three things?

21.4 Calculate standard time to polish a pair of shoes. Use 15% for allowances.

Element	Time/unit, hr	Rating, %	Occurrence/ cycle
Get equipment	.01	90	1.0
Polish one shoe	.02	110	2.0
Put equipment away	.025	100	1.0

21.5 Why should unusually low times in a time study be investigated?

THOUGHT-DISCUSSION QUESTIONS

1. If you have an organization with 1000 employees and 10 people in the industrial engineering department (clerks, technicians, and engineers), what proportion of the IE department time should be allocated to improving productivity and what proportion to time study?

2. Time standards, on the average, could be accurate to within 1%, 5%, 10%, 15%, or 20%. When should each percent be used?

REFERENCES

1. Fein, M. *Wage Incentive Plans.* WM and ME Division Publication 2. Atlanta, Ga. American Institute of Industrial Engineers, 1970.

2. Hicks, C., and Young, H. A comparison of several methods for determining the number of readings in a time study. *Journal of Industrial Engineering,* Vol. 13, No. 2 (1962): 93-96.

3. Parks, G. Extreme value statistics in time study. *Journal of Industrial Engineering,* Vol. 16, No. 6 (1965): 351-355.

4. Peer, S., and Kennedy, W. Video methods analysis. *Industrial Engineering,* Vol. 5, No. 2 (1973): 16-20.

5. Quick, J.; Duncan, J.; and Malcomb, J. *Work Factor Time Standards.* New York: McGraw Hill, 1962.

6. Rice, R. Survey of work measurement and wage incentives. *Industrial Engineering,* Vol. 9, No. 7 (1977): 18-31.

7. Salvendy, G., and McCabe, G. Auditing standards by sample. *Industrial Engineering,* Vol. 8, No. 9 (1976): 25-29.

8. Salvendy, G. Effects of equitable and inequitable financial compensation on operator's productivity, satisfaction and motivation. *International Journal of Production Research,* Vol. 14, No. 2 (1976): 305-310.

9. Schantz, J. Video tape recording saves time for IE's. *Industrial Engineering,* Vol. 13, No. 7 (July 1981): 48-54.

10. Shaw, A. Time Study Manual of the Erie Works of the General Electric Co. Erie, Pa., 1978.

11. Westinghouse Electric Corp. *Work Measurement Techniques and Application,* R-131-Rev. Pittsburgh, Pa.: Westinghouse Electric, 1953; p. 57.

22

RATING

NORMAL PACE

As was discussed in chapter 21, the determination of the amount of time for a task requires four steps: (1) design the proper work method (discussed in most of the chapters of this book), (2) using a sufficient statistical sample, record the amount of time taken by an experienced worker (chapter 21), (3) adjust the measured time to that which would be required at a "normal" pace (this chapter), and (4) add allowances (chapter 23). The result is *standard* time. (In England, normal time is called basic time.)

$$(\text{Measured time})(\text{Pace}) = \text{Normal time} \tag{22.1}$$

$$\frac{\text{Normal time}}{1 - \text{Allowance, percent}} = \text{Standard time}$$

To define a normal time, you must first define a *normal* pace. This is not a trivial problem. Unless normal is defined, any time standard is useless. It is analogous to recording the temperature as 100 without specifying whether the Fahrenheit or Celsius scale is being used.

Fein recommends defining normal pace in relation to motivated pace (see Figure 22.1).[6,7,8]

Motivated Productivity Level *(MPL):* The work pace of a motivated worker, possessing sufficient skill to do the job; physically fit to do the job; after adjustment to it; and working at an incentive pace that can be maintained day after day without harmful effect.

Normal or Acceptable Productivity Level *(APL):* The work pace which is established by management, or jointly by management and labor, at a level which is considered satisfactory; it is established at a given relationship to Motivated Productivity Level.

MPL is a function of human work capacity. It might be capable of being determined by engineers using such criteria as energy expenditure for physical work (see chapter 12) or information processing capacity (see principle 8 of chapter 14) for psychomotor work. *APL* is whatever proportion of *MPL* that is jointly satisfactory to management and labor. Since this proportion is *negotiated*, the definition of *APL* is a political decision rather than a technical decision.[11] *Negotiated* may not be the proper word as in many cases management presents a defined value of *APL* and workers accept it. Once the political decision has been made, then the engineers can evaluate other jobs in relation to the defined value of *APL*.

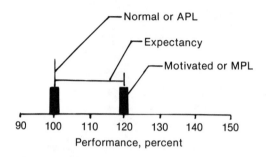

Figure 22.1 *Motivated Productivity Level (MPL)* is established first as the "anchor." Then, a negotiated distance away, (Incentive Expectancy), Acceptable Productivity Level *(APL)* is located. In common usage, however, *APL* is "normal" and has the numerical value of 100%—which implies *APL* is the anchor when, in fact, *APL* isn't. Boepple says Work-Factor and British Standard 3138 (both at *MPL*) are equal and that there is a 20% difference between them and *MTM* (at *APL*).[3] However, Table 17 (Introduction to Work Study, 1978) gives expectancy for British Standard as 33%; daywork rate is defined as walking 3 miles/hr and 75%; incentive rate is defined as walking 4 miles/hr and 100%.[12]

A key decision is what percent of nonincentive experienced workers are expected to work at the 100% pace. Is it 10%, 50%, or 90%? My recommendation is for a "tight" definition of 100% so that only a few achieve it. The others, due to group psychology, will tend to strive for it. The standard for retaining the job would be lower than standard performance. On the other hand, if 100% is easily achieved, many workers will restrict their output to 100%, since they are paid by the hour.

Although the scale actually is "anchored" to the *MPL* value, because of common usage and custom the value of *APL* is defined as 100%. The difference between the two usually is called *incentive expectancy* or *expectancy*.

$$APL = MPL - \text{Expectancy} \tag{22.2}$$

where   *APL* = Acceptable productivity level (normal), percent

       MPL = Motivated productivity level, percent

  Expectancy = Incentive expectancy, percent

The values for expectancy are political decisions, not ranges of human work capacity or "scientific" decisions. Fein says typical expectancy values were 10%-15% during the 1930s and 20%-25% during the 1940s.[8] During World War

II, the US War Labor Board ruled that 30% was fair and equitable. Rice, survey-ing 1500 US firms, reported target for day work averaged 101% while target for incentive was 119%—a difference of 18%.[22] Actual performance for day work averaged 93% and for incentive averaged 123%—a difference of 30%. Fein reports 30% is the most common expectancy value in the USSR while China uses 25%.

Methods Time Measurement *(MTM)* times are *APL* times with a 25% expec-tancy scale while Work-Factor are *MPL* times.[8]

Expectancy also is related to allowances; the lower the allowances, the higher the expectancy tends to be and the higher the allowances the lower the expec-tancy tends to be.[6] The value of expectancy, being a political decision, depends on the political strength of management and labor. Expectancy should not be expected to be constant from country to country, from firm to firm, or even within the multiple plants of a specific firm. Fein comments that both expectancy and base wage rates/hr are a part of collective bargaining; they are beyond the province of engineers.

RATING TECHNIQUES

Gilbreth broke tasks down into therbligs (see Table 8.1). Using *MTM* nomen-clature we speak of moves, reaches, grasps, positions, etc. The essential problem of rating is that each of these therbligs changes its proportion of the total task as the pace changes.[23]

For example, at a 100% pace, the total task may take 1.0 min, composed of .1 min (10%) for reach, .2 min (20%) for grasp, .3 min (30%) for move, and .4 min (40%) for position. However, assume the worker speeds up so that the total task only takes .8 min (i.e., overall pace is 1.0/.8 = 125%). Then the reach may increase to a pace of 150% and take .067 min (8.4% of the total), grasp and move may increase at the same 125% as the task as a whole and take .160 min and .240 min (keeping their percentages at 20% and 30%), while position may increase only to a pace of 120% and take .333 min (increasing its proportion to 41.6% of the total).

The rater's problem is which therblig to watch. The low-skill (difficulty) ther-bligs, such as reach, generally change more than the overall task; the high-skill therbligs (such as position and sometimes grasp) change less than the overall task. In addition, the proportion of low-skill to high-skill therbligs varies from element to element, so the rater cannot rate just one element and assume the rating is valid for the entire task.[13]

The simplest rating system is called *pace rating*—sometimes *speed rating* or *tempo rating*.[17] The observer estimates the *pace;* that is, primarily concentrates on the dynamic therbligs such as reach and move rather than the stationary ther-bligs such as position and grasp. This single-factor technique is the most com-mon approach.

The second approach, sometimes called *objective rating,*[19] is to rate in two steps of rating and a third of calculation. *First,* the observer rates the speed; *second,* the observer estimates the task difficulty, also called *effort* (i.e., in effect estimates what proportion of the total time is composed of easy, average, and difficult therbligs); and *third,* the observer multiplies the speed factor by the diffi-culty factor to get the overall factor, pace. This second approach is a descendant of the original Westinghouse system, first used in 1925. The original Westing-house system had four components: skill—proficiency at following a given method; effort—will to work; conditions; and consistency—high variability went with low pace.

Smalley gives the following example to distinguish speed and difficulty[25]:

An unburdened man walks 8 m across a level, unobstructed floor in 6 s; his pace is rated at 110%. After picking up a heavy load, he carries it 8 m up an inclined, obstructed ramp in 6 s. When walking up the ramp the speed does not change but difficulty does, so the pace increases. If, instead of picking up the load, he had returned the 8 m in 5 s instead of 6, the speed would increase and the difficulty would be constant, so pace increases.

Table 22.1 shows how methods influence performance. As has been mentioned previously, the analyst should specify an efficient method before recording the time necessary to do the method. *Method*, however, can be defined in different levels of detail. Level 1, management-controlled, gives design details such as tools and equipment, fixtures and containers, workplace layout, and material flow. Level 2, which management tries to control, gives the general motion pattern.

Table 22.1 Martin Divided Methods into Level 1, Management-Controlled; Level 2, Which Management tries to Control; and Level 3, Operator-Controlled.[17]Although Level 3 Greatly Influences the Operator's Proficiency, Most Time Study Personnel Have Difficulty in Detecting Changes in Level 3.[10]

Level	Detail of Methods Description
1. Management-controlled	Tools and equipment
	Fixtures and parts containers
	Workplace layout
	Material flow
2. Management attempts to control	General motion pattern
3. Operator-controlled (influenced by training)	Specific motion pattern, including hand coordination
	Unnecessary motions
	Types of motions (undesirable or highly refined)
	Amount of fumbling
	Eye-hand coordination
	Delay intervals between motions

SOURCE: Reprinted with permission from *Industrial Engineering*, August 1970. Copyright American Institute of Industrial Engineers, Inc., 25 Technology Park/Atlanta, Norcross, Ga. 30092.

Level 3, controlled by the skill and training of the worker, gives hand-hand coordination, number of unnecessary motions, number of fumbles and hesitations, and eye-hand coordination. *Objective rating* tries to consider level 3 (or for low-volume operations, level 2); level 1 is outside the scope of performance rating and within the scope of methods design. The fine points of these "micromethods" of a specific worker are not detected by the usual rater even though they are critical to the time taken by the worker.[10]

REDUCING RATING ERRORS

To improve rating, consider some of its problems.

First, rating is a skill, a technique—inaccurate for specific time studies, although hopefully rating is sufficiently accurate when averaged over many studies. Raters who say they are perfectly accurate on each and every time study either do not understand what they are doing or are lying.

Second, rating is a subjective judgment vs. a mentally remembered scale. There must be an objective scale (see the previous section on normal pace); the rater must be familiar with it.

Third, the rater's subjective scale must be calibrated vs. the objective scale. The calibration (training) should be frequent and effective.

What are the calibration problems?

One "problem" is not an industrial problem but is a problem of academic studies of pace rating. Investigators usually have a large number of raters (say 100) from a number of different companies rate a number of different operations. When they calculate the variability of the total group, approximately half of the variability is within-company error and half is between-company error.[16,18] That is, raters have been trained for their company's APL and the APL varies between companies. Excluding the between-company variability, Fein reports the Society for Advancement of Management study (1200 USA raters) had a within-company mean deviation of 6%; Moores (100 British raters) had a within-company standard deviation of 7%.[8]

A possible problem is the duration of observation time. Andrews and Barnes demonstrated that there was no appreciable difference in rating accuracy whether the observer watched a movie of an operation for 5, 10, or 15 s.[1] Gambrell, however, recommends a movie be shown at least 45 s.[9] Murrell reported that, on a study by Bevis of an actual operation, the highly skilled operator had a mean time of .056 for a job with a standard time of .087; that is, performance *averaged* 155%.[20] Yet individual cycles ranged from 87% to 218% of standard! The distribution of times was the typical positive skew distribution with 90% of the times between -1σ and $+3\sigma$. Thus, since the rater is tracking a "moving target," sufficient cycles should be studied so the sample is representative (see chapter 21). Certainly the rating should not be on individual cycles. Rate on an overall basis for each element.

One real problem is that raters tend to rate "flat"[2,4,15,16,18,24,27]; see Figure 22.2. Flat rating probably is due to overemphasis on the motion speed of reaches—a minor component of total time—and neglect of the skill demonstrated in positioning—a major component of total time.

Training films for rating should have a separate filming of each pace rather than merely speeding up or slowing down the camera speed since the composition of the elements changes with the pace.

Gershoni found, both in England and in Israel, that raters generally were unable to detect very important differences in micromethods.[10] His recommended solution was to use a predetermined time system notation to describe the method on the time study form, assuming that the system forces the rater to pay more attention to the exact motion pattern.

Gershoni also pointed out that raters both in England and in Israel were unable to detect untrained operators when studying films and so increased their rating errors. This may not be a problem in a "real situation" as presumably no time study is taken unless the operator is "experienced."

Das and Reuter looked at the training procedure in detail.[4,5,21] They recommended modifying the Society for Advancement of Management (SAM) films. See Table 22.2 for a list of SAM rating films. The SAM films mix operations (a scene of Form Rug Cups, then Cut Cork Tube, then Deburr, etc., each at a different pace). Das says this is good for testing but, for training, a number of paces should be shown for Form Rug Cups, then a number of paces for Cut Cork Tube, etc. After each scene the rater should be given the true pace immediately after putting down the estimate; if the estimate is in error by more than 5%, the film should be backed up and the scene shown over. In training terminology, apply the principles of knowledge of results and immediate feedback.

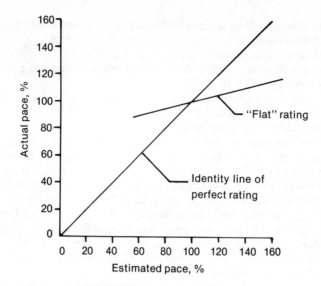

Figure 22.2 *Perfect pace rating* would have each estimate fall on the identity line. If the estimates had no systematic bias, they would fall randomly above and below the identity line. Unfortunately, raters tend to have a systematic bias—being "tight" when the pace is fast and "loose" when the pace is slow. Physiological cost as a function of speed (such as watts of human energy expended/kg-m of output) has a general "U" shape with the curve increasing steeply at the extremes.[14,30] The bottom of the U curve is not necessarily at *APL;* for walking, for example, it is at 4.04 km/hr.

Gambrell said raters trained to estimate to the nearest percent (97%, 109%) had less variability than those who rounded to the nearest 5% (95%, 110%).[9]

SUMMARY

Pace rating, a necessary step in determining the time for a job, can be done although there are inaccuracies on any specific study. Well-trained raters have standard deviations of about 6%-7% so 95% of their estimates on a specific study would be within 12%-14% of the true pace.

Normal pace (acceptable productivity level or *APL*) is a political decision which depends on the relative strength of labor and management. *APL* is not the same from country to country, firm to firm, or even within the multiple plants of a firm.

There are two standard rating techniques. Speed rating evaluates only the speed of working. Objective rating also evaluates the difficulty of the task.

Rating variability can be reduced to "reasonable" levels by careful definition of *APL* and periodic training of the raters.

SHORT-ANSWER REVIEW QUESTIONS

22.1 Give the formula for determining the amount of time for a task.

22.2 What is the relation between Motivated Productivity Level, Acceptable Productivity Level, and Expectancy?

22.3 What is the difference between pace rating (also called speed rating and tempo rating) and objective rating?

Table 22.2 The Society for Advancement of Management (SAM) has 18 Time Study Rating Films; They Are Subdivided into Three Series. For Each Operation a Worker (Who Actually did the Task) is Shown in an Introductory Scene plus 5 Scenes at Various Paces.

Series	Reel	Operations
5201	1	Deal cards, toss blocks, transport marbles
5201	2	Dink tile squares, fold gauze, pack gaskets
5201	3	Countersink, kick press, shear rubber tile
5201	4	Cut cork tiles, deburr, form rug cups
5201	5	Feed rolling mill, shovel sand, stack cartons
5201	6	Pack cans, seal cartons, tape boxes
5201	7	Bolt pipe flange, check tires, fill radiator
5201	8	Collate papers, staple papers, tear bills
6302	1	Collate papers, comptometer, typing-electric
6302	2	Ditto, key punch, posting
6302	3	Calculator-multiplication, filing, staple papers
6302	4	Card sorter, mail sorting, tear bills
6503	1	Filing, mail sorting, typing-nonelectric
6503	2	Check writing, comptometer, key punch
6503	3	Drill press, punch press, turret lathe
6503	4	Engraving, form rug cups, mechanical assembly
6503	5	Sweep floors, wash windows, wet mop floors
6503	6	Pack cans, seal cartons, stack cartons

22.4 What is flat rating? What probably causes it?

22.5 Within an organization, raters' standard deviations are about 6%-7%. This means 95% of rating estimates are within what percent of true pace?

THOUGHT-DISCUSSION QUESTIONS

1. Assume you have started a new business and wish to start setting time standards. How are you going to define 100% pace to your raters?

2. When a standard time is set for an operation, some managements consider 100% as a goal which will be achieved by only a few operators, others as a typical value for operators, and others as a minimum and cause for dismissal if not achieved. Which definition do you think best? Why?

REFERENCES

1. Andrews, R., and Barnes, R. The influence of the duration of observation time on performance rating. *Journal of Industrial Engineering*, Vol. 18, No. 4 (1967): 243-247.

2. Anson, C. Accuracy of time-study rating. *Engineering* (1954): 301-304.

3. Boepple, E. Coordinator of Research and Development of Wofac Co. Personal communication, May 17, 1977.

4. Das, B. A statistical investigation of the effect of pace and operation on performance rating. *International Journal of Production Research*, Vol. 3, No. 1 (1964): 65-71.

5. Das, B. Applying programmed learning concepts to instruct in performance rating. *Journal of Industrial Engineering*, Vol. 16, No. 2 (1965): 94-100.

6. Fein, M. A rational basis for normal in work measurement. *Journal of Industrial Engineering*, Vol. 18, No. 6 (1967): 341-346.

7. Fein, M. Work measurement today. *Industrial Engineering*, Vol. 4, No. 8 (1972): 14-20.

8. Fein, M. Work measurement: concepts of normal pace. *Industrial Engineering*, Vol. 4, No. 9 (1972): 34-39.

9. Gambrell, C. The independence of pace rating versus weight handled. *Journal of Industrial Engineering*, Vol. 10, No. 4 (1959): 318-322.

10. Gershoni, H. An analysis of time study based on studies made in the United Kingdom and Israel. *American Institute of Industrial Engineers Transactions*, Vol. 1, No. 3 (1969): 244-252.

11. Gottlieb, B. A fair day's work is anything you want it to be. *Journal of Industrial Engineering*, Vol. 19, No. 12 (1968): 592-599.

12. International Labour Office, *Introduction to Work Study*, Geneva: International Labour Office, 1978.

13. Jinich, C., and Niebel, B. Synthetic leveling—how valid? *Industrial Engineering*, Vol. 2, No. 5 (1970): 34-37.

14. Kerkhoven, C. The rating of performance levels of SAM films. *Journal of Industrial Engineering*, Vol. 14, No. 4 (1963): 170-174.

15. Lifson, K. Errors in time-study judgments of industrial work pace. *Psychological Monographs General and Applied*, No. 355, Vol. 67, No. 5 (1953).

16. Mansoor, E. An investigation into certain aspects of rating practice. *Journal of Industrial Engineering*, Vol. 18, No. 2 (1967): 184-190.

17. Martin, J. A better performance rating system. *Industrial Engineering*, Vol. 2, No. 8 (1970): 36-41.

18. Moores, B. Variability in concept of standard in the performance rating process. *International Journal of Production Research*, Vol. 10, No. 2 (1972): 167-173.

19. Mundel, M. *Motion and Time Study*. New York: Prentice Hall, 1960.

20. Murrell, H. Performance rating as a subjective judgement. *Applied Ergonomics*, Vol. 5, No. 4 (1974): 201-208.

21. Reuter, V. New way to learn pace rating. *Industrial Engineering*, Vol. 9, No. 7 (1977): 16-17.

22. Rice, R. Survey of work measurement and wage incentives. *Industrial Engineering*, Vol. 9, No. 7 (1977): 18-31.

23. Sakuma, A. New insight into pace rating. *Industrial Engineering*, Vol. 7, No. 7 (1975): 32-39.

24. Schell, H. A study on effort rating. *Modern Management*, Vol. 9 (April 1949): 9-20.

25. Smalley, H. Another look at work measurement. *Journal of Industrial Engineering*, Vol. 18, No. 3 (1967): 202-218.

26. Society for Advancement of Management. *A Fair Day's Work*. New York: New York: Society for Advancement of Management, 1954.

27. Sury, R. A study of time study rating research. *Production Engineering*, Vol. 41 (1962): 24-31.

28. Sury, R. A comparative study of performance rating systems. *International Journal of Production Research*, Vol. 1, No. 2 (1962): 23-38.

29. Thompson, D., and Applewhite, P. Objective effort level estimates in manual work. *Journal of Industrial Engineering*, Vol. 17, No. 2 (1968): 92-95.

30. van der Walt, W., and Wyndham, C. An equation for prediction of energy expenditure of walking and running. *Journal of Applied Physiology*, Vol. 34, No. 5 (1973): 559-563.

23

ALLOWANCES

RELATION TO NORMAL PACE

To review:

$$STDTIM = (MESTIM * PACE)/(1\text{-}SHIALL) \qquad (23.1)$$

where $STDTIM$ = Standard time for a task, hr/unit

           $MESTIM$ = Measured time for a worker, hr/unit

           $PACE$ = Rating of a worker, percent

           $NORTIM$ = Normal time for an average experienced worker, hr/unit

           $SHIALL$ = Allowance for a task expressed as percent of shift, percent

Thus, if an element by a worker has a measured time, $MESTIM = .010$ hr/unit, and the person making the time study gave the specific worker a rating, $PACE = 120\%$, then $NORTIM = .012$ hr/unit. If the task allowance, $SHIALL = 10\%$, then the standard time $STDTIM = .012/.9 = .0133$ hr/unit.

Equation 23.1 gives allowances as a percent of the total shift. That is, 10% for a 480-minute shift means that 48 min are allowed.

In England, however, the allowance is expressed as a percent of work time. Thus the British Standard formula is:

$$STDTIM = (MESTIM * PACE)(1 + WTALL) \qquad (23.2)$$

$WTALL$ = Allowance for a task expressed as a percent of work time, percent

Equation 23.2 gives allowances as a percent of working time, not shift time. If

48 min of allowances were desired per shift, then $480 - 48 = 432$ min of work. Then $432(x) = 480$ and $x = 1.11$; thus $WTALL = 11.1\%$.

Allowances for a task usually are subdivided into personal allowances, fatigue allowances, and delay allowances; see the remainder of this chapter for more detail on each.

However, there is an important distinction concerning allowances in different organizations. What is the base? Does every task get an allowance or are allowances only given for *cause*? If, for example, in organization A a minimum allowance of 15% was given to all tasks, then a difficult task might receive an allowance of 25%; in organization B, a minimum allowance of 5% means the difficult task should receive an allowance of 15%; in organization C, a minimum allowance of 0% means the difficult task should receive an allowance of 10%. See Table 23.1.

Table 23.1   Standard Time for a Task Depends Upon a Combination of the Definition of *Normal Pace* and the Base Allowance Given Every Task

	Organization		
	A	B	C
Measured time, hr/unit	.0100	.0100	.0100
Rating (if all three use the same definition of *normal pace*), %	120	120	120
Normal time, hr/unit	.0120	.0120	.0120
Base allowance, %	15	5	0
Allowance for the same difficult task, %	25	15	10
Standard time, hr/unit	.0160	.0141	.0133
Measured time, hr/unit	.0100	.0100	.0100
Rating (if all three use the same definition of *standard time*), %	120	136	152
Normal time, hr/unit	.0120	.0136	.0152
Base allowance, %	15	5	0
Allowance for the same difficult task, %	25	15	5
Standard time, hr/unit	.0160	.0160	.0160

If all three organizations use the same concept of normal pace, then normal time is the same in all three but standard time is .0160, .0141, or .0133 hr/unit. If a worker completed one unit in .0150 hr, in organization A this would be considered as $.0160/.0150 = 107\%$ performance while in B it would be $.0141/.0150 = 94\%$, and in C it would be $.0133/.0150 = 89\%$.

On the other hand, all three organizations might use the same concept of standard time/unit. Then their time study personnel would be trained to a different concept of normal so that someone doing the job in .0100 hr/unit would be rated as 120% in A, as 136% in B, and 152% in C.

Since organizations have different concepts of normal and different allowance bases, comparisons between organizations must be made very carefully. Unfortunately some people don't realize the connection between allowances and the concept of normal pace and make comparisons between organizations that are not valid.

In some situations, work involves machines which work unattended for part of the operator's work cycle (see Figure 23.1). It may be possible to take some of the personal, fatigue, or delay allowances during the unoccupied time. These situations can become quite complicated—especially with multiple machines—and each case must be considered on its own merits. However organizations should set guidelines.

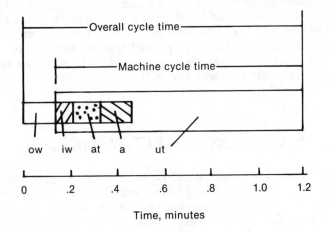

Overall cycle time

Machine cycle time

ow   iw   at   a   ut

| | | | | | |
0   .2   .4   .6   .8   1.0   1.2

Time, minutes

Figure 23.1 *Machine time allowance terms* are shown in the figure of one worker and one machine. OW = Outside Work = work which must be done outside the machine (process) controlled time. *IW* = Inside Work = work which can be done within the machine time. *UT* = Unoccupied Time = operator not engaged in inside work, attention time, or taking authorized rest. (*UT* usually has 0% allowance.) A = Allowances. Personal and Environmental Fatigue allowances usually must be taken away from the machine but physiological and psychological fatigue allowances, in some cases, may be taken at the machine. It is difficult to use incentive pay when the outside work time is a small proportion of the total time. Generally the goal is a 1% increase in pay for a 1% increase in output. But if the machine cycle dominates, then even very good performance on the outside time has little effect on the total time.

In many organizations there are standardized break periods—coffee breaks—say 15 minutes in the first part of the shift and the same in the second part. The organization can consider this 30 min/shift as part of the allowances and pretend each day has an 8-hour shift or they can consider allowances as separate from the organized breaks and pretend each day has 7.5 hours of work. Williams reports that the 15% rest allowance of the International Labour Organization and of Imperial Group (his company in England) includes all breaks including lunch; thus 15% × 480 min = 62 min; "by taking a 30 min meal break and two 10 min tea breaks, he would still have 12 min at his disposal."[11] My personal preference is to keep the standardized breaks (and especially meal breaks) out of allowances since, when breaks are in the allowances, the allowances become distorted when shorter or longer shifts are worked. That is, 15 min in a 240 min segment = 6.7% while 15 min in a 300 min segment = 5%. However, the contrary argument can be made that "coffee" and lunch breaks certainly both are rest times even though lunch time normally is not paid.

WHO DETERMINES THE ALLOWANCE?

Due to the relation between base allowance and the definition of normal pace, the determination of the base allowance is a political decision, not a scientific decision. Management, for example, may decide that *MTM* standards are their standards or 90% of *MTM* is their 100%. Managements may use an industry-wide standard or even a nationwide standard such as British Standard 3138. Remember, however, if two British companies both use BS 3138 as 100% but one gives 10% personal allowance to everyone and the other gives 15%, then they require different standard times for the same job. Once the base allowance and the definition of normal pace have been determined, then the engineers have the responsibility of determining the work pace of specific individuals and the allowances for specific tasks. (Then some managements will consider 100% for each person as a goal and others as a minimum!)

PERSONAL ALLOWANCES

Personal allowances are given for such things as blowing your nose, coughing, going to the toilet, getting a drink of water, smoking, etc. There is no "engineering" or "scientific" basis for the percent to give. They are for the individual and do not vary with the task. Lazarus' study reported that the mean of 23 industries (235 plants) was 5.6%.[7] With the exception of 10% for the lumber and wood industry and 7.6% for primary metals, all the others were between 4.6% and 6.5%. The International Labour Office reports common numbers are 5% for males and 7% for females.[5] Different allowances for men and women are no longer legal in the USA or England.

FATIGUE ALLOWANCES

Fatigue allowances are given in most industries. Lazarus' survey in the USA gave a mean of 5.1% with an industry range of 0 to 10%.[7]

There is very, very little published evidence supporting the few values or even which factors to consider. See Broadbent for a review.[2] I will use three subcategories of physiological, psychological, and environmental and will put the factors of Williams, Cornman, and ILO into these subdivisions (see Table 23.2).[3,5,10]

Authors don't consider the same factors; they divide the problem differently (e.g., Temperature and Humidity in one combined factor vs. two separate factors), they give the same factors different names (Monotony and Repetition of Cycle); and the allowances, for the same degree of a factor, differ drastically. In addition, the ILO recommends higher allowances for women than men.

Williams' values are an amalgam of industrial values in England; no evidence is given for their validity although "the results now are in use for the Imperial Group." All his values assume a "standard fatigue allowance of 10% which includes personal allowances" is given to everyone. Cornman's values are from an American consulting firm. His procedure is to add allowances from all his factors and then subtract 25%. That is, if total from the tables was 35%, then the fatigue allowances equal 10%. The ILO values were supplied by a British consulting firm in 1956. They assume a personal needs allowance (for needs away from the workplace) of 5% for men and 7% for women and a "basic" fatigue allowance for everyone of 4% plus "variable" fatigue allowances from the tables.

Belbin points out that many fatigue allowances are given for factors which do not influence fatigue or performance but simply reflect an unpleasant job.[1] Examples are monotony, odors, and noise. My recommendation is to give allowances only when they affect performance; if they just reflect unpleasant situations, don't give allowances (changing the time/unit) but change the wage rate per hour.

Table 23.2   Factors Considered for Fatigue Allowances[3,5,10]
Some Factors Have Been Renamed or Combined for This Table

PHYSIOLOGICAL	Williams	Cornman	ILO
Weight, Force, or Pressure	Table 23.3	Table 23.4	Table 23.3
Posture, Position	Table 23.5	Table 23.6	Table 23.7
Restrictive Clothing	Table 23.8		
PSYCHOLOGICAL			
Discipline, Concentration, Mental Demand, Visual Demand	Table 23.9, Table 23.10	Table 23.11	Table 23.12
Monotony, Repetition of Cycle	Table 23.13	Table 23.14	Table 23.15
Duration		Table 23.16	
ENVIRONMENTAL			
Thermal and Atmospheric, Temperature, Humidity, Air Supply, Air Conditions	Table 23.17	Table 23.18, Table 23.19, Table 23.20	Table 23.21
Noise, Aural Strain	Table 23.22	Table 23.23	Table 23.21
Vibration	Table 23.24		
Light, Visual including Lighting	Table 23.25	Table 23.26	Table 23.21

Physiological Fatigue Allowances

Weight lifted or force or pressure applied is considered in Table 23.3 (Williams and ILO) and Table 23.4 (Cornman). The ILO gives different allowances for men and women, but Williams and Cornman do not. As was commented on in chapter 12, different weight-lifting allowances for men and women can be justified physiologically. However, due to political pressures for "equality," different allowances for men and women probably shouldn't be used. The ILO approach does not consider the important factor of rest time within the cycle. However, its concave curve seems more logical than the straight lines of Williams or the convex curve of Cornman. Rohmert discusses allowances for muscular fatigue.[8] See chapter 12 for more material on manual material handling. Posture or body position is considered in Table 23.5 (Williams), Table 23.6 (Cornman), and Table 23.7 (ILO). My personal opinion is that the allowances do not rise steeply enough; see Principles 1, 2, and 3 of chapter 14.

The last factor is Restrictive Clothing (Table 23.8). In most cases, restrictive clothing probably should not be an allowance as the effect of the clothing is considered in the measured time. Williams' values are for the fatigue in addition to the increase in measured time.

Psychological Fatigue Allowances

The effect of discipline, concentration, mental and visual demand, and mental or visual strain is considered in Tables 23.9, 23.10, 23.11, and 23.12. For surveys

Table 23.3 Weight, Force, or Pressure Allowance (%)[5,10]

Weight, Force or Pressure, kg		Williams		ILO	
		Short Duration (up to .1 min)	Long Duration (over .1 min)	Males	Females
0 to 2.5		0	0.3	0	1
2.6	5.0	1.1	2.0	1	2
5.1	7.5	2.2	3.6	2	3
7.6	10.0	3.3	5.7	3	4
10.1	12.5	5.2	7.9	4	6
12.6	15.0	6.8	10.1	6	9
15.1	17.5	8.5	12.3	8	12
17.6	20.0	10.1	14.9	10	15
20.1	22.5*	11.8	17.6	12	18
22.6	25.0	13.4	20.4	14	—
25.1	30.0	15.9	24.5	19	—
30.1	40.0	20.8	31.8	33	—
40.1	50.0**			58	

* ILO recommended maximum = 22.5 for females
** ILO recommended maximum = 50 for males

Source: Reprinted with permission from *Industrial Engineering* Magazine, December 1973. Copyright American Institute of Industrial Engineers, Inc., 25 Technology Park/Atlanta, Norcross, GA 30092.

Table 23.4 Physical Demand Allowances (%) (Cornman)[3] Add 3% if Work is Done in a Difficult Work Position

Level of Effect, kg	Time Effort Is Applied			
	Up to 15%	15% to 40%	40% to 70%	Over 70%
Up to 2.2	0	0	3	3
2.2 to 11	0	0	3	7
11 to 27	0	3	7	10
Over 27	3	7	10	13

Source: Reprinted with permission from *Industrial Engineering* Magazine, Vol. 2, No. 4, 1970. Copyright American Institute of Industrial Engineers, Inc., 25 Technology Park/Atlanta, Norcross, GA 30092.

of the literature see Konz and Smith.[6,9] Machine-paced work and inspection are common tasks where these allowances may be necessary. A high allowance here may reflect a poor job design; the cure is not the giving of an allowance but a redesigned job. See especially chapter 13 for Principle 3 (decouple tasks) and chapter 16 for Principle 6 (give short breaks often).

Most allowance systems give some allowance for boredom, monotony, lack of a feeling of accomplishment, etc. (see Tables 23.13, 23.14, 23.15, and 23.16). My opinion is that boredom, etc., are unlikely to cause fatigue and thus lack of performance; they primarily reflect unpleasantness which should be reflected in the wage rate per hour rather than in the time/unit.

Table 23.5   Posture and Motion Allowances (%) (Williams)[10]

Category	Allowance, %
Sedentary work, no muscular strain (light assembly, packing, or inspection)	0
Sitting, slight muscular strain (operating low-pressure foot pedal)	1
Standing or walking; body erect and supported by both feet on ground	2
Standing on one foot (operating press foot control)	3
Unnatural postures (kneeling, bending, stretching, or lying down); light shoveling; holding unbalanced loads	4
Crouch; working with manual restraint (restricted tool movement in confined areas)	5
Awkward posture in conjunction with heavy work (bending or stooping in lifting heavy weights; carrying heavy loads on level ground or down slopes; one or both hands as convenient)	7
Carrying or moving awkward and heavy loads over rising ground; climbing stairs or ladders with heavy loads; working with hands above shoulder height (painting ceilings)	10

Source: Reprinted with permission from *Industrial Engineering* Magazine, December 1973. Copyright American Institute of Industrial Engineers, Inc., 25 Technology Park/Atlanta, Norcross, GA 30092.

Table 23.6   Position Allowances (%) (Cornman)[3]

Category	Allowance, %
Sit or combination of sit, stand, and walk where change of position is not more than 5 minutes apart; arm and head positions at normal working height	2
Stand or combination of standing and walking where sitting is allowed only during rest periods; also for situations where arms and head are out of normal working range for periods of less than 1 minute	3
Workplace requires constant stooping or standing on toes; also for work requiring extension of arms or legs	5
Body is in cramped or extended positions for long time periods; also for where attention requires motionless body	7

Source: Reprinted with permission from *Industrial Engineering* Magazine, Vol. 2, No. 4, 1970. Copyright American Institute of Industrial Engineers, Inc., 25 Technology Park/Atlanta, Norcross, GA 30092.

Table 23.7   Posture Allowances (%) (ILO)[5]

Category	Males	Females
Seated	0	0
Standing	2	4
Slightly awkward		
Abnormal position (given in addition to standing allowance)		
Slightly awkward	0	1
Awkward (bending)	2	3
Very awkward (lying, reaching up)	7	7

Table 23.8   Restrictive Clothing Allowances (%) (Williams)[10]

Category	Allowance, %
Clothing or apparatus not restrictive	0
Additional weight for protection vs. elements restricts movement; gloves affect handling; face mask restricts breathing	1 to 5
Heavy breathing apparatus, cumbersome clothing (asbestos suit restricting movement)	6 to 15

Source: Reprinted with permission from *Industrial Engineering* Magazine, December 1973. Copyright American Institute of Industrial Engineers, Inc., 25 Technology Park/Atlanta, Norcross, GA 30092.

Table 23.9   Discipline Allowances (%) (Williams)[10] Discipline Considers the Need for Exact Timing or Conformance to a Machine Cycle. Cycle Durations are for Guidance; Allowances Need not be Given for all Cycles of Less Than .2 Min

Discipline Demand	Cycle Length, Minutes	Allowance, %
Negligible	Over .20	0
Low	.16 to .20	0 to 2
Medium	.09 to .15	2 to 3
High	Below .09	3 to 5

Source: Reprinted with permission from *Industrial Engineering* Magazine, December 1973. Copyright American Institute of Industrial Engineers, Inc., 25 Technology Park/Atlanta, Norcross, GA 30092.

Table 23.10   Concentration Allowance (%) (Williams)[10] Allowance Considers the Effects of Relaxing Attention and Disturbing Influences in Relation to Personal Responsibility for Safety, Precision, Quality, Security, and Cost

Category	Allowance, %

No concentration required
    Normal tasks with automatic hand and eye coordination
    Driving truck or fork lift under normal conditions
	0

Moderate
    Adding columns of figures
    Fork lift driving in very congested aisles
    Band sawing
	1 to 4

High
    Truck driving in very heavy traffic
    Fine electrical wiring
    Engraving using an eye glass
	4 to 8

Source: Reprinted with permission from *Industrial Engineering* Magazine, December 1973. Copyright American Institute of Industrial Engineers, Inc., 25 Technology Park/Atlanta, Norcross, GA 30092.

Environmental Fatigue Allowances

The effect of climate is considered in Tables 23.17, 23.18, 23.19, 23.20, and 23.21. See Hancock and Chaffin for a more rational approach to heat stress allowances.[4] If the environmental problem is seasonal, give the environmental

Table 23.11   Mental (Visual) Demand Allowance (%) (Cornman)[3] Considers the Degree of Mental and Visual Fatigue Sustained Through Mind-Eye Concentration and Coordination

Category	Allowance, %
Only occasional mental or visual attention Operation practically automatic or attention required only at long intervals	2
Frequent mental and visual attention; intermittent work or operation involves waiting for a machine or process to complete a cycle (some checking)	3
Continuous mental and visual attention for either safety or quality reasons; usually repetitive operations requiring constant alertness or activity	5
Concentration or intense mental and visual attention in laying out or doing complex work to very close accuracy or quality or coordinating a high degree of manual dexterity with close visual attention for sustained periods of time; also all purely inspection operations (checking quality is prime object)	8

Source: Reprinted with permission from *Industrial Engineering* Magazine, Vol. 2, No. 4, 1970. Copyright American Institute of Industrial Engineers, Inc., 25 Technology Park/Atlanta, Norcross, GA 30092.

Table 23.12   Visual and Mental Strain Allowances (%) (ILO)[5] No Distinction is Made Between Males and Females

Category	Allowance %
Visual strain	
Fairly fine work	0
Fine or exacting (reading a micrometer or slide rule; ring frame tenting with light yarn)	2
Very fine or exacting	5
Mental strain (remembering a long and complicated process sequence or attending to a number of machines simultaneously)	
Fairly complex process	1
Complex process or wide attention span	4
Very complex	8

Table 23.13   Monotony Allowance (%) (Williams)[10] Allowance Considers the Absence of Mental Stimulation Due to Task Tediousness, Lack of Variation in the Work, Absence of Competitive Spirit or Companionship, and Degree of Repetition

Category	Allowance, %
Long cycle repetitive work (above .10 min/cycle)	0
Average repetitive work or nonrepetitive work	0 to 3
Highly repetitive work (less than .05 min/cycle)	3 to 5

Source: Reprinted with permission from *Industrial Engineering* Magazine, December 1973. Copyright American Institute of Industrial Engineers, Inc., 25 Technology Park/Atlanta, Norcross, GA 30092.

512

Table 23.14   Repetition of Cycle Allowance (%) (Cornman)[3] Considers the Hypnotic, Monotony Effect of Short-Cycle Repetitive Operations

Category	Allowance, %
Operator varies operation pattern or schedules own work; operations vary from day to day or operations may not be done daily	3
Reasonably fixed operation pattern or where deadlines or pressure to complete are present; task is regular although operator can vary operations from cycle to cycle	7
Periodic completion of operations scheduled; operations regular in occurrence; thought and motion patterns made at least 10 times/day	10
Completion of thought and motion patterns more than 10 times/day; machine-paced operations; most piece-rated operations; operators suffer boredom and lack of control	13

Source: Reprinted with permission from *Industrial Engineering* Magazine, Vol. 2, No. 4, 1970. Copyright American Institute of Industrial Engineers, Inc., 25 Technology Park/Atlanta, Norcross, GA 30092.

Table 23.15   Mental and Physical Monotony Allowances (%) (ILO)[5] No Distinction is Made Between Males and Females for Mental Monotony but Females are Considered Less Susceptible to Physical Monotony than Males

Category	Males	Females
Mental monotony (more likely to occur in the office than the shop)		
• Low	0	0
• Medium	1	1
• High	4	4
Physical monotony (repeated use of fingers, arms, hands, legs)		
• Rather tedious	0	0
• Tedious (cycle of 5-10 seconds)	2	1
• Very tedious	5	2

Table 23.16   Duration Allowances (%) (Cornman)[3] Consider the Amount of Time to Complete a Job and Obtain a Feeling of Accomplishment or Being Finished

Category	Allowance, %
Operation or suboperation completed in 1 minute or less	3
Operation or suboperation completed in 15 minutes or less	7
Operation or suboperation completed in 60 minutes or less	10
Operation or suboperation completed in over 60 minutes	13

Source: Reprinted with permission from *Industrial Engineering* Magazine, Vol. 2, No. 4, 1970. Copyright American Institute of Industrial Engineers, Inc., 25 Technology Park/Atlanta, Norcross, GA 30092.

Table 23.17 Thermal and Atmospheric Allowances (%) (Williams)[10] Consider the Demands Despite Presence of Protective Clothing or Equipment

Ventilation and Circulation and Humidity	Dry Bulb Temperature, C			
	Below −1	−1 to 13	13 to 24	24 to 38
Adequate ventilation and circulation Normal climatic humidity	10 to 20	1 to 10	0	1 to 10
Inadequate ventilation and circulation Nonstandard climatic conditions, causing some discomfort	20 to 25	5 to 10	0 to 5	5 to 15
Very poor ventilation and circulation, fumes, dust, steam, causing irritation to eyes, skin, nose, throat	20 to 30	10 to 20	5 to 10	10 to 20

Source: Reprinted with permission from *Industrial Engineering* Magazine, December 1973. Copyright American Institute of Industrial Engineers, Inc., 25 Technology Park/Atlanta, Norcross, GA 30092.

allowance only during the relevant season, not year around. Many times environmental allowances are given when the problem really is unpleasantness (which should be reflected in the wage rate/hr) rather than decrement in performance. See chapter 20 for more on climate.

Table 23.18 Temperature Allowances (%) (Cornman)[3] Consider Average Temperature in Performing Daily Duties

Category	Allowance, %
Mechanical or electrical control of temperature for comfort (for inactive or office personnel, usually from 22 to 24 C. Normally active or plant, usually from 20 to 21C)	1
Temperature controlled by job requirements (heat from machines, ovens, materials); for inside work, from 24 to 30 C; for outside work, from 27 to 32 C; normal circulation	2
Temperatures controlled by job requirements (heat from machines, ovens, materials); for inactive or office personnel, below 18 C or above 27 C; for outdoor work or where normal air circulation, below 4 C or above 32 C	2
When normal air circulation is available, temperature below 2 C and above 35 C; when normal air circulation is not available, above 32 C	3

Source: Reprinted with permission from *Industrial Engineering* Magazine, Vol. 2, No. 4, 1970. Copyright American Institute of Industrial Engineers, Inc., 25 Technology Park/Atlanta, Norcross, GA 30092.

Noise and vibration are considered in Tables 23.21, 23.22, 23.23, and 23.24. (See chapter 18 for more on noise.) In general, studies have been unable to detect a difference in performance as a function of noise level (if auditory communication is not part of the job); most noise allowances probably reflect unpleasantness rather than fatigue.

Table 23.19 Humidity Allowance (%) (Cornman)[3] Considers Extra Concentration Due to Perspiration, Wiping of Brow, Pulling at Clothing, Etc.

Category	Allowance, %
Normal, comfortable humidity supplied by air conditioning or heating systems; no sensation of dryness or humidity; for temperatures of 21 C to 24 C, humidities of 40% to 55%	1
Unusually dry conditions (after 30 minutes have skin sensation or burning nostrils); less than 30% humidity or high humidity, noticeable upon entrance to area by clammy skin sensation (60% to 85% humidity)	2
Unusually high humidity; clothing becomes damp quickly (humidity over 80%)	2
Humidity or wetting conditions such as steam rooms or outdoors in rain where special clothing must be worn	3

Source: Reprinted with permission from *Industrial Engineering* Magazine, Vol. 2, No. 4, 1970. Copyright American Institute of Industrial Engineers, Inc., 24 Technology Park/Atlanta, Norcross, GA 30092.

Table 23.20 Air Supply Allowances (%) (Cornman)[3] Consider Oxygen Availability or Repulsion of Human Body to the Surroundings

Category	Allowance, %
Normal operations outdoors or in air-conditioned area where filtering or washing of air supplies fresh, odor-free air	1
Normal nonair-conditioned plant or office; occasional stuffiness; movement of air supplied by movement of personnel or from machines; no air filtration	2
Extremely small and enclosed surroundings; air movement nil; or dusty conditions caused by the job (regardless of dust type); limited smoke (either foreign or operator generated)	3
Extremely smoky, toxic, or dusty conditions; nauseating or mentally disturbing fumes (although not injurious to health); air movement or exhausting does not remove effects	5

Source: Reprinted with permission from *Industrial Engineering* Magazine, Vol. 2, No. 4, 1970. Copyright American Institute of Industrial Engineers, Inc., 24 Technology Park/Atlanta, Norcross, GA 30092.

Table 23.21 Environmental Allowances (%) (ILO)[5] No Distinction is Made Between Males and Females

Category	Allowance, %
Air conditions (excluding climatic factors)	
Well ventilated, fresh air	0
Badly ventilated but not toxic or injurious fumes	5
Work close to furnaces, etc.	5-15
Noise (loud noise at irregular intervals, as in riveting; testing where listening for changes in pitch, tone, or level)	
Continuous	0
Intermittent loud (testing of auto engines; press shop)	2
Intermittent very loud or high pitched, loud	5
Light (level below IES recommendations, excessive glare, poor contrast)	
Slightly below recommended level	0
Well below	2
Quite inadequate level	5

Table 23.22 Physical Environment Including Noise Allowances (%) (Williams)[10] Considers Discomfort Caused by Dirt, Oil, Grease, Water, or Other Liquids, Ice, Chemicals, etc., and Noise Irritation (Pitch, Volume, Irregularity)

Category	Allowance, %
Clean, bright, dry surroundings; normal "machine" and "human" noise	0
Dirty, wet, greasy, contaminated surroundings	0 to 3
Uncomfortable noise	0 to 4
Combination of several factors	0 to 8

Source: Reprinted with permission from *Industrial Engineering* Magazine, December 1973. Copyright American Institute of Industrial Engineers, Inc., 24 Technology Park/Atlanta, Norcross, GA 30092.

Table 23.23 Noise Allowances (%) (Cornman)[3] Consider Fatigue to the Nervous System (Changes in Noise as Well as Loudness)

Category	Allowance, %
Normal noise level in average office or in industrial plant making light-weight products (30-60 dBA); intermittent music easily heard and enjoyed	1
Unusually quiet area where noise is almost absent (library; less than 30 dBA); also for constant, loud noise (tin shop, knitting room, city street); 60-90 dBA of constant noise; music may not be heard with pleasure	2
Normally quiet surroundings with intermittent loud or annoying noises (nearby riveter, elevated train, punch press); noises of sharp nature and above 90 dBA; also nonintermittent noises above 100 dBA (boiler factory)	3
High frequency or otherwise annoying noise, intermittent or constant	5

Source: Reprinted with permission from *Industrial Engineering* Magazine, Vol. 2, No. 4, 1970. Copyright American Institute of Industrial Engineers, Inc., 24 Technology Park/Atlanta, Norcross, GA 30092.

Table 23.24 Vibration and Instability Allowances (%) (Williams)[10] Consider Physical Movement as it Affects Work Demand, Precision, or Safety

Category	Allowance, %
Very low and steady (hand drill, saw)	0 to 1
Medium; significant but predictable vibration	2 to 3
High, distressing, difficult to control; floor vibration, pneumatic drilling, unevenly moving surface	4 to 7

Source: Reprinted with permission from *Industrial Engineering* Magazine, December 1973. Copyright American Institute of Industrial Engineers, Inc., 24 Technology Park/Atlanta, Norcross, GA 30092.

Table 23.25 Visual Including Lighting Allowances (%) (Williams)[10] Consider the Closeness of the Work as it Affects Visual Attention and Strain

Category	Adequate Lighting	Inadequate or Disturbed Lighting
No special eye attention required	0	0
Close eye attention intermittently; or continuous eye attention with varying focus	0	1
Continuous eye attention with continuous focus	1 to 3	4 to 8

Source: Reprinted with permission from *Industrial Engineering* Magazine, December 1973. Copyright American Institute of Industrial Engineers, Inc., 24 Technology Park/Atlanta, Norcross, GA 30092.

Lighting is considered in Tables 23.21, 23.25, and 23.26. See chapter 17 for more on the eye and vision and illumination.

Taking an overview of the allowances, most seem to have an inadequate range. Cornman, for example, gives a 1% allowance to excellent lighting and 3% to working in the dark. In addition, remember that Cornman has "25% deductible."[3]

Table 23.26 Light Allowances (%) (Cornman)[3] Primarily Consider Eye Strain to Focus Unless Light is so Poor as to Require Extra Body Motions

Category	Allowance, %
"Normal" lighting (200 to 500 lux in most industries; 500 to 1000 in offices and inspection); absence of glare is apparent	1
Occasional glare is inherent part of job or where substandard or special lighting is required	2
Continual glare is inherent part of job; also for work requiring constant change from lighted area to darkness (less than 50 lux); also work requiring "venetian-blind" effect (shiny and dull surface in lathe turning)	2
Work in absence of light or where sight is obstructed (noticeable by feel of fingers or feet); eyes not used or are straining (photo darkroom, mechanic under machine, etc.)	3

Source: Reprinted with permission from *Industrial Engineering* Magazine, Vol. 2, No. 4, 1970. Copyright American Institute of Industrial Engineers, Inc., 24 Technology Park/Atlanta, Norcross, GA 30092.

Taking an overview of the three subcategories, for Physiological, Williams has a minimum of 0% and a maximum (for a 20 kg load) of 40%. Cornman's range is from 2% to 17% while the ILO range is from 0% (for males) to 17%. For Psychological, Williams' range is from 0%-18%, Cornman's is from 8%-34%, and the ILO is from 0% to 13%. For Environmental, Williams' range is from 0%-53% while Cornman is 5%-19% and ILO is 0%-30%. Remember also Cornman's "25% deductible."

DELAY ALLOWANCES

Lazarus reported an all-industry mean in the USA of 5.3% with industry ranges from 0% to 10%.[7]

Delays occur from a wide variety of reasons such as machine breakdown, interrupted flow of materials, conversations with supervisors, etc. If the delay is long (say 1 hour), then the operator "clocks out" and works on something else. For short delays an extra allowance is given. The allowance depends upon the job and the working situation at that time and so should not be expected to be the same from plant to plant or time to time. If a delay allowance is given to every job or if the same delay allowance has been given to a job for many years, this probably reflects a redefined *normal* pace.

"Engineered" delay allowances can be determined either from occurrence sampling (see chapter 9) or an all-day time study.

SUMMARY

Allowances are intimately related to the definition of normal pace. If allowances are not given for "cause," then they, in effect, reflect a redefined "normal pace."

Allowances usually are given as personal allowances, fatigue allowances, and delay allowances. There is very little scientific basis for most of the allowances now used. Perhaps the best procedure at the present time is to use allowances that at least are rational and consistent. Williams', Cornman's, and the ILO's all are rational and consistent. None, however, can be considered "scientific."

A large allowance indicates a large potential for a job redesign.

SHORT-ANSWER REVIEW QUESTIONS

23.1 Allowances are related to the definition of normal pace. Explain.

23.2 Give examples of what personal allowances are for. What are typical values?

23.3 What are the three subdivisions of fatigue allowances? What is a typical fatigue allowance for the USA?

23.4 Should allowances be given for an unpleasant job? Explain your answer.

23.5 Give examples of what delay allowances are for. What are typical values for the USA?

THOUGHT-DISCUSSION QUESTIONS

1. Should allowances be included in the standard by dividing normal time by (1 − Allowances) or by multiplying normal time by (1 + Allowances)? Why?

REFERENCES

1. Belbin, R. Compensating rest allowances—part 1 and 2. *Work Study and Industrial Engineering* (now called *Management Services*), Vol. 1 (1957): 416-426; Vol. 2 (1958): 40-46.

2. Broadbent, D. Is a fatigue test now possible? *Ergonomics*, Vol. 22, No. 12 (1979): 1277-1290.

3. Cornman, G. Fatigue allowances—a systematic method. *Industrial Engineering*, Vol. 2, No. 4 (1970): 10-16.

4. Hancock, W., and Chaffin, D. A practical method for industrial heat stress allowance determination. *AIIE Transactions*, Vol. 9, No. 2 (1977): 144-154.

5. International Labour Office. *Introduction to Work Study*. Geneva: International Labour Office, 1971; chapter 17.

6. Konz, S. Endurance and rest for work with concentration and attention. *Proceedings of the Human Factors Society*, (1979): 210-213.

7. Lazarus, I. Inaccurate allowances are crippling work measurements. *Factory* (April 1968): 77-79.

8. Rohmert, W. Problems in determining rest allowances. *Applied Ergonomics* (1973): Vol. 4, No. 2, 91-95 and Vol. 4, No. 3, 158-162.

9. Smith, W. A review of the literature relating to visual fatigue. *Proceedings of the Human Factors Society*, (1979): 362-366.

10. Williams, H. Developing a table of relaxation allowances. *Industrial Engineering*, Vol. 5, No. 12 (1973): 18-22.

11. Williams, H. Personal communication, 1976.

24

PROGRESS CURVES

CONCEPT OF TWO LOCATIONS FOR LEARNING

Individual Learning

Individual learning is improvement with a *constant* product design and *constant* tools and equipment.

Consider Joe playing golf at the Riverview course. He has a driver, a 5 iron, a 9 iron, and a putter, and some slightly used golf balls. For his first round of 18 holes he shoots 137. His next round is 131. Then 127, 115, 118, 112, 114, 110. Eventually he levels out at a score of 100.

All during the time that his score was improving he played the same course; his required output was the same. He used the same clubs, tees, and balls; his tools and equipment were the same. He modified the same work method by reducing jerks, getting his feet set just right, and holding the club slightly tighter.

In the same way, Joe might go to his job and turn out more and more widgits/hr as he gained more experience even though the design of the widgits did not change and his tools and equipment did not change. The improvement is due to better eye-hand coordination, fewer mistakes, reduced decision time, etc.

Organization Learning (Manufacturing Progress)

This is improvement with *changing* product design, *changing* tools and equipment, and *changing* work methods. It is individual learning plus organization learning; often it is called *manufacturing progress*.

Using the golf example, Joe might switch golf courses to the Sunnyside course.

There he can average 96 since there are not as many sand traps. This is similar to a change in product design; less is required to complete a unit.

He could modify his tools and equipment. One step might be to get a complete set of clubs; another might be to get new golf balls.

He could modify his work method further. Perhaps he could use a different putting stance. He could improve his technique for estimating distance.

For an occupational example, consider the waitress Maureen serving coffee and doughnuts. During the individual learning, Maureen learned where the coffee cups are, the coffee pot, the prices of each product, etc. The amount of time it took her to serve a customer declined to a plateau. Now let's consider some organizational changes which might occur. Management might set a policy to serve coffee in cups without saucers and to furnish cream in sealed one-serving containers so that the container need not be carried upright. These changes in product design (learning by the organization) reduce time for Maureen and all other waitresses. Tools and equipment also can be modified. Put one coffee pot at each end of the counter. Redesign the sales slip to speed computations and reduce errors. Buy a coffee pot with a better handle so less care is required to prevent burns. These changes in tools and equipment also help all waitresses reduce time per customer. The organization may even decide to leave the bill with the customer when the last food item is served so as to further reduce time per customer. Organization progress includes all the changes—not just individual learning.

For an industrial product, say an airplane, manufacturing progress includes the increased skill of the assembly operator plus the reduced supervision needed by the supervisor as more and more planes are built; the reduced time needed by the fork truck driver since the driver knows exactly when to bring the materials; the reduced repairs required due to incorrect drawings as the initial errors on drawings are eliminated; reduced need for the operators to look up information as by the fiftieth airplane they remember that the red part goes above the green part

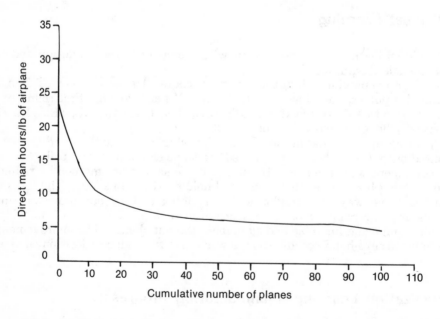

Figure 24.1 *Practice makes perfect.* As more and more units are produced the fixed cost is divided over more units so fixed cost/unit declines. In addition, however, variable cost/unit declines as fewer mistakes are made, less time is spent looking up instructions, better tooling is used, etc. The variable cost data usually can be fitted with an equation of the form, $y = ax^b$.

instead of having to look it up, etc. It includes better material usage as parts are trimmed with less scrap and better materials are used.

Organization progress comes from three factors: (1) operator learning with existing technology, (2) influence of new technology, and (3) economics of scale.

Point 1 was just discussed. Examples of new technology are the subsurface bulblike nose on the front of tankers (a nose that increased tanker speed at very low cost), improved turbine blade design which increased turbine efficiency, and solid state electronics which improved the performance of TV sets and computers.

Economy of scale occurs as equipment with twice the capacity costs less than twice as much; capital cost/unit is reduced. Operator costs are reduced as fewer work hours are needed/unit of output.

QUANTIFYING IMPROVEMENT

Log-Log Concept

Practice makes perfect has been known for a long time. Wright made a key step when he published manufacturing progress curves for the aircraft industry.[38] Wright made two major contributions. First he quantified the amount of manufacturing progress for a specific product. It's not too useful to a businessman to tell him that time to assemble an airplane is going down; he wants to know how much it is going down. Wright told him. See Figure 24.1 for a hypothetical example of airplane cost/unit vs. cumulative airplanes produced.

This was quite a big step but giving a manager a complex equation, $y = ax^b$, was too much in the 1930s; in fact, engineers should be careful even today in presenting formulas to managers. Managers like simplicity. So Wright took the curve out of the data and presented the data in a straight line. The managers hardly noticed that he had put the curve in the axes. Figure 24.2 has the same data plotted on the log-log scale. See Figure 24.3 for Cartesian coordinates, semilog coordinates, and log-log coordinates.

On Figure 24.2, look at the "cumulative number of airplanes" axis. Call the physical distance between the 1st and 2nd airplane "Δx"; it is equal to 1 airplane. Mark off this same physical distance from 2; it will reach to 4. Mark off the same physical distance from 4; it will reach to 8. The same physical distance on a log scale is represented by a series of 1, 2, 4, 8, 16, . . . We also can say that every time we move the same distance to the right we double the quantity (1, 2, 4, 8, 16, . . .).

Next consider the straight line fitted to the data. The rate of decline of the line, the slope, is $\Delta y/\Delta x$. If we pick the Δx as the distance between doubled quantities, then Δx will be the same no matter whether we are dealing with 8 airplanes to 16 airplanes, 16 airplanes to 32 airplanes, or 25 airplanes to 50 airplanes.

Wright gave learning curves in terms of new cost when the quantity doubled. For example, assume that the manufacturing cost/lb of the 8th airplane was 12 hours and cost of the 16th airplane was 9.5. Then this is considered to be "a 78% curve"; that is, for $x_2 = 2 x_1$, $y_2 = .78 y_1$. If cost at unit 16 was 10 then it would be 10/12 or an 83% curve. (If you wish to determine the b in the equation $y = ax^b$, use the conversion from Table 24.1.)

Determining a Curve

The formula just recommended uses the straight line $\log y = \log a + b (\log x)$. Scholarly articles often advocate fitting more complicated formulas to learning

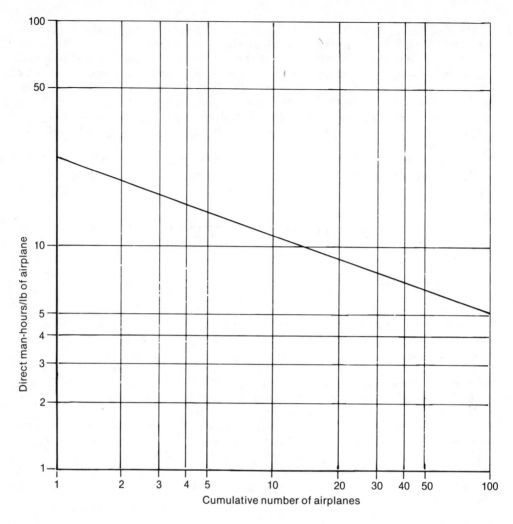

Figure 24.2 *Managers* like straight lines. Plotting $y = ax^b$ on log-log paper gives a straight line. The key piece of information desired by managers is the rate of improvement—the slope of the line. The convention is to refer to reduction with doubled quantities. If quantity $x_1 = 8$, then quantity $x_2 = 16$. Then if cost at x_1 is $y_1 = 100$ and cost at x_2 is $y_2 = 80$, this is an "80% curve."

data. (A statistician has been defined as one who draws a mathematically precise line from an unwarranted assumption to a foregone conclusion!) Carlson, for example, recommends adding two more terms to the polynomial to get $\log y = \log a + b (\log x) + c (\log x)^2 + d (\log x)^3$. Others recommend other curves. The trade-off is accuracy of estimate vs. difficulty of obtaining the estimate and ease of use. Lipka shows how to convert various types of curves to straight lines.[23] In brief, for $y = ax^b$, use $\log x$ and $\log y$; for $y = ae^{bx}$ use x and $\log y$; for $y = a + bx^n$, where n is known, use a plot of x^n vs. y; for $y = x/(a + bx)$, plot x vs. x/y or $1/x$ vs. $1/y$. See chapter 25. The simple formula also facilitates comparisons among references. The sheer simplicity of log-log plots has another advantage: they are done by hand. Since computers are available, people no longer plot the data and fit a curve by eye; they keypunch some numbers and dump them into a computer curve-fitting routine. The unfortunate result is that keypunch errors, transcrip-

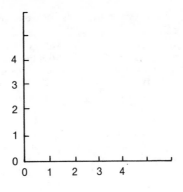

Cartesian coordinates

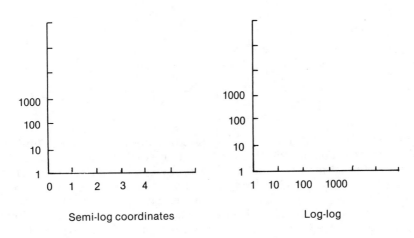

Semi-log coordinates   Log-log

Figure 24.3 *Axes* can differ.

tion errors, accounting errors, etc., are no longer detected as "garbage in becomes Bible out" as the computer prints all numbers to six decimal points.

Conversely, you might have some very ordinary data that are not very convincing in themselves; make them more convincing by having them printed by a computer; even more convincing to the uninitiated are the same data *plotted* by a computer. The consensus has been that the simple straight line on log-log paper is best.[14]

Table 24.2 shows how these complicated equations are obtained. During the month of March various people wrote down on charge slips a total of 410 hours against this project's charge number. The average work hours/unit during March then becomes 29.3. The average x coordinate is $(1 + 14)/2 = 7.5$. (Because the curve shape is changing so rapidly in the early lots, some authors recommend plotting the first lot at the 1/3 point $[(1 + 14)/3]$ and points for all subsequent lots at the midpoint. What they are really saying is that one straight line for the early lots and a second less-steep line would be most accurate but one "adjusted" line is simpler.)

During April, 9 units passed final inspection and 191 hours were charged against the project. Cumulative hours of 601 divided by cumulative completed output of 23 gives average hours/unit of 26.1; the 26.1 is plotted at $(15 + 23)/2 = 19$ in Figure 24.4 using log-log paper. Complicated curve-fitting routines can be

Table 24.1   Factors for Various Improvement Curves

Improvement Curve, % Between Doubled Quantities	Learning Factor, b, for Curve $y = ax^b$	Multiplier to Determine Unit Cost if Average Cost is Known	Multiplier to Determine Average Cost if Unit Cost is Known
70	−.515	.485	2.06
72	−.474	.524	1.91
74	−.434	.565	1.77
76	−.396	.606	1.65
78	−.358	.641	1.56
80	−.322	.676	1.48
82	−.286	.709	1.41
84	−.252	.746	1.34
85	−.234	.763	1.31
86	−.218	.781	1.28
88	−.184	.813	1.23
90	−.152	.847	1.18
92	−.120	.877	1.14
94	−.089	.909	1.10
95	−.074	.926	1.08
96	−.059	.943	1.06
98	−.029	.971	1.03

used but the accuracy of the data rarely justifies anything beyond a line fitted by eye. The purpose of using progress curves is to predict the future. Calculating the future to 6 decimal points is foolish. Fitting by eye has an advantage in that it requires a plot of the data; outliers and abnormal points are obvious; computer routines will fit a straight line (or whatever curve you specify) whether the straight line is valid or not. For these data the percent is about 79. A simple way to calculate it is to read the cost from the line at a quantity, the quantity doubled, and the quantity quadrupled. For example, 24.9 for 20, 19.7 for 40, and 15.5 for 80. Then 15.5/24.9 = .6225 and the square root of .6225 = .79. If errors in reading are not so important, just use one doubling. For example, 19.7/24.9 = .79. If someone reports a curve as 78.898%, have that person gather the data next time.

Table 24.2   Time and Completed Units as They Might Be Reported for a Product

Month	Units Completed (Pass Final Inspection)	Month's Direct Labor Hours Charged to Project	Cumulative Units Completed	Cumulative Work Hours Charged to Project	Average Work Hours/Unit
March	14	410	14	410	29.3
April	9	191	23	601	26.1
May	16	244	39	845	21.7
June	21	284	60	1129	18.8
July	24	238	84	1367	16.3
August	43.	401	127	1708	13.4

Although average cost/unit is what is usually used, you may wish to calculate cost at a specific unit. Conversely, the data may be for specific units and you want average cost. Table 24.1 gives the multiplier for various slopes. The multi-

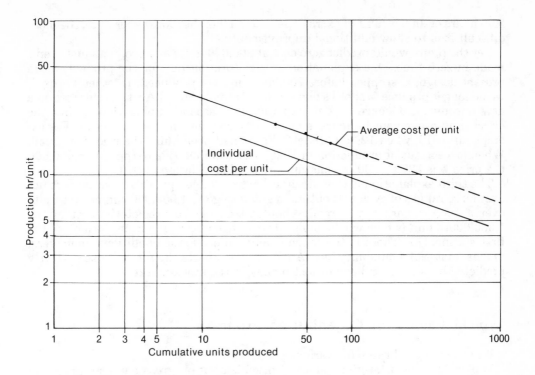

Figure 24.4 *Average cost/unit* from Table 24.2 gives a "79% curve." Cost/unit can be estimated by multiplying average cost/unit by the factor from Table 24.1. Thus the average cost of the first 20 units is estimated as 25.9 from the fitted line; the cost of the 20th unit is 25.9 (.658) = 17.0 hours.

plier for a 79% curve is estimated as (.641 + .676)/2 = .658. Thus if we wished to estimate the cost of the 20th unit, it would be (24.9 hr) (.658) = 16.4 hr. This multiplier is based on the fact that the average cost curve and the unit cost curve are parallel after an initial transient. "Initial transient" usually is 20 units and sometimes as few as 3.

Cost/unit is especially useful in scheduling. For example, if 50 units are scheduled for August, then work-hr/unit (77% curve) at unit 127 = (13.4) (.656) = 8.8 and at unit 177 = 7.8. Thus between 390 and 440 hours should be scheduled.

Price/unit often is used for between-company comparisons since price/unit is available while cost/unit probably isn't. If cost/unit is calculated from price/unit, the additional assumption must be made that profit margins do not change.

Looking at the curve in Figure 24.4 you can see how the extrapolated line would predict average cost/unit at 200 to be 11.4 hr, at 500 to be 8.3, and at 1000 to be 6.6. If we add more cycles on the paper the line eventually reaches zero cost at cumulative production of 200,000 units. Can cost go to zero? Can a tree grow to the sky? No.

The log-log paper has helped understand improvement but it also deceives. Note that cost for unit 20 was 24.9 hours. When output was doubled to 40 units, cost dropped to 19.7; doubling to 80 dropped cost to 15.5; doubling to 160 dropped cost to 12.1; doubling to 320 dropped cost to 9.6; doubling to 640 dropped cost to 7.6. Now consider the improvement for each doubling. For the first doubling from 20 units to 40 units, cost dropped 5.20 hours or .260 hr/unit of extra experience. For the next doubling from 40 to 80, cost dropped 4.2 hours or .105 hr/unit of extra experience. For the doubling from 320 to 640, cost dropped

2.00 hours or .006 hr/unit of extra experience. Thus the more experience, the more difficult it is to show additional improvement.

Yet the figure would predict zero cost at 200,000 units and products just aren't made in zero time. One explanation is that total output for the product, in its present design, is stopped before 200,000 units are produced. In other words, if we no longer produce Model Ts and start to produce Model As, then we start on a new improvement curve at zero experience. A second explanation is that the effect of improvement in hours is masked by changes in labor wages/hr. For example, in 1910, when 12,300 Model T Fords had been built, the price was $950. When it went out of production in 1926 after a cumulative output of 15,000,000 the price was $270; $200 in constant prices plus inflation of $70.

The third explanation is that straight lines on log-log paper are not perfect fits over large ranges of cycles. If output is going to go to 1,000,000 cumulative units over a 10-year period, you really shouldn't expect to predict the cost of the 1,000,000th unit (which will be built 10 years from the start) from the data of the first six months. There is too much change in economic conditions, managers, unions, material availability, etc. Anyone who expects the future to be perfectly predicted by a formula has not lost money in the stock market.

Typical Values for Organization Progress

What are typical rates for various products?

Tables 24.3 and 24.4 give values of organization progress curves reported in the literature. These include all progress: better designs, more efficient factory layouts, economies of scale, individual learning, improved technology from supplier industries, etc. Those values that represent sales prices also include the decline in cost/unit of initial investment in engineering, buildings, machinery, etc.

In summary, the rate of improvement depends on the amount that can be learned: the more that *can* be learned, the more that *will* be learned. The amount that can be learned depends on two factors: amount of previous experience with the product and degree of mechanization.

Titleman gives an explanation, shown with Figure 24.5, which he says predicts the learning curve ratio "with less than 2.5% error."[35] He requires the estimator to divide total time for an operation into two categories: manual time and mechanical time. Then calculate the manual ratio which is manual time over total time. A job requiring 10 minutes manual time and 20 minutes mechanical (machine) time would have a manual ratio of 33% and an estimated learning curve of 95%.

Percent of Task Time		Manufacturing Progress, %
Manual	Machine	
25	75	90
50	50	85
75	25	80

Thus, if you were dealing with a product which required 1000 hr/unit, which was scheduled for total production of 70 units, and required about 75% manual time, you might estimate an 80% curve. If this same product was scheduled for total production of 2000 units, you would plan to purchase more machines, set up a better production line, select raw material sizes more carefully, etc., and less learning would occur during this production—85% might be appropriate. Then if cumulative production over a 15-year period reached 10,000 units, a rate of 87,

Table 24.3 Manufacturing Progress Rates Reported in the Literature with Their References

Rate, %	Number of Cycles	Examples
60		Production man hours/cumulative units of steel produced since 1867[18]
68	1,000	Test and adjust time for product B[9]
70	10,000,000 to 3,000,000,000	Average revenue/unit for silicon transistors for industry accumulated volume[8]
70	3,000,000	Price/unit for integrated circuits for industry total accumulated volume (1964-1968)[8]
70	300,000,000 to 3,000,000,000	Price/lb of polyvinyl chloride (constant dollars) for industry accumulated volume; from 30,000,000 to 300,000,000 it was 95%[8]
70	50,000,000 to 80,000,000	Average price/unit (constant dollars) for free-standing gas ranges for cumulative industry volume (1952 to 1967)[8] Overhead cost/airplane[38]
72	1,000,000 to 2,000,000,000	Price/unit of integrated circuits (1963-1972)[29]
73	11,000	Electronic assembly of product A[9]
75	8,000	Electro-mechanical assembly of product B[9]
75		\$/ton for US electric utility coal 1948 to 1971 (in 1970 prices). 1.5 to 5.5 billion tons. Sharp increase in price upon passage of Clean Air Act of 1970.[12]
75		Cents/kw-hr for electricity price in USA from 1926 to 1970 in 1970 dollars. 0.4 to 22 kw-hr $\times$ 10^{12}. Price rose above trend during 1930s. Discontinuity in curve in 1971 due to effect of Clean Air Act of 1970.[12]
76		Man hours/barrel of petroleum refined in US since 1888[18] Assembly labor/unit on 15 different models of machine tools[17] Maintenance man hours/shutdown in a General Electric plant
78-84		Production hours/unit for Liberty ships, Victory ships, tankers and standard cargo vessels in USA during World War II[17]
79	1,000	Service time—IBM electronic machine[20]
80		Cost/barrel of catalytic cracking unit capacity (inflation has changed actual dollar cost to 94%)[18]
80		Man hours/airframe in USA during World War II[17] Labor cost/airplane[38]
80		\$/barrel of retail gasoline processing cost for USA 1919 to 1969. 1970 dollars. .2 to 37 C units where C = 10^{16} BTU[12]
81	20,000	Service calls/150 machines (IBM electro-mechanical)[20]
82		Total labor/unit on 16 different models of machine tools[17]
84	8,000	Final assembly labor/unit on product B[9]
85	2,000	Labor hours/gun barrel on boring machine[2]
86	8,000	Repetitive final assembly labor hours/unit on product B[9]
86	12,300 to 15,000,000	Cost/unit for Ford Model T. Price in 1910 when 12,300 had been built was \$950. In 1926, when 15,000,000 had been built, price was \$270 (\$200 in constant prices)[18]
86	1,000	Service time—IBM electronic machine[20]
87	20,000	Service time—IBM electro-mechanical machine[20]
88		Machining labor/unit on 15 different models of machine tools[17] Cost of purchased subassemblies on airplanes[38]
90		Output of a fluid catalytic cracking unit
95		Lbs of raw material/airplane[38] \$/barrel of price for crude oil at the well in USA 1869 to 1971. 1970 prices. .01 to 80 C units where C = 10^{16} BTU[12]
96	11,000 to 120,000	Hours/unit for electronic assembly of product A[9]

90, or even 93% might be appropriate for this last cycle (1,000 to 10,000 units) of the log-log paper.

Allemang recommends a procedure using product design stability, a product characteristics table (complexity, accessibility, close tolerances, test specifications, and delicate parts), parts shortage, and operator learning to calculate an allowance.[1] Advantages cited are reduced judgment and more documented estimates (especially important for government auditors).

Typical Values for Learning

Table 24.5 gives some learning values reported in the literature.

Smyth used a different approach, a job evaluation-type plan.[34] In place of Titleman's two factors, he used six: length of cycle, education required, rules and regulations that needed to be known, rhythm and dexterity, mental dexterity, and physical demand. For a spin and tin wire operation in a radio factory, he assigned (for the factors in order) "over 240 units/hr," "equivalent to 8th grade," "rules and regulations at minimum," "medium amount of rhythm and dexterity," "minimum," and "light physical effort." The factors had a total of 69 points which was equivalent to 40 hours of learning time. The most complicated job was allowed 384 hours of learning time. Clifford and Hancock point out that age of the operator affects the rate; older workers time/move and time/position declined at a slower rate than for younger workers. Turban reports inexperienced workers (either new workers or veteran workers learning new skills) take twice as long to learn as experienced workers.[36]

The improvement takes place through reduction and elimination of fumbles and delays rather than greater movement speed.[4,5] Stationary motions such as position and grasp improve the most while move and reach improve little.[31] It is reduced "information processing" time rather than faster hand speed that makes the reduction. Salvendy and Pilitsis reported kcal/cycle declined with experience as well as time/cycle.[32]

Carlson and Row report interruptions in work cause forgetting, with the amount of forgetting depending on the length of time between "batches."[7]

Work-Factor learning factors (see Table 24.6) can be approximated by two learning rates—around an 80% curve for cycles 1-50 and around 90% for cycles 50-500.[37] The table permits estimation of the mean time/unit required for quantities less than 500. For example, if the Work-Factor time was 1.5 min/unit and 100 units were to be made in this release to the shop, then the estimated mean time for 100 units would be 1.5 (1.3) = 1.95 min/unit; total time would be 1.95 (100) = 195 min. This estimate does not include allowance time so if allowances were 15%, the total time would be 195/.85 = 229 min.

The effect of similar work can be considered:

$$ESTIME = WFTIME\,(SMFACT\,[QMULT \text{-} 1] + 1) \qquad (24.1)$$

where $ESTIME$ = Estimated time/unit including learning allowances

        $WFTIME$ = Work-Factor time/unit

        $SMFACT$ = Similarity factor

               = 1.0 for 0 to 30% similar

                0.5 for 31 to 60% similar

                0.25 for 61 to 90% similar

                0 for over 90% similar

        $QMULT$ = Quantity multiplier from Table 24.6

Table 24.4 Japanese Organization Progress Curves.[25] Konz has Estimated Cumulative Units, Cumulative Work Time, Early Times, and Late Times From the Japanese Figures and Tables to Give the Reader a "Feel" for Each Situation.

Industry	Firm & Product	Operation	Learning %	Cumulative Units	Prod. Period, Days	"Early" Time/Unit	"Late" Time/Unit
Airline	V1	Maintenance/flying hr	91	396,000	2,000	4.45	2.44
Automobile	B1	Total	80		800	200.	45.
	B2		78		1,100	160.	95.
	J	Assembly	84				
	J	Lot work	87				
	J	Mechanical milling	93				
	J	Drilling	85				
	J	Lathe	87				
	Q	Assembly	83	3,892	300	1.52	0.24
	U1	Total, workers/unit	84	900,000		2.3	0.7
	U2	Engine	82		70	700.	150.
Chemical	D1	Paper (days/10,000 tons)	74	350,000 tons		800.	70.
	D2	Paper (days/10,000 tons)	75	1,300,000 tons		800.	40.
	F	Tire	89	1,146	21	276.	150.
Construction	E	Aluminum welding touchup, hr	86	153	30	53.1	17.7
Electrical and Electronic	C1	Assembly line, min	86		400	0.400	0.135
	C2	Assembly line, min	88		300	0.090	0.037
	C3	Assembly line, min	89		280	0.080	0.037
	C7	Assembly line, min	82		28	0.15	0.055
	C8	Automatic lead insert machine, min	86		100	0.1	
	C9	Automatic lead insert machine, min	87		100	0.1	
	C4	Defect percent	81		100	0.65	0.17
	C5	Defect percent	90		100	0.45	0.18
	C6	Defect percent	70		70		0.15
	01	Total	83	26,200	220	9.2	4.4
	01	Total	83	26,200 to 106,800	180	4.4	3.1
	P	Tape recorder total	86	652	300	84.	16.
	Y1	Inspection	89	22,500	210	288.	164.
	Y2	Inspection	88	5,600	210	198.	117.
	Y3	Inspection	83	1,921	105	48.	30.
	Y4	Inspection	85	787	105	43.	33.
Food Processing Machine	T	Total	93	2,500	22	1.37	0.96
	A	Lathe	84				
	A	Drilling	90				
	A	Milling	87				
	H1	Grinding for casting	94	up to 8,000			
	H1	Grinding for casting	83	8,000 to 16,000	250	2.27	1.61
	H2	Grinding for casting	96	up to 8,000			
	H2	Grinding for casting	87	8,000 to 16,000	250	2.25	1.74
	L1	Mechanical cutting	93				
	L2	Assembly	87				
	N1	Total for cooling products	88	370			
	N2	Total for cooling products	91	350		505.	
	N4	Total for cooling products	91	400			
	N5	Total for cooling products	90	390			
	N6	Total for cooling products	85	400			
	N7	Total for cooling products	87	470			
	N8	Total for cooling products	91	450			
	N9	Total for cooling products	92	450			
	N10	Total for cooling products	89	650		970.	
	N11	Total for cooling products	93	500			
	Z	Machine assembly	87	700		350.	150.
Metal	X	Stainless Steel	87		800	15.	11.
	W	Metal grinding	97	15,200		0.72	0.53
Precision	I2	Daily loss time, min (job 2)	76	110		250.	50.
	I2	Daily loss time, min (job 20)	83	20		11.	6.
	I2	Daily loss time, min (job 79)	79	80		200.	70.
	I2	Daily loss time, min (job 40)	79	80		38.	10.
	I3	Assembly (cash register A), hr	96	40		190.	150.
	I3	Assembly (cash register B), hr	93	35		170.	85.
	I3	Assembly (cash register C), hr	92	90		50.	30.
	K1	Nut making	80	1,000		1.	
	K2	Nut making	81	1,000		1.	
	K3	Nut making	82	1,000		1.	
	K4	Nut making	86	1,000		1.	
	K5	Nut making	90	1,000		1.	
	K6	Nut making	95	1,000			
	K7	Nut making	98	1,000			
	S2	Thread	83	10,000		0.9	0.4
	S4	Thread	90	1,700		1.8	1.
Shipping	G	Total (small ships), hr/ship	76	48,200 (53 ships)		3,320.	667.
	R1	Machining, %	86	460 (7 ships)		100.	66.
	R2	Boiler, %	88	480 (7 ships)		100.	69.
	R3	Tank and pipe, %	84	430 (7 ships)		100.	61.
	R4	Stringer, %	84	423 (7 ships)		100.	60.

Table 24.5 Learning Rates Reported in the Literature With Their References

% Rate	Number of Cycles	
68		Truck body assembly[15]
70	12	Bag-molded aircraft cowls[33]
72	50	Complex, 300 hr/unit assembly[9]
74		Machining and fitting of small castings[15]
78	212	Complex cored large radrome[33]
80	400,000	Keyboard entry on business machines[19]
		Precision bench assembly[26]
	6,800	Press molded housings[33]
82		Burring, sanding, and hand forming[24]
		Shearing plates[35]
		Grinding[26]
83		Fitting[35]
		Power sawing[26]
	40	Sorting cards into compartments[10]
	4,000	Substituting letters for symbols[10]
		Radio tube assembly[15]
		Servicing automatic transfer machines[15]
84	2,000,000	Cigar making (90% for cycles 2,000,000 to 10,000,000)[10]
		Lathes[26]
85		Gas cutting, thin plates-machine[35]
		Man attendance hours—washing machines[15]; effective hours were 88%, wasted hours rate was 80%
87	1,000	Drill, ream and tap[22]
		Drilling[26]
88		Welding-manual[35]
89	10,000	Punch press[9]
		4 sec. cycle assembly[5]
		Washers on pegs (initial cycle time = 20 min)[11]
		Milling[26]
90		Punch press[24]
		Bench inspection[26]
91	8,000	Adding pairs of digits[10]
92		Assembly with jig[35]
	14,000	Assembly (70% from 14,000 to 25,000; 95% from 25,000 to 600,000 when put on piecework)[39]
		Gas cutting, thick plates-machine[35]
		Welding[26]
		Deburring, cleaning[26]
94	300	Pegs in pegboard[39]
		Welding, submerged arc[35]
95	450	Countersink[22]
	10,000	Punch press—average of 5 operators on 5 operations (range 89 to 98)[3]
	450	Screwdriver work (85% for cycles 450 to 3500)[4]
95		Reduction of running speed (m/s) vs. distance (km) for world-class male runners (.94 for females)[30]
96.5		Milling—no jig[35]
98.5		Grinding—manual[35]
		Chipping—pneumatic[35]
		Blast cleaning[35]
		Milling—with jig[35]
		Assembly—no jig[35]

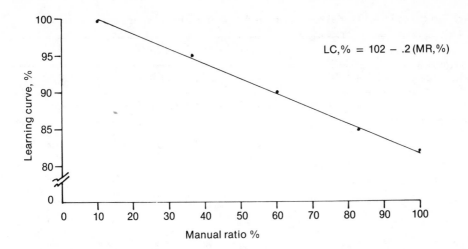

Figure 24.5 *Predict learning percent* from the manual ratio for an operation (Titleman).[35] For example, a job with 10 minutes "manual" time and 5 minutes "mechanical" time has a 67% manual ratio and learning percent is predicted as 88%.

Source: Reprinted with permission of *Product Engineering Magazine* (a Morgan-Gampian Publication).

For example, if $WFTIME = 1.5$, the quantity released is 100, and 80% of the operations were fully familiar for this operator, then $SMFACT = .25$ and $QMULT = 1.3$. Then $ESTIME = 1.5 \,(1.075) = 1.61$ min/unit.

Table 24.6 also allows estimation of the specific unit at which 100% of Work-Factor time should be achieved. For a 90% curve, the unit curve is about 85% of the mean curve (see Table 24.1). A line on log-log paper parallel to the $QMULT$ values and at 85% of their values cuts the axis between 200-400 units (depending on the cycle time and the similarity). Thus a worker is expected to achieve Work-Factor standard after 200 to 400 cycles. Remember, however, that Work-Factor assumes the worker is experienced at the start and only familiarity with a specific unit is needed.

APPLICATIONS

As discussed in chapter 21, use of time/unit has four applications in all organizations: cost allocation, scheduling, acceptable day's work, and evaluation of alternative methods. In some organizations there is a fifth application as wages are based on output. For more examples, see Nanda and Adler.[27]

The fact that labor hours/unit decline as output increases makes computations for all the applications more complex than if it could be assumed constant. Ah, for the simple life!

Cost Allocation

For cost allocation, many authors have pointed out the necessity of using the projected lower costs in bidding on new contracts or make-buy decisions. For example, assume average cost of the first 100 units is 250 hr/unit, you have a 80% manufacturing progress curve, and you must estimate manufacturing costs for a reorder of 100 units. Then you should estimate average cost of .8 (250) =

Table 24.6   Work-Factor Learning Allowances for Individual Operators[37]

Quantity of Pieces to be Produced	Work-Factor Time Cycle			
	−3.0 Min.	−6.0 Min.	−12.0 Min.	>12.0 Min.
	Quantity Class Multipliers (Learning Allowances Included)			
1	5.0	5.0	5.0	5.0
2	4.1	4.2	4.3	4.4
3	3.6	3.7	3.9	4.0
4	3.3	3.4	3.6	3.8
5	3.0	3.2	3.4	3.6
6	2.9	3.0	3.2	3.4
7	2.7	2.9	3.1	3.3
8	2.6	2.8	3.0	3.2
9	2.5	2.6	2.9	3.1
10	2.4	2.6	2.8	3.0
15	2.1	2.2	2.5	2.7
20	1.9	2.0	2.3	2.5
30	1.7	1.8	2.0	2.2
40	1.6	1.7	1.9	2.0
50	1.5	1.6	1.7	1.9
60	1.4	1.5	1.7	1.8
70	1.4	1.5	1.6	1.7
80	1.3	1.4	1.5	1.7
90	1.3	1.4	1.5	1.6
100	1.3	1.4	1.5	1.6
110	1.2	1.3	1.4	1.5
120	1.2	1.3	1.4	1.5
130	1.2	1.3	1.4	1.5
140	1.2	1.2	1.4	1.5
150	1.2	1.2	1.3	1.4
160	1.2	1.2	1.3	1.4
170	1.2	1.2	1.3	1.4
180	1.1	1.2	1.3	1.4
190	1.1	1.2	1.3	1.4
200	1.1	1.2	1.3	1.3
250	1.1	1.1	1.2	1.3
300	1.1	1.1	1.2	1.2
350	1.1	1.1	1.1	1.2
400	1.1	1.1	1.1	1.2
450	1.06	1.1	1.1	1.2
500	1.05	1.1	1.1	1.1

200 hr/unit for the cumulative output of 200. Thus the first 200 will take 40,000 hours; the first 100 took 25,000 hours so this reorder of 100 will require 15,000 hours. It is not necessary to charge a lower price due to the improvement—you may wish to have higher profits—but it certainly is desirable to know your true costs when making your bid.

Shared costs are important. That is, if a component is used on more than one final product (standardization), it can progress much faster on the curve because its sales come from multiple sources. Thus the costs of all the final products are pulled down. Remember, however, that low costs do not guarantee success if the product is obsolete. That is, products have "life cycles" of birth, growth, decay, and death.

Scheduling

For scheduling, changing labor hours/unit is important. If you wish to keep units/week constant, then you should expect to decrease your labor force for the project as time passes. If you wish to keep your labor force constant, then your output/week should increase week by week. Consider also the effect of different lot sizes and labor turnover. For example, consider the effect on economic lot size for spare part components. Say, for example, that a unit has a standard manufacturing cost of $11.33 based on a lot size of 100 units, setup cost of 3 hr, run cost of 1 hr/unit, labor cost of $10/hr (including overhead), and material cost of $1/unit. Standard cost then is:

Material	$ 1.00
Labor	
Setup (3/100) ($10)	0.33
Run 1 ($10)	10.00
	$11.33

Assume that due to poor scheduling or a customer emergency it is decided to run a special lot of 6. Assume a learning curve of 95%. Then run time $= 1.0/(.95)^4 = 1.23$ hr/unit and total cost becomes:

Material	$ 1.00
Labor	
Setup (3/6) (10)	5.00
Run 1.23 (10)	12.30
	$18.30

Glover has a good discussion of scheduling problems.[14]

Evaluation of Alternatives

Evaluation of alternative methods is simpler when labor hr/unit are assumed to be constant; we want accurate predictions, however, so use a changing cost/unit. People often are disappointed by the high cost of a new method or model early in its curve when compared to the old method or model which had years of experience.

For example, consider a decision whether to manufacture an item in a factory (manufacturing)—or in the field (construction). Assume that construction at the site is a "little factory." Assume the real factory produces 10,000 units/yr and has an 80% progress curve. The decision is whether to make a 200 unit order in the factory or in the field.

The production ratio is 10,000/200 = 50. Call unit production cost in the factory = 1. Then, for the .8 curve, cost is multiplied by 1/.8 = 1.25 for each doubling. There are 5.65 doublings in 50 ($2^{5.65} = 50$) so cost is $(1.25)^{5.65} = 3.53$ times higher to manufacture in the field than in the factory. Fisher used this example to explain the high cost of nuclear power plants.[12] Due to economies of scale for heat recovery, it is desirable to build very large nuclear power plants; because the plants are so large, components are built on the site rather than in centralized factories and costs are very high.

Although the example is simplified, the calculations help explain the advantages of manufacturing vs. construction.

Acceptable Day's Work

For acceptable day's work evaluations, units produced/standard time should increase with experience. If a new operator's output is plotted vs. the typical learning curve for that job, it is possible to predict quite early whether an operator will ever make standard. See Knowles and Bell and Ghormley for examples of learning curves used to select operators[13,21]; see Nissley for a form used to give a daily target for operators as they progress along the curve.[28] Glover says good predictions can be made after two weeks.[16]

Assume a time standard, y, is set at experience level, x. Assume further, for ease of understanding, that $y = 1.0$ minute and x is 100 units. That is, a time study technician, Bill, made a time study on the first 100 units produced by Roger, calculated the average time and got 1.0 min/unit. What happens after Bill leaves and Roger continues working? See Table 24.7. Bill started the study at 7:30 a.m. and studied 100 units, getting an average time of 1.0 minutes. He stopped the study when Roger started his coffee break. Let's assume that 95% is the appropriate rate for this job for the first 5000 units. Roger completes his 200th piece shortly before lunch and his 400th before the end of the day. His average time/unit for the first day is about .90; he has performed at 111% of the "standard" time of 1.0 min/unit. The second day Roger completes his 800th unit early in the afternoon; he will complete his 3200 unit within a week of the time study. His average time/unit of .77 min/unit during the week was 129% of standard.

No time standard is accurate unless accompanied by a run quantity.

There are a number of points which can be made from the example.

First, assume that this unit is produced in lots of varying sizes depending on sales. Roger had a requirement of 3200 units. At .77 min/unit, the 3200 units took 41.1 hours; he turned in a reported time for the lot of 44 hours; since standard was 53.3 hours, his efficiency was 53.3/44 = 121%. The supervisor was happy and Roger was happy. Four months later, an order for 400 units is released to Bill O'Dell. Bill, who is trying to become a supervisor, worked very hard and produced them in .87 min; he reported 400 × .87/60 or 5.8 hours for the job; the

Table 24.7 Demonstration of the effect of learning on performance vs. standard. Even a small learning rate can have a major effect on performance. X = experience level of the operator when time study was taken; say 50, 100, or 500 cycles. The table gives time/unit based on a time standard of 1.0 min/unit; therefore if actual time standard was 5.0 min/unit, then time/unit at 98% and $2X$ would be .98 (5.0) = 4.9.

Learning curve, %	Time/unit at			Percent of standard at		
	2X	4X	32X	2X	4X	32X
98	.98	.96	.90	102	104	111
95	.95	.90	.77	105	111	129
90	.90	.81	.59	111	123	169
85	.85	.72	.44	118	138	225

foreman compares 5.8 with 6.67 standard hours and calculates his efficiency as 6.67/5.8 = 115% and says to himself good but not as good as Roger. For accuracy the foreman should have used a standard time of .77 min/unit for Roger and a standard time of .90 for Bill. Roger's real efficiency was 41.1/(.77 × 3200/60) = .94 while Bill's was 5.8/(.90 × 400/60) = .97. Thus, unless learning curves are considered, supervisors will have an incorrect idea of whether a specific operator is doing an acceptable day's work.

Second, the magnitude of the learning effect should be emphasized. Criticism is made of the accuracy of the rating procedure used in time study. Considerable effort and training are used to reduce errors to ±5%; that is, a person's pace might be estimated as 95% when in actuality it might be between 90% and 100%. However, this 10% range is minor compared to the errors which can occur if a standard time is used regardless of the run quantity.

The major cause of inaccurate time standards is failure to consider learning. Thought: If none of the operators in your plant ever turn in a time that improves as they gain experience, does that mean you have obtuse operators? Might it be stupid supervisors?

Incentives

If you pay by results, failure to recognize that labor time changes with output causes a number of problems. One is dissatisfaction of the workers as some get "easy" jobs (N is large) while others get "tough" jobs (N is small). Another effect is restriction of output. Rather than report output of 200% of standard (resulting in complaints of high wages from fellow workers and possible changing of the standard by the organization), the employees just increase their leisure time.

SUMMARY

Practice makes perfect. Both organizations and individuals improve with practice. Learning has been found in many different industries in a variety of countries. The amount of learning can be predicted for various situations. Time standards make the assumption that time/unit is constant vs. cumulative units produced. This incorrect assumption can cause major errors in cost allocation, scheduling, evaluation of alternatives, acceptable day's work, and incentive pay.

SHORT-ANSWER REVIEW QUESTIONS

24.1 How did Wright make a curve into a straight line?

24.2 What is meant by "garbage in, Bible out?"

24.3 Assume a Work-Factor time was .5 min/unit and 200 units were to be made on this release. If allowances are 20%, what is estimated standard time?

24.4 Assume a time standard was set as .05 hr/unit after observing the first 50 cycles. If the task has a 98% learning curve, what would be the average time/unit at 100 cycles, 200 cycles, 400 cycles, 800 cycles, and 1600 cycles? What percent of standard could be expected at each number if the operator worked at full capacity?

24.5 Is learning or inaccurate rating a greater potential source of incorrect time standards? Explain your answer.

THOUGHT-DISCUSSION QUESTIONS

1. Learning reduces the time/unit with experience so a time standard without a run quantity is soon likely to be in error. Why don't operators complain about time standards without a run quantity?

2. Discuss the relative effect of learning and pace rating as a cause of inaccurate time standards.

REFERENCES

1. Allemang, R. New technique could replace learning curves. *Industrial Engineering*, Vol. 9, No. 8 (1977): 22-25.

2. Andress, F. The learning curve as a prediction tool, *Harvard Business Review*, Vol. 32 (Jan.-Feb. 1954): 87-97.

3. Bapu, H. Learning curves for punch press operators. Project report to S. Konz, May 1967.

4. Barnes, R., and Amrine, H. The effect of practice on various elements used in screwdriver work. *Journal of Applied Psychology* (1942): 197-209.

5. Barnes, R.; Perkins, J.; and Juran, J. A study of the effect of practice on the elements of a factory operation. University of Iowa Studies in Engineering Bulletin 22, No. 387, 1940.

6. Carlson, J. Cubic learning curves: precision tool for labor estimating. *Manufacturing Engineering and Management* (November 1973): 22-25.

7. Carlson, J., and Rowe, A. How much does forgetting cost? *Industrial Engineering*, Vol. 8, No. 9 (1976): 40-47.

8. Conley, P. Experience curves as a planning tool. *IEEE Spectrum* (June 1970): 63-68.

9. Conway, R., and Schultz, A. The manufacturing progress function. *Journal of Industrial Engineering*, Vol. 10, No. 1 (1959): 39-54.

10. Crossman, H. A theory of the acquisition of speed skill. *Ergonomics,* (1959): 153-165.

11. Daniels, R. Factors affecting industrial learning on interrupted production schedules. MS thesis, Kansas State University, 1966.

12. Fisher, J. *Energy Crises in Perspective.* New York: Wiley-Interscience, 1974.

13. Ghormley, G. The learning curve, *Western Industry*, Vol. 17, No. 9 (1952): 31-34; Vol. 17, No. 10 (1952): 37-39; Vol. 17, No. 12 (1952): 47-49; Vol. 18, No. 2 (1953): 61-65.

14. Glover, J. Manufacturing progress functions I. An alternative model and its comparison with existing functions. *International Journal of Production Research*, Vol. 4, No. 4 (1965): 279-300.

15. Glover, J. Manufacturing progress functions II. Selection of trainees and control of their progress. *International Journal of Production Research*, Vol. 5, No. 1 (1966): 43-59.

16. Glover, J. Manufacturing progress functions III. Production control of new products. *International Journal of Production Research*, Vol. 6, No. 1 (1967): 15-24.

17. Hirsch, W. Manufacturing progress functions. *The Review of Economics and Statistics,* Vol. 34 (May, 1952): 143-155.

18. Hirschman, W. Profit from the learning curve. *Harvard Business Review*, 42 (January-February 1964): 125-139.

19. Kilbridge, M. Predetermined learning curves for clerical operations. *Journal of Industrial Engineering,* (1959): 203-209.

20. Kneip, J. The maintenance progress function. *Journal of Industrial Engineering*, Vol. 16, No. 6 (1965): 398-400.

21. Knowles, A., and Bell, L. Learning curves will tell you who's worth training and who isn't. *Factory Management & Maintenance* (June 1950): 114-115.

22. Konz, S. Learning curves for drill press operations. MS thesis, University of Iowa, August 1960.

23. Lipka, J. *Graphical and Mechanical Computation.* New York: Wiley, 1918.

24. McCambell, E., and McQueen, C. Cost estimating from the learning curve. *Aero Digest* (October 1956): 36-39.

25. Morooka, K., and Nakai, S. Learning Investigation, Report by Committee of Learning Investigation to Japan Society of Mechanical Engineers, February 1, 1971.

26. Nadler, G., and Smith, W. Manufacturing progress functions for types of processes. *International Journal of Production Research*, Vol. 2, No. 2 (1963): 115-135.

27. Nanda, R., and Adler, G. (Ed.) *Learning Curves Theory and Application,* Monograph 6. Norcross, Ga.: American Institute of Industrial Engineers, 1977.

28. Nissley, H. The importance of learning curves in setting job shop standards. *Mill and Factory,* Vol. 44 (May 1949): 119-122.

29. Noyce, R. Microelectronics. *Scientific American,* Vol. 237, No. 3 (1977): 63-69.

30. Riegel, P. Athletic records and human endurance. *American Scientist,* Vol. 69 (May-June 1981): 285-289.

31. Rohmert, W., and Schlaich, K. Learning of complex manual tasks. *International Journal of Production Research,* Vol. 5, No. 2 (1966): 137-145.

32. Salvendy, G., and Pilitsis, J. Improvements in physiological performance as a function of practice. *International Journal of Production Research,* Vol. 12, No. 4 (1974): 519-531.

33. Sheffler, F. Estimating for reinforced plastics. *Modern Plastics* (May 1957): 135-150, 243.

34. Smyth, R. How to figure learning time. *Factory Management and Maintenance,* (March 1943): 94-96.

35. Titleman, M. Learning curves—key to better labor estimates. *Product Engineering,* (November 18, 1957): 36-38.

36. Turban, E. Incentives during learning. *Journal of Industrial Engineering,* Vol. 19 (1968): 600-607.

37. Work-Factor learning time allowances. Ref. 1.1.2, Wofac Company, Moorestown, N.J., 1969.

38. Wright, T. Factors affecting the cost of airplanes. *Journal of Aeronautical Sciences,* Vol. 3 (February 1936): 122-128.

39. Youde, L. A study of the training time for two repetitive operations. MS thesis, State University of Iowa, 1947.

2

25

STANDARD DATA SYSTEMS

REASONS FOR STANDARD DATA

Rather than measure new times for each and every new operation, often it is desirable to *reuse* previous times—to use *standard data*. The reasons for reuse are cost, consistency, and ahead of production.

Cost

The cost of looking up a number in a table or solving an equation is less than the cost of taking a time study to determine the same number. For example, if we wished to know the time to reach 300 mm to a washer in a bin, grasp it, and move it to an assembly, we could make a time study of someone doing the task, analyze the data, and determine a number. Or, we could use *MTM* or Work-Factor (see chapter 8). The time study technique would take much more time and thus more money. The cost of using standard data, however, is (1) the cost of setting up the standard data system (a capital cost, a fixed cost) plus (2) the cost of using the system (an operating cost, a variable cost). To have economical total costs, the capital cost must be divided by many uses. Thus standard data are economically justifiable only for repetitive tasks or elements.

Consistency (Fairness)

Standard data are more consistent than one specific time study since they come from a bigger data base (that is, data from multiple studies on multiple operators). In statistical terms, $\sigma_{\bar{x}} = \sigma_x / n$. That is, the means of n individual time studies (represented by the σ_x) vary more than the means of the average of the studies (represented by the $\sigma_{\bar{x}}$). On any one time study, there is a rating and thus a possible error. But random rating errors tend to cancel over many studies. Thus, use of standard data for setting a specific standard minimizes any variation due to a specific operator or observer.

Although it seems heresy, in setting time standards it is more important to be consistent (small random error) than to be accurate (small constant error). (Error = "actual" time minus standard time.) It's nice, of course, to be both. For example, assume actual time for job A and B is 1.0 min/unit. A *consistent* system might set time for A at .90 and for B at .91, or time for A at 1.05 and B at 1.04. An *accurate* system might set time for A at .96 and for B at .99, or time for A at .99 and B at 1.03. See Figure 25.1 for a sketch of random and constant errors.

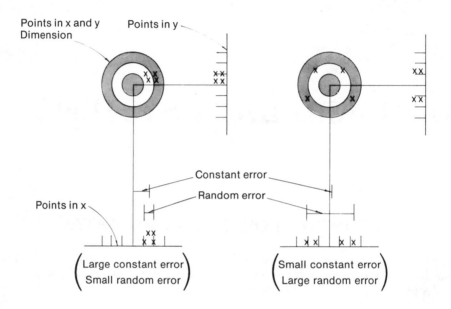

Figure 25.1 *Random and constant errors* are shown on a two-dimensional bullseye (Error = Bullseye—Hit location). The lower figures show random and constant errors in one dimension. Constant errors can be corrected by calibration. Random errors can be reduced by improved techniques, better devices, and training. Random error can be compensated for by increasing sample size; that is if you "fire enough shots," you can "hit the bullseye" since error approaches zero as *n* approaches infinity.

Constant errors are much easier to correct or handle administratively. For example if, for the whole department, the standard averaged .90 min and performance averaged 1.00 min, department efficiency is 90%. The supervisor can adjust production schedules, the cost accounting department can adjust costs, etc. But if standards varied randomly, then the supervisor would have a difficult time knowing which products would take more time, the cost accounting department would not know which product costs were wrong, and the supervisor would think some workers were working slower or faster than they really were. Perhaps even more important is that the workers would lose faith in the fairness of the time standards.

Ahead of Production

Standard data have a third advantage; they can be determined ahead of production. Timing with a stopwatch requires something to be timed. There must be an experienced operator, a work station with tools and equipment, and parts to be worked on. But times often are needed for cost estimates, for determining which alternative work method to use, or for determining how many machines to buy or workers to hire.

SYSTEM STRUCTURE

Levels

There are three levels of detail: micro, elemental, and macro. Table 25.1 gives an example of each level.

Table 25.1   Standard Time Systems Can Be at Three Levels of Detail

Micro system (typical component time range from .01 to 1 s)

Element	Code	Time		
Reach	R10C	12.9 *TMU*		
Grasp	G4B	9.1		
Move	M10B	12.2	27.8 *TMU* = 1 s	
Position	P1SE	5.6		
Release	RL1	2.0	1 s = .036 *TMU*	

Elemental system (typical component time range from 1 to 1000 s)

Element	Time
Get equipment	1.5 minutes
Polish shoes	3.5
Put equipment away	2.0

Macro system (typical component times vary upward from 1000 s)

Load truck	2.5 hours
Drive truck 200 km	4.0
Unload truck	3.4

Micro level systems are typified by *MTM* and Work-Factor. Times of the smallest component range from about .01 to 1 s. Components usually come from a predetermined time system rather than a time study.

Elemental level system components vary from about 1 to 1000 s. Components come either from micro level combinations or time study.

Macro level system component times range upward from about 1000 s (about .25 hr). Components come from elemental level combinations, from time study, from occurrence sampling, or even from employee activity logs.

Constant Vs. Variable Elements

Elements either can be constant (yes-no) or variable. Table 25.2 has examples of both constant and variable elements.

Accuracy

In some applications "estimated standards" (random error of ±20%) may be more cost-effective than "engineered standards" (random error of, say, ±5%).[2]

Table 25.2   Elements are of Two Types. Constant Elements Either Occur or Not; They Are "Go" or "No-Go." Variable Elements Vary; the Amount of Time Allowed Depends Upon the Amount of a Variable That Occurs.

Constant elements	Time, min
Get supplies	4.5
Put away supplies	8.0
Lubricate machine	5.0
Use hoist	2.0

Variable elements	
Walk	Time = f (distance, difficulty)
Insert pegs	Time = f (number of pegs)
Drill hole	Time = f (depth of material, drill rev/min)
Repair TV set	Time = f (number of defective components, complexity of defective components)

Estimated standards can use such crude input data as historical data and employee logs of time spent on various activities.[9,12] Analysis can be by multiple regression in which sophisticated statistical techniques are used to determine the *average* effect of various variables.[1,3,10]

CURVE FITTING

Data often can be presented concisely with one of the following formula forms (the summary comes from Lipka).[5]

The data may be a function of one parameter *(a)*; that is, a function can be fitted by a straight line through the origin (see Figure 25.2). If the line passes through the origin, the equation has the form $y = bx$ where b is the slope.

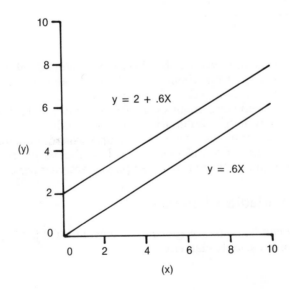

Figure 25.2 *Straight lines* of the form $y = bx$ go through the origin while $y = a + bx$ has a y intercept = a.

The second possibility is that the data are a function of two parameters *(a, b)*: There are five "simple" possibilities.

1. If the straight line does not pass through the origin, it has the form $y = a + bx$ where $a =$ the value of the y intercept.

2. The curve may be of the form $y = ax^b$ and is called a geometric curve. See Figure 25.3. If data can be approximated by an equation of the form $y = ax^b$, then the plot of (log x, log y) approximates a straight line. Generally you would plot *(x, y)* on logarithmic coordinate paper rather than plotting (log x, log y).

3. The curve may be of the form $y = ae^{bx}$, an exponential curve where $e =$ approximately $2.718281 \ldots$ is the base of natural logarithms. It also can be written as $y = ar^x$ where $e^b = r$. See Figure 25.4. If data can be approximated by an equation of the form $y = ae^{bx}$, then the plot of (x, log y) approximates a straight line.

4. The curve may be of the form $y = a + bx^n$ where n is known or suspected. If data can be approximated by an equation of the form $y = a + bx^n$, where n is known, than the plot of *(x^n, y)* approximates a straight line.

5. The data may be of the form $y = x/(a + bx)$ or $x/y = a + bx$; that is, a hyperbola with asymptotes $x = -a/b$ and $y = 1/b$. If the data can be approximated by an equation of the form $y = x/(a + bx)$ or $x/y = a + bx$, then the plot of *(x, x/y)* or of $(1/x, 1/y)$ approximates a straight line.

The third possibility is that the data are a function of three or more parameters *(a, b, c)*.

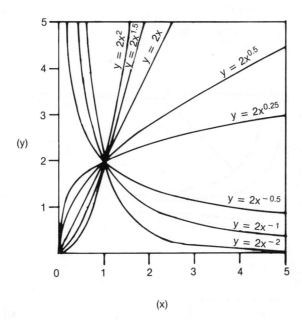

Figure 25.3 *Geometric curves* have the form $y = ax^b$. All six curves drawn have $a = 2$, but b has different values. The geometric curves with b positive all pass through the points (0,0) and (1, a); as one variable increases so does the other. The curves with b negative all pass through the point (1, a); they have $x = 0$ and $y = 0$ as asymptotes, and as one variable increases, the other decreases.

The equation $y = ax^b$ may be modified into $y = ax^b + c$. If b is positive, the new equation represents a curve with y intercept c; if b is negative the new equation represents a curve with asymptote $y = c$. See Figure 25.5.

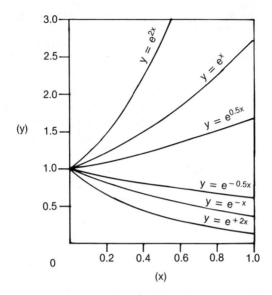

Figure 25.4 *Exponential or logarithmic curves* have the form $y = ae^{bx}$. All six curves drawn have $a = 1$, but b has different values. The curves all pass through the point $(0, a)$ and have $y = 0$ for an asymptote.

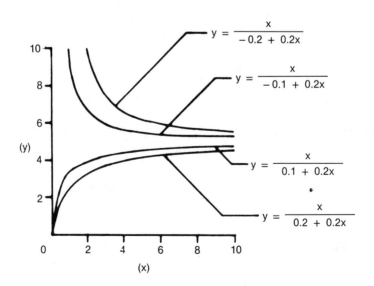

Figure 25.5 *Hyperbolas* with the form $y = x/(a + bx)$ or $x/y = a + bx$ have asymptotes $x = -a/b$ and $y = 1/b$. The four curves all have the value of $b = .2$ but a has different values.

The simple exponential $y = ae^{bx}$ may be modified into $y = ae^{bx} + c$. In the new equation, the asymptote $y = c$. See Figure 25.6

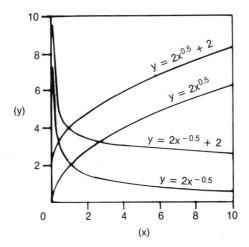

Figure 25.6 *Parabolas or hyperbolas* with a third constant have the form $y = ax^b + c$. All four curves drawn have $a = 2$, but b and c have different values. If b is positive, the parabolic curve has a y intercept $= c$. If b is negative, the hyperbolic curve has asymptotes $x = 0$ and $y = c$.

The equation of the straight line $y = a + bx$ may be modified into a parabola, $y = a + bx + cx^2$. This, in turn, is just the first three terms of a polynomial, $y = a + bx + cx^2 + dx^3 + e x^4 \ldots$ (You may recognize the form more easily if you mentally insert a $x^0 = 1$ as the term for the parameter a.) Each term beyond two gives one bend to the curve. That is, for just two parameters (a, b), the curve is a straight line; for three (a, b, c), the curve has one bend; for four (a, b, c, d), the curve has two bends, etc.

The equation $y = x/(a + bx)$ may be modified into the hyperbola, $y = c + x/(a + bx)$. In the new equation, $x = 0$ gives $y = 0$ while formerly $x = 0$, $y = c$.

A fourth possibility is that the data may be fitted by one equation over one range of values and another equation over a different range.

To analyze experimental data: (1) plot the data, (2) guess the appropriate curve shape, and (3) determine the constants for the selected curve shape. Computer routines generally are used to determine the actual constants.

Use a criterion minimizing the sum of the deviations (from the actual data value $[Y]$ to the predicted value $[Y_{est}]$) squared—least squares. Although it is possible to fit equations "by eye," the cost of computation now is so reasonable that the improved prediction accuracy is worthwhile. A strong advantage of the least-squares technique is that it reduces the amount of judgment involved. That is, once you have a given set of data and a given equation form, you always get the same answer. Unions like that.

Table 25.3 and Figure 25.7 give some example data fitted by a least-squares straight line, by a least-squares parabola, and by a least-squares exponential.

To fit a least-squares straight line,

$$Y = A + BX \tag{25.1}$$

solve the "normal equations":

$$\Sigma Y = AN + B\Sigma X$$
$$\Sigma XY = A\Sigma X + B\Sigma X^2 \tag{25.2}$$

Rather than solve two equations with two unknowns, an alternate computational format is:

$$A = \frac{(\Sigma Y)(\Sigma X^2) - (\Sigma X)(\Sigma XY)}{N\Sigma X^2 - (\Sigma X)^2} \tag{25.3}$$

$$B = \frac{N\Sigma XY - (\Sigma X)(\Sigma Y)}{N\Sigma X^2 - (\Sigma X)^2}$$

To determine how good the fit of the equation is to the points, calculate the standard error, the correlation coefficient, and the coefficient of determination. The standard error of estimate of Y on X is:

$$S_{Y.X} = \sqrt{\frac{\Sigma(Y - Y_{est})^2}{N}} \tag{25.4}$$

For $N < 30$ and straight lines, use the modified standard error

$$^sY.X = \frac{N}{N-2}\left(S_{Y.X}\right) \tag{25.5}$$

For parabolas, use N-3, for cubic, use N-4, etc.

Table 25.3 Calculations for three different equations to fit the data shown in Figure 25.7. The first step is to make the summations shown below the table. The second step is to calculate the As and Bs. The third step is to calculate the estimated value of Y for each X and the standard error and the coefficient of determination. Be sure to retain sufficient digits in the calculations.

Data			Estimated Y			Error ($Y - Y_{est}$)		
X, Points	Y, Min	$Z = \log_e X$	Straight line	Parabola	Exp.	Straight line	Parabola	Exp.
5	1.05	1.609	1.48	1.13	.96	−.43	−.08	+.09
9	1.70	2.197	1.67	1.55	1.62	+.03	+.15	+.08
14	2.10	2.639	1.92	1.94	2.11	+.18	+.16	−.01
19	2.20	2.944	2.16	2.28	2.46	+.04	−.08	−.26
27	2.70	3.296	2.55	2.77	2.85	+.15	−.07	−.15
36	3.10	3.584	2.99	3.19	3.18	+.11	−.09	−.08
38	3.40	3.638	3.09	3.27	3.24	+.31	+.13	+.16
46	3.40	3.829	3.48	3.53	3.45	−.08	−.13	−.05
52	3.80	3.951	3.78	3.67	3.59	+.02	+.13	+.21
57	3.70	4.043	4.02	3.74	3.69	−.32	−.04	+.01
					$S_{Y.X}$	.21	.11	.13

$\Sigma N = 10$

$\Sigma X = 303$ $\Sigma X^3 = 554{,}744$

$\Sigma X^2 = 12{,}201$ $\Sigma X^4 = 26{,}816{,}820$

$\Sigma Y = 27.15$ $\Sigma X^2 Y = 41{,}746.7$ $\Sigma Z = 31.507$

$\Sigma Y^2 = 81.3925$ $\Sigma Z^2 = 105.9405$

$\Sigma XY = 970.35$ $\Sigma YZ = 92.5792$

An easier alternate computational form for equation 25.4 (for straight lines only) is:

$$S^2_{Y.X} = \frac{\Sigma Y^2 - A\Sigma Y - B\Sigma XY}{N} \qquad (25.6)$$

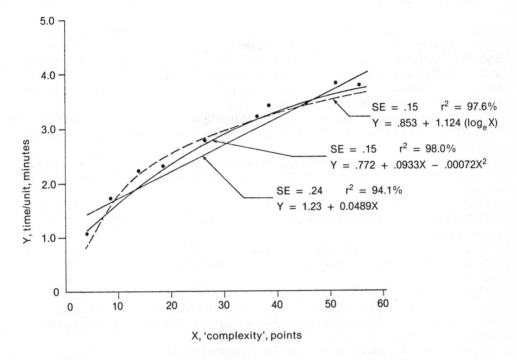

Figure 25.7 *Three curves* (a straight line, a parabola and an exponential) are plotted through the same data. The standard error is smallest for the parabola.

$$r = \pm \sqrt{\frac{\text{Explained variation}}{\text{Total variation}}} = \pm \sqrt{\frac{\Sigma(Y_{est} - Y)^2}{\Sigma(Y - Y)^2}} \qquad (25.7)$$

An alternate computational form (usable only for straight lines) is:

$$r = \frac{N\Sigma XY - (\Sigma X)(\Sigma Y)}{\sqrt{[N\Sigma X^2 - (\Sigma X)^2][N\Sigma Y^2 - (\Sigma Y)^2]}} \qquad (25.8)$$

The coefficient of determination is r^2; it gives the proportion of the total variation that is explained. By definition, r, the correlation coefficient (and thus also r^2, the coefficient of determination) cannot exceed 1. Use r^2 or standard error to determine which equation gives the best fit. With additional terms in the equation, more of the variation must be explained so r^2 must increase. Standard error, however, might decrease with additional terms as the degrees of freedom (the divisor) increases for each additional term. You must decide whether a more complicated equation is worth the increased accuracy.

To fit a least-squares parabola,

$$Y = A + BX + CX^2 \qquad (25.9)$$

solve the "normal equations":

$$\Sigma Y = AN + B\Sigma X + C\Sigma X^2 \qquad (25.10)$$

$$\Sigma XY = A\Sigma X + B\Sigma X^2 + C\Sigma X^3$$

$$\Sigma X^2 Y = A\Sigma X^2 + B\Sigma X^3 + C\Sigma X^4$$

Equations 25.10 can be remembered by noting they can be obtained by multiplying equation 25.9 by 1, X, and X^2, respectively, and summing on both sides. This least-squares polynomial can be extended to a least-squares cubic equation (add a DX^3 term to equation 25.9), quartic equation (add DX^3 and EX^4 terms), etc. For polynomials beyond a straight line, calculate standard error from equation 25.4 and correlation from equation 25.7.

To fit a least-squares exponential, approximate it by making a log (base e or base 10) transformation of the X values and then use equations 25.3, 25.6, and 25.8.

When a straight line is fitted to the data, the equation is $Y = 1.23 + .0489\ X$ with a modified standard error of .24 minutes and a coefficient of determination of 94.1%.

When a parabola is fitted to the data, the equation is $Y = .772 + .0933\ X - .00072\ X^2$ with a modified standard error of .15 minutes and a coefficient of determination of 98.0%.

When an exponential is fitted to the data, the equation is $Y = -.853 + 1.124\ Z$ with a modified standard error of .15 minutes and a coefficient of determination of 97.6%.

Therefore if the parabolic equation is used, the equation will predict the Y value (the time) within $\pm$ .15 minutes 68% of the time and within $\pm$ .30 minutes 95% of the time, assuming the sample continues to represent the universe! The parabola explains 98% of the total variability of the data.

Be careful not to tell the computer the wrong curve shape. If, for example, you tell the computer to fit a least-squares straight line to the data, the computer will fit a straight line to the data, even if a parabola, hyperbola, or exponential curve would be better. As a general policy you should examine the residuals between the actual points and the predicted points; if the residuals have a pattern, consider another equation form. For example, consider 10 points and a fitted straight line. If the deviations are randomly + or −, the line is satisfactory. But if the first 2 deviations and the last 2 deviations are + but the middle 6 are −, then a curve would give a better fit.

BUILDING THE STRUCTURE

The structure ("building") has three levels:

1. *Details of elemental standards* ("bricks" and "boards"). Method: Detailed descriptions such as *MTM* or Work-Factor motion analyses or time study method description. Time: Detailed time studies or predetermined time analyses.

2. *Elemental standards* ("walls" and "floors"). Method: Detailed description of work to be done. Time: Detailed elemental time tables or formulas from which time was obtained.

3. *The work standard* ("building"). Method: General statement of work to be done with location, part number, etc. Time: Totaled time. For purposes of

this chapter, it will be assumed that material exists at level 1; it is to be consolidated to levels 2 and 3.

- The first step in developing elemental standards is to classify the data in terms of similar types such as types of machine, types of material, or types of product. Later, when you are very familiar with the data, it may be possible to use data from diverse activities, but first get related items together.

- Second, break the job into elements; determine typical elements; get data for typical elements. If some jobs have elements that occur rarely, it may be worth setting a standard for these unusual elements. For example, replacing a muffler may take .5 hours on all cars except 1974 Zasmobiles or cars with over 150,000 km. One alternative is to give a blanket allowance to other jobs (say 34%) to "cover" the odd jobs.[8] Another is not to give a standard at all for these miscellaneous jobs.

- Third, list the major factors that you think affect the time. Try to use variables that are counted easily (pages of drawings, meters of welding wire used, number of pieces made, tons melted). If possible use a variable that will be known ahead of production. Then subdivide into constant elements and variable elements. Steffy and Darby recommend dividing variable elements into categories such as dimensional (part length, thickness), process (machining rev/min, distance from pan to machine), and material properties (tensile strength, difficulty).[11]

- Fourth, classify each factor as having a major or minor effect on time. Plot each factor vs. time (see the earlier section on curve fitting). See how well the data points can be fitted by a straight line, by a curved line, or by two straight lines with different slopes. Another possibility may be a "family" of curves. For example, cutting time on a lathe depends on part diameter and on cutting tool width. Put time on the y axis and part diameter on the x axis. Then plot the time for each cutting tool width as a separate curve. Several authors recommend the use of stepwise multiple regression on a computer as a preliminary screen to identify the important variables. You would list possible variables, such as part diameter and part diameter squared, log of part diameter, cutting tool width, cutting tool width squared, and type of tool holder as possible variables. (Using squares and logs as well as the variable itself allows for nonlinear curves.) The program then computes which variables are important in which combinations. The criterion is r^2, the ratio of explained/total variance.

- Fifth, examine the deviation of individual data points from the curve. If a point deviates too much (Steffy and Darby say 5% is too much), investigate it to see if it is an outlier. Outliers caused by reversals of digits and clerical mistakes can be eliminated when the mistake is found. If no mistake is found, it still may be possible to eliminate a point statistically (see Table 21.4).

- Sixth, make the decision. Does the equation have a sufficiently small standard error for your purposes? The allowable error depends on the cost of reducing the error and the cost of making a bad decision due to an inaccurate time. Before the advent of computerized curve-fitting routines and hand-held calculators to solve equations, it was necessary to keep most equations as straight lines. Giving an equation that required raising a value to the 2.47 power would get you laughed out of the office. Now it can be solved in 5 s on a hand calculator! On the other hand, don't get carried away with complex equations. If you can't put it on a graph, it probably is too complicated.

STANDARDS FROM THE STRUCTURE

Before the advent of hand calculators and computers, standard data formulas usually were solved for typical values and the results presented in a table. Now, due to these improvements in computing technology, results should be presented in formulas that can be solved for specific values. (The formulas should give ranges of values to avoid extrapolation beyond the data.)

Rather than just using the computer's power to solve equations, it also is possible to use its ability to make decisions (if given the decision rules), add and subtract values, add allowances, etc.; in short, to calculate a standard.[4,6,7] The computer program also can have diagnosis and error-checking routines built into the standard data program. Although there is a large capital cost involved in writing the original programs, these programs are being written and are available from consulting firms.

SUMMARY

To save money, reuse previous data for standard operations rather than obtaining new data (a new time study). Standard data tend to be consistent, even if not accurate, and are available for use ahead of production. With reductions in the cost of computation—calculators and computers—it now is economic to fit least-squares equations to data and to consider much more complex situations without an excessive cost of setting the standard. Although standard data times rarely are "perfectly" accurate, neither are time studies and a "quick and dirty" answer may be satisfactory.

SHORT-ANSWER REVIEW QUESTIONS

25.1 Give the three reasons for reuse of previous times—the use of standard data.

25.2 Give an example of random errors and constant errors. How do you correct random errors? How do you correct constant errors?

25.3 Give an example of a component from micro system, an elemental system, and a macro system.

25.4 Give the equation form for the dependent variable as a constant, as a straight line, as a parabola, and as a polynomial.

25.5 Give the definition of the coefficient of determination. Give the definition of the coefficient of correlation.

THOUGHT-DISCUSSION QUESTIONS

1. It is more important for time standards to be consistent than to be accurate. Discuss.

2. When fitting an equation to data, what should you use to determine which equation to use? Equation complexity? Correlation? Standard error?

REFERENCES

1. Bertoni, C., and Martin, J. Work measurement of purchasing. *Industrial Engineering*, Vol. 8, No. 1 (1976): 31-36.

2. Boyer, C. Work measurement: the flap over MIL-STD-1567. *Industrial Engineering,* Vol. 8, No. 11 (1976): 14-25.

3. Brown, J. How to measure group performance. *Industrial Engineering,* Vol. 1, No. 9 (1969): 14-16.

4. Johnson, T. Computer-generated wage incentive rates. *Industrial Engineering,* Vol. 5, No. 6 (1973): 10-13.

5. Lipka, J. *Graphical and Mechanical Computation.* New York: Wiley & Sons, 1918.

6. Martin, J. The 4M data system. *Industrial Engineering,* Vol. 6, No. 3 (1974): 32-38.

7. Murphy, R. A computerized standard data system. *Industrial Engineering,* Vol. 2, No. 10 (1970): 10-17.

8. Shurtz, S. Maintenance standards in a public utility. *Industrial Engineering,* Vol. 7, No. 11 (1975): 14-17.

9. Smerilson, H. Standards for engineers. *Industrial Engineering,* Vol. 7, No. 10 (1975): 12-17.

10. Steffy, W., and Ashby, D. A new approach to foundry production standards. *Foundry,* 93 (September 1965): 66-69.

11. Steffy, W., and Darby, D. Computer-generated time standards. Technical Report, Industrial Development Division. Ann Arbor: University of Michigan, 1968.

12. Van Kirk, R. Work standards for highway design engineers. *Industrial Engineering,* Vol. 5, No. 1 (1973): 34-39.

PART VIII
IMPLEMENTATION
OF DESIGN

26

JOB INSTRUCTION

PROBLEM

Use job instruction whenever the job and the workers' knowledge of the job do not match. In instruction-aided work, the operators always have a detailed procedure (to reduce memory requirements) or information aids (to reduce calculations and improve decisions) before them; in trained work the operator memorizes the procedure or calculates the information. With aids and with training, the goal is to change behavior—to cause specific responses to specific stimuli. You might train a secretary to run a duplicating machine or a mechanic to adjust the brake pedal; you might use instruction aids to fix a broken duplicator or to assemble a unit with 200 components. An instruction aid does not eliminate the need for training, it just reduces the training required.[29] Using a computer memory to store ticket reservations and calculations reduces the training required for the clerk, but some training is still necessary. An important point is that investment in an aid is not lost if an individual changes jobs. The usual solution is some training and some aids rather than all one or the other.

Instruction and training requirements often are underestimated because they are considered necessary only for organizations expanding the total number of employees. However, they can be required if (a) the job changes (e.g., a different machine speed is used, a different raw material is used, or the work content is allocated to a different work station), or (b) the job does not change but the worker changes.

Change in the worker occurs with a static total work force as individual employees quit, go on vacations, get promoted, get demoted, are fired, or are absent for the day.

Change in the worker also occurs if the total work force is expanded. Since the new jobs are at all work levels but the new employees usually begin with the bottom job levels, the existing employees begin a chain reaction of job shifting so that 10 new positions may require 40 workers to learn new jobs.

Change in the worker occurs even with reduction in the work force. Considerable instructional requirements are created by *bumping*. The bumping concept is that the specific individual laid off depends on the time worked in a specific department (department seniority), plant (plant seniority), or even company (company seniority). If a company eliminates the second shift in a plant, a drill press operator on the second shift with high seniority may bump a drill press operator on the first shift. The first shift operator bumps an inspector; the inspector bumps a material handler, and the material handler is laid off. A typical value for the ratio of people in new jobs to persons laid off is 3; the exact number depends on the extent of the seniority (lower ratio for department seniority) and the job proficiency requirements (lower ratio for stiffer requirements). Permitting an employee a period of time (say 90 days) before proficiency on the new job must be demonstrated gives a very high ratio (as well as very low work quality during this period).

Unfortunately, management may not realize that instruction is inadequate, because poor cost accounting systems conceal poor job instruction. Poor job instruction primarily affects quality rather than gross output. Cost accounting systems usually do not detect changes in quality costs—especially since no supervisor likes being "chewed out" for poor quality. That is, supervisors hide quality problems from the cost accountants and charge quality costs to a wide variety of "innocent" accounts. Thus no "flag" shows the poor quality resulting from inadequate instruction.

INSTRUCTION AS A COMMUNICATION CHANNEL

Figure 26.1 shows job instruction as a communication channel.

RECEIVER	SENSOR	MESSAGE	TRANSMITTER	SOURCE
Radio	Electronic circuit	Music	Broadcast transmitter	Tape
Trainee	Eye, ear	"2 + 2 = 4"	Book, voice	Teacher

MEDIUM
Radiowaves
Light & sound waves

Figure 26.1 *The comparison* of two communication systems shows (top) broadcasting music to a radio vs. (bottom) a teacher showing a trainee that "2 + 2 = 4."

Receiver-Sensor

Should the employee receive the message with audio, visual, or kinesthetic sensors?

Learning by doing (primarily kinesthetic input) has a long-established reputation for success. "Hands on" experience results in learning with long retention. For example, most teachers feel that students learn more when they write notes rather than when they underline or just read words. Motor movements, such as riding a bicycle, are very resistant to extinction. Use kinesthetic input for highly repetitive psychomotor movements in which "mental" processing becomes automatic.

Visual input is best for complex "mental" information, due to the ability of the eyes to recognize a wide variety of patterns in both time and space. Reading is an example of recognizing spatial patterns. The eyes interpret very complex time and space interrelationships, e.g., in vehicle driving.

Use auditory input for simple "mental" information such as *stop* or *go, right* or *left;* it is unreliable for complex situations. (AVO: Avoid Verbal Orders.) By use of pitch as well as duration and intensity, a sound can be used to represent up to 10 different situations, but the flexibility and information content of auditory signals are far less than for visual signals. If verbal orders must be used, have the receiver repeat them back to the sender, preferably in different words. Table 26.1 shows that, for a specific laboratory, purely verbal communication (the telephone) was used 7% while visual (reading and writing) was used 26%. Face-to-face, at 35%, is a mixture of visual and auditory. Perhaps most striking is that 68% of "work time" was spent communicating.

For most job-instruction situations use the eye.

Table 26.1   Percent of Time Spent by Technical, Supervisory, and Clerical Personnel in a Research and Development Laboratory; 3100 Observations[15]

Communicating	68	
Face-to-Face		35
Telephone		7
Reading		12
Writing		14
Equipment use	16	
Laboratory		13
Office		3
Other	16	
	100	

Medium

There is a variety of media for both the auditory and visual channels.

Audio messages are transitory; the message does not remain for inspection but "decays." Nonrecorded messages can be misunderstood through momentary lapses of attention, environmental or electronic noise, or just too rapid an information-presentation rate. Tape recording permits replaying, but most messages are not recorded. Even if a recording is available, employees may not replay it since (1) they may not realize that they misunderstood the message or (2) they do not understand how to use the recording equipment. Recorded audio messages have a fixed rate of presentation that forces the listener to receive the information only at that speed. Constant speed causes problems for those with poor command of the language. On the other hand, more proficient listeners become bored with slow rates of information presentation. For psychomotor tasks such as typing or assembly, work will be slowed down due to the slow rate of information presentation. For problem solving involving two people communicating,

Chapanis demonstrated that solution time averaged 15 minutes when the voice was one of the communication methods; without the voice, solution time was about 30 minutes.[5]

In most job-instruction situations the medium should be visual: demonstrations, slides, movies, TV, photographs, microfilm, models, or printed instructions.

Moving or still? Moving pictures (TV or movie) show sequences such as hand-arm motions in both real time and slow motion. Still pictures have the advantage that the workers can view them at their own pace; in addition, referring to the previous picture is simple. Both backing up and freezing the image are possible with TV and movies, but viewers may not do either due to reluctance to use the equipment or because they are in a group. Moving pictures stimulate bored people more than a series of stills.

Using a physical model of the object rather than a slide of the object or a photograph of the object does not give an information transmission advantage—if the photograph or slide gives all the relevant information.[7] Poor focusing, poor reproduction, or omission of an important detail make the physical model superior to the abstraction because a model can be turned upside down, be viewed in three dimensions, be brought closer to the eyes for a better view, and even be disassembled. When using a physical model in several assembly stages, conserve space by mounting the models on a turntable. However, photographs permit views that are difficult to obtain with models (i.e., magnification, views with important features emphasized). Photographs use little physical space and can be duplicated.

Color has more informational content that does black and white. Color stimulates bored viewers. Yet the extra expense of color is not necessary for many messages such as layout of a work station or a printed message such as "use acid flux" because the brightness contrast is sufficient and color adds no additional information. In fact, "artistic" use of color (such as yellow letters on a brown background or green letters on blue) degrades contrast and reduces legibility. Poor legibility also is disliked.[17]

Multiple channels for instruction-aided work are difficult to justify on a cost-effective basis. Dickey and Konz found that audio supplementation of visual work instructions for an assembly task caused less successful performance than purely visual instructions.[7] The problem was that the tape-recorded audio instruction was presented at a fixed rate. For the initial cycles, it was too fast in relation to the worker's ability. After practice the workers had to wait for the messages. For the visual message without audio supplementation, productivity was not restricted by the fixed pace of the audio instruction.

The reader should be careful not to overemphasize mechanical devices. Our culture puts a heavy emphasis on mechanical techniques and de-emphasizes investigation of communication without "hardware"[8]; that is, we emphasize "hardware" not "software," the "medium" not the "message."

Message

Good instructional material starts with identification of what needs to be communicated. Then consider who the audience is; consider the most-competent user and, especially, the least-competent user.

All messages are codes; codes must be decoded. Since many codes are easy to decipher (and for instructional purposes that is the way it *should* be), the decoding (translation) problem is often forgotten. A pictorial message requires the least translation. (A word equals .0001 picture.) For example, show an operator the correct location of solder joints by showing a picture of the assembly with circles about the solder joints; teach how to load a camera with a picture showing the loading operation. Translation from the message to the task is simple and direct.

Be more visual than verbal. Instruction in words rather than in pictures requires the words to be translated. *Employez le tournevis.* A French reader knows this means "Use the screwdriver." There are many situations in which workers may speak or read a language other than the primary language of the country (Spanish in many sections of the USA, English in the French-speaking areas of Canada or French in the English-speaking areas, and Turkish, Arabic, Greek, or Italian by many of the "guestworkers" in Northern Europe). "They haven't spoken English in America for years." Write sentences in active, affirmative ways; avoid negatives and passives. Keep the main topic of the instruction early in the sentence.[4]

The vocabulary of university graduates is not the same as the vocabulary of most workers. Most industrial workers are not comfortable with polysyllabic vocabulary typified by "subsequent, prior, chartreuse, incorporate, and simultaneously"; they like little words such as "after, before, blue-green, put-in, and at the same time." (Eschew obfuscation.) See Table 26.2.

Calculate the years of schooling required to read your writing from the following formula developed by the Gunning-Mueller Clear Writing Institute. People prefer to read below their level. *TV Guide,* the nation's biggest circulation magazine, has a value of 6; the *Wall Street Journal, Time,* and *Newsweek* average 11.

$$GL = .4\,(A + P) \tag{26.1}$$

where GL = Grade level, years

A = Average words per sentence. Treat independent clauses as separate sentences.

P = Percent of words with 3 or more syllables. For P, omit (a) capitalized words, (b) easy combinations such as screwdriver and guestworker, and (c) verbs that reach 3 syllables with the addition of es or ed.

A simplified Flesch reading ease formula is[14]:

$$GL = .4\,A + 12\,S - 16$$

where S = Syllables/word

Table 26.2   Use Simple Words

Bad	Better
parameters	values, variables
scrutinize	look at
incorporate	put in
verification	check
precede	before
prior	before
facilitate	help, make easy
subsequent	after
simultaneously	at the same time
via	by
inquire	ask
equivalent	equal
simulate	pretend

Tabular formats are superior to narrative formats; flowcharts give fewer errors than do narrative formats[13]; see Table 26.3 for an improvement over the following text.[28] Paper is cheap so don't condense the tables or force the user to make calculations.

When time is limited, travel by Rocket, unless cost is also limited, in which case go by Space Ship. When only cost is limited, an Astrobus should be used for journeys of less than 10 orbs, and a Satellite for longer journeys. Cosmocars are recommended, when there are no constraints on time or cost, unless the distance to be travelled exceeds 10 orbs. For journeys longer than 10 orbs, when time and cost are not important, journeys should be made by Super Star.

Keep information in specific behavioral terms ("With the left hand, hold the latch on the left side. With the right hand, insert key 147 and turn clockwise") rather than generalities ("Unlock box"). A poor instruction is "Inspect nameplate to ensure that it has been properly installed." Better is "Inspect nameplate. Location should be 75 mm from top and 100 mm from left side of cover. Nameplate should not be able to be pried up with fingernails."

Transmitter-Source

Expensive equipment often greatly increases job instruction cost.

Table 26.3 Decision Structure Tables Communicate With Fewer Errors Than Text or Decision Diagrams (Boxes Connected by Lines). Use of the Connective Words (if, and, then) and Verbs (is) Aid Communication. Single and Double Lines Help Also.[28]

If TIME is	and COST is	and JOURNEY LENGTH is	then TRAVEL MODE is
limited	unlimited	any	Rocket
limited	limited	any	Spaceship
unlimited	unlimited	less than or equal to 10 orbs	Cosmocar
unlimited	unlimited	more than 10 orbs	Superstar
unlimited	limited	less than or equal to 10 orbs	Astrobus
unlimited	limited	more than 10 orbs	Satellite

Mechanical equipment (TV recorders, slide projectors, movie projectors, tape recorders, and filmstrip projectors) are most suited for classroom-type presentations to groups. The equipment requires electrical power, considerable space, and very considerable maintenance.

For individualized instruction, it is very hard to surpass the printed or written page and the photograph. These communication aids are especially valuable on the shop floor because multiple copies can be made for each worker, no power is needed, storage is compact, and no instruction is necessary for equipment use.

In learning, there is considerable evidence over the last 70 years that, for the same total hours of practice, distributed practice (short training sessions) produces faster learning than massed practice (long sessions). Braddeley and Longman, for example, emphasize the importance of the amount of training/day and found one session of one hour/day was better than one session of two hours or two sessions of two hours.[3] It is easier to learn a single standard procedure (even if it is longer and more complex) than a variety of short ones for various circumstances. Ideally, equipment could be operated both with the long standard procedure and the short one (used by skilled operators).

TRAINER

Who

Let us assume that the correct work method has been developed. However, just because the engineer or supervisor know what is to be done does not mean that the operator knows. Assume we wish to have the worker memorize a specific procedure rather than continually consult job instructions such as those in a technical manual. Transfer of knowledge—training—can be done by a number of different people and methods.

Transfer of knowledge requires: (1) knowledge of the subject by the trainer, (2) proper teaching materials and technique, (3) sufficient time for training, and (4) a trainee able and willing to learn. Assume condition 4 is not a problem. We will consider four possible trainers.

One possibility the British call "sit by Nellie"; that is, learn by observing a fellow worker. But "Nellie" may not know the proper work method or have proper teaching materials, proper teaching techniques, or sufficient time to devote to training a fellow worker.

Second, let the supervisor do the training. Many of the same problems remain with the special problem of lack of time. Supervisors often do not have enough free time to devote to training. In theory, training is one of the primary responsibilities of every supervisor; in practice, supervisors usually do not have sufficient time, and training is omitted.

Third, use in industry the technique used in public education. Get a teacher knowledgable in the subject; give the teacher the proper teaching materials and allow sufficient time for the transfer. In public education, students are grouped in classes so that the teacher's time is prorated over a large number (say 25) of students to cut the cost of the teacher. The student's time is assumed to have a low value. In industry, however, the student's time costs the organization wages. In addition, the need usually is for one or two trainees for a specific skill within a short time period. In some industrial situations teacher costs can be prorated. A session on company medical benefits can be given to all new employees regardless of the specific skill they will use at work. Another example is upgrading skills of existing employees after working hours (typists might be trained in dictation). After learning the new skill, they have the potential to be promoted once a job opening is available.

The fourth approach is not to prorate the teacher's time over more students but to automate the teacher's job—eliminate the teacher. "Self-service" is well accepted in many facets of our life. To make self-instruction work, use unambiguous material and clear transmission, as there is no person to answer questions.

Programmed instruction attempts to solve these problems of self-service.

How?

Programmed instruction began with B.F. Skinner, a psychologist who wanted to help his country during World War II. What can the military do with a Ph.D.

psychologist who trained rats to go through a maze? Well, it seems that there was a group with a problem. They had determined that the reason that bombs dropped above ships didn't hit the ships very often was that the ships turned after the bomb was dropped. The solution: have the bomb turn also. The field of electronics was not yet developed sufficiently to do this remotely, so the "pilot" had to ride the bomb.

Although the Japanese used people as pilots, the Americans decided to use pigeons. The pigeon would look at a TV picture of the target ship, sent by a camera in the nose of the bomb. Then the pigeon would peck at the ship and, by sensing where the screen was being pecked, the bomb could be steered! But who was going to train the pigeons? Enter Skinner!

The war ended before the project succeeded. Then one day in 1947 Skinner came home furious after visiting his daughter's kindergarten class. "I give my pigeons better training than my daughter gets." Skinner then began to develop programmed learning for humans.

How had Skinner trained pigeons?

Teach a pigeon to turn a circle clockwise. First, watch until it turns part of a circle clockwise, then give it a piece of grain. Continue giving grain for turning but gradually require more and more of the circle for the reward (i.e., shape its behavior).

Embedded in this example are a number of principles:

1. Specifically define the exact behavior you wish the trainee to do (turn a circle clockwise).

2. Present the information in small steps, increments, or modules (small arcs first).

3. Present information at a pace determined by the trainee.

4. Give immediate feedback of the correct answer—knowledge of results—to the trainee (grain for the correct movement).

5. Observe behavior of the trainee. If not satisfactory, modify program.

Research indicates that, when applied to humans, steps 2 through 4, although desirable, are not essential.[20]

Step 1 means that we must specifically define exactly which motor movements we want. Glittering generalities such as "know the job" or "understand the process" are replaced by "if the temperature on gauge 7 rises above 115 C, turn the red switch on column 7 to 'off' and call your supervisor." Give training programs the same level of attention to detail as you do to computer programming; much poor teaching is simply insufficient attention to detail.

Step 2 says that information is more digestible in small bites. The best size of the bite may depend on the appetite.

Step 3, self-pacing, is a key concept. Let trainees receive the message at their own rate rather than at some rate appropriate for the "average." Since, by definition, only 20% of any group is within the 40th to 60th percentile, a message delivered at an average rate is at an inappropriate rate for 80% of the trainees.

Step 4, immediate feedback to trainee, is desirable though not essential. Trainees with greater memory abilities may be able to absorb considerable information and then later match the answer with the question. Delay of the correct answer means that the trainee might miss a key early point and thus be mislead on a considerable amount of subsequent material.

Step 5, immediate feedback to the trainer, is very important (feedback distinguishes an efficient "closed loop" system from an "open loop" system in education just as it does in electronics).

Yet the key concept has not yet been mentioned. There are no stupid students, just terrible teachers. It stems from Skinner's belief that he was smarter than any pigeon. If the pigeon did not learn, it was not the pigeon's fault, it was up to Skinner to develop a better teaching method. This concept, that poor learning is the teacher's fault, is hard for many teachers to accept. They retort that some people could never learn nuclear physics no matter how long they studied; they have neither the capability nor the motivation. True. Use the "no stupid students, just terrible teachers" concept to keep yourself on your toes and doing the best possible job rather than looking for excuses.

INSTRUCTION-AIDED WORK

Instruction Aids

Instruction aids can be subdivided into procedural and psychomotor.

Present procedural ("big picture") information in decision structure tables such as Tables 26.3, 26.4, and 26.5 or in a routing sheet such as in Table 26.6. These tables give various decisions that have been made about processes, such as feed and speed to use. It is important that these decisions be made in advance by experts rather than by inexperienced operators under time pressure and without knowledge of all the relevant information. The tables also serve as a convenient storage location for tool numbers, standard costs, etc. and also provide a good training aid. In CAM (Computer-Aided Manufacturing), the decision structure tables, routing sheets, and other information are stored in a computer. The computer permits an easily updated central reference point, complex decision structure tables, and facilitates comparison with other similar parts (group technology).

Table 26.4   One Explicit Form of Written Standard
Instruction is a Decision Structure Table. This Specific Example Gives the
Check-Cashing Policy of a Grocery Store.

If TYPE OF CHECK is	And AMOUNT OF CHECK is	And BANK OF CHECK is	And CUSTOMER ADDRESS is	Then CUSTOMER IDENTIFI-CATION REQUIRED is	And ON BAD CHECK LIST	Then DECISION is
Two-Party	Any	Any	Any	—	—	Reject
Company	Up to $25	Any	—	0	—	Accept
	$25.01 to $200	Any	Any	1	—	Accept
	$200.01 and up	—	—	—	—	Reject
Personal	Up to $25	Local	Local	1	yes / no	Reject / Accept
			Out of Town	1	yes / no	Reject / Accept
		Out of Town	Local	1	yes / no	Reject / Accept
			Out of Town	2	yes / no	Reject / Accept
	Over $25	Any	Any	—	—	Reject

Table 26.5 An Example of a Decision Structure Table As Applied to Select Either the Drill Size Before Tapping or the Clearance Drill Size When You Wish a Hole Large Enough so That the Drill Will not Touch the Bolt Threads

National Special Thread Series			
If BOLT DIAMETER, INCHES is	And THREADS PER INCH is	Then DRILL SIZE FOR 75% THREAD is	Then CLEARANCE DRILL BIT SIZE is
1/16	64	3/64	51
5/64	60	1/16	45
3/32	48	49	40
7/64	48	43	32

Table 26.6 Routing Sheets (Also Called Operations Charts) Vary Depending on the Organization, But the Table Below is Typical. Operation Numbers Are Given in Multiples of Ten So That Operations Originally Omitted Can be Added Without Changing the Numbers of the Other Operations. Some Routing Sheets Also Include Bill of Material Information.

Part name Punch   Part number 541-675   Raw material 1040 10 mm round

Operations SK   Date 20 Jan 75   Used on Model 80

Operation Number	Operation Name	Machine	Tooling	Feed, mm/rev	Speed, rev/min	Hr/unit	Remarks
10	Turn 4 mm dia Turn 3 mm dia	J & L T. Lathe	# 642 Box	.225	318		
	Cut off to length	J & L	# 6 cutoff	Hand	318	.008	
20	Mill 5 mm radius	#1 Milwaukee	Tool 84	Hand		.004	
30	Heat-treat	#4 furnace				.006	
35	Degrease	Vapor degreaser					
40	Measure hardness	Rockwell tester				.002	
50	Store						

Present psychomotor ("fine-grain picture") information in pictorial instruction sheets such as in Figure 26.2 or present this information on videotape.[2,22] Emphasize Grasp and Position elements because these are the "skill" elements; Move and Reach are less important. In general, pictures without supplementary words are better than words without supplementary pictures but pictures plus words are best of all. Words can explain the "why" as pictures give the "how." Element breakdowns (as with *MTM*) communicate concepts to other engineers but don't give the worker the necessary details. If you're using videotape, have the operators add the supplementary words later when viewing their own actions. If you're using photographs, use instant photographs stapled on a sheet of paper. If using a sketch, be sure it is realistic; many people have difficulty in understanding drafting conventions so isometric is best.[23] See Table 19.6. Have the operators write their supplementary remarks on the paper; that is, the people who use the technical manual should help write it. Make a photocopy for the office records,

Figure 26.2a *When cutting a two-crust pie* the first step is to stack enough plates to fill the tray.

Figure 26.2b *Holding* a marker by the edges makes it easier to center:
1. More accurate
2. Hands don't obscure center of pie

Figure 26.2c *Use the edge* of the pan as a guide to position marker in center of pie.

Figure 26.2d *Press* marker lightly on crust. Only guide marks are needed.

but keep the best copy at the machine. Laminate or put in a loose-leaf folder to keep off grease, etc. Leave space for changes. Avoid the "gold-plating" of having professional photographers, typed instructions, elegant mounting, etc. If an elegant copy is desired, wait until the instructions have been polished by the users, and count on several drafts (most first drafts are poor). These more polished instructions can be justified if the procedure is a "standard manufacturing process" and the instruction widely distributed. Especially important is that the instruction sheet remain at the point of use—not in a file cabinet in the office.

Human judgments are most accurate when they involve direct comparisons rather than use of the memory. Harris and Chaney demonstrated that photographic limit aids were useful in reducing inspector errors.[12] They recommend the following steps: (1) establish a panel of experts to judge the quality, (2) have the experts evaluate quality on a sample of items near the borderline (include various types of possible defects), (3) mount or photograph (inspector's eye view) the items with emphasis on the borderline items, (4) give each inspector a set.

DESIGN OF GRAPHS, FORMULAS, TABLES, AND PRINT

The following recommendations can be used for training (where the operator memorizes the procedure) or for work instruction aids (where the operator depends on the aid without help from a teacher). Without the feedback from a teacher, good design is even more important.

Information can be presented as graphs, formulas, tables, or in print.

Graphs

Graphs and bar charts give general relationships but are quite poor for detailed values. Figure 26.3 shows walking time as a function of distance. Figure 26.4 is an alternate form; the relationship between the two variables is obscured but conversions from one scale to the other are made more accurately.

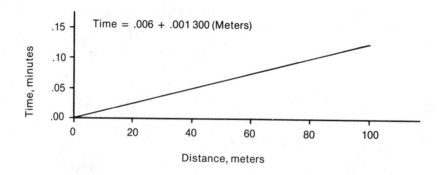

Figure 26.3 *General relationships* such as up or down are shown by a graph or curve but they are difficult to use for accurate values. Graphs (giving the line or data points) are superior (for both time and errors) to bar charts (where the data point is the end of a column from the axis).[25]

Characteristics of good graphs are:

1. Give units for both axes.
2. Mark subdivisions of scales (tick marks) on axes—especially important for scales with non equal intervals as log scales or probability scales.
3. Make scales in units of 5 or even units of 2. Use 2, 4, 6, 8 . . . or 5, 10, 15, 20 . . . ; avoid 1, 3, 5, 7 . . . or 1, 4, 7, 10 . . . Worst of all are computerized scales such as 1.31, 2.62, 3.93, etc. resulting from a programming decision to make the figure use the full paper size.
4. Use brief labels rather than keys.[18]
5. Indicate data points with open circles or triangles. Not as good are filled circles and triangles. Worst are dots, crosses, and Xs.

Formulas

Formulas permit exact calculations; use them for multiple-variable information (such as the modified DuBois formula for surface area for the human body):

$$SA = .208 + .945 (.007\ 184\ H^{.725}\ W^{.425}) \tag{26.1}$$

where SA = body surface area, m²

H = height, cm

W = weight, kg

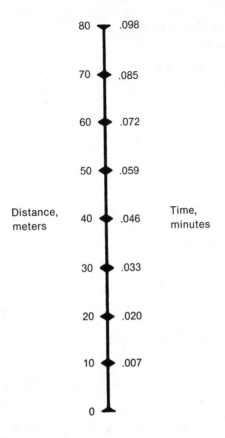

Figure 26.4 *Conversion lines* reduce the distance from the scale to the line that occurrred in Figure 26.3; thus it should improve accuracy but at the loss of whether the relationship is a straight line or a curve.

In the past people avoided formulas since calculations required pencil and paper, slide rules, or logarithms. The computer, desk calculator, and personal calculator have reduced these problems; consider the formulas rather than automatically rejecting them.

To reduce computational errors, present the formula in the units that the user will enter. The formula for Figure 26.3, for example is:

$$T = .006 + .001\ 300\ (D) \tag{26.2}$$

where T = time, minutes

D = distance, m

If the users want their answer in minutes, don't give the following formulas since there is a chance for conversion errors:

$$T = .000\ 100 + .000\ 063\ (D) \tag{26.3}$$

where T = time, hours

D = distance, feet

or

$$T = .000\,100 + .000\,021\;(D) \tag{26.4}$$

where T = time, hours

D = distance, yards

Reduce decimal point errors by presenting numbers in groups of three with an intervening space:

$$T = .006 + .003\,800\;(D) \tag{26.5}$$

rather than

$$T = .006 + .003\,8\;(D) \tag{26.6}$$

or

$$T = .006 + .0038\;(D)$$

Tables

In most situations, tables are best—whether numerical (e.g., time tables) or non-numerical (store directories indicating the floor for merchandise). Following are seven principles of good table design.

1. Use explicit tables, not implicit tables. Explicit tables give all the information directly and do not require the user to make calculations. Implicit tables require the user to make calculations. Table 26.7 is an explicit table of the normal distribution. In the implicit form only the upper half of the table is printed; the reader is assumed to know that the distribution is symmetrical and to be able to "mirror" calculations. Table 26.8 shows explicit and implicit forms for converting F to C; in the implicit form you must make additions. In other implicit forms, you must interpolate.

   Implicit tables save paper but take more time to use and have a higher error rate. The general public (that is, a casual user of the table) and people with less intelligence (by definition, half of the population is below average) find implicit tables very difficult to use.[26,27] Use explicit tables.

2. Avoid matrix tables, especially matrix tables on a diagonal. Table 26.9 is a travel chart giving the distance between various departments. If a matrix must be used, present information vertically and horizontally as in Table 26.10. Better is a linear (one-dimensional) table as Table 26.11. Save space by "folding" linear tables. The telephone book is an example. For the general public, use explicit linear tables.

3. Arrange the items in a column rather than a row to reduce search time.[26]

4. Divide items in a column into groups. Groups of 5 are better than groups of 10 which are better than no groups at all.[24] "Logical" grouping as in Table 26.7, if possible, is desirable.

5. Search left to right (the normal direction of reading); Figure 26.8 gave degrees C in the left column and degrees F in the right. In indices (such as parts lists which might include part number, part name, next higher assembly number, quantity used, etc.), put the indexed item (such as part number) on the left. Then put the next columns in order of importance to index user. If next higher assembly number was what was looked for, put it next; if part name was the least-used information, put it in the far right column.[11] Search direction is not critical if the same person uses the same table repetitively.

Table 26.7   An Explicit Table of the Normal Distribution; in an Implicit Table Only the Negative or Positive Values of z Would Be Given

Number of standard deviations, z	Cumulative area from negative infinity
−3.0	.001 300
−2.33	.01
−2.0	.022 800
−1.96	.025
−1.64	.05
−1.28	.10
−1.00	.159
−0.80	.212
−0.60	.274
−0.40	.345
−0.20	.421
0	.5
+0.20	.579
+0.40	.655
+0.60	.726
+0.80	.788
+1.00	.841
+1.28	.90
+1.64	.95
+1.96	.975
+2.0	.977 200
+2.33	.99
+3.0	.998 700

Table 26.8   Examples of Implicit and Explicit Tables; the Less Desirable Implicit Table Saves Space But Requires the User to Make Calculations

Implicit format				Explicit format			
Degrees, C	Degrees, F	Degrees, C	Degrees, F	Degrees, C	Degrees, F	Degrees, C	Degrees, F
0	32.0	1	1.8	0	32.0		
10	50.0	2	3.6	1	33.8	11	51.8
20	68.0	3	5.4	2	35.6	12	53.6
30	86.0	4	7.2	3	37.4	13	55.4
40	104.0	5	9.0	4	39.2	14	57.2
50	122.0	6	10.8	5	41.0	15	59.0
		7	12.6	6	42.8	16	60.8
		8	14.4	7	44.6	17	62.6
		9	16.2	8	46.4	18	64.4
		10	18.0	9	48.2	19	66.2
				10	50.0	20	68.0

6. Omit redundant abbreviations within the body of a table to reduce search time.[27] Abbreviations distinguish between columns; column distinction, if necessary, is better accomplished by variations in type style or darkness. That is, print one column darker or print it in a different type face. Tinker

Table 26.9   Don't Use Matrix Tables With Diagonals; If a Matrix Table Must be Used, Present the Information Horizontally and Vertically as Table 26.10

Office
Punch Press
Drill Press
Sheet Metal
Die Storage
Shipping Dock

Table 26.10   Matrix Tables are Difficult for the General Public to Use Without Error; Use Linear Explicit Tables Such as Table 26.11 if Possible Even Though They Do Use More Space

	Office	Punch Press	Drill Press	Sheet Metal	Die Storage	Shipping Dock
Office	—	50	50	75	70	25
Punch Press	50	—	100	80	30	50
Drill Press	50	100	—	75	80	60
Sheet Metal	75	80	75	—	30	30
Die Storage	70	30	80	30	—	25
Shipping Dock	25	50	60	80	25	—

Table 26.11   Fewest Errors as Well as Least Time are the Advantages of a Linear Explicit Table; They Require More Space than Matrix Tables. "Fold" (Have Multiple Columns) to Save Space.

Office to
Punch Press	50
Drill Press	50
Sheet Metal	75
Die Storage	70
Shipping Dock	25

Punch Press to
Office	50
Drill Press	50
Sheet Metal	75
Die Storage	70
Shipping Dock	25

Drill Press to
Office	50
Punch Press	100
Sheet Metal	75
Die Storage	70
Shipping Dock	25

Sheet Metal to
Office	75
Punch Press	80
Drill Press	75
Die Storage	30
Shipping Dock	80

Die Storage to
Office	70
Punch Press	30
Drill Press	80
Sheet Metal	30
Shipping Dock	25

Shipping Dock to
Office	25
Punch Press	50
Drill Press	60
Sheet Metal	80
Die Storage	25

reported equal legibility for columns of figures separated by a pica of space whether or not there was a vertical line (a rule); a pica space was better than just a rule.[24]

7. Split codes (part numbers, zip codes, telephone numbers, voucher numbers, charge numbers) into 2, 3, or 4 digits. Use GL4-6717 not GL46717; use 399-28-7456 not 399287456; use 660 50 009 not 66050009; use RL J2042 not RLJ2043.[16] To minimize errors, use short numeric codes; avoid alphabetic or alphanumeric codes.[21] When using mixed letters and numbers, omit the letters l and O. O often is confused in talking; are you referring to the letter or the number? If possible, say "zero" for the number, not "oh."

Print

The following summarizes the work of Tinker, who summarized the literature (including 58 of his own articles) in *Legibility of Print.*[24]

1. Use "reasonable" type styles. Readers not only find extreme styles less legible; they consider them less attractive. Terry Faulker of Eastman Kodak made the same point in Figure 26.5. An English group preferred typefaces that were Press Roman, Theme, or Unives over Baskerville, Bodoni Book, Aldine Roman, and Century.[6] Text with only the first letter of the sentence

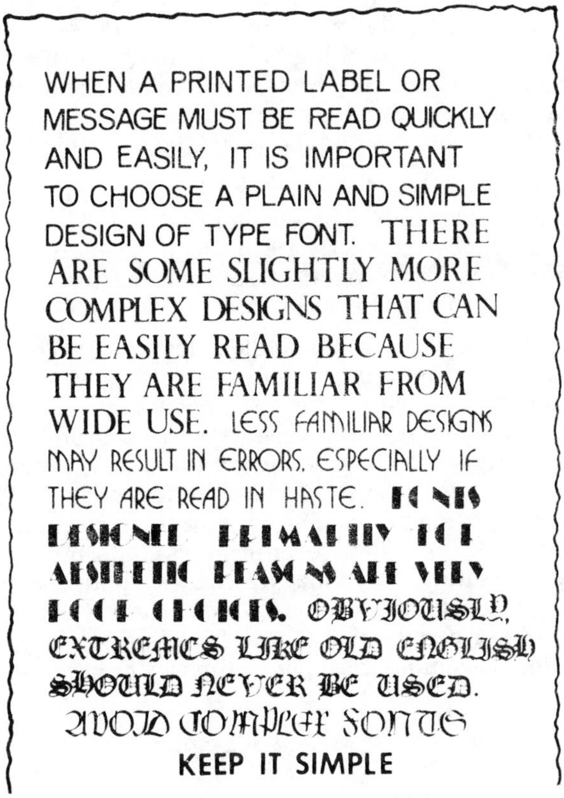

WHEN A PRINTED LABEL OR MESSAGE MUST BE READ QUICKLY AND EASILY, IT IS IMPORTANT TO CHOOSE A PLAIN AND SIMPLE DESIGN OF TYPE FONT. THERE ARE SOME SLIGHTLY MORE COMPLEX DESIGNS THAT CAN BE EASILY READ BECAUSE THEY ARE FAMILIAR FROM WIDE USE. LESS FAMILIAR DESIGNS MAY RESULT IN ERRORS, ESPECIALLY IF THEY ARE READ IN HASTE. IT IS NOT ESTABLISHED THAT LEGIBILITY AND AESTHETIC APPEAL ARE VERY HIGHLY RELATED. OBVIOUSLY, EXTREMES LIKE OLD ENGLISH SHOULD NEVER BE USED. AVOID COMPLEX FONTS KEEP IT SIMPLE

Figure 26.5 *Readers* vote legible fonts to be more attractive.[9]

capitalized is more legible than text with each word capitalized or with the initial letter of every word capitalized. Tables 26.12 and 26.13 give standards for handwriting.

To minimize computer coding problems, have: (1) a loop on the top of the letter *o* to distinguish it from zero, (2) the digit 4 with an open top to distinguish it from 9, (3) the letter *D* with a straight back to distinguish it from zero, (4) the letter *U* with a flat bottom to distinguish it from *V*, (5) the number one as a straight line without a hook or base to distinguish it from I, and (6) the comma with a distinct tail to distinguish it from the period and decimal point.

2. Don't use Roman numerals. This applies everywhere—chapter numbers in books, volume and table numbers, dates, and so forth. Perry reported that Roman numbers took 50% to 100% more time to read and produced from 3 to 30 times as many errors.[19]

3. Type size, length of the line, and the space between the lines are interrelated in a complex fashion. Figure 26.6 gives changes in reading speed for 10-point type. The point of a type in this case describes the space from the top of an ascending letter such as *t* to the bottom of a descending letter such as *y*. Printers use different units. For width 1 pica = 1/6 inch = 4.233 mm; for height 1 point = 1/72 inch = .353 mm. Elite typewriter type is 14.4 points (10 characters per linear inch, 5/vertical inch); pica typewriter type is 12 points (12 characters per linear inch, 6/vertical inch). This book is set in 10-point type with 1-point leading. Table 26.14 gives Tinker's comparisons among the optimum for each type size. Tinker's general recommendation is to use 11-point type. It is more legible than 10-point type; readers prefer it, and it gives greater flexibility in line width and leading. (Leading is the space between the lines. "Leading" is pronounced with a short e as it comes from the pieces of lead formerly used.) Although readers generally prefer some leading, Tinker says readers prefer a 10-point type "set solid" (minimum spacing) over an 8-point type with 2-point leading.

4. Leave adequate margins. The common printing practice of printing on only 50% of a page is probably wasteful for flat material. However, pages in bound files, books, and folders are difficult to read in the curved inner margin (the gutter). Photocopies often omit the top or bottom 10 mm of material as well as material in the gutter.[10]

5. Use double columns rather than single columns. As Tinker says: "a) More words can be printed on a larger paper which saves total margin area per 100 words. b) Fewer running heads are needed. c) It is possible to make more economic use of space devoted to figures, tables and formulas, many of which can be confined to one of the columns without reducing legibility. d) Relatively large tables, cuts and lengthy formulas can be printed across both columns. In a single column arrangement it might be necessary to print the cuts and tables sidewise, or print a long formula on two lines which would break up its unity. e) In many instances, double-column composition avoids tipping in large tables and cuts by virtue of the large page available. In addition, both printing experts and general readers express a strong preference for a double-column arrangement over single column."[24]

SUMMARY

The first problem of training is to realize there is a training problem. Many, if not most, quality problems are due to management failings rather than to poor worker attitudes. Of management failings affecting quality, failure to communi-

Table 26.12 Recommended Handwriting Characters and Guide for Understanding of the Nuances of Each Character and of the Rationale for Selection[1]

Letter	Guide	Letter	Guide
A	Use of squared top not supported by sufficient evidence of confusion.	N	Parallel legs.
B	Overhang top and bottom is used to reduce possibility of confusion with numeral 8 or 13. Distinct center division required to avoid similarity to letter D.	O	Loop added at top by arbitration to avoid virgule, now too confusing.
		P	Overhang at top added for consistency with letters B, D, and P.
C	No evidence of confusion; there is some similarity to left parenthesis if curve is not deep enough.	Q	No special convention.
		R	Overhang at top added for consistency with letters B, D, and P.
D	Overhang top and bottom is used to reduce possibility of confusion with numeral zero. This convention is similar to that for letter B.	S	Serif added at top only for ease of preparation and to distinguish from numeral 5 and special character dollar sign.
E	Rounded left side is to be avoided to reduce confusion with ampersand.	T	No special convention.
F	Similar to letter E above.	U	This convention adopted to distinguish from letter V and lowercase letter u.
G	Strong, emphasized serif reduces possibility of confusion with letter C or numerals 6 and 0.	V	No special convention required if the letter U has an identifying characteristic.
H	Parallel sides.	W	Center division extends to top of letter. Rounded bottom should be avoided.
I	Serifs top and bottom are de facto standards.		
J	Top serif reduces confusion with letter U.	X	No special convention.
		Y	Vertical leg bisects angle formed by top legs to avoid confusion with numeral 4.
K	Slanting legs are joined at center.		
L	No special convention.	Z	Horizontal bar is de facto standard.
M	Legs spread at bottom; center division extends to bottom of letter. Rounded tops should be avoided.		

Table 26.13 Recommended Handwriting Numbers and Guide for Understanding of the Nuances of Each Number and of the Rationale for Selection[1]

Numbers	Guide
0	Closed circle with no added identifying characteristic
1	Single vertical bar, no added identifying characteristic
2	No loop at bottom
3	Curved lines, no straight top line
4	Open top to reduce confusion with 9
5	Vertical and top lines joined at right angles
6	Loop closed at bottom to avoid confusion with zero or lower case b
7	Crossbar used in Europe considered confusing with letter Z, and does not have support in USA
8	Made with two circles adjoining vertically to avoid confusion with special characters ampersand and dollar sign
9	Straight leg from common usage

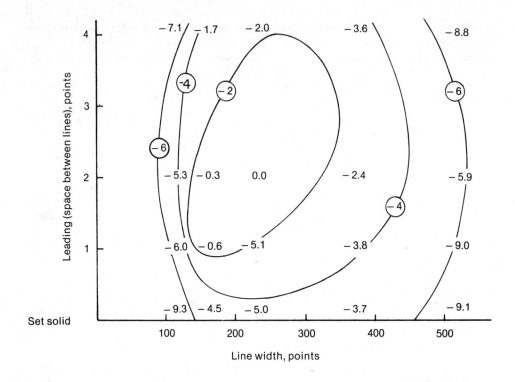

Figure 26.6 *Changes in reading speed (%)* are shown for 10-point type (1 point = ¹⁄₇₂ inch = .353 mm). 100% is defined as reading 2-point leading and 16-pica length. Experiments were at 9, 14, 19, 31, and 43 pica (103, 168, 228, 372, and 516 points).[24]

Table 26.14 For Each Type Size a "Near Optimum" Line Length Was Used; the Type Was Granjon With 2-Point Leading (504 Subjects)[24]

Type size, points	Line length, pica	points	Reading speed, percent
12	24	288	100
11	22	264	99
9	18	216	98.7
10	20	240	97.3
8	16	192	95.6
6	14	168	94.0

cate precisely what is to be done is very common. (Management usually blames the worker, though, rather than itself.)

Second, decide what to communicate. The knowledge must be transferred from the engineer's brain to the worker's brain to become useful. Your goal is a specific motor response.

Third, decide on the medium to transmit the message. In general, visual is better than audio; visual material that includes pictures is better than visual material without pictures.

Fourth, decide who will do the communicating. The more automated the communication, the more care is needed. Eliminating feedback to the writer invites problems.

Fifth, decide how to communicate using Skinner's five principles and remembering that there are no stupid students—just terrible teachers.

Use the last section of the chapter to make yourself a better communicator.

SHORT-ANSWER REVIEW QUESTIONS

26.1 Why does "bumping" cause training requirements?

26.2 Why does assembly to tape-recorded instructions tend to decrease productivity?

26.3 Using pigeon training for an illustration, give the five steps of programmed learning.

26.4 What are limit aids? Why are they used?

26.5 Sketch the recommended handwriting style for the following capital letters: D, O, U, and Z. Sketch the recommended handwritten numbers 2 and 4.

THOUGHT-DISCUSSION QUESTIONS

1. For a specific job with which you are familiar, should there be more job aids or more memorization? Make specific recommendations.

2. If programmed learning is so good, why isn't it used everywhere?

REFERENCES

1. Association for Computing Machinery. *Communications of ACM,* Vol. 12, No. 12 (December 1969): 697-698.

2. Booher, H. Relative comprehensibility of pictorial information and printed words in proceduralized instructions. *Human Factors,* Vol. 17, No. 3 (1975): 266-277.

3. Braddeley, A., and Longman, D. The influence of length and frequency of training session on the rate of learning to type. *Ergonomics,* Vol. 21, No. 8 (1978): 627-635.

4. Broadbent, D. Language and ergonomics. *Applied Ergonomics,* Vol. 8, No. 1 (1977): 15-18.

5. Chapanis, A. Interactive human communication. *Scientific American,* (March 1975): 36-42.

6. Cooper, M.; Daglish, H.; and Adams, J. Reader preference for report typefaces. *Applied Ergonomics,* Vol. 10, No. 2 (1979): 66-70.

7. Dickey, G., and Konz, S. Manufacturing assembly instructions: a summary. *Ergonomics,* Vol. 12, No. 3 (1969): 369-382.

8. Eames, R., and Starr, J. Technical publications and the user. *Human Factors,* Vol. 7, No. 4 (1965): 262-269.

9. Faulkner, T. Keep it simple. *AIIE Ergonomics News* (April 1972).

10. Goodey, J., and Matther, K. Technical information for architects. *Applied Ergonomics,* Vol. 2, No. 4 (1971): 198-206.

11. Hamill, B. Experimental document design: guidebook organization and index formats. *Proceedings of Human Factors Society,* Rochester, N.Y., 1980; 480-483.

12. Harris, D., and Chaney, F. *Human Factors in Quality Assurance.* New York: Wiley, 1969; 155-165.

13. Kammann, R. The comprehensibility of printed instructions and the flowchart alternative. *Human Factors,* Vol. 17, No. 2 (1975): 183-191.

14. Kincaid, J., and Fishburne, R. Readability formulas for military training manuals. *Human Factors Society Bulletin*, July 1977.

15. Klemmer, E., and Snyder, F. Measurement of time spent communicating. *The Journal of Communication*, Vol. 22, No. 2 (June 1972): 142-158.

16. Konz, S.; Braun, E.; Jachindra, K.; and Wichlan, D. Human transmission of numbers and letters. *Journal of Industrial Engineering*, Vol. 19, No. 5 (1968): 219-224.

17. Konz, S.; Chawla, S.; Sathaye, S.; and Shah, P. Attractiveness and legibility of various colours when printed on cardboard. *Ergonomics*, Vol. 15, No. 2 (1972): 189-194.

18. Milroy, R., and Poulton, E. Labelling graphs for improved reading speed. *Ergonomics*, Vol. 21, No. 1 (1978): 55-61.

19. Perry, D. Speed and accuracy of reading Arabic and Roman numerals. *Journal of Applied Psychology*, Vol. 36 (October 1952): 346-347.

20. Schramm, W. The Research on Programmed Instruction. Washington, D.C.: US Department of Health, Education and Welfare, Office of Education 34034, 1964.

21. Stanhagen, J., and Carlson, J. Identifying and controlling coding errors in information systems. *Ergonomics*, Vol. 22, No. 4 (1970): 441-452.

22. Sumbingco, S.; Middleton, R.; and Konz, S. Evaluating a programmed text for training food service employees. *Journal of the American Dietetic Association*, Vol. 54, No. 4 (1969): 313-316.

23. Szlichcinski, K. Telling people how things work. *Applied Ergonomics*, Vol. 10, No. 1 (1979): 2-8.

24. Tinker, M. *Legibility of Print*. Ames, Iowa: Iowa State University Press, 1963.

25. Verhagen, L. Experiments with bar graph process supervision displays on VDUs. *Applied Ergonomics*, Vol. 12, No. 1 (1981): 39-45.

26. Wright, P., and Fox, K. Presenting information in tables. *Applied Ergonomics*, Vol. 1 (1970): 234-242.

27. Wright, R., and Fox, K. Explicit and implicit tabulation formats. *Ergonomics*, Vol. 15, No. 2 (1972): 175-187.

28. Wright, P., and Reid, F. Written information: some alternatives to prose. *Journal of Applied Psychology*, Vol. 57, No. 2 (1973): 160-166.

29. Wulff, J., and Berry, P. Aids to job performance (chapter 8) in *Psychological Principles in System Development*, Gagne, R. (ed). New York: Holt, Rinehart and Winston, 1963.

27

RESISTANCE TO CHANGE

CHALLENGE OF CHANGE

Most of the chapters of this text discuss obtaining a technically excellent design. Material is given concerning hand movement patterns, handtool design, workstation design, noise, toxicology, and so forth. However, when the engineer has decided upon a technically correct concept, the concept then must be translated into practice. Translation of a concept into practice involves the "sociopolitical" world as well as the "technical" world. Engineering students, raised on a diet of mathematics and physical science, often have difficulty adjusting to this "sociopolitical" world, which treats their technical "solution" as merely a "proposal."

$$\left(\begin{array}{l} \text{Technical quality} \\ \text{of a proposal} \end{array} \right) \left(\begin{array}{l} \text{Acceptance} \\ \text{of a proposal} \end{array} \right) = \text{Amount of Improvement}$$

Youth also tend to emphasize authority and lack the patience for mediation.[23] F. Leisman, when reviewing this chapter, commented[17]:

"Without acceptance, nothing has happened. The technique or technology may be a thing of beauty, the analysis may be perfect and even foolproof, the proposal may be well written and well presented and overwhelmingly justified with excellent payback. But it is nothing, zero, a total waste of time, unless it is acceptable to those who are impacted.

"Your skill at implementing change will determine how far and how fast you will be recognized by your supervision and your peers as having the potential for increased responsibility in your organization. This skill in toto consists of 85% approach and 15% technical application.

"The bottom line of engineering is implementation—there is no other."

Staff units value change because that is how they prove their worth; line units value stability because change inconveniences them or reflects unfavorably on them. On the other hand, staff units are strongly committed to preserve control and rule systems while line units prefer flexible interpretation of control systems.[7] The statement that "where you stand depends upon where you sit" expresses the concept that whether you are for or against a change may depend upon which job you presently are holding.

PROCESS OF CHANGE

Munson and Hancock break the change process down into five steps[21]; see Table 27.1. The situation starts with stability, "unfreezes," "moves," "refreezes," and then is stable again. The technical change that, sooner or later, arouses no resistance must be extremely trivial, so how can you unfreeze, move, and refreeze a situation?

Table 27.1   Stages of the Change Process. The More Individuals Who Have to Be "Unfrozen," "Moved," and "Refrozen" the More the Challenge! The Longer a Situation is "Frozen" the More Difficult it is to "Unfreeze".[20]

Stage	Description
Frozen	stability
Unfreeze	rethink goals; consider alternatives
Move	try new methods; feelings of insecurity
Refreeze	"I like it;" "it really works"
Frozen	stability

As an example, consider three workers (Joe, Pete, and Sam) rotating daily among jobs A, B, and C of the subassembly of product X. They are paid by the piece. An engineer proposes that Joe always do A, Pete always do B, and Sam always do C. The engineer computes that their pay will increase 25% (due to increased specialization and thus more pieces/hr) and the organization's cost/unit will drop (due to lower burden cost/unit even though labor cost/unit remains constant). However, the workers refuse to change. What they resist usually is not technical change but social change.[17]

Figure 27.1 gives a "force diagram" of this situation. The "pointer" can be pulled toward the present (the status quo) or toward what is proposed. Facts are shown as solid arrows and emotions-attitudes are shown as dashed arrows. Note: (1) the many arrows (i.e., how complex even small problems can be), and (2) the preponderance of emotions-attitudes over facts.

The pointer can be made to go to the proposed side by increasing the forces toward the proposed, by decreasing the forces toward the present, or by a combination of the two.

As a general strategy for dealing with emotions-attitudes, reduce the negative ones (pulling toward status quo) rather than increase the positive ones (pulling toward change).[19] Increasing positive emotions-attitudes may be counterproductive. For example, increasing the employees' fears of management reprisal (for not making the change) may just set up counter emotions to resist the change ("it's us or them"; "solidarity forever").

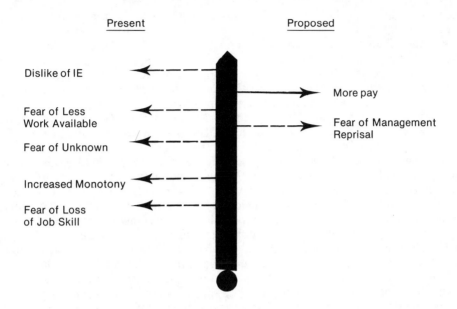

Present   Proposed

Dislike of IE

More pay

Fear of Less
Work Available

Fear of Management
Reprisal

Fear of Unknown

Increased Monotony

Fear of Loss
of Job Skill

Figure 27.1 *Emotions-attitudes* are very important factors in a potential change situation. Engineers tend to overlook the critical importance of the worker's social circle breakup, fears of the unknown, etc. One common technique of reducing financial fears is to have the organization guarantee no financial loss to the worker. If labor requirements are reduced, the worker is transferred to some other job (with no loss in pay) and the organization "eats" the cost of the surplus worker until attrition reduces the work force. However, even with no financial loss, the workers may not approve of the social losses, the changes in power, and the disruptions and so prevent the change.

Another important concept is that of negotiation between the outside experts (the technical experts) and the inside experts (the people who must do the changing). The outside experts tend to focus on "facts" while the inside experts tend to focus on emotions-attitudes. Technical experts with scientific backgrounds expect many failures in experiments. Local experts, however, interpret failures as proof that the technicians do not know their business.[2] Sometimes, however, the inside experts come up with a good modification to obtain an "organizationally best" solution and then the ego involvement of the technical experts in *their* "technically best" solution becomes a problem. The negotiation process has a number of advantages: (1) negotiation takes time and thus permits people to change their concepts gradually; (2) clearly unacceptable designs will be eliminated (these bad designs generally occur because the technical experts didn't understand the problem or all the consequences of their proposal); (3) negotiation improves communication between the groups. The communication gives the technical experts better facts for their proposal and reduces unnecessary fears among the inside experts. It helps the technical experts understand the (usually sound) reasons of inside experts for opposing changes so the necessary modifications in the proposal can be made. What is the interval between paying the costs and receiving the benefits?

Acceptance of a technical proposal needs to come from both the management and the workers. Thus, negotiations may be done in two stages: once between the technical experts and management and a second time for the revised proposal between management and the workers.

An alternate technique is have the technical experts and the workers prepare the proposal, which they then present to management. Examples of this are the Japanese *Quality Circle* (Jishu Kanri) programs (see the fish diagrams in chapter 7 and the discussion in chapter 10) and the Israeli Productivity Circles.[8] The Productivity Circles emphasize quick direct financial benefits for the workers when a change is made. (Although improved productivity aids the workers through their wages and retention of the plant's competitive position, these benefits seem too vague to many workers; they want something *now*.) Technical expert-worker projects also have been used in the US for many years; for published examples see Coch and French.[3,5] Other examples are the Scanlon plans and Evolutionary Operation of Processes (see chapter 4) and the famous work simplification program of Allan Mogensen.[9] Quality of Work Life, abbreviated *QWL*, attempts to replace the typical adversary relationship between management and employees with a cooperative one. Its greatest implementation seems to be in the automotive industry—which has a long history of adversary relationships. Key words of *QWL* are cooperation, involvement, trust, and mutual rapport. The key problem is to convince management to change its autocratic approach; this takes time and *QWL* programs take 5 years or more. Participation, to be of value, must be based on a search for relevant ideas; the shallow notion of participation as a manipulative tool needs to be debunked.[17]

The "area of freedom," see Figure 27.2, limits the freedom to make changes.[20] In coordinating local experts and technical experts, be sure both groups under-

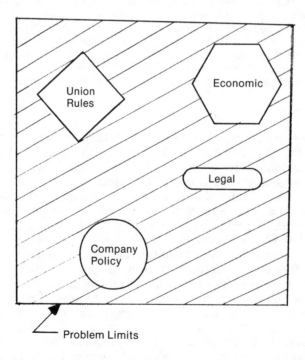

Figure 27.2 *Decisions are limited* by the "area of freedom." Limitations might be economic (keep capital cost less than $10,000), union rules (all maintenance work must be done by in-plant workers), company policy (all reductions in the work force must be done by attrition), legal (noise level must be less than 90 dBA), etc.

stand what is not permissible (spend over $10,000, require over 50 hr work/week, reduce product quality, etc.).

Knowledge from the local experts and technical experts is depicted schematically in Figure 27.3. Note: (1) the technical experts tend to be "long on information and short on emotions" and (2) not all facts are known to the technical experts. Thus coordination between the two sides may improve solution quality as well as solution acceptance.

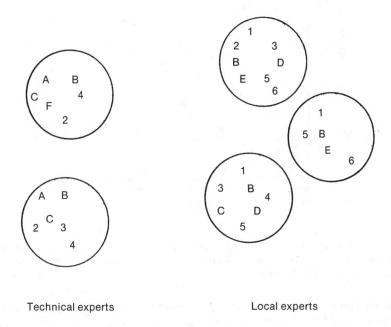

Technical experts   Local experts

Figure 27.3 *Knowledge* concerning a problem is spread through the members of both the local experts and the technical experts. "Facts" are depicted by letters (e.g., A = optimum work height is at the elbow, B = the machine has a bad bearing, C = capital cost of a roller conveyor is $2000, etc.). Emotions-attitudes are depicted by numbers (e.g., 1 = we are underpaid, 2 = productivity in the department is unsatisfactory, 3 = that engineer is a stuck-up snob, etc.).

The "Taylorism" concept of workers blindly following the orders of their supervisors has been made obsolete by the increasing complexity of society and the increased education of workers; see Figure 27.4.

QUALITY CIRCLES

Quality Circles began in Japan in 1963, developing from the needs of Japanese industry and the characteristics of Japanese society and management. Needs concerned the long-range industrial goals of Japan. Previously they had competed on the basis of low labor costs. However they wanted to improve the Japanese standard of living and this required higher wages. In addition, there were millions of competitors in Asia willing to work for very low wages. Thus the Japanese decided to compete on the basis of quality and technology. To get the quality they decided to get "everyone" to help rather than just the elite group of engi-

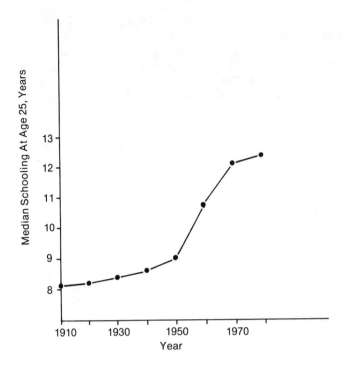

Figure 27.4 *Education levels* have increased since the time of Taylor. Workers had only a grade school education when Taylor formulated his policy of job design by experts without worker contribution.

neers and managers—i.e., for quality, use quantity. This concept of mass participation was aided by the management style of participation in which consensus was highly valued and by the long-range outlook of management.

The standard approach to job design in developed societies since Taylor has been to have technical experts design the job and the workers follow the instructions; this policy often is called "Taylorism." It has been extraordinarily successful in improving productivity and the standard of living in many societies; see chapter 3. The Japanese, however, decided that with their highly educated work force the workers could contribute some ideas on parts of the job. The elite groups of engineers and managers would still work on the major problems but "the masses" would work on the many minor problems. The emphasis would be on quality problems. Since the ordinary blue collar workers had not been given technical training on how to improve quality, it would be necessary to train them and then give them a mechanism to use the training.

The result is that, 20 years later, Japanese products have a quality reputation. Almost all of Japan's 54,000,000 workers have some quality training, approximately 8,000,000 have had technical courses in quality from the Union of Japanese Scientists and Engineers, and membership in *Quality Circles* has risen from 80,000 people in 10,000 Circles in 1966 to 6,000,000 in 600,000 Circles in 1980.[1] By the late 1970s the *Quality Circle* concept had spread to Korea, Taiwan, Brazil, and the USA.[14] The US Quality Circle association is the International Association of Quality Circles (PO Box 30635, Midwest City, OK 73140). Founded in 1978, by March 1982 it had 3600 members and 35 local chapters.

As now structured, Circles range from 3 to 25 members, but since part of the concept is to get everyone to participate, 10 or so seems to work best; membership is voluntary. (If there are more volunteers than places, tell the others to wait their turn or form another Circle.) Meetings typically are once a month but may be as often as one a week; an hour is a typical length. Normally they are within working hours; if outside working hours (such as for assembly line workers), they are paid extra. In Japan, rewards for Circle members are not financial, but praise, publicity, and self-satisfaction. Cole feels, "Monetary incentives must play a larger role in the US Circles than the Japanese Circles since the Japanese have the lifetime employment system and very large annual bonuses (5 months salary)."[6] Projects are nominated by the workers and by the staff. Safety and housekeeping projects are done as well as quality. Direct labor cost reduction projects should be done by others—they are not the "mission" of Quality Circles. Gryna recommends letting the workers nominate more in the first couple of years as they have many minor workplace frustrations to clear away; the project to work on is selected after discussion within the Circle followed by agreement between the Circle leader and the management.[12]

The following example shows how Miss F. Hashimoto's QC team at Matsushita Electric analyzed switches used for volume control on stereos.[15]

Step one was to select the project. Among the defects from the assembly line, the largest percentage was attributed to the volume switch. After consultation with management, the Circle decided to study switch defects.

Step two was to analyze present conditions. Figure 27.5 shows their Pareto diagram for defects over a 3-month period. The Y axis is percent defective. The X axis is a series of bars, arranged in descending magnitude, of the various causes. They cross-hatched the major cause for emphasis. Then they plotted cumulative defects (line with dots). "Rotation" was 70% of the defects. The Pareto distribution, also called the "insignificant many and the mighty few," encourages "fighting giants," so switch rotation was picked as the project.

Figure 27.6 shows an analysis of the causes of rotation defects: 87% were uneven rotation. Then the Circle members used a cause-effect, or fish diagram, Figure 27.7, to organize the problem and improve communication. Development of the diagram required a series of Circle meetings. The fish head contains the goals, the major bones the major causes, and minor bones the contributors to the major causes. With Figure 27.7 as a guide, they collected defect data from the in-process inspectors, sorted it by cause, and developed Figure 27.8.

The third step was to establish goals. At their next Circle meeting they established three goals: reduce rotation defect rate from 1.3% (as of January 1965) to 0.5% by December 1965, while introducing a more stable control system; develop an overall Circle implementation plan, Figure 27.9; selectively attack the problems for improvement using the Pareto chart order.

The fourth step was to promote control activities. One activity was an *np* chart, Figure 27.10. For their production, a sample ($n = 400$) taken every hour was appropriate. When an out-of-control condition developed (point beyond control limits), the Circle had a meeting. Depending upon the nature of the problem, they either determined the problem cause or a measure to prevent reoccurrence of the problem. In order to control common deficiencies in previous operations, check lists were developed for critical control points. The operators used these check lists to check their own work. After improved procedures were put into practice and found workable, the standard procedure was revised to ensure continued use of the new method, and sample size was changed from 400 to 1600.

As a result of these activities the defect rate was reduced from 1.3% to 0.3% for an annual saving of 400,000 Y (about $1000 at the exchange rate in 1965). In her paper, Miss Hashimoto concluded that although they had achieved their goal, they had not completely eliminated the problems. They were determined to continue improvement to achieve a still better result. In addition, she commented, that as an inexperienced Circle leader, she had not worked enough with people from other departments.

A number of comments can be made about the example project. First, this was a small project with a $1000/year saving. Yet a considerable amount of work was necessary to achieve the saving. In the many examples of Japanese *Quality Circles,* I have not seen return on investment reported. Typically, the number reported is annual savings or percent defects. This implies that the engineering cost and capital cost of making the change are not considered to be worth reporting. However, this engineering cost may be quite low as it is done during normal work time; meeting costs probably are charged to training or general quality costs.

Second, the Japanese blue collar Circle members used techniques that, by American standards, are sophisticated. Pareto diagrams, fish diagrams, and *np* control charts are not even known to many American engineers—much less used. The key here is management's emphasis on training members in the use of these techniques and encouraging their use. Gryna, in discussing American *Quality Circles,* emphasizes the need for training.[12] Example training modules are:

- Introduction to Quality Circles
- Brainstorming
- Fish diagrams
- Histograms
- Check lists and data recording
- Case study
- How to make graphs
- How to make a presentation

Another training need is for interpersonal skills by the Circle leader.

Third, quality problems were considered to be technical problems—not motivational problems. Improvements were use of modified hand tools, modified assembly procedures, modified operator training, and the use of check lists by the operators.

Fourth, many problems are communication problems. Fukuda of Sumitomo Electric summarized the results of 87 groups at his firm in Table 27.2.[10,11] Figure 27.11 is a modified version of one of his figures showing that information must be known to both the operator and the technical staff in order to be practiced. Knowledge alone is not sufficient as both groups must want to apply the knowledge. Fukuda summarized the figure in the saying "defense before offense." In "defense," try to move from categories B, C, and D to A. Defensive examples are to put warnings and suggestions into easily readable visible form, to clearly define standard operations, and to develop tools enabling less effort yet more skill to be put forth. Only when you are "farming as well as you know how" do you take the offense and make changes in equipment and manufacturing conditions.

One way to emphasize communication is to *post* the fish diagrams on the wall next to the machines and invite *everyone* to make comments and suggestions. (You will have many "detectives looking for clues" rather than just one person.) The goal is to get everyone to *"think* beyond what they are *told* to do."

The Circle then takes the best of these ideas for further investigation. For investigation, a good technique is to use a control chart (such as Figure 27.10). The *active* approach is to make a change in the cause and note the result on the control chart; the *passive* approach is to watch the control chart, and, when the chart changes, try to go back and find what changed in the process. The active

Table 27.2   Types of Countermeasures
Found Effective by QC Circles at Sumitomo[10]

Countermeasure type	Type of result (%)			
	Limited		Considerable	
	Slow	Quick	Slow	Quick
Warnings and suggestions put into easily readable visual form—defensive	41	25	57	68
Clearly defining standard operations—defensive	24	35	24	53
Developing tools enabling less effort yet more skill to be put forth—defensive	12	5	37	32
Making improvement in equipment—offensive	5	5	30	12
Making changes in manufacturing conditions—offensive	0	5	20	5

Quick = within 3 months
Slow = over 3 months

Limited = Reduction of defects of less than 40%
Considerable = Over 40%

approach is preferable since it produces more knowledge in a shorter time period with less effort. Remember, however, as was pointed in the EVOP discussion in Chapter 4, to make evolutionary changes rather than revolutionary changes.

PROPOSAL FOR CHANGE

The proposal should have the following general format. On the cover sheet should be the project title, the data, your name, and the word *Recommendation* followed by 50 to 100 words giving the recommendation. At the top of the second page should be the word *Problem* followed by a 10 to 50 word statement of the problem. Next should come the word *Analysis* followed by the analysis. This section may take three to ten pages. If it takes over ten pages, cut the material down by putting details in appendices. Finally, put the word *Conclusions* and briefly give them.

Proposals should go through several drafts. Supervisors should never accept handwritten drafts—too many errors hide in handwriting. It is important that the writer know for whom the report is finally intended and what the report's purpose is. Establish the writing style wanted (informal active voice such as "We measured background noise levels" or formal passive voice such as "Background noise levels were measured"). Strunk and White, in 85 pages popular since 1959, tell how to write clearly with style.[22]

The contents of the proposal should include annual costs of the present and proposed methods, one-time costs of the proposal (such as new equipment, training costs, installation costs), expected life of the proposal, and expected savings. See Figure 4.3 for an example form. Assumptions should be stated specifically (i.e., product life = 3 years at 8,000 units/yr, no changes in product design, change will not affect spare parts sales, quality will be unchanged, etc.). The proposal also should list a "timetable" of the sequence of executive actions required to implement the proposal (1 May: change drawings; 15 May: stop production of existing design; 15 May: change customer's service manual; 20 May: change standard cost of product, etc.).

Will the project generate income, reduce costs, or reduce risks? Know who in the audience makes the decisions and what they like. Some decision makers tend to be risk-avoiders and others emphasize potential economic gain. Generally they will be primarily interested in the economic aspects of the proposal rather than the technical aspects. Since they accept the proposal as technically sound, avoid "technical overkill." Remember that the decision maker will be reviewing many proposals and not all will be accepted; why should yours be?

In addition to the written proposal, proposals usually must be presented orally. Try to take only 15-30 minutes in your formal presentation; additional detail can come from questions or the written report. Use transparencies or flip-charts (one for every 60-90 seconds is a good rule). See Table 27.3. Rehearse your talk before your colleagues. Clawson even recommends rehearsing the question and answer period.[4]

If possible, pick the best time and place to present ideas. For example, if the decision maker is impulsive, present ideas in the presence of others so it is more difficult for an idea to be rejected arbitrarily. If a proposal affects several departments, informally "clear" the proposal with each department before formally bringing it up for discussion. This "no surprises" policy reduces the automatic rejection of the unknown and also may improve the technical merit of the proposal (as shown in Figure 27.3).

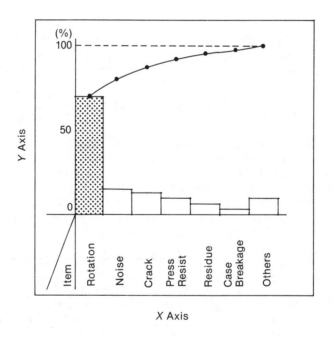

Figure 27.5 *A Pareto analysis,* using 3-months data, showed that switch rotation accounted for 70% of the defects. Thus it was selected for further analysis.

IMPLEMENTATION OF CHANGE

Once the proposal is accepted, the final step is implementation. *Implementation is not automatic.* One common way of "sinking" a proposal is to "smile it to death." Managers not wanting the proposal just ignore implementing it by "losing" change forms, promising to do something "when I have time," "forget-

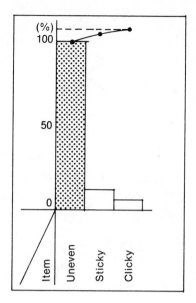

Figure 27.6 *Another Pareto analysis,* analyzing rotation defects by phenomena, showed that 87% of the probems were uneven rotation.

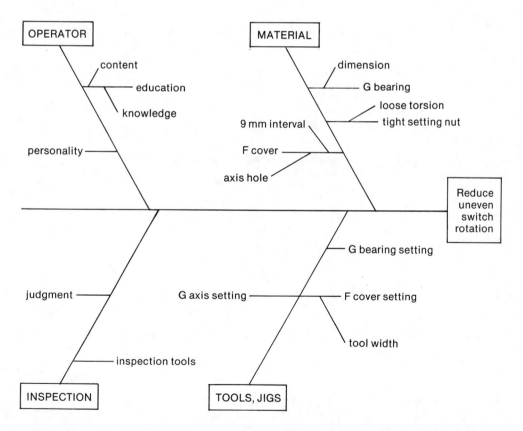

Figure 27.7 *Uneven rotation errors* were the target of a cause-effect diagram. The fish head is the goal, major bones are the major categories, and minor bones are the subdivisions of the major categories.

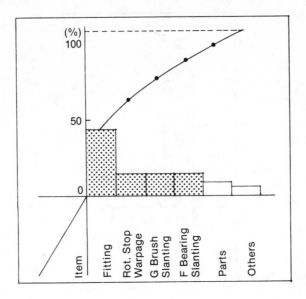

Figure 27.8 *Using the fish diagram* in consultation with the in-process inspectors, the Circle developed a new Pareto diagram of causes of uneven rotation.

	O J F M A M J J A S O N D
1. Analysis of Present Conditions	△─O
2. Dev. and Imple. of Improvements	
• Improper Fitting	△─O─▢
• F Rot. Stop Warpage	△─O─▢
• G Brush Slanting	△─O─▢
• F Brearing Slanting	△─O─▢
• Parts	△─O─▢
• Others	△─O─▢
3. Introduction of More Stable Control System	△─O─▢
	△────────O─▢────▢

Figure 27.9 *Goals and a schedule* were set by the Circle at their next meeting.

ting" to do things. Workers can sabotage by saying "the parts don't fit," by making the parts not fit, by neglecting maintenance.

Installation can be divided into five stages[13]:

1. Gaining acceptance of the change by the department supervision

2. Gaining approval of the change by plant and general management

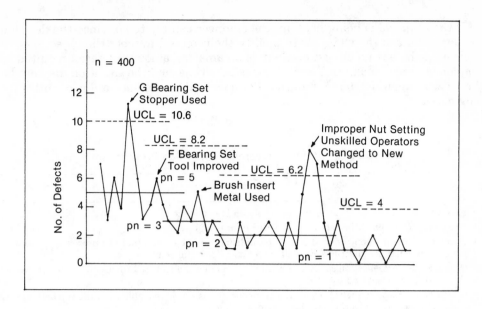

Figure 27.10 *An np chart* helped identify when defects occurred so their causes could be eliminated. The first control limits describe the original proccess. The second set of limits describes the results after the second and third Circle meetings. The third set describes results after the fourth, fifth, and sixth meetings, and the fourth set after the seventh meeting.

	Managerial-Technical Staff	Known		Unknown
Operator		Want to apply	Don't want to apply	
Known	Want to apply	Practiced A	Not Practiced B	Not Practiced D
	Don't want to apply			
Unknown		Not Practiced C		Not Practiced E

Figure 27.11 *Many problems are communication problems.* Area C, for example, shows that if information is known to the technical-managerial staff but not to the operator, the technique is not practiced.

3. Gaining acceptance of the change by the workers involved and their representatives

4. Retraining the workers to use the new methods

5. Maintaining close contact with the job progress until satisfied that it is running as intended

If the proposal is being held up, the engineer can try to convince the decision maker to "crack the whip" or to modify the proposal to meet the objections.

Even if there is no objection, there is an amazing amount of detail required to implement most changes. These changes take time, and the situation may not be "refrozen" again for many months. Patience and persistence are necessities for engineers.

Table 27.3   Effective Overhead and Slide Presentations

Overheads

- Stay within 7½ x 9½ format as projector platen is not 8½ x 11.
- Use color to "outline" the talk as well as for figures and overlays.
- Don't put too much onto one sheet. Two guidelines are: 1 sheet/minute of presentation and 45 words maximum per sheet (6-7 words/line; 5 lines; 3 vertical columns).
- Organize your presentation so it has a beginning, middle, and end. Tell what you will tell them, tell them, tell them what you told them.
- Use high-contrast originals; be sure the transparency is easily readable from the farthest viewer position; check by projection, not hand-held viewing.
- Use, for graphs, grids or graph paper under paper originals to get proper scales and relationships.
- Keep projector on your left if you point with your left hand; on your right if you point with your right. Point with a pointer or pen, not your finger.
- Have material prewritten on the film. Handwriting speed is about .4 words/s, speaking is 2 to 4 words/s, and reading is 3 to 9 words/s.

Slides

- Use color, not black and white. Make color slides from black and white text by adding a yellow or *light* pastel overlay to the slide. Reverse slides (light letters on dark background) can be made in color from black and white text.
- Keep information/slide to one idea; no more than 3 curves/graph.
- Make material *easily* readable from the farthest viewing position. Graphs and tables that are satisfactory in print need to be simplified and lines emphasized for slides. Leave space (at least height of a capital letter) between lines of text.
- Use duplicate slides rather than back up during a presentation.
- Practice your talk. Do it *early* so you can make changes in the slides.

SUMMARY

A technical proposal is useless unless it is accepted *and* implemented. Change involves the sociopolitical world as well as the world of facts. Reduce negative emotions-attitudes to improve acceptance. Worker participation helps.

Proposal presentation is important. Once the proposal is accepted, it must be implemented. Implementation requires attention to detail and possibly additional compromise.

SHORT-ANSWER REVIEW QUESTIONS

27.1   Give the five stages of the change process, starting with frozen.

27.2   Give three advantages of negotiation between the outside experts and the inside experts.

27.3 Give names of four different "plans" or techniques which emphasize cooperation of technical experts and workers before presenting proposals to management.

27.4 List the four sections into which a technical report should be divided.

27.5 On an oral technical presentation, how much time should each transparency or flip chart take?

THOUGHT-DISCUSSION QUESTIONS

1. Acceptance of a proposal must come both from the workers and from management. Discuss some things that can improve acceptance of a proposal by management without decreasing acceptance by workers.

2. Describe a situation in which change is proposed. Make a force diagram (Figure 27.1) showing facts as solid arrows and emotions-attitudes as dashed arrows. What do you recommend to increase movement toward the proposed solution?

REFERENCES

1. Arai, J. Japanese productivity: what's behind it? *Modern Machine Shop*, Vol. 52, No. 4 (1979): 117-125.

2. Beals, R. Resistence and adaptation to technological change; some anthropological views. *Human Factors*, Vol. 10, Nov. 6 (1968): 579-588.

3. Chaney, F. Employee participation in manufacturing job design. *Human Factors*, Vol. 11, No. 22 (1969): 101-106.

4. Clawson, R. *Value Engineering for Management, 100-110.* New York: Auerback Publishers, 1970.

5. Coch, L., and French, J. Overcoming resistance to change. *Human Relations*, Vol. 1, No. 4 (1948): 512-532.

6. Cole, R. Will QC circles work in the U.S.? *Quality Progress*, Vol. 12, No. 7 (1980): 30-33.

7. Dalton, M. *Men Who Manage.* New York: Wiley & Sons, 1959.

8. Dar-El, E., and Young, L. Systems incentives: three ways to better productivity. *Industrial Engineering*, Vol. 9, No. 4 (1977): 24-29.

9. Frost, C.; Wakely, J.; and Ruh, R. *The Scanlon Plan for Organizational Development.* Ann Arbor: Michigan State University Press, 1974.

10. Fukuda, R. The reduction of quality defects by the application of a cause and effect diagram with the addition of cards. *International Journal of Production Research*, Vol. 16, No. 4 (1978): 305-319.

11. Fukuda, R. Introduction to the CEDAC. *Quality Progress*, Vol. 14, No. 11, (Nov. 1981): 14-19.

12. Gryna, F. *Quality Circles.* New York: American Management Association, 135 W. 50th St., N.Y., N.Y. 10020, 1981.

13. ILO, Define, Install, Maintain (chapter 12) in *Introduction to Work Study*, Geneva: International Labour Office, 1969.

14. Irving, R. QC payoff attracts top management. *Iron Age*, Vol. 222, No. 3 (1979): 64-65.

15. Konz, S. Quality Circles: Japanese success story. *Industrial Engineering*, Vol. 11, No. 10 (1979): 24-27.

16. Konz, S. Quality circles: an annotated bibliography. *Quality Progress*, Vol. 13, No. 4 (1981): 30-35.

17. Lawrence, P. How to deal with resistance to change. *Harvard Business Review*, Vol. 1 (1969): 4-12, 166-176.

18. Leisman, F. An Engineer-in-charge of Manufacturing Development, General Motors Corp., personal communication, June 2, 1977.

19. Maier, N. A human relations program for supervision. *Industrial and Labor Relations Review,* Vol. 1 (1948): 443-464.

20. Maier, N. *Principles of Human Relations.* New York: Wiley & Sons, 1952.

21. Munson, F., and Hancock, W. Problems of implementing change in two hospital settings. *American Institute of Industrial Engineers Transactions,* Vol. 4, No. 4 (1972): 258-266.

22. Strunk, E. W., and White, E. *The Elements of Style,* 3rd ed. New York: MacMillan, 1979.

23. Von Lazar, A., and Mikesell, D. Science, technology and political development, *Human Factors,* Vol. 10, No. 6 (1968): 599-606.

28

JOB EVALUATION

GOAL OF JOB EVALUATION

Job evaluation is a method for determining the relative worth of jobs. It is concerned with providing hard data on which to come to conclusions about the rate for the job. The "formula for fairness" offers system and stability if planned for and developed on the bases of concensus, cooperation, and concord. It is a rational consideration of the more important differences between the inputs of human work. The crux of the problem is assessing and agreeing on the fairness of differential pay—the problems of equity and acceptability. The systems are not perfect—at root they are judgments, a thread in a total fabric, a skeleton framework for a system of equitable payment.[3]

One administrative problem is that it is difficult to justify different wage rates for *each* job—thus similar jobs are grouped into 10 to 20 "labor grades." See Figure 28.1. That is, similar jobs might all be paid $6.40/hr instead of some at $6.20, some at $6.30, some at $6.50, etc. Another problem is how to separate the worth of the job and the worth of the individual doing the job. In Japan, for example, the worth of the individual (evaluated by the person's age) is given a high weight and the job relatively little. In the USA, however, the job is the primary factor although some allowance is made for the holder of the job. The individual allowance usually is some combination of merit and seniority.

JOB DESCRIPTION

Job evaluation has four steps: (1) Describe the class of job (e.g., Clerk-Typist II), (2) specify the minimal hiring requirements, (3) arrange the jobs in order of worth, and (4) describe a specific job at a specific location—a "position description" (Clerk-Typist for Electrical Engineering). See Figure 28.2.

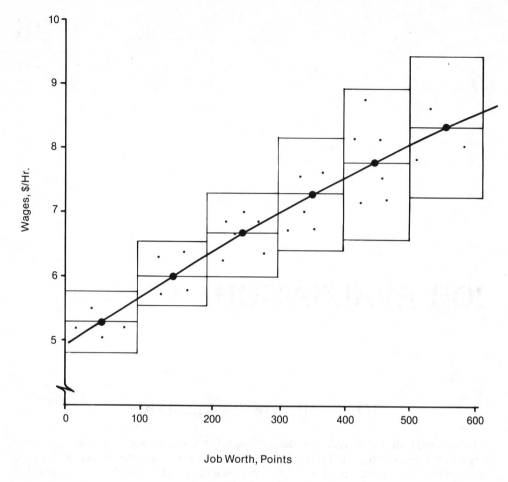

Figure 28.1 *Labor grades* (the boxes) summarize the equation of wage vs. worth. All points (jobs) within a box are considered of equal worth for administrative purposes. The vertical dimension of the box permits adjustment for different individual contributions to the same job; it should be greater for more complex jobs. The boxes generally have a vertical overlap so a person at "the bottom of the bracket" in a higher grade may not earn as much as a "top of the bracket" person in a lower grade. Both the overlap and the vertical spread from the mean wage for the grade should increase with complexity so individual jobs can be modified with minimal effect on an incumbent's pay and so there is more scope for merit rating.

Describe the job class

This step has two substeps: (1) Securing the factual information and (2) writing the information in standardized format. The information to obtain includes:

- duties and the percent of time devoted to each
- responsibilities
- knowledge and skills needed
- performance standards to be met
- what is used
- working conditions, including hazards

Securing the information generally is done by someone from the personnel department using some combination of questionnaires, interviews, and observation of the jobs. Questionnaires, if used, need to be checked and supplemented by worker interviews for better worker acceptance and improved accuracy. Interviews with the workers are the most common approach; use a structured interview. Observation is a good check but generally is not sufficient by itself.[5]

Although the information gathering and writeup should be done by a trained person, the results should be checked for clarity and brevity by a committee of interested parties. For accuracy in the third step, ranking of the jobs, all job descriptions should be about the same length and contain about the same level of detail. Raters will be more consistent if the person writing the job description knows what types of information they want and includes it in the description.

Keep the number of job descriptions (and jobs) reasonably small. The most common problem is overspecification of jobs and resulting lack of flexibility as workers "work to rule" and refuse to do work that is not in their job descriptions. For example, a carpenter might refuse to turn on the power as that would be an electrician's work. Thus all job descriptions should have a "miscellaneous duties" section; in Figure 28.2 it is "Perform other duties as assigned."

Specify minimal hiring requirements

This step, job specification, has become a very difficult task with the increased emphasis on equal employment opportunity, nondiscrimination, and hiring the handicapped. In many situations, organizations are sued if they don't make a "reasonable effort" to modify the job to fit a handicap (such as a bad back). If you put too rigorous requirements on a job, you may be challenged to prove that they are relevant in a court case. For example, can you prove in court that a high school diploma is necessary for job X? Desire to hire well-qualified employees is not proof! On the other hand, putting no requirements on a job doesn't help much either. Hiring a person and then allowing that person 30-90 days to prove they can do the job results in poor quality and may result in worker injury if the person isn't physically qualified. If the entry-level job is part of a job sequence (such as helper, apprentice, electrician), specify enough qualifications for the entry job or soon you will have people in the entry job who cannot complete the normal sequence.

Arrange jobs in order of job worth

This major step will be discussed in detail, after the next paragraph, in "Arranging Jobs in Order."

Describe a position

After the general categories of jobs are described (clerk typist I, clerk typist II, secretary I, etc.) it is necessary to describe individual jobs at individual locations (the job in Electrical Engineering). Then it can be determined that this specific job should be a clerk typist II. Sometimes the positions are described first and then the combinations are used to form a job description.

SECTION A: Position Purpose:

Explain concisely why the duties and responsibilities assigned to this position are essential to agency operations.

The person occupying this position functions as the office receptionist, typist, and clerk for the Electrical Engineering Department. Since this position is the only one of its type in the Department, it is obviously critical to the operation of the Department. This person is expected to meet students, staff, and the public, provide general information about the Department, and direct the visitor to the proper place or person for more information or help. Duties of this position include the typing of research manuscripts, daily correspondence, and other material submitted by faculty and staff.

SECTION B: Duties and Responsibilities:

Instructions: (1) Number each duty and indicate approximate percent of time spent on each major duty or group of duties. (2) Include specific data as to responsibility for direction of work of other employees; position numbers and class titles of employees supervised; degree of responsibility for funds or actions, decision making, and program and policy planning; nature, purpose, and level of contacts within and outside the agency. (3) Indicate how independently of supervision this position functions, or conversely, how closely and directly the position is supervised.

Duty No. and Percent of Time	Duties
1. 25%	Act a office receptionist, answer general questions about the Department, direct visitors, answer telephones, take messages. Person must have general knowledge about operation of the Department and University. Visitors include faculty, staff, students, and the general public.
2. 45%	Type research papers, proposals, correspondence, examinations, and forms. Most work is submitted to this person in the form of hand-written drafts. Work is checked by person submitting the material. Little supervision is required; work checked when completed.
3. 3%	Employee uses duplicating machine and copying machine. Employee is responsible for meeting deadlines set by faculty, staff, and supervisor.
4. 25%	Employee maintains records of all students enrolled in Electrical Engineering. Related tasks include assigning advisors, transferring students, posting semester grades, recording of drop-add slips, and filing all student-oriented transactions. Employee maintains line schedule, textbook lists, and course outlines. Little supervision is given. Results checked as files accesed.
5. 2%	Handle mail and maintain office supplies. Assist other staff members with clerical work. Perform other duties as assigned.

SECTION C: MINIMUM QUALIFICATIONS: (Education and Experience, Certificates, Licenses, Degress, Skills Required)

Good oral and written abilities, Vo-Tech graduate with course emphasis on secretarial science, one-year experience.

Figure 28.2 Job description for a clerk-typist II. Job descriptions should be accurate, brief, and clear statements of what the worker is expected to do. Begin sentences with verbs. Be specific. State duties rather than qualifications of the incumbent.

ARRANGING JOBS IN ORDER

Background

There are many different jobs: lion tamer, astronaut, cook, secretary, engineer, corporate executive, teacher, welder, truck driver, assembly operator, etc. For-

Table 28.1   A Comparison of 21 Point Rating Plans
Showing the Number of Factors Used and the Number of Components into
Which Each General Factor is Broken Down.[5]

Company	Skill	Effort	Respon-sibil-ity	Work-ing Con-ditions	Total Factors
General Foods Corporation	7	1	2	0	10
General Electric Company	2	2	1	1	6
Revere Copper and Brass, Inc.	2	2	3	2	9
Westinghouse Electric Corp.	3	3	4	0	10
American Optical Company	1	1	1	1	4
Cheney Bros. Company	1	0	1	1	3
R. G. Le Tourneau, Inc.	8	2	4	2	16
Inland Steel Company	3	2	5	3	13
Carnegie-Illinois Steel Corp.	4	2	4	2	12
Baldwin Locomotive	4	2	3	3	12
Western Tablet & Stationery Co.	4	2	3	2	11
Cessna Aircraft Co.	3	2	4	2	11
Lockheed Aircraft Corp.	2	2	1	2	7
Greenfield Tap & Die Corp.	4	2	2	2	10
An Underwear Manufacturer	6	2	4	3	15
A Textile Mill	2	2	2	2	8
National Electrical Mfrs. Assn.	3	2	4	2	11
Industrial Management Society	5	1	3	2	11
Stigers & Reed*	14	5	1	15	35
George S. May Co.**	6	4	3	4	17
McClure, Hadden, & Ortman*	7	6	6	5	24

*"The Theory and Practice of Job Rating"
**Management Consultants

tunately, this variety can be reduced as job evaluation has as a goal the arrangement of jobs within an organization, not between organizations. However, there still may be a great diversity. This diversity is reduced to a practical level by setting up multiple plans—for example, one plan for clerical/office, one for managerial/technical, and one for shop/maintenance. Different plans for the different groups are justified two ways:

1. Factors vary in importance by group. For example, effort and working conditions have relatively little importance in distinguishing among office jobs but considerable importance in shop jobs. Responsibility and initiative might vary considerably among professional/technical jobs and little among clerical jobs.

2. Wage curves are the goal. Wages for the work are, to some extent, determined by competition vs. other employers. The relative wages of shop jobs generally are higher than office jobs since so many people are willing to exchange the "prestige and status" of an office job for the higher wages of a shop job. In addition, professional/technical jobs have to compete on a national rather than a local basis. An engineer may look at wages all over the country while shop people may compare wages only within the community; thus the level of competition is higher for jobs in which people are more mobile and organizations have to offer relatively higher wages to meet the competition. Thus, except for monopolies such as the government, organizations generally have a local wage curve (rather than a national curve) for both clerical and shop personnel.

There are four approaches to arranging jobs: (1) ranking, (2) classification (such as the Civil Service grades of the US Government), (3) factor comparison, and

Table 28.2 Scoring System for the MIMA Job Evaluation Plan for Shop Jobs[6]

Job Factors	Degrees				
	1st	2nd	3rd	4th	5th
SKILL					
1. Education or Trade Knowledge	14	28	42	56	70
2. Experience	22	44	66	88	110
3. Initiative and Ingenuity	14	28	42	56	70
EFFORT					
4. Physical Demand	10	20	30	40	50
5. Mental and/or Visual Demand	5	10	15	20	25
RESPONSIBILITY					
6. Equipment or Process	5	10	15	20	25
7. Material or Product	5	10	15	20	25
8. Safety of Others	5	10	15	20	25
9. Work of Others	5	10	15	20	25
JOB CONDITIONS					
10. Working Conditions	10	20	30	40	50
11. Hazards	5	10	15	20	25

Grade Ranges

Score Range	Grades	Score Range	Grades
139	12	250-271	6
140-161	11	272-293	5
162-183	10	294-315	4
184-205	9	316-337	3
206-227	8	338-359	2
228-249	7	360-381	1
Maximum Points			500

(4) point systems. This chapter will discuss only the most popular approach, point systems. For information on the others (and more detail on point systems), consult the references at the end of the chapter.

Point plans. The general concept of a point system is to analyze the "levels" of the "factors" relative to a job. Each factor has a number of levels with points allocated for each level. Then the analyst just adds up the points to get job worth. Although simple in concept, judgment is required.

The first judgment step is in the selection of the factors and levels for the plan. As a general consensus over the last 50 years, the various factors have been grouped into categories of skill, effort, responsibility, and job conditions. Typical subdivisions of skill are factors of "education," "experience," and "initiative and integrity." Another breakdown of skill is "preemployment training," "employment training and experience," "mental skill," and "manual skill." Typical factors of effort are "physical demand" and "mental or visual demand." Typical factors of responsibility are for "equipment or process," "material or product," "safety of others," and "work of others." Another breakdown of responsibility is for "materials," "tools and equipment," "operations," and "safety of others." Typical factors of job conditions are "working conditions" and "hazards." Another breakdown is "surroundings" and "hazards."

Table 28.1 gives the number of factors within each of these four categories. Most plans use 10-15 different factors. Since some factors are more important than others and wider ranges of points are given for some factors than others, a smaller number of factors probably would be sufficient from a technical viewpoint. However, from an administrative (political) viewpoint, the larger number of factors permits the administrator to say "our plan covers everything."

Table 28.2 gives the factors and points assigned for the Midwest Industrial Management Association plan for shop jobs; there is another plan for office jobs. Table 28.3 gives the levels of factor 10 (Working Conditions) and Table 28.4 gives the levels of factor 11 (Hazards).

If, for example, the job of a lathe operator were being evaluated, for working conditions it probably would be 2nd degree or 20 points; for hazards it probably would be 3rd degree or 10 points. In the interests of accurate job evaluation and acceptance of the job evaluation, an evaluation of any specific job should never be made by a single individual. Each allocation of points should be checked by a committee of diverse backgrounds and interests and a consensus reached. Jobs should be reevaluated periodically (every 5 or 10 years) as jobs change considerably over time. The success of job evaluation plans has been founded on such analytical, detailed, consensus-building techniques to replace opinionated, general, whimsical approaches.

Table 28.3   Factor 10; Working Conditions[6]

This factor measures the surroundings or physical conditions under which the job must be done and the extent to which those conditions make the job disagreeable. Consider the presence and relative amount of exposure to dust, dirt, heat, fumes, cold, noise, vibration, wetness, etc. When working conditions vary with specific work assignments, such as found in maintenance jobs, the degree selected must represent the weighted average of all the conditions encountered.

1st DEGREE
Excellent working conditions with absence of disagreeable conditions

2nd DEGREE
Good working conditions; may be slightly dirty or involve occasional exposure to some of the elements listed above

3rd DEGREE
Somewhat disagreeable working conditions due to exposure to one or more of the elements listed above to the extent of being objectionable; may be exposed to one element continuously or several elements occasionally, but usually not at the same time

4th DEGREE
Disagreeable working conditions where several of the above elements are continuously present to the extent of being objectionable

5th DEGREE
Continuous and intensive exposure to several extremely disagreeable elements; working conditions *particularly disagreeable*

SUMMARY

Job evaluation determines the relative worth of jobs. It is an analytical, consensus-building approach, not a procedure based on science. The first step is the job description. Then the jobs are arranged in sequence and a wage value is assigned. Labor grades simplify administration and permit paying for individual merit as well as the value of the job.

Table 28.4   Factor 11; Hazards[6]

This factor measures the hazards, both accident and health, connected with or surrounding the job, considering safety clothing, devices or equipment that have been installed or furnished the work location, the material being handled, the machines or tools used, the work position, and the probable extent of injury in case of accident.

1st DEGREE
Accident or health *hazards negligible;* remote probability of injury

2nd DEGREE
Accidents improbable, outside of minor injuries, such as abrasions, cuts or bruises; health hazards negligible

3rd DEGREE
Exposure to lost-time accidents possible, such as severe injuries to hand or foot, loss of finger or eye injury, etc., some exposure to health hazards, not incapacitating in nature

4th DEGREE
Exposure to incapacitating accident or health hazards, such as loss of arm or leg, impairment of vision

5th DEGREE
Exposure to accidents or health hazards which may result in *total disability or death*

SHORT-ANSWER REVIEW QUESTIONS

28.1   How is pay based both on the worth of the job and the worth of the person holding the job?

28.2   What are the three ways of obtaining information for a job description?

28.3   Why is it bad to have too many different jobs?

28.4   Why do organizations have different plans for different groups?

28.5   List the four major categories for the factors in the point systems.

THOUGHT-DISCUSSION QUESTIONS

1. If you thought your organization was biased against females, how might the bias be reflected in the job evaluation system?

2. Should a job evaluation plan be used that covers all jobs within an organization rather than different ones for shop, offices, and managerial?

REFERENCES

1. Bartley, D., *Job Evaluation,* Reading, MA: Addison-Wesley, 1981.

2. Henderson, R., and Clarke, K., *Job Pay for Job Worth,* Atlanta, GA: Business Publishing Div., Georgia State University, 1981.

3. Livy, B., *Job Evaluation: A Critical Review,* London: George, Allen and Unwin, 1975.

4. McCormick, E., *Job Analysis: Methods and Applications,* New York: American Management Association, 1979.

5. Patton, J.; Littlefield, C.; and Self, S. *Job Evaluation Text and Cases,* Homewood, IL: Irwin, 1964.

5. Treiman, D., *Job Evaluation: An Analytic Review,* Office of Publications, National Academy of Sciences, Washington, DC, 1979.

INDEX

614